DEVELOPMENTS IN
EARTH SURFACE PROCESSES, 13

NATURAL HAZARDS AND HUMAN-EXACERBATED DISASTERS IN LATIN AMERICA

Special Volumes of Geomorphology

DEVELOPMENTS IN EARTH SURFACE PROCESSES

1. PALEOKARST: A SYSTEMATIC STUDY AND REGIONAL REVIEW
   P. BOSÁK, D. FORD, J. GLAZEK and I. HORÁCEK (Editors)
   [OUT OF PRINT]
2. WEATHERING, SOILS & PALEOSOLS
   I.P. MARTINI and W. CHESWORTH (Editors)
3. GEOMORPHOLOGICAL RECORD OF THE QUATERNARY OROGENY IN THE HIMALAYA AND THE KARAKORAM
   JAN KALVODA (Editors)
   [OUT OF PRINT]
4. ENVIRONMENTAL GEOMORPHOLOGY
   M. PANIZZA
5. GEOMORPHOLOGICAL HAZARDS OF EUROPE
   C. EMBLETON and C. EMBLETON-HAMANN (Editors)
6. ROCK COATINGS
   R.I. DORN
7. CATCHMENT DYNAMICS AND RIVER PROCESSES
   C. GARCIA and R.J. BATALLA (Editors)
8. CLIMATIC GEOMORPHOLOGY
   M. GUTIÉRREZ
9. PEATLANDS: EVOLUTION AND RECORDS OF ENVIRONMENTAL AND CLIMATE CHANGES
   L.P. MARTINI, A. MARTINEZ CORTIZAS and
   W. CHESWORTH (Editors)
10. MOUNTAINS WITNESSES OF GLOBAL CHANGES RESEARCH IN THE HIMALAYA AND KARAKORAM: SHARE-ASIA PROJECT
    RENATO BAUDO, GIANNI TARTARI and
    ELISA VUILLERMOZ (Editors)
11. GRAVEL-BED RIVERS VI: FROM PROCESS UNDERSTANDING TO RIVER RESTORATION
    HELMUT HABERSACK, HERVÉ PIÉGAY and
    MASSIMO RINALDI (Editors)
12. THE CHANGING ALPINE TREELINE: THE EXAMPLE OF GLACIER NATIONAL PARK, MT, USA
    DAVID R. BUTLER, GEORGE P. MALANSON, STEPHEN J. WALSH and
    DANIEL B. FAGRE (Editors)

Developments in Earth Surface Processes, 13

NATURAL HAZARDS AND HUMAN-EXACERBATED DISASTERS IN LATIN AMERICA

Special Volumes of Geomorphology

Edited by

Edgardo M. Latrubesse
University of Texas at Austin, Department of Geography and the Environment, Austin, USA

Amsterdam – Boston – Heidelberg – London – New York – Oxford
Paris – San Diego – San Francisco – Singapore – Sydney – Tokyo

Elsevier
Radarweg 29, PO Box 211, 1000 AE Amsterdam, The Netherlands
Linacre House, Jordan Hill, Oxford OX2 8DP, UK

First edition 2010

British Library Cataloguing-in-Publication Data
A catalogue record for this book is available from the British Library

Library of Congress Cataloging-in-Publication Data
A catalog record for this book is available from the Library of Congress

ISBN: 978-0-444-53117-9
ISSN: 0928-2025

For information on all Elsevier publications
visit our website at books.elsevier.com

Printed and bound in Great Britain

09 10 10 9 8 7 6 5 4 3 2 1

Contents

Editorial Foreword

This new book edited by Edgardo Latrubesse concerns natural hazards and their associated processes and disasters in many parts of Latin America. Most people can recall reading or hearing about a catastrophe in that part of the world, but are usually vague about where it or how it occurred, or about what caused the problem in the first place. This thirteenth volume in our book series on Developments in Earth-Surface Processes sets the stage for a new look at these phenomena in this part of the world, where mainly scientists skilled in Iberian languages have much information.

The natural hazards of earthquakes and volcanic eruptions are, of course, directly related to plate-tectonic translocations and collisions in the lithosphere. In the uppermost portions of this lithosphere, different kinds of slope-failure hazards reflect the many actions, interactions, and reactions between rock materials and climatic variations, particularly precipitation, although seismic or volcanic accelerations can also play major roles. In addition, the interactions and translocations of the hydrosphere and the atmosphere with the lithosphere and its processes generate tsunamis, hurricanes, and floods. These complex or even chaotic processes cause great hardship, the study of which can enable better scientific understandings of the many processes, about which we need far greater information to deal with more effectively.

In these times of accelerating global change, as climates shift worldwide and human populations increase exponentially without much constraint, it is clearly incumbent on us to pay close attention to these natural hazards. To know the problems is to move closer to solutions to them. Many countries of Latin America, with their limited infrastructure and weak economies, are highly vulnerable to environmental and political problems brought on by these natural hazards, especially as they are exacerbated by the actions of humankind.

Another serious world issue is that some forms of attempted predictive geomorphology, such as flood-hazard mapping and recurrent landslides require long-term monitoring, good record keeping, and analysis of old writings to find comparable events and to translate them into modern equivalents. Such data can be used to produce statistical probabilities of the future occurrence of similar events in the same areas. But as global change shifts climatic thresholds, such statistical treatments will perhaps not be so feasible and so unpredicted hazards will lead to greater damage to infrastructure and to greater loss of life. Thus this book is a useful beginning in the assessment of hazards and catastrophes on one of the most vital continents of the Southern Hemisphere. Because the people of this area are vulnerable to the vagaries of forces from plate-tectonic, atmospheric, hydrospheric, and cryospheric motions, regional assessments provide an important analytic tool.

In the first millennium CE in South America, the Tiwanaku population on the Altiplano around Lake Titicaca grew beyond the carrying capacity of their land, as a result of which they were unable to cope when a catastrophic 400-year drought arrived. Ultimately, it meant the end of their empire (Thompson, et al., 1988; Kolata, 2001). Similarly, the El Niño–Southen Oscillation (ENSO) affected irrigation capabilities so profoundly in coastal Peru that it apparently brought the Moche civilization to an end there as well (Fagen, 1999; Bowen, 2005). The fate of many of these prehistoric civilizations in Latin America therefore has strong implications for the modern world, illustrating the potentially devastating effects of similar natural hazards. This book, therefore, can serve as a useful warning for the future about possible environmental cataclysms and other Earth-surface problems in Latin America. In producing this volume in the English language, which ensures greater distribution of knowledge about natural hazards and human-exacerbated disasters in Latin America, Edgardo Latrubesse has done a significant service.

REFERENCES

Bowen, M. 2005. *Thin Ice*. Henry Holt Books, New York

Fagen, B. 1999. *Floods, Famines and Emperors: El Niño and the Fate of Civilizations*. Basic Books, New York.

Kolata, A. L. 2001. Environmental thresholds and the "natural history" of an Andean civilization. In *Environmental Disaster and the Archaeology of Human Response*, eds. G. Bawden and R.M. Reycraft, pp. 163–178. University of New Mexico, Albuquerque, Maxwell Museum of Anthropology.

Thompson, L.G., Davis, M. E., Mosley-Thompson, E., and Liu, Kam-Biu, 1988. Pre-Incan agricultural activity recorded in dust layers in two tropical ice cores. *Nature, 336*, pp. 763–765.

John F. Shroder Jr.
Editor-in-Chief

Foreword

The International Association of Geomorphologists (IAG) welcomes this book on the Geomorphology of Natural Hazards and Human-Exacerbated Disasters in Latin America, for it is a truly international compilation that deals with a very important and timely issue. I congratulate its editor, Professor Edgardo Latrubesse, on bringing it to fruition. The IAG has recognized the significance of geomorphology for understanding, predicting and mitigating hazards and disasters by establishing a working party on Geomorphological Hazards, and hazards will be the theme of the Regional Meeting we are holding at Brasov in Romania in September 2008.

As a recent report by the Centre for Research on the Epidemiology of Disasters (CRED) has shown (CRED, 2007), natural disasters have devastating impacts. In 2006, for example, there were 427 reported natural disasters that globally killed more than 23,000 people, affected almost 143 million others, and were the cause of more than US$ 34.5 billion in economic damages. Floods and various types of windstorms were the two major causes of economic damage. Not all natural disasters are essentially geomorphological, but many of them are, including avalanches, landslides and other types of mass movements, desertification and soil erosion, floods, coastal storm surges, earthquakes, tsunamis, and volcanic eruptions (Alćantara-Ayala, 2002).

Hydrometeorological disasters appear to be increasing in occurrence and tend to be more frequent than geological disasters. The occurrence and consequences of disasters may be changing because of increasing human impacts on the environment, because of changes in climate (especially a tendency to more intense storm activity in many parts of the world), and because more people and economic activities are being located in hazardous places. Whether one is talking about new housing estates located on flood-prone areas in the United Kingdom, or favelas constructed on very steep slopes in Brazil, or the position of New Orleans, it is clear that much economic activity is located in illogical sites.

Geomorphologists cancontribute to the study of natural disasters and hazards in many different ways (Rosenfeld, 2004). Identifying potentially susceptible locations is important and can involve locating those places where disasters have occurred in the past, and at the same time assessing their past frequency and magnitude. A huge number of techniques are available, including sequential photography, archival material, and dendrochronology. Geographical Information Systems (GIS) and analysis of geotechnical properties of materials can be used to determine where disasters such as landslides might occur in the future. Geomorphological maps can show areas that may be at risk and may form the basis for land zonation and planning decisions. It is also crucial to understand the causation and mechanics of geomorphological change if appropriate mitigation procedures are to be undertaken successfully. It is equally crucial to understand some of the geomorphological and environmental consequences of engineering solutions that have been proposed for natural disaster mitigation and to evaluate the relative merits of "hard" and "soft" solutions. Geomorphologists also need to be involved in debates about the future significance and impacts of global climate changes and to identify those ?hot spots? where systems may be tipped beyond some crucial threshold. For example, sea-level rise and increased storm surge activity may both combine to make coastal lowlands highly vulnerable to flooding; glacier retreat may remove buttressing from slopes and make them more prone to mass movements; permafrost decay may cause shorelines and riverbanks to become more prone to catastrophic retreat; and lower soil moisture levels in drylands may stimulate dune reactivation and dust storm generation.

The chapters in this volume indicate the wide range of geomorphological disasters that afflict the different parts of Latin America, the role of climate change and anthropogenic factors, and the methods employed by geomorphologists in their attempts to understand and mitigate their worst effects, not least in developing countries.

Andrew Goudie
University of Oxford and President of the IAG

Preface

Disasters have been a main concern at the global scale, and many of them are essentially geomorphologic. Developing countries are among the countries most profoundly affected by these disasters, and Latin American countries are not the exception. In addition to the typical disasters that affect people in a direct way and produce economic loss, Latin America is also suffering environmental disasters, notably, massive rates of deforestation, desertification, or glacier recession in which geomorphologic processes have a main role. Large disasters have happened in every decade: for example, the eruption of Nevado de El Ruiz (1985), Hurricane Mitch (1998), the killer floods and landslides in Vargas-Venezuela (1999), and the Pisco earthquake (2007). These disasters shocked society and reminded us that, with minor exceptions, our region has advanced little in disaster mitigation and prevention because of the general failure of public policies during these last decades.

In spite of this not very hopeful perspective, some links between Latin American researchers have started to appear through some broad scientific programs such as CYTED (Science and Technology Program among Ibero-American Countries) and between South American researchers through the pioneer PROSUL program implemented by the National Council of Research of Brazil (CNPq), which offers some hope of obtaining a broad scientific interaction among South American researchers in the future. In each South American country, geomorphologists have been trying to provide results and interactions with political and social actors, with varying success. Often geomorphologists have been relegated to a secondary role because of politicians' preference for "hard" engineering solutions after disasters rather than "soft" prevention and mitigation policies. On the other hand, the International Association of Geomorphologists (IAG) has been trying to increase the participation of geomorphology in society as well as to create mechanisms for strengthening the links between Latin American researchers. During the International Conference on Geomorphology, in Tokyo, Japan, 2001, I was elected to participate on the Executive Committee of the International Association of Geomorphologists. Between 2001 and 2005, with the support of the past IAG president, Mario Panizza, my main tasks related to promoting Latin American research, increasing the participation of young scientists from developing countries, fostering the creation of new national associations/groups in geomorphology, and improving the participation of scientists in developing countries in geomorphologic research. Because of these novel initiatives of IAG, three national groups of geomorphology have been created in Venezuela, Colombia and Bolivia. In addition, the participation of the Argentinean Group of Geomorphology in the IAG has become regularized. All this was possible because of our effort and work in collaboration with Latin American colleagues. A very good start in reactivating interest in geomorphology and disasters in Latin American countries was the Regional Conference on Geomorphology Geomorphic Hazards: Towards the Prevention of

Disasters, which was organized by Mexican colleagues from the Universidad Autónoma de México in 2003, such as Irasema Alcántara Ayala (one of the contributors to this book).

During the last few years, the IAG under the current president, Andrew Goudie (2005–2009), has maintained that our specialists should have a more active role in society, implementing the appropriate means to increase the awareness of other scientists, decision makers, stakeholders, planners, authorities, and the general public about the role of geomorphology in the understanding of hazards and disaster prevention. To reach these goals, several specific meetings and activities have been organized, and a new working group was created.

In September 2006, together with my colleague Selma Simões de Castro (one of the contributors to this book), we organized the IAG Regional Conference on Geomorphology and the VI SINAGEO (Brazilian Symposium of Geomorphology, official activity of the Brazilian Union of Geomorphology). Our main objective was to strengthen the links between South American researchers and to enable our students and young researchers to access results from researchers with particular experience in tropical areas across the world. The Symposium involved more than 700 participants from 21 countries and proved successful. The meeting was the largest conference specifically devoted to geomorphology ever organized in South America, in terms of both the number of national and international participants. Geomorphologic hazards in tropical countries were an important concern of the meeting. A new successful Latin American Symposium on Geomorphology was organized in 2008 in Belo Horizonte, Brazil by the Brazilian Union of Geomorphologists in association with the VII Symposium of Geomorphology. This exemplifies the increasing academic interest in geomorphology. The quantity of results on applied geomorphology presented at the meetings is an indicator that geomorphology's contribution to society has been growing substantially during the last years in several Latin American countries, especially Brazil. This book combines both policies of the IAG in a single product: the participation of Latin America researchers in a "hot" topic such as hazards and disasters.

Natural hazards and human-exacerbated disasters occur particularly in developing countries, and geomorphologic processes are one of the main agents. In this book the geomorphology of natural hazards and human-exacerbated disasters in Latin America is addressed by experts with a long history of research on the continent and includes results from Mexico, Central America, Venezuela, Colombia, Ecuador, Peru, Bolivia, Chile, Argentina, and Brazil. The book contains 21 chapters written by 39 researchers from over 28 academic or research institutions from 11 countries.

The main objective of the book is to offer a vision of the geomorphologic dynamics of the main disasters, describing their mechanisms and consequences for South American societies. Major disasters produced by volcanoes, earthquakes, floods, landslides, and tsunami hazards are a main focus as well as the description of the main hydrometeorological and geological drivers. Human-induced environmental disasters are also included, such as desertification in Patagonia and soil erosion in Brazil. The recession of South American glaciers as a response to recent climatic trends and sea-level scenarios is also discussed.

The approach is broad, analyzing causes and consequences and including social and economic costs, discussing environmental and planning problems, but always describing the geomorphologic/geologic-involved processes. This particular approach differentiates this book from others which are devoted to assessing the more "social" aspects of disasters, thus emphasizing mainly economic and social effects.

I believe that this book can be a main reference for a variety of graduate and undergraduate students and professionals interested in geomorphology, physical geography, environmental sciences, environmental geology, and ecology, as well as for planners, environmental engineers, and environmental consultants. I am hopeful and expectant that mitigation and prevention policies in Latin America encouraged by the active participation of geomorphologists in society and as a consequence of better interaction among the different social and politic actors can help to decrease the impact of disasters in the future.

My sincere thanks are due to Andrew Goudie, president of the International Association of Geomorphologists, for contributing the Foreword. I would also like to acknowledge the colleagues who acted as referees, helping us to obtain a better result. I also especially thank John (Jack) Shroder, Editor-in-Chief of the Developments in Earth Surface Processes Series, for his fundamental help in the editorial process and to the staff of Elsevier that helped me through the edition of this book.

Finally, I am totally grateful to the authors for participating in this book, thereby transmitting their research results and their social understanding of the countries where many of them have been living their entire lives or where they have acquired much of the scientific experiences that have shaped their professional careers.

Edgardo M. Latrubesse
University of Texas at Austin, Department of Geography and the Environment, GRG 334, A 3100, Austin TX, 78712, USA

List of Contributors

Ana Luiza Coelho-Netto
GEOHECO/Geo-Hydroecology Laboratory – Geography Department, Federal University of Rio de Janeiro - UFRJ, Brazil

Adriano Cecioni
University of Concepción, Concepción, Chile, Departamento de Ciencias de la Tierra, Casilla (P.O. Box) 160–Concepción – Chile

André de Souza Avelar
GEOHECO/Geo-Hydroecology Laboratory – Geography Department, Federal University of Rio de Janeiro - UFRJ, Brazil

Andrea Coronato
Centro Austral de Investigaciones Científicas (CADIC-CONICET). B. Houssay 200 (9410) Ushuaia, Argentina

Andrew S. Goudie
St. Cross College, Oxford, OX1 3LZ, United Kingdom

Blanca León
Museo de Historia Natural, Av. Arenales 1256, Apartado 14-0434, Lima-14, Peru

Carlos Benavente
INGEMMET, Av. Canadá 1470, San Borja, Lima, Perú.
cbenavente@ingemmet.gob.pe

Carlos Ferrer
Universidad de los Andes, Geographical Institute, Mérida Venezuela.
e-mail: carlosferrerve@yahoo.com

Carlos H. Costa
Departamento de Geología, Universidad Nacional de San Luis, Chacabuco 917, 5700 San Luis, Argentina.
costa@unsl.edu.ar

Daniel Brea
Instituto Nacional del Agua, Programa Hidráulica Fluvial, Universidad Nacional de La Plata, Área Departamental Hidráulica, La Plata, Argentina

Edgardo M. Latrubesse
University of Texas at Austin, Department of Geography, GRG 334, A3100, Austin TX, 78712, USA

Elizabeth Mazzoni
Universidad Nacional de la Patagonia Austral, Unidad Académica Río Gallegos Lisandro de la Torre 1070, Río Gallegos, Santa Cruz, Argentina
E-mail: gis@uarg.unpa.edu.ar

Enrique J. Schnack
Comisión de Investigaciones Científicas de la Provincia de Buenos Aires – CIC- and Laboratorio de Oceanografía Costera, Facultad de Ciencias Naturales y Museo, Universidad Nacional de La Plata, 64 y 120, 1900 La Plata, Argentina.
eschnack@netverk.com.ar

Fausto O. Sarmiento
Department of Geography, The University of Georgia, Athens, GA. 30602. USA

Federico I. Isla
Consejo Nacional de Investigaciones Científicas y Tecnológicas – CONICET- and Centro de Geología de Costas, Facultad de Ciencias Exactas y Naturales, Universidad Nacional de Mar del Plata. Complejo Universitario, Funes 3350, 7600 Mar del Plata, Argentina. fisla@mdp.edu.ar

Franck A. Audemard M
Departamento de Ciencias de la Tierra, FUNVISIS, Apartado Postal 76.880, Caracas 1073-A, Venezuela. faudemard@funvisis.gob.ve

Irasema Alcántara-Ayala
Instituto de Geografía, Universidad Nacional Autónoma de México (UNAM), Circuito Exterior, Ciudad Universitaria, 04510 Coyoacán, México, D.F., Tel: (+5255) 56 22 43 41, Fax: (+5255) 56 16 21 45,
E-mail: irasema@igg.unam.mx.

Jaime Argollo
Universidad Mayor de San Andrés, Instituto de Geología y Medio Ambiente, La Paz, Bolivia

Jaime Laffaille
Universidad de los Andes, Geophysical Laboratory and Foundation for the Prevention of Seismic Risk (FUNDAPRIS). Mérida, Venezuela.
e-mail: jaime@ula.ve

Jorge Rabassa
CADIC-CONICET, and Universidad Nacional de la Patagonia-San Juan Bosco, C.C. 92, 9410 Ushuaia, Tierra del Fuego, Argentina

Jose Candido Stevaux
Universidade Estadual de Maringá-GEMA, Campus, Avda. Colombo 5790, Maringá, 87020-900, Brazil

José Pereira de Queiroz Neto
Universidade de Sao Paulo, Department of Geography,
AV. Prof. Lineu Prestes 338
Butanta, 05508-900 - Sao Paulo, SP - Brasil

Kenneth R. Young
Department of Geography and the Environment, University of Texas at Austin, Austin TX 78712, U.S.A.

Klaudia Laffaille
Centro de Investigación en Gestión Integral de Riesgos, Mérida, Venezuela. e-mail: k.laffaille@gmail.com

Laura P. Perucca
Consejo Nacional de Investigaciones Científicas y Técnicas (CONICET) - Gabinete de Neotectónica. INGEO. Facultad de Ciencias Exactas, Físicas y Naturales. Universidad Nacional de San Juan

Laurence Audin
LMTG-UMR5563, CNRS/IRD/Université Tolouse 3, Casilla 18-1209, Lima 18, Perú. Laurence.audin@ird.fr

Maria Lucia de P. Hermann
Departamento de Geociências, Universidade Federal de Santa Catarina, Florianópolis, Campus Universitario (88040-900), SC, Brazil

Maximiliano Bezada
Universidad Pedagógica El Libertador-UPELD, Departamento de Geociencias, Caracas, Venezuela. mbezada@cantv.net

Michel Hermelin
Departmento de Geología, Universidad EAFIT, Carrera 49-7 Sur 50, Medellín, Colombia.

Mirian Vazquez
Universidad Nacional de la Patagonia Austral, Unidad Académica Río Gallegos. Lisandro de la Torre 1070, Río Gallegos, Santa Cruz, Argentina.
E-mail: gis@uarg.unpa.edu.ar

Natalia Hoyos
Laboratorio de Sistemas de Información Geográfica, Universidad EAFIT, Carrera 49-7 Sur 50, Medellín, Colombia.

Norberto O. Garcia
Facultad de Ingeniería y Ciencias Hídricas, Universidad Nacional del Litoral, CC495-3000 Santa Fe, Argentina

Paul A. Baker
Duke University, Division of Earth and Ocean Sciences, Durham, NC 27708, USA

Rosana Nieto Ferreira
Department of Geography, East Carolina University, Greenville, NC 27858-4353, U.S.A

Samia Aquino
University of Texas at Austin, Department of Geography and the Environment, GRG 334, A3100, Austin TX, 78712

Selma Simões de Castro

Laboratorio de Geologia e Geografia Fısica-Universidade Federal de Goiás, Campus II-74001-970, Goiania-GO, Brazil

Stella M. Moreiras
Consejo Nacional de Investigaciones Científicas y Técnicas (CONICET). Instituto Argentino de Nivología, Glaciología y Ciencias Ambientales
(IANIGLA) - CRICYT
e-mail:moreiras@lab.cricyt.edu.ar

Verónica Pineda
University of Concepción, Concepción, Chile, Departamento de Ciencias de la Tierra, Casilla (P.O. Box) 160–Concepción – Chile

Willy Alvarenga Lacerda
Civil Engineering Program – Geotechnical Department, Federal University of Rio de Janeiro - UFRJ, Brazil

CHAPTER 1

Climate and Geomorphologic-related Disasters in Latin America

Norberto O. Garcia[1], Rosana Nieto Ferreira[2], *and* Edgardo M. Latrubesse[3]

[1]*Facultad de Ingeniería y Ciencias Hídricas, Universidad Nacional del Litoral, CC495-3000 Santa Fe, Argentina*
[2]*Department of Geography, East Carolina University, Greenville, NC 27858-4353, U.S.A.*
[3]*University of Texas at Austin, Department of Geography, GRG 334, A3100, Austin TX, 78712, USA*

Contents

Corresponding author: *Norberto O. García, Facultad de Ingeniería y Ciencias Hídricas, Universidad Nacional del Litoral, CC495-3000 Santa Fe, Argentina, nogarcia@fich1.unl.edu.ar*

Developments in Earth Surface Processes, Volume 13
ISSN 0928-2025, DOI 10.1016/S0928-2025(08)10001-3

1. Introduction

Latin America extends from the northern border of Mexico at 32°N to Cape Horn at 56°S in the southernmost tip of Chile, and from Cape Branco at 35° 30′W to Cape Fariñas at 81° 20′W. With a landmass that extends over 20,340,000 km^2, mostly in the Southern Hemisphere, Latin America represents 16% of the Earth's landmass and is much more extensive in the tropics than in the higher latitudes. The Andes Mountains run along the entire western coast of South America (Fig. 1.1) forming an effective meteorological barrier between the Pacific Ocean and the continent. Similarly, along Mexico and Central America the Sierra Madre Mountains act as a barrier to the trade wind easterlies. In Latin America the topography, geographic position, and contrasting surrounding oceanic conditions create great climatic diversity. Section 2 of this chapter offers an overview of Latin America's

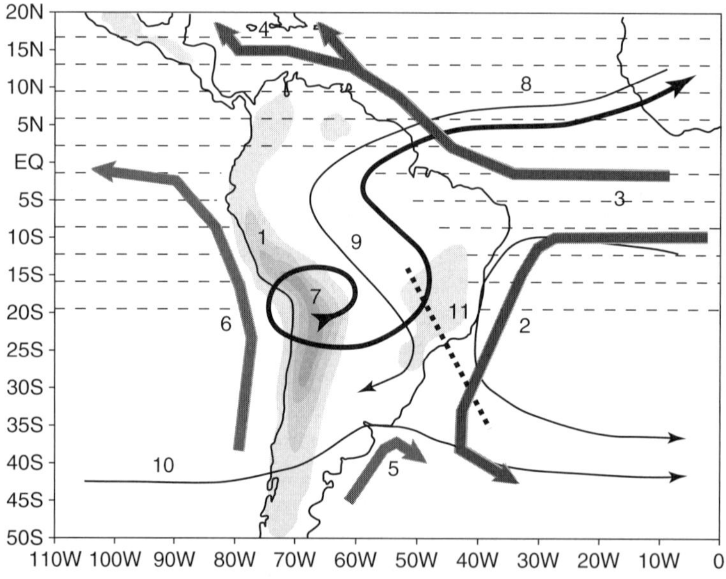

Figure 1.1 Schematic of the summertime circulation in Latin America (adapted from Zhou and Lau, 1998).

1. Andes Mountains
2. Warm Brazil Current
3. Equatorial Current
4. Caribbean Current
5. Malvinas Current
6. Humboldt Current
7. Bolivian Anticyclone
8. Trade wind easterlies
9. South American Low-Level Jet
10. Midlatitude Westerlies
11. South American Convergence Zone

complex climate variability from the monsoon cycle to the effects of the El Niño-Southern Oscillation (ENSO) phenomenon, as well as from the effects of tropical storms and hurricanes in the tropical regions of Latin America, to the effects of extratropical cyclones in the midlatitudes. Section 3 ties the climate and weather features to their geomorphological effects over many regions of Latin America.

2. Latin American Climate

The combination of the wide range of latitudes covered by Latin America, the existence of major orographic features, and significant oceanic influences has produced extremely diverse, complex climate patterns. This section discusses the main climate patterns in Latin America.

2.1. Ocean Currents

Ocean currents play an important role in modulating local and global climate patterns. Latin American climate is strongly influenced by oceanic currents in both the Atlantic and Pacific oceans. The main oceanic currents that influence Latin America are as follows.

The Antarctic circumpolar current influencing the southern continental extreme is situated between 50° and 60° south latitude. The present narrowness of the water connection between South America and Antarctica, however, causes this great oceanic current to suffer perturbations and modifies its sweep. The main discharge passes through Drake Strait, and a smaller discharge is deflected north up the south Chilean coast. The northern deflection, called the Humboldt Current (Fig. 1.1), travels along the western coast of South America to the vicinity of the equator, giving unique characteristics to the coastal climates.

The Humboldt Current transports cold waters northward along the Chilean coast. The shape of the coast, the depth of the sea, and the Coriolis force, together with the southeast trade winds, cause a deflection in its northerly reaches, and thus the current separates from the Peruvian coast. This causes upwelling of colder deep waters that rise to the surface, conditioning the principal climatic features in the coastal zone between the equator and 20°S.

On the other hand, the eastern coast of South America has notably different characteristics, showing a seasonal behavior that is not observed with the same intensity along the Pacific coast. On the eastern coast, the Atlantic equatorial current is of vital importance and meets the South American continent near Cape Branco. As it flows from east to west, the Atlantic equatorial current (Fig. 1.1) bifurcates into two warm currents that travel along the South American coast, the northward-flowing Northern Brazil and Caribbean Currents, and the southward-flowing Brazil Current.

The Brazil Current (Fig. 1.1) travels southwesterly, reaching Cape Corrientes on the Argentine coast. At the end of the summer months in the Cape Corrientes zone, the warm Brazil Current converges with the Malvinas Current that flows past the Patagonian coast, carrying cold waters toward the north. At the end of winter, the Malvinas Current (Fig. 1.1) displaces the Brazil Current away from the coast and extends northward to Cape Frio (7°S) along the Brazilian Atlantic coast. The warm waters, thus separated from the continental margin, then bathe the Brazilian coast only between Cape Branco (7°S) and Cape Frio (20°S, García, 1994). This direct effect of the shifting oceanic conditions can be clearly observed by comparing the monthly mean temperatures in the warmest and coldest months of both continental coasts (Table 1.1 and Fig. 1.2).

Another indication of the effect of the adjacent seas on the occidental and oriental coasts of South America is the thermal contrast between the shores (Fig. 1.2). During the austral summer, the oriental coast of South America is everywhere warmer than the occidental coast. This characteristic, however, does not prevail throughout the year because during the winter the cool Malvinas Current influences the oriental coast and the monthly temperatures are practically the same along both coasts from the southern extremity to 30°S.

The Northern Brazil Current bathes the northern coast of Brazil and becomes the Caribbean Current as it flows northwestward between the northern coast of South America and the Caribbean Islands. The Caribbean Current ultimately flows into the Gulf of Mexico through the narrow strait that separates the Yucatan Peninsula and the Island of Cuba. The seasonal variability of these currents causes a marked seasonal variability of sea-surface temperatures (SSTs) in the Caribbean Sea and Gulf of Mexico, with the warmest SSTs occurring during the boreal summer in the Caribbean Sea. The seasonal changes in SSTs in the Caribbean

Table 1.1 Warmest and Coldest Monthly Mean Temperatures on the Oriental and Occidental South American Coasts (in °C)

Latitude	Locality	Temp. Warmest Month (°C)	Temp. Coldest Month (°C)	Annual Amplitude (°C)
8° S	Lambayeque (west)	24.2	17.5	6.7
	Recife (east)	26.3	22.9	3.4
23° S	Antofagasta (west)	20.3	13.4	6.9
	Rio de Janeiro (east)	26.1	20.8	5.3
33° S	Valparaiso (west)	19.7	11.7	6.2
	Curitiba (east)	20.1	12.6	7.5
40° S	Valdivia (west)	16.5	7.6	8.9
	Bahía Blanca (east)	22.8	7.6	15.2
50° S	Punta Arenas (west)	10.7	2.3	8.4
	Río Gallegos (east)	13.1	1.9	11.2

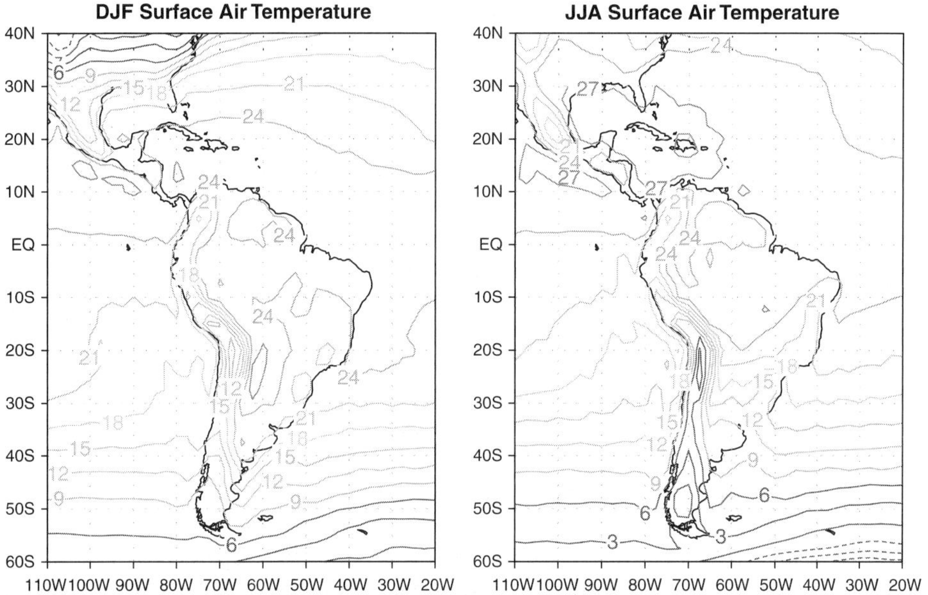

Figure 1.2 Average (1979–2006) surface air temperature (°C) from the National Centers for Environmental Prediction (NCEP). Reanalysis dataset during (a) December–January–February, and (b) June–July–August.

Sea and Gulf of Mexico are reflected in the seasonal changes of surface air temperature shown in Figure 1.2. During the boreal summer, the presence of warm ocean SSTs in the Caribbean Sea and Gulf of Mexico contribute to the maintenance, intensification, and formation of hurricanes that affect Central America and the Caribbean.

2.2. Air Temperature Patterns

The surface air temperature patterns in Latin America show well-established seasonal characteristics, with the land quickly responding to the Sun's apparent movement with respect to the equatorial plane, so that the maximum temperature occurs in South America on the summer of southern hemisphere. During the austral summer (December, January, and February—DJF), the continent's warmest area is situated in southcentral Brazil, Bolivia, and northern Argentina (Fig. 1.2). In addition, over the occidental coastal zones the effect of the cold Humboldt Current is notable. To a lesser extent, the effect of the cold Malvinas Current is also felt in the temperatures of the Argentine coast as far north as the mouth of the Rio de la Plata.

During the austral winter (June, July and August—JJA), the highest temperatures are present in Central America, Mexico, and the Caribbean. During the austral winter, cooler temperatures occur in southern South America, lessening the temperature differences between land and sea. The effect of the Humboldt

Current on the adjacent coast is noticeable even during these months. Autumn and spring are the transition seasons and have similar temperature patterns.

2.3. Pressure and Wind Patterns

In Latin America, the pressure field trough, marked by the intertropical convergence zone (ITCZ) and the convergence of the southeast and northeast trade winds, oscillates from north to south following the seasonal shifts in solar declination. In April, solar declination is at a minimum over the terrestrial equatorial plane, generating the lowest air pressures of the year in the equatorial region. As the solar declination increases northward, the equatorial trough is also displaced northward and reaches its maximum northern latitude during the austral winter (Fig. 1.3a), which produces a zonally elongated low-pressure pattern that extends throughout the continent from the Isthmus of Panama to northern Patagonia.

As October approaches, the minimum solar declination returns to the equatorial plane during its southward migration, causing the equatorial trough to return to the equator. In the austral summer, the trough reaches its southernmost latitude over the southern and central Amazon (Fig. 1.3b).

Over the oceans, the equatorial trough migrates almost imperceptibly, always staying within the northern hemisphere in the Pacific coastal areas, in the latitudes

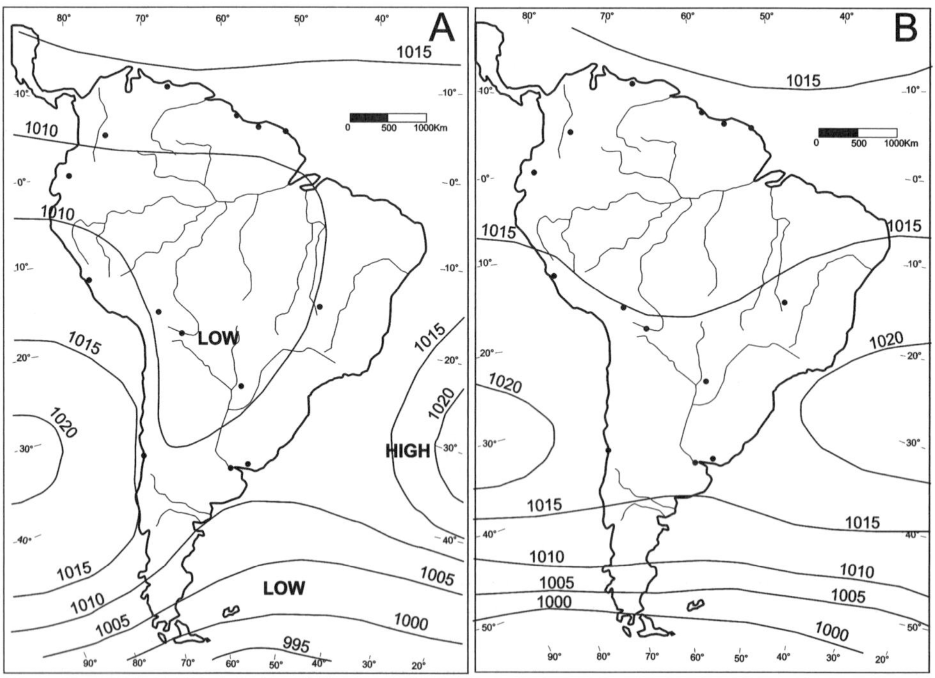

Figure 1.3 Sea-level pressure (mb) during (a) December–January–February, and (b) June–July-August.

of Panama and north Colombia. Along the Atlantic coast the trough migrates only a few degrees of latitude with the seasonal cycle.

In austral winter, the subtropical high-pressure centers present near 30°S in the Pacific Ocean and at 27°S over the Atlantic Ocean are connected across the South American continent. During the austral summer, the belt of high pressure in the subtropics is interrupted by heating of the landmass, generating lower pressure over land that divides the semipermanent subtropical anticyclones.

Thanks to the characteristic cloud cover, the ITCZ is easy to identify in the rainfall patterns shown in Figure 1.4. It crosses the western coast of Colombia at approximately 5°N, and curves around the Amazon Basin east of the Andes Mountain Range, reaching as far as the northern part of Paraguay. It curves northeastward over the eastern part of the Amazon Basin and extends into the Atlantic Ocean near 5°N.

Northeast trade winds prevail to the north of the ITCZ (Fig. 1.4), blowing with moderate intensity on the northern coast of the continent between the Amazonas and Panama. To the south of the ITCZ, along the oriental coast extending to 15.5°S, the trade winds blow from the southeast in the oceanic areas and swing gradually to the east and northeast as they penetrate inland. Further south, toward the Drake Strait, midlatitude westerly winds are intense and persistent.

During the austral winter, the location of the ITCZ coincides with the warmest SSTs and reaches its northernmost position (Fig. 1.4b). Northeast winds are felt only over the Caribbean coasts of Venezuela and Colombia. The southeast trade winds are located on the southern side of the ITCZ, over the Atlantic between 5°N and 15°S. Once again during this season, midlatitude westerly winds are intense and persistent near the Drake Strait. The wind systems are subject to strong local influences, particularly due those of topography in mountainous regions.

2.4. Rainfall Patterns

The wind, pressure, and temperature patterns described in the previous sections give rise to the observed seasonal variability of precipitation in the Americas. This complex seasonal cycle is strongly influenced by the surrounding oceans, topography, land surface type, thermodynamic conditions, and large-scale circulations. Together, the seasonal cycle of winds, precipitation, temperature, and sea-level pressure fields in tropical South, Central, and North America share general characteristics with classical monsoon climates in other parts of the world (Zhou et al. 1988) and can therefore be referred to as the American Monsoon System.

2.4.1. The South American Monsoon System

Figure 1.4 shows the climatological low-level winds and the Global Precipitation Climatology Project (GPCP) precipitation patterns in South America during the austral and boreal summers. The peak rainy season in South America occurs during December through February (Fig. 1.4a) and is referred to as the South American Monsoon System (SAMS). The SAMS is characterized by a precipitation pattern that

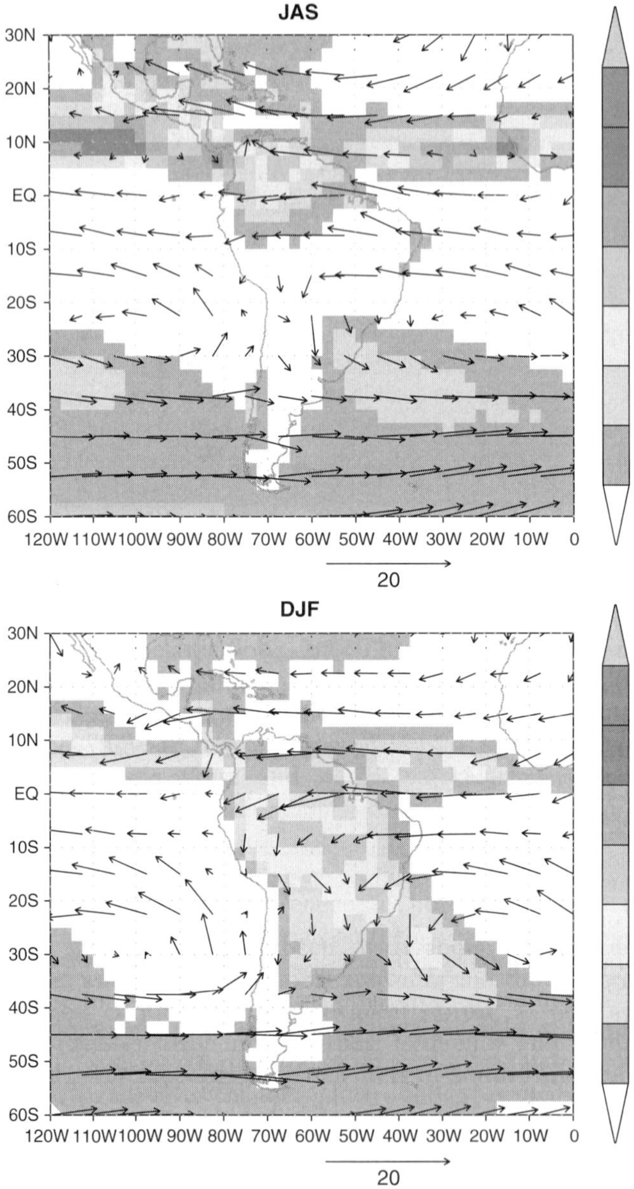

Figure 1.4 Average rainfall (mm/day) from the GPCP dataset and average low-level winds (m/s) from the NCEP Reanalysis during (a) JAS (July–August–September), and (b) DJF (December–January–February).

is dominated by the presence of the ITCZ to the south of the equator, bringing plentiful precipitation to most of South America. Another characteristic feature of the SAMS is the presence of the South Atlantic Convergence Zone (SACZ). The SACZ is a northwest-southeast-oriented band of enhanced precipitation that extends from the Amazon Basin to the Atlantic Ocean (Fig. 1.1 and Fig. 1.4b). The SAMS is also characterized by strong easterly winds that carry warm, moist air from the tropical Atlantic into the Amazon Basin. Upon reaching the Andes Mountains, this easterly air current turns southward, flowing along the Andes Mountains (Gandu and Geisler 1991) forming a circulation feature known as the South American low-level jet (SALLJ; Fig. 1.1 and Fig. 1.4b). The SALLJ provides the warm humid air that fuels precipitation along the SACZ. During the SAMS season Central America, Venezuela, Chile, and southern Argentina remain notably dry.

Important local maxima of annual rainfall amounts occur at the mouth of the Amazon River and in the western Amazon Basin (Fig. 1.5). The wettest region in all of Latin America occurs near the border of Colombia and Panama. The Colombian town of Andagoya (5N, 76W) is among the wettest places on Earth, with annual rainfall amounts of over 6800 mm. In contrast, one of the driest places on Earth is the town of Arica, Chile, where years can go by without a drop of rain falling from the sky.

The main dry regions of South America can easily be seen in the mean annual rainfall shown in Figure 1.5. In a typical year, the sinking air motion produced by the South Pacific high-pressure system allows little rain to fall in the coastal areas of Peru and Chile from the equator to 30°S. The presence of the Andes Mountains

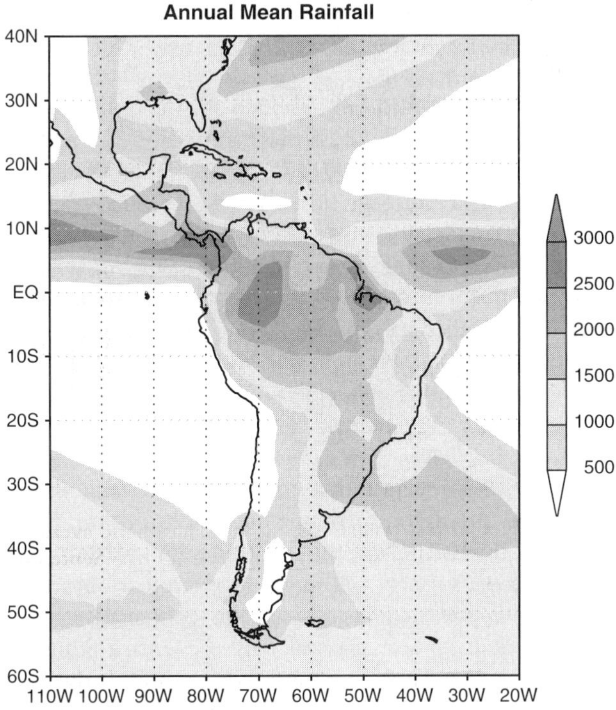

Figure 1.5 Average total annual rainfall (mm) from the GPCP dataset.

contributes to the formation of a rain shadow, or an area of small rainfall amounts in southeastern Argentina. The northeast region of Brazil usually receives very small amounts of annual rainfall (Fig. 1.5) because of the sinking air motion of the South Atlantic high-pressure system (Fig. 1.3). These dry South American regions are characterized by desert and steppe landscapes.

2.4.2. Central America and the Caribbean

During the boreal summer when the ITCZ is centered to the north of the equator (Fig. 1.4b), rainfall is plentiful from the northern Amazon Basin through Venezuela, Colombia, Caribbean and Central America all the way to southern Mexico and parts of the southwestern United States, marking the peak of the North American monsoon season. North of the equator, Latin America's driest region is the Sonoran desert in northern Mexico.

Over the East Pacific and Atlantic oceans, the central latitude of the ITCZ and associated precipitation band remains fairly constant throughout the year, reflecting the fact that the underlying sea-surface temperature maxima in these oceanic regions does not migrate between the two hemispheres.

A unique annual cycle of precipitation that displays a bimodal distribution with maxima in June and September–October and a distinctive relative minimum during July and August occurs over central-southern Mexico, most of Central America, and parts of the Caribbean (e.g., Magaña et al. 1999). This unique annual cycle of precipitation is called the *midsummer drought, veranillo*, or *canícula*, depending on the region where it occurs, and it poses challenges to agriculture and hydroelectric energy production in those regions.

During the boreal summer and fall, Mexico is affected by tropical storms and hurricanes that form in both the Eastern Pacific Ocean and Atlantic Ocean (Fig. 1.6). The number of Atlantic Ocean hurricanes that affect Mexico each year varies greatly from no strikes during some years (e.g., 1991, 1994) to the eight tropical storms and hurricane strikes that occurred during the 2005 Atlantic hurricane season (http://maps.csc.noaa.gov/hurricanes/viewer.html). Because the Yucatan Peninsula protrudes into the Gulf of Mexico, it receives the brunt of the tropical storms and hurricane activity in that country. On average one to two Eastern Pacific tropical storms and/or hurricanes affect the west coast of Mexico each year. Central America is also affected by hurricanes that form in both the Eastern Pacific and Atlantic Ocean, with one or two tropical storms or hurricane strikes per year.

Although the seasonal cycle is the strongest mode of variability in Latin America, the circulation and precipitation patterns of this region undergo important variability in timescales that span from diurnal to interdecadal timescales. Along the Nordeste (northeast) coast of Brazil, the sea-breeze circulation often combines with a low-level easterly jet to produce large westward-propagating squall line systems. These Amazonian squall lines sometimes cross the entire Amazon Basin over a period of two days, producing copious amounts of rainfall along the way (Silva Dias and Nieto-Ferreira 1992, Cohen et al. 1995). Midlatitude frontal systems pass through South America every one to two weeks (Garreaud 2000), bringing precipitation mostly to the subtropical portions of the continent but occasionally also into the tropical portions of the continent (Kousky 1979). Variability of the SACZ also produces strong variability in the precipitation and wind patterns in South America on intraseasonal timescales (Nieto-Ferreira et. al. 2003). On these intraseasonal timescales, the Madden-Julian Oscillation causes 30- to 50-day oscillations in precipitation and wind patterns in the eastern Amazon Basin and SACZ regions during the austral summer (Carvalho et al. 2004) as well as in Central America and the Caribbean during the boreal summer. On interannual timescales, ENSO produces circulation and precipitation variability that affect most of Latin America. On interdecadal timescales, the variability of rainfall in the drought-prone Nordeste region of Brazil is strongly affected by the variability of SSTs in the tropical Atlantic (Moura and Shukla 1981). The number of Atlantic ocean tropical storms and tropical cyclones that affect Mexico is also affected by the interdecadal variability of tropical Atlantic SSTs (Landsea and Gray 1992).

2.5. ENSO

The ENSO is a coupled ocean–atmosphere phenomenon that occurs in the tropical Pacific Ocean and affects weather and climate all over the Earth. El Niño episodes occur every two to seven years and are characterized by a warming of the SSTs along the eastern tropical Pacific Ocean that occurs around Christmas time, hence the name El Niño, or Christ Child in Spanish. El Niño events last about 12 to 15 months, though sometimes El Niño conditions persist for up to two years. La Niña events are characterized by warm SST anomalies in the western tropical Pacific Ocean and cold SST anomalies in the eastern tropical Pacific Ocean. Typical atmospheric conditions during El Niño events include weakened trade wind easterlies over the equatorial Pacific, more rainfall than normal in the eastern Pacific Ocean (Fig. 1.6) where SSTs are warmer than usual, and less rainfall than normal in the western Pacific Ocean, Australia, and

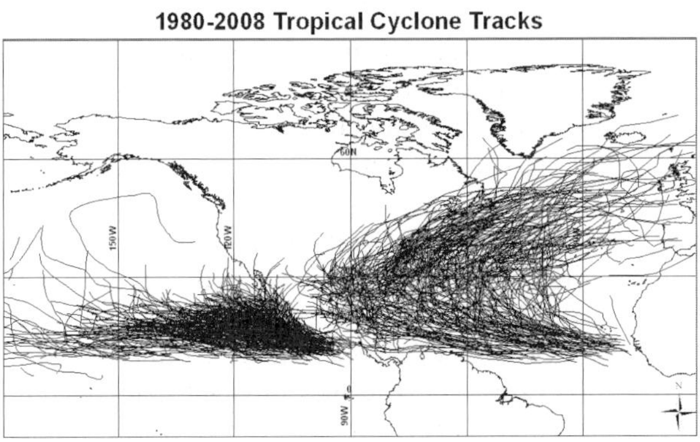

Figure 1.6 Hurricane tracks in the Eastern Pacific and Atlantic oceans from 1851 to 2007. Image courtesy of the National Hurricane Center–National Oceanic and Atmospheric Administration (NOAA).

Indonesia. On interannual timescales, ENSO produces circulation and precipitation variability that affect most of Latin America. For instance, when compared to La Niña events, strong El Niño events bring increased summertime (DJF) rainfall to coastal Peru and parts of Uruguay, Argentina, and southern Brazil, while a decrease in summer (DJF) rainfall amounts is observed in the Amazon Basin (Fig. 1.7) (Aceituno 1988). Central America and parts of the Caribbean experience drier than usual summers (JJA) during El Niño events than during La Niña events (Fig. 1.7). However, it is important to keep in mind that not all El Niño events are created equal and that different El Niño events can produce very different rainfall anomalies in Latin America.

3. Geomorphologic consequences of climate

Some of the climate patterns described in the previous section play an important role in triggering natural disasters in Latin America.

3.1. ENSO

Much of Peru's coastal areas and parts of the Pacific coast of Ecuador and Colombia are subject to the two- to seven-year ENSO cycle, causing alternating periods of floods and droughts. Large recent El Niño events occurred in 1982/1983, 1991/1992, and 1997/1998. Northwestern Peru is arid to semiarid, but during a strong El Niño year it receives several times its typical annual rainfall. An extended land-sea breeze system brings Pacific Ocean moisture over the continental area, while trade winds carry in additional moisture from the Amazon (Bendix 2000).

Flash floods occur in the desert, with associated damage to houses and infrastructure. Many of the surface features of the dry coastal areas of Peru resulted from infrequent fluvial events associated with the sudden flow of ephemeral streams, with overbank flooding and mud flows during El Niño events (see Chapter 9 by Young and Leon; Fig. 1.8). For example, in San Cristobal, Ecuador, with a mean monthly rainfall of 14 mm for December, the El Niño event of 1982 discharged 548 mm of rainfall. Bolivia and the northern part of Chile can also be affected by extraordinary floods (Latrubesse et al., Chapter 10, and Cecioni and Pineda, Chapter 18).

In the Andean countries (Venezula, Ecuador, Peru, and Bolivia), the El Niño of 1997/1998 was estimated to have caused an economic loss of ~$7,500 million. In contrast, the Peruvian Amazon is subject to droughts during El Niño events. For example, the El Niño of 1997/1998 generated drought conditions affecting the Amazon Basin and triggering tree mortality (see, for example, Laurance et al. 2001). El Niño also produces a marked decrease in rainfall in the Caribbean, Andean and Orinoquia zones of Colombia, affecting hydroelectric power generation and causing a variety of socio-economic effects.

García and Mechoso (2005) studied the influence of ENSO on South American rivers. They divided the time series of stream flows into three categories: warm

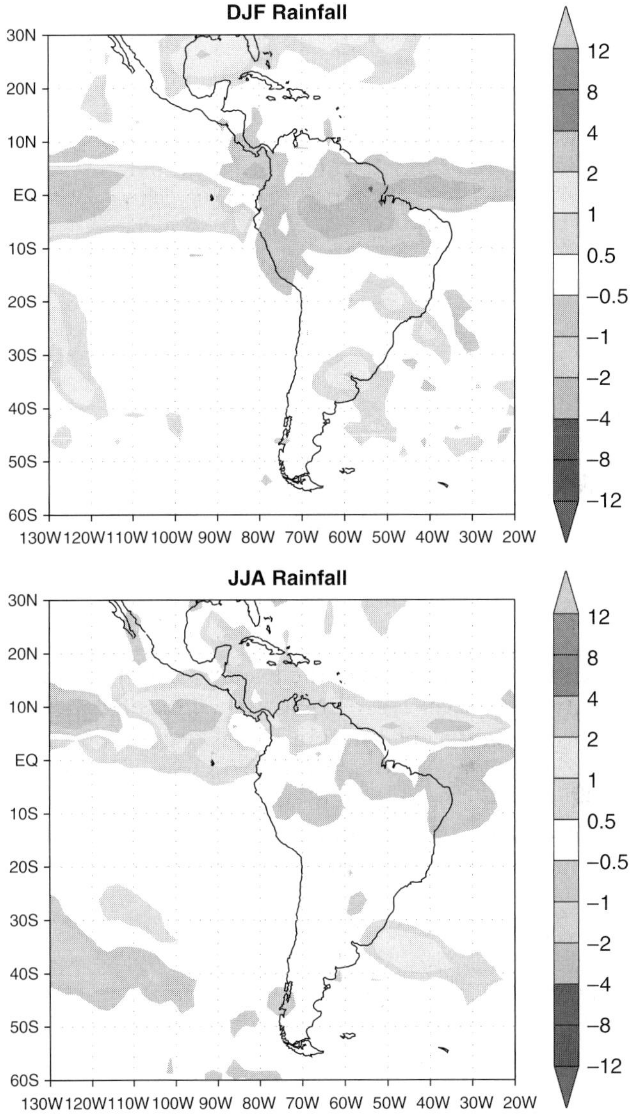

Figure 1.7 ENSO impacts on rainfall. El Niño minus La Niña rainfall (mm/day) in (a) December–January–February, and (b) June–July–August. Data from the Global Precipitation Climatology Project (1979–2008).

events, cold events and neutral years. They found that the tropical rivers in central and western South America (Amazon and Orinoco) showed a tendency of slightly increased stream flows during cold events, but suffered very little impact from warm events. The tropical rivers in northeastern South America (the Tocantins and San Francisco) showed the opposite behavior, with decreased stream flows during cold

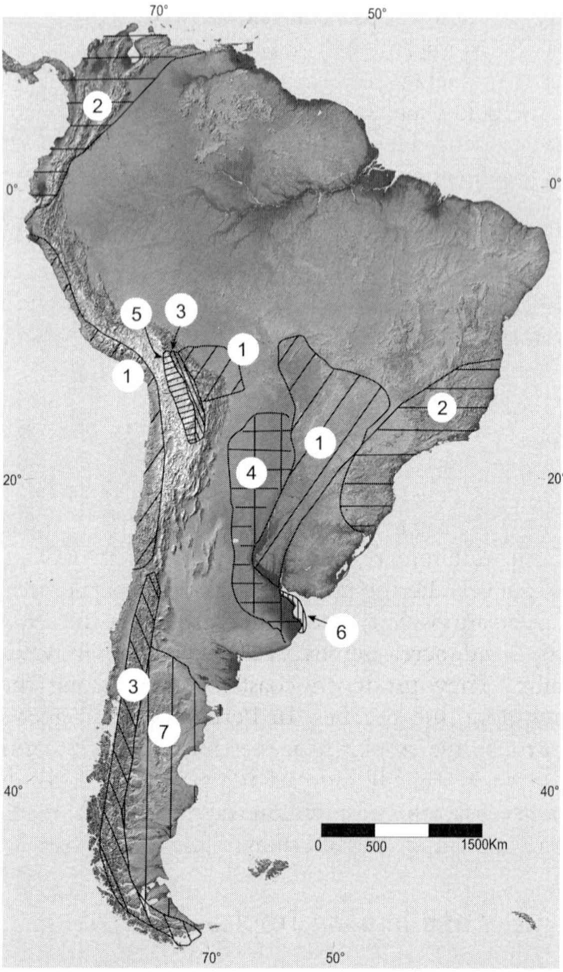

Figure 1.8 Main areas affected by hazards and disasters produced by geomorphologic processes.

1. Floods and landslides related to El Niño–ENSO events
2. Mountains/ranges in tropical areas affected by landslides and floods driven by intense tropical rainfall
3. Andean receding glaciers
4. Floods in flat lands
5. Droughts related to ENSO events
6. Coastal floods driven by SE winds (*sudestada*)
7. Desertification driven predominantly by overgrazing (wind and hydric erosion)

events and practically no differences during the warm events. However, none of the differences were statistically significant. The La Plata Basin rivers (the Paraná and Uruguay), on the other hand, showed significant stream flow increases during warm events and small differences during cold events.

These results are broadly consistent with the current view of the region's climate. Over tropical South America during El Niño events, the rising motion tends to be weaker than average and rainfall over the eastern Amazon and northeast Brazil tends to be below average, with broadly the reverse situation during La Niña events (Marengo and Hastenrath 1993). One would therefore expect a tendency of stronger rainfall and higher stream flows during cold events and the opposite during warm events. In general, however, ENSO only explains a small fraction of the interannual variance of rainfall in the Amazon (Marengo et al. 1993, 2001; Rao et al. 1996; Dettinger et al. 2000). Therefore, the consequences for stream flows have small statistical significance. The Tocantins and San Francisco rivers are in the northeastern part of South America, where the regional climate is also influenced by sea-surface temperature (SST) and intertropical convergence zone (ITCZ) anomalies in the Atlantic. The La Plata Basin's climate, however, is significantly affected by ENSO through atmospheric teleconnections (Cazes et al., 2003).

At the same time, as discussed in this book by Islas and Schnack (Chapter 16), the South American coast is particularly exposed to ENSO-triggered effects. The effects are different in different regions. In Colombia, ENSO produces seasonal increases in mean sea level, altering the dynamics of the barrier islands. In Ecuador, coastal and island populations were badly affected during the 1997–1998 El Niño season by wave action and increasing sea levels, due to Kelvin waves from storms in the northern Pacific. They produced coastal erosion along the shoreline and destroyed structures along the beaches. In Peru, ENSOs trigger strong inputs of sediment to the coast, while similar processes impact the estuarine complexes of Lagoa dos Patos (southern Brazil) and cause Paraná River/Rio de la Plata floods (Argentina–Uruguay). In northern Brazil, on the other hand, dry conditions during El Niño favor the migration of coastal dunes.

3.2. Mountain Areas and Intense Tropical Rainfall

High and intense rates of tropical rain falling on mountain landscapes with steep slopes are a deadly combination, generating disasters. As analyzed in the chapters devoted to Colombia (Hermelin and Hoyos, Chapter 7), Venezuela (Bezada, Chapter 6), and Brazil (Coelho Netto and Avelar, Chapter 12 and Stevaux et al., Chapter 13), landslides and flash floods full of detritus flowing along mountain river valleys can attain catastrophic dimensions during periods of intense rainfall, especially for densely populated regions. Extreme landslide events produce excessive sediment yields, which converge into the channel network that drains the watersheds, causing immediate mass sedimentation of drainage systems in the river valleys, the piedmont plains, or coastal flatlands.

In Brazil, the majority of these phenomena occur in the steep mountainous lands of the *serras* and coastal ranges along the Brazilian Atlantic region (São Paulo, Rio de Janeiro, Minas Gerais, Espirito Santo, Rio Grande do Sul, Santa Catarina, and Paraná states). The relief is higher than 1200 m and close to the coast, but inland the mountainous escarpments can reach altitudes of up to 2000 m and relief amplitudes of more than 700 m. The combination of the steep relief, plus the existence of saprolite produced by tropical chemical weathering and high rates of rainfall, which can surpass 5000 mm (the highest pluviometric rates in Brazil), favor landslide processes. Most landslides occur during the summer (January to March) and are usually induced by less frequent, intense, and spatially nonuniform rainfall events (Coelho Netto and Avelar, Chapter 12; Stevaux et al., Chapter 13).

In Colombia, mountainous regions occupy about one-third of the country's territory, but these areas are home to most of the population. Rivers with headwaters in the Cordillera tend to receive high rates of precipitation. Rainfall can be extremely intense in the mountain areas, reaching values of more than 200 mm in a few hours. Heavy tropical rain acts as the triggering factor for erosion and produces generalized undercutting along stream banks and the reactivation of mass movements on the slopes. Because of the shifting of the ITCZ, landslides and floods occur predominantly during the rainy season, although part of the country has a bimodal distribution of rains, with floods occurring in April–May and October–November. Several examples of catastrophic floods and landslides are discussed in this book by Hermelin and Hoyos (Chapter 7).

In Venezuela, more than 50% of the population lives in the coastal-mountain regions, as most of the country's economic activities are concentrated in these areas of extreme geomorphological hazard, which are triggered by hydro-meteorological events.

Several tragic episodes of destruction caused by debris flow in alluvial fans along the Caribbean Cordillera and mountain rivers are analyzed by Bezada (Chapter 6). The debris flows and floods on alluvial fans at the mouths of the Caribbean coastal mountain drainage network in the state of Vargas on December 15 and 16, 1999, resulted in a catastrophic death toll of approximately 15,000 people, inundating coastal communities and causing severe property destruction estimated at more than $2 billion (Larsen and Wieczorek 2006). The catastrophic floods had been induced by extraordinary rains, which were caused by the extratropical synoptic atmospheric conditions associated with the polar front activity that affected the north of Venezuela during November–February. These particular conditions are produced with very variable annual frequency and intensity. The atmospheric instability is strengthened by the barrier created by the Caribbean Mountain Range along the north Venezuelan coast, which increases the amount and intensity of these extraordinary rains (Goldbrunner 1961, 1976, 1984). In just a few days, the rainfall reached values greater than the annual regional average precipitation (525 mm). The accumulated rainfall of 293 mm during the first two weeks of December was followed by an additional 911 mm of rainfall from December 14 to 16 (Larsen and Wieczorek 2006).

3.3. Floods in Tropical Rivers

Floods in the tropical rivers of South America affecting large populations are not as severe as those in other tropical regions of the world, such as Asia. This is a paradox considering that of the 34 largest tropical rivers in the world, in terms of water discharge, 14 flow entirely through South American countries, mainly Brazil (Latrubesse et al. 2005). Nevertheless, because of the high rates of rainfall recorded in tropical regions, together with the human occupation of dangerous and vulnerable floodplain environments, flood damage can be considerable in urban areas.

Of the large South American fluvial basins, the most dramatic floods take place in the Parana River Basin. The floods affect a large portion of the population in Brazil and Argentina, and produce significant economic losses and damage to infrastructure (Fig. 1.8). With a huge drainage area of 2,605,000 km^2 and a mean annual water discharge in the middle reach of 18,860 m^3/s in Corrientes, this giant river—one of the ten largest in the world in terms of water discharge—drains a variety of landscapes and climatic regions in Brazil, Bolivia, Paraguay, and Argentina. Almost all of the water discharged originates from two subbasins: the Upper Parana River Basin and the Paraguay River Basin (Table 1.2). With a drainage area of 1,150,000 km^2, lying mainly in Brazilian territory, and an absolute maximum rainfall of more than 2250 mm (in the middle Iguaçu River tributary basin), the Upper Paraná River Basin contributes 70% of the water discharged, largely due to the rainfall produced in southwestern Brazil. The Paraguay River, with a drainage area of 1,095,000 km^2, contributes ~22.4% of the water discharged, mainly originating from the Upper Paraguay River Basin, where the Pantanal of Mato Grosso is located.

The Upper Parana River Basin encompasses the most populated states in Brazil, including the industrial and agricultural cores of the country. With 24 million inhabitants, the Metropolitan Region of São Paulo (MRSP), located in the Tiete River Basin, a tributary of the Parana River, is one of the largest urban agglomerates in the world and the most important industrial center in South America.

Strong episodic rainfalls on small catchment areas flowing from the Serra do Mar and Cantareira Mountain Ranges to the Tietê River and its tributary, the Tamanduatei River, can produce dramatic floods in the São Paulo area. The problems of inundation on the Tiete River and its tributary began at the end of nineteenth century, when their floodplains started to be urbanized, especially owing to the construction of railroads and factories. The main problem was the human modifications to the original fluvial system. The Tiete was formerly characterized by a meandering pattern flowing over a broad natural floodplain. However, this channel has been totally modified by canalization, with large tracts having been cemented, accelerating the response of the systems to floods but eliminating the floodplain's capacity of absorption. The Upper Tietê River Basin in the MRSP has an area of 5985 km^2 and has a mean discharge of 104 m^3/s, with 30 m^3/s^{-1} of this being effluent. In some months, river discharge is only one-sixth of effluent discharge (see Chapter 13, for further details). The urban area of the Upper Tiete Basin increased from 190 km^2 in 1930 to 1900 km^2 in 2001—that is, a tenfold increase in 70 years. The most intensive and lasting floods occurred in 1924, 1983, 1991, and 2005.

Floods in the Parana River can produce remarkable disasters in Argentina as well. Along its middle reach, the Parana River flows on a wide and complex floodplain, 13 to 40 km in width, that is completely inundated during extraordinary floods. Fluvial stages and discharges have been recorded on the Parana River since the nineteenth century (1891 to present). In 116 years, the river has reached the critical stage, necessitating the evacuation of endangered populations 22 times. Very large floods are considered "exceptional" because of their magnitude. Nevertheless,

Table 1.2 General information for different sub-basins in the Rio de la Plata basin (1931–2002)

		Drainage Area (km^2)	Length (km)	Average Slope (m/km)	Mean Annual Discharge (m^3/s)	Period of Floods
Upper Paraná	**Parana**	1,510,000	3780	0.2381	**18858**	Jan-Mar
	Grande		1050	0.6898	1930	Jan-Mar
	Tiete		700	0.4143	752	Jan-Mar
	Paranapanema		750	0.6529	1384	Jan-Mar
	Iguazu		650	1.0189	1409	Sept-Nov
Paraguay		1,095,000	1850	0.1162	**4469**	May-July
Uruguay		365,000	1650	0.5875	**5122**	June-Oct

more floods than just the exceptional ones are, in reality, severe floods, often associated with El Niño-ENSO events. In contrast to the Parana River Basin, the Amazon Basin, the largest and most complex fluvial network in the world, is characterized by large rivers in a rainforest area with a very low population density. The largest Amazon cities are concentrated along the main rivers, but they are strategically located in areas that are not drastically affected by floods. A few cities in the southwestern Brazilian Amazon, where floods affect some urban areas because of a lack of urban planning, can be considered exceptions. For example, lowland rivers produce floods in Rio Branco, the capital city of the State of Acre, which affect thousands of people, as described in this book by Stevaux et al. (Chapter 13). However, the recurrence and magnitude of floods are very regular, and river management is relatively easier to solve. The main problem here is not the catastrophic characteristics of exceptional floods but insufficient urban planning, as the cities have grown very quickly, with humans occupying the floodplains (risk areas).

3.4. Tropical Cyclones and Disasters

The relatively narrow portion of Latin America that extends from Central America to Mexico is in the unenviable position of being affected by tropical storms and hurricanes that form both to their west in the eastern Pacific Ocean and to their east in the Atlantic Ocean (Fig. 1.6). The worst hurricane disaster in Central America was caused by Hurricane Mitch in 1998, which formed in the Caribbean Sea on October 22, 1998, and made landfall in Honduras on October 29, 1998. The hurricane moved very slowly through Honduras and Nicaragua, producing enormous amounts of rainfall. Some towns received in one day the amount of rainfall they usually receive in one year, leading to catastrophic flooding and landslides. Hurricane Mitch is one of the deadliest hurricanes in history, with nearly 11,000 people killed and over 8,000 left missing. In addition to the loss of human life, flooding and landslides caused an estimated $6 billion in damage. Floods related to hurricanes and cyclonic circulations had been affecting Central America on a large scale and part of Mexico as well (Alcántara Ayala, Chapter 4).

Tropical cyclones seldom affect South America, and when they do, they are confined to the Atlantic Basin, at low latitudes on the coast of Colombia and Venezuela. About 15 tropical cyclones have affected Colombia and Venezuela in the past 100 years, with only three making landfall (Tropical Storm Alma in 1974, Tropical Storm Bret in 1993 and Hurricane Joan in 1988). All three had tropical-storm intensity at the time of landfall in Colombia. Hurricane Joan hit Colombia and Venezuela as a tropical storm in October 1988, causing flooding and landslides that killed 36 people and left an estimated 27,000 people homeless. No precise data on Joan's economic damage or human impact were estimated (Pielke et al., 2003).

The first hurricane ever reported in the South Atlantic was Catarina, which reached category 2 on the Saffir-Simpson scale just before making landfall in Santa Catarina State on the southern coast of Brazil on March 28, 2004. Hurricane Catarina damaged some 36,000 homes, destroyed 1600 homes, and caused severe damage to agriculture in small coastal cities (Coelho Netto and Avelar, Chapter 12).

3.5. Drought and Floods in the Plains of the Middle Latitudes

The Chaco-Pampean Region, with an area of ~950,000 km^2 (which is 35% of the territory of Argentina) is the most densely populated and richest agriculture- and livestock-producing region of that country. It is characterized by a very flat landscape, where a local relief can be an impediment for water drainage, having even more impact than the regional slope.

Because of a heritage of widespread inactive quaternary aeolian landforms, a large part of the Pampa can be considered a paleodesert, which today is frequently flooded. In these extreme conditions, fluvial systems have great difficulty organizing well-structured drainage networks and frequently lack defined basin limits.

Historical floods and droughts have been recorded since 1575, and the area has been suffering an alternation of dry and wet episodes until the present day (Moncau 2001). However, since 1970, increasing precipitation has been recorded, causing an elevation of the water table in many regions and favoring flood events (Fig. 1.8).

3.6. A Bad Combination in Patagonia: Inappropriate Land Use in a Windy Land

Patagonia covers the southernmost part of the American continent, from 37°S almost to Cape Horn, at 56°S. With an area of 790,000 km^2, Patagonia lies between the subtropical high-pressure belt and subpolar low-pressure areas. As previously discussed, the area is entirely dominated by westerly winds. The Andes act as a topographic barrier, intercepting humid winds from the Pacific Ocean and defining a narrow western band (windward), with hyper-humid to humid climates, and a wider eastern area (leeward), where subhumid, semiarid, and arid climates prevail. The arid Extra-Andean Patagonia in Argentina covers an area of approximately 550,000 km^2.

Annual rainfall in Extra-Andean Patagonian territory is less than 250 mm a year, with rainfall prevailing during the winter. Permanent strong winds coming from the west and southwest blow across the entire region, with a mean annual velocity of more than 6 m/s. The windiest area is located between 47^0 and 49^0, where mean annual velocities reach more than 10m/s (Barros et al. 1997). Desertification processes started at the end of the nineteenth century with the settlement of woolgrowers in different areas (Fig. 1.8). Between 1880 and 1950, sheep stock increased rapidly from 1,790,000 head in 1895 to more than 25,000,000 head in 1952 (Huerta 1991).

As explained in detail in this book by Mazzoni (Chapter 17), overgrazng has been the main factor causing desertification. Of the 73.5 million ha of Patagonia that has been analyzed, 93.6% shows some signs of desertification (Mazzoni, Chapter 17). Huge areas have been classified into the severe or very severe categories of desertification. This means that they have highly degraded lands where the environmental damage is irreversible in terms of development for most economic activities.

Water and wind erosion are the geomorphological processes that are most strongly evidenced. Water erosion occurs in the mountain environments, especially those located in the western portion of Extra-Andean Patagonia, whereas in the plateau environments, gullies, usually several meters in depth, appear and the terrain suffers soil salinization. Wind action is characteristic of the area, with mounds, nebkhas, thickening deflation basins, and desert pavements being typical features. Dune fields appear only in isolated patches.

3.7. Extratropical Storms and the "Sudestada"

Many storm surges, with a duration varying from a few hours to two or three days, have been recorded along the Argentine coast in association with the northward-traveling tidal wave (Fig. 1.8). The storm surges are produced by the combined action of an anticyclone located to the west of Argentina (the semipermanent South Pacific anticyclone) and a cyclone located over the South Atlantic moving toward the east or northeast. Because of this situation, strong winds from the south or southwest and high water levels affect the whole Argentine coast, as well as the Rio de la Plata shores in Uruguay, and southern Brazil. The most hazardous floods affect the Rio de La Plata shores and the eastern coast of the Province of Buenos Aires due to the persistence of southeastern winds. These winds, if very strong, are often caused by an anticyclone located over southern Argentina and the adjacent ocean. The sea achieves an extraordinarily high level when, among other factors, a depression forms to the north of Buenos Aires, over Uruguay and southern Brazil. In addition, flooding is exacerbated by the Coriolis force piling up water as the storm surge travels along the Argentine coast, producing remarkably hazardous effects in the Rio de La Plata. This phenomenon is known regionally as *sudestada*.

3.8. Snow/Ice Storage in the Andes and Climatic Change

Global climate change (GCC) triggers various consequences, such as rising mean annual or seasonal temperature, rising or diminishing precipitations at the regional level, rising global sea level, and an increase in the frequency of extreme meteorological events, among others. In South America, especially in the Andean Mountain Range, Patagonia, Tierra del Fuego, and the Antarctic Peninsula, the impact of these changes has been observed in the glaciers and permafrost areas at least since 1978 and, particularly, in the last decade of the twentieth century (Rabassa, Chapter 19) (Fig. 1.8).

The main consequences of climatic change have been a fast glacier margin recession, the thinning of the ice cover, elevation of the regional snow line, and a reduction of Andean areas under permafrost. These negative effects have also been recorded in small mountain glaciers and ice all over the Andean Mountain Range, from Venezuela to Tierra del Fuego.

Rabassa (Chapter 19) has estimated that if the present rate of ice recession is maintained, most, if not all, of the cirque glaciers in Patagonia and Tierra del Fuego will disappear during the next two decades. In addition, both valley glaciers and the Patagonian ice sheets will be severely reduced. This could cause great damage to environmental and landscape heritage, stocks of water resources, and tourism resources.

Rabassa (Chapter 19) also estimates that the collapse of the ice barrier in the Antarctic Peninsula would create a great supply of large icebergs, affecting navigation routes and increasing sailing risks.

3.9. Climatic Trends over Recent Decades

Climatic trends over recent decades have been studied in the La Plata Fluvial Basin and the Pampean region, over an area of ~3,500,000 km^2, across a variety of geological settings and environmental conditions, and covering several countries (Brazil, Bolivia, Paraguay, Uruguay, and Argentina). In general terms, two anomalous precipitation patterns have occurred during the last 70 years: dry conditions from the mid 1930s to late 1960s and hyperhumid conditions from the late1960s onward (García and Vargas 1998).

The spatial–temporal structures of trends and interannual oscillatory components of precipitation over the whole La Plata Basin have been investigated by Krtepper and García (2004), using multivariate singular spectrum analysis. They presented evidence that the precipitation has cycles in the interannual frequency band of about 6 and 3.5 years, and a quasi-biennial oscillation, together with a trend component that explains less than 5% of the total variance. The relative importance of each oscillatory mode can be appreciated by the percentage of explained slower-than-annual variance. In particular, the 6-year, 3.5-year, and quasi-biennial cycles together account for 30.1% of the total. The 6-year oscillatory component (Fig. 1.9) shows a spatial pattern of extreme differences between opposite phases in annual precipitation, with two clear maxima (more than 50 mm) over the Iguazu River Basin and the sources of the Paranapanema and Tiete Rivers in Brazil. The

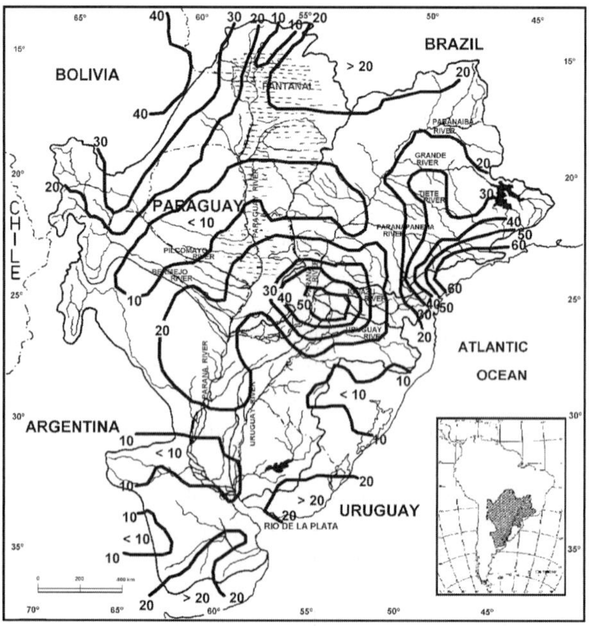

Figure 1.9 Differences (mm) between extreme high and low phases of the 6-year oscillation over the La Plata Basin.

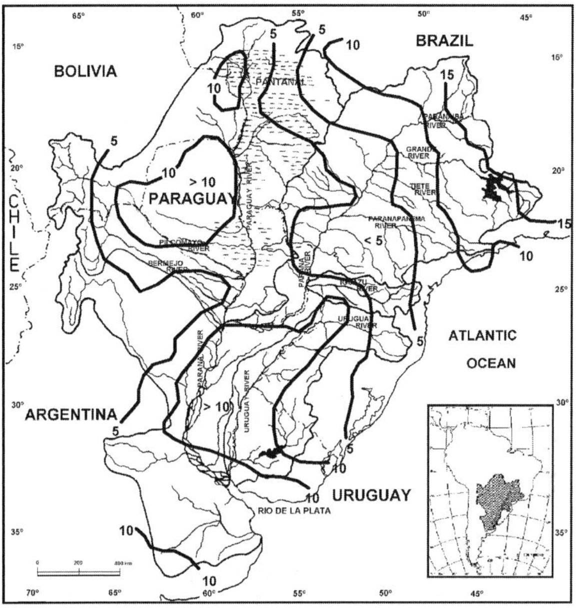

Figure 1.10 Differences (mm) between extreme high and low phases of the 3.5-year oscillation over the La Plata Basin.

map of opposite phases for the 3.5-year component (Fig. 1.10) shows a lower spatial variability compared to the 6-year pattern, and the largest values in the Brazilian Highlands are around 25% of those of the 6-year oscillation.

Larger values for increases in annual precipitation between 1950 and 1984 were recorded over the Pantanal (Brazil), and extreme decreases in values were recorded over the Pilcomayo and Bermejo River basins (two tributaries coming from the Andes). This had a direct consequence on the large-scale behavior of the major rivers of the basin, especially for the Paraguay River, where the largest contributions to the total variance, for periods longer than a year, corresponded to an upward trend.

Between 1956 and 1991, in the northeast of Argentina, the increases in the mean annual precipitation were greater than 10% in wide zones and up to more than 30% in other zones (Castañeda and Barros, 1994; García, 2000). The existence of changes in the trends and sudden increases in the means was shown by García and Vargas (1998).

There is no exact uniformity in the dates of changes in the tendencies of annual precipitation values among the scarce and widely spread rain-gauging stations. However, a notable similarity of behavior exists, as the stations, almost without exception, showed tendencies for changes in the same, or almost the same, years, around 1917–1918, 1943–1944 and 1970–1971 (see Table 1.3). The trends in annual precipitation have been positive throughout the entire set of stations, without exception, starting from the year 1970–1971.

Table 1.3 Temporal Behavior of the Annual Mean Precipitation in Northeastern Argentina (from García et al., 2007)

N^{er}	Gauging Station	Province	Change of Trend	Jump of the Mean
1	POSADAS	MISIONES	1944/1960/1977	1979
2	PARANA	ENTRE RÍOS	1942/1958/1969	1970
3	GUALEGUAYCHÚ	ENTRE RÍOS	1919/1942/1970	1973
4	CONCORDIA	ENTRE RÍOS	1917/1938/1968	1970
5	LABOULAYE	CORDOBA	1918/1952/1971	1977
6	MARCOS JUÁREZ	CORDOBA	1917/1943/1971	1970
7	RECONQUISTA	SANTA FE	1921/1944/1971/1993	1974
8	PASO DE LOS LIBRES	CORRIENTES	1920/1937/1971	1973
9	ROQUE SAENZ PEÑA	CHACO	1943/1960/1978	1980
10	CERES	SANTA FE	1917/1939/1960/1970	1970
11	CORRIENTES	CORRIENTES	1916/1945/1971	1973
12	CORDOBA	CORDOBA	1918/1939/1971	1972
13	FORMOSA	FORMOSA	1917/1943/1962	1962

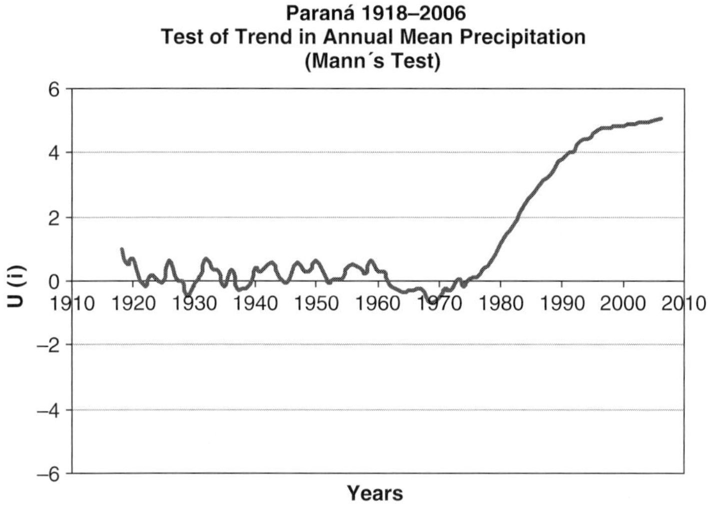

Figure 1.11 Graph of U(i) vs. Years (Mann's Test) for precipitation at Paraná station (center of the NEA region) (adapted from García et al., 2007).

The sudden increases in the means that marked a change in the climate were not always coincident with the trend changes, although most of them happened in the 1970s. This seems to have been a key date for climatic change because, starting in that period, the amount of water in the region, of both pluvial and fluvial origin, has increased. At the same time, the phreatic levels have risen dramatically. This positive trend embraced the whole region of northeastern Argentina as well as the south of Paraguay (Fig. 1.11).

In this region, one area, where the greatest positive trends have taken place, stands out: the south of the Province of Corrientes, where the increases in the mean annual values were more than 400 mm. As in the region of northeastern Argentina, the isohyets are approximately meridional, and the increase in the bi-annual precipitation over the whole region results in a displacement of the isohyets toward the west. Because of this, and also because of the implementation of new technologies, the agricultural frontier expanded toward the west. However, the negative side of these changes was the continuous or repeated flooding.

This increase in precipitation, which was not simultaneous across the whole of the Plata Basin, is probably linked to the greatest frequency and intensity of the El Niño phenomenon and to the strengthening of the South American low-level jet. These positive trends took place simultaneously with a remarkable heating of the polar areas (Barros and Doyle 1996; Barros et al. 2000).

Another indication of behavior change was seen through the recording of the greatest frequency of daily precipitations of more than 100 mm in almost all of the rain-gauging stations in the region (Canziani 2003, García et al. 2007). This is manifested by a greater decadal frequency of convective events, which were almost certainly mesoscale convective systems (MCS), although this was not demonstrated as starting from 1970 but from the first part of the twentieth century.

Changes in the frequency of droughts and floods have also been recorded in the Pampean region of Argentina. The frequency of annual droughts has been observed to decrease over the whole Humid Pampas region since 1970, with an average of one drought every three years until 1969 and one every five years thereafter. Considering the relative degrees of precipitation, it was concluded that the more intense dry periods occur during autumn, winter, and spring rather than summer (Venencio and García 2005). Venencio and García found that the spatial and temporal behavior of the dry periods showed that the amplitude and phase values of the three most significant harmonics in the series corresponded to the dry years. Both the summer rainfall and the winter droughts stress the amplitude of the second harmonic, which contains information about the amount of precipitation and represents the annual precipitation cycle, whereas the summer droughts diminish it. The second spectral component is the most significant, while the first is related to the ENSO signal in the equatorial Pacific Ocean (Venencio and García, 2005). These results indicated that, on average, both the north and the south-central regions of the Pampa have homogeneous behavior with regard to the intensity and time of occurrence of the dry events, whereas the extreme southern region has unique behavioral characteristics during the dry events.

4. Final Remarks

With most of its territory located in the tropical and equatorial areas, intense rainfall events are the main cause of natural disasters in Latin America. Intense tropical rainfall in mountain areas is responsible for some of the most catastrophic floods and landslides on the continent, such as those produced in northern Venezuela, Colombia, and southeastern Brazil. Central America and Mexico suffer severe flooding and landslides due to the intense rainfall, winds, and storm surges of landfalling hurricanes on their Atlantic and Eastern Pacific coasts.

On interannual timescales, ENSO events trigger a variety of geomorphological hazards across the continent, such as floods, landslides, and coastal erosion in the desert areas of the Pacific coast, floods and landslides in southeastern Brazil, and floods in the La Plata Basin. On decadal and longer timescales, climate trends that consist of increased precipitation and floods as well as an alarming recession of the Andean glaciers have had detrimental effects on one of the most productive agricultural areas in the world, the Pampas region.

Many of the terrible consequences of the natural hazards and disasters described in this chapter have been aggravated by the increased human-induced vulnerability of various Latin American regions. For instance, intensive and inappropriate land use in the past has produced environmental disasters and damage such as desertification in Patagonia and fluvial and sheet wash erosion in some states of southeastern Brazil. The scarcity or nonexistence of hazard and environmental planning, and the lack of disaster prevention and mitigation programs in Latin American countries, further aggravate the vulnerability of the region's populations to natural hazards and disasters. As public perception of these phenomena increases, so does the public sensitivity to the results of climate-driven disasters. Catastrophic environmental scenarios proposed by climate models within the framework of global climate change and Latin American governments will be pushed to put in place mechanisms to mitigate the effects of natural disasters in the region.

CHAPTER 2

Geomorphology as a Tool for Analysis of Seismogenic Sources in Latin America and the Caribbean

Carlos H. Costa[1], Franck A. Audemard M[2], Laurence Audin[3], *and* Carlos Benavente[4]

[1]*Departamento de Geología, Universidad Nacional de San Luis, Chacabuco 917, 5700 San Luis, Argentina. costa@unsl.edu.ar*
[2]*Departamento de Ciencias de la Tierra, FUNVISIS, Apartado Postal 76.880, Caracas 1073-A, Venezuela. faudemard@funvisis.gob.ve*
[3]*LMTG-UMR5563, CNRS/IRD/Université Tolouse 3, Casilla 18-1209, Lima 18, Perú. Laurence.audin@ird.fr*
[4]*INGEMMET, Av. Canadá 1470, San Borja, Lima, Perú. cbenavente@ingemmet.gob.pe*

Contents

Developments in Earth Surface Processes, Volume 13
ISSN 0928-2025, DOI 10.1016/S0928-2025(08)10002-5

1. Introduction

Seismicity, volcanism, and related processes have constituted a significant hazard in most regions of Latin America and the Caribbean since the Spanish arrival, and catastrophic events have been recorded in the pre-Hispanic period of America as well. The identification and analysis of most volcanic hazardous sources are quite direct, owing to their prominent geomorphic signature and precursory effects, such as gas emission and local seismicity. However, the recognition of many potential seismic sources commonly expressed at surface as faults and folds is not so straightforward.

In the outer lithosphere, an earthquake implies a sudden release of elastic energy through one or several rupture areas that may remain blind, without affecting the Earth's surface. However, depending on the size of the ruptured area and the depth of the hypocenter, the earthquake-related rupture may reach the surface, causing deformation as faulting or folding phenomena. These ruptured areas are located at major dynamic lithospheric borders as subduction zones and other interplate boundaries. Earthquakes tend to cluster in space and time at these critical zones, being envisaged as seismogenic sources, and therefore these zones are considered to be prone to future hazardous earthquakes.

A significant part of the deformation resulting from the interacting lithospheric plates is accommodated along plate boundaries at rates commonly ranging from 50 to 90 mm/yr. However, part of this deformation at a crustal scale is also transmitted through the continental crust away from these boundaries, elastically stressing the continental interiors at a much lower deformation rate. This fact turns many anisotropies and weaknesses of the continental crust (i.e. faults) into potential seismogenic sources, even those located at areas considered stable continental interiors (e.g., the Mississippi valley in the United States and Marryat and Tennant Creek in Australia). Because of the low slip rates (commonly <1 mm/yr), these structures may not show records of historical seismicity, and usually they cannot even be imaged by seismicity. Consequently, the significance of these structures as potential seismogenic sources can be missed or underestimated if the data for assessing the seismic hazard are based solely on the historical and instrumental seismicity of a region or individual structure.

Active fault studies have demonstrated a consistent relation between the slip rate of geological structures and the recurrence interval for destructive earthquakes along them (Audemard and Singer, 1996; Sieh, 1996; McCalpin, 1996; Villamor and Berryman, 1999; and many others). These results suggest that repetition of large earthquakes may be witnessed at a human temporal scale in structures accommodating significant deformation at interplate boundaries. For these cases, seismicity commonly corresponds to the spatial location of seismic sources, which most frequently correlates at surface with geological structures that show evidence of recent or even historic deformation. These seismic sources may be located onshore (i.e., the Mérida Andes of Venezuela and Tierra del Fuego in South America, the Polochic-Motagua area in Central America), or they may remain offshore such as those represented by shallow earthquakes along subduction zones.

The recurrence pattern for most seismogenic structures in the long term (10^4-10^5 years) could be periodic, clustered or random. However, is common that those structures with significant slip rates (> 5mm/yr) have already produced a large earthquake during the historical record, which in the Americas barely covers the last 500 years. This time span may allow the recording of at least one large earthquake along these structures, which can be regarded as representative for its seismic potential. However, it is well known that elastic energy can be stored for a considerable period of time (10^3–10^5 years) along structures or crustal anisotropies at very slow strain rates, which do not necessarily mean a lower seismogenic capability. Although large earthquakes at intraplate regions are less common, they can produce substantial damage, not only because of the earthquake itself, but also because people are much less prepared and structures are generally not designed to withstand strong ground motion.

When evaluating a seismic source's capability for producing destructive earthquakes in the future, it is important to estimate its maximum seismogenic potential. This is because the ground peak acceleration, which is related to the size of the seismic event among other characteristics, determines the safety parameters for building design and construction in general. Therefore it is a basic input in territorial planning.

Traditionally, seismic hazard assessments, even in intraplate regions, have relied on the seismic catalogue to identify areas where damaging earthquakes might occur in the near future. But based on contemporaneous experiences and on the scientific knowledge of mainly the last three decades, it is now widely accepted that there is a need to widen the possible seismic scenarios by expanding the time window of the seismic record into the past. In other words, it has been demonstrated that seismic hazard cannot be properly assessed with the data illuminating only the last 500 years (or frequently less than that). For areas where the seismic cycle of a source involves a time span beyond historic records, there are significant chances that they are characterized by moderate to long recurrence intervals (10^3–10^5 years). In the Andean region, for example, the geologic structures that show evidence of movements during the Quaternary (<1.8 Ma) are thought to be the ones that exhibit the highest likelihood of experiencing seismic events with social impact in the future.

The scientific approach to expanding the time frame of earthquake records into prehistoric times is based on the wide consensus that earthquakes of magnitude M > 6,5 with depths shallower than 30 km commonly deform the Earth's surface and/or lead to other secondary effects such as liquefaction and slope instability (Slemmons, 1977; Wallace, 1981; Bonilla, 1988; Wells and Coppersmith, 1994, McCalpin, 1996; and many others). Evidence of this sort can be preserved in the landscape and in the Quaternary stratigraphic record at deformation zones, with the analysis and interpretation of such evidence undertaken in the fields of earthquake geology and paleoseismology (Wallace, 1981, 1986; Yeats and Schwartz, 1990; McCalpin, 1996; Yeats and Prentice, 1996, among many others). Geomorphology plays an important role in this multidisciplinary approach because terrain analysis constitutes one of the most important steps in identifying and evaluating geomorphic assemblages potentially related to recent deformation of seismic origin. Although paleoseismic studies have less accuracy than the seismic catalogue, they help to provide more realistic assessments of the seismogenic capability for many structures, particularly in intraplate areas (e.g., Audemard, 2005).

Some case histories of structures with different tectonic regimes from the Andean region have been selected here to illustrate the role of terrain analysis in recognizing evidence of Quaternary activity. In some cases they have led to more detailed paleoseismic studies for seismic hazard assessments.

2. The Seismotectonic Setting of Latin America and the Caribbean

Western South America and most parts of Central America and the Caribbean lie close to dynamic areas of the Earth's crust, where earthquakes are common and often destructive. Figure 2.1 outlines major lithospheric plate boundaries and

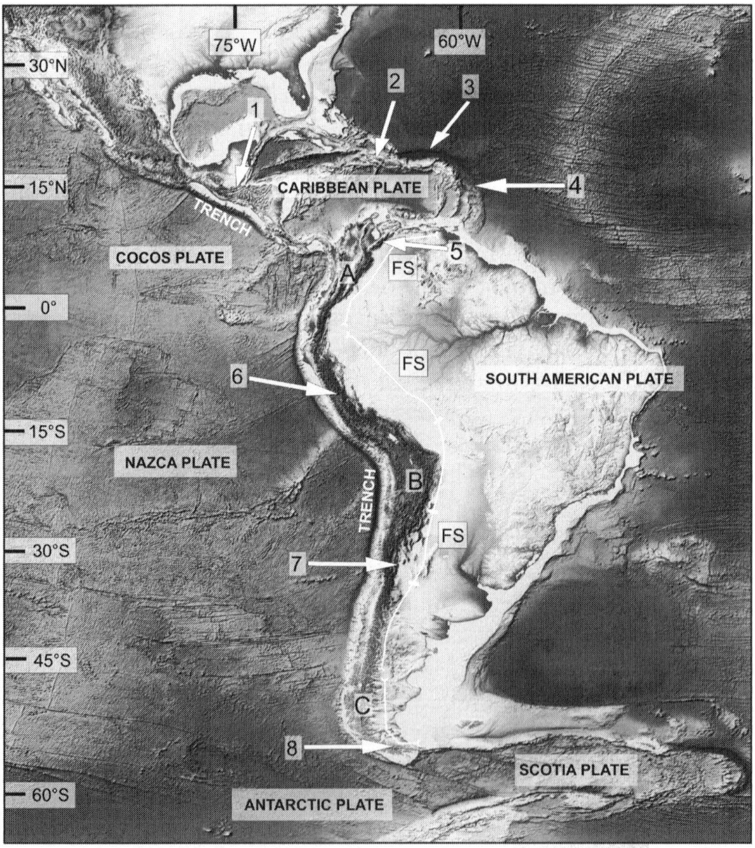

Figure 2.1 Major plate domains of South/Central America and the Caribbean. A, B, C: Northern, Central, and Southern Andes, respectively. 1. Motagua-Polochic deformation zone, 2. Hispaniola island, 3. Puerto Rico trough, 4. Lesser Antilles subduction zone, 5. Mérida Andes of Venezuela and Boconó fault, 6. Cordillera Blanca, Perú, 7. Argentina Precordillera, 8. Magallanes-Fagnano transform zone. FS: Flat-slab subduction segments. DEM downloaded from http://www.ngsg.noaa.gov/.

provides a suitable base for understanding the occurrence of earthquakes in space and time, particularly those related to interplate areas.

2.1. The Caribbean and Central America

Most of the Caribbean plate corresponds to oceanic crust and island arcs separating the major plates of North and South America (Fig. 2.1). The main historic earthquakes, current seismicity, and active geological structures are concentrated along plate margin interactions, linked to subduction zones and transform boundaries. In some cases, these transform zones do not exhibit a well-defined margin, but instead constitute a diffuse area characterized by higher seismic activity and slip rates of first-order geological structures.

The boundary with the North America plate (where almost all the territory of Mexico lies) is dominated by a left-lateral regime, expressed onshore near the Guatemala–Honduras border as the Motagua-Polochic deformation zone (Malfait and Dinkelman, 1972). The Motagua fault was the source for a M_s 7.5 earthquake in 1976 (Ms: Surface wave magnitude), accompanied by 230 km of left-lateral slip (Plafker, 1976). An example of the geometric array accompanying the surface rupture is shown in Figure 2.2. Other faults related to this plate boundary are the Oriente-Septentrional fault system in the island of Hispaniola, with active seismicity and documented evidence of prehistoric earthquakes (Prentice et al., 1993). The wrenching interaction between plates changes east of Hispaniola into subduction at the Puerto Rico trench.

The Lesser Antilles subduction zone constitutes the eastern boundary of the Carribbean plate, where an active volcanic arc is the main feature above sea level. Earthquakes and volcanism are related to subduction processes, and although minor

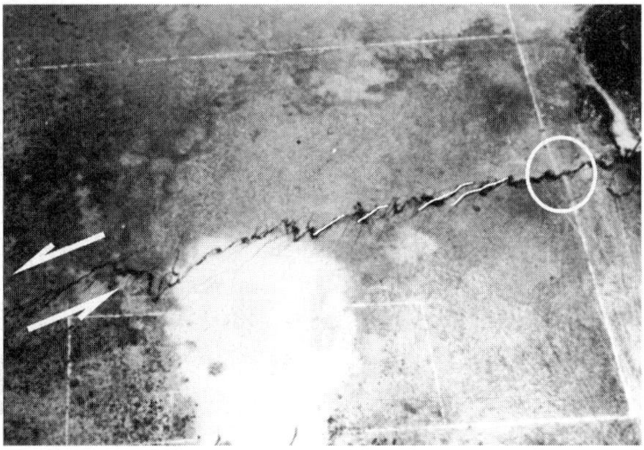

Figure 2.2 Surface ruptures along the Motagua fault at a soccer field, related to the Guatemala 1976 earthquake. Note the en echelon pattern (partly highlighted by white traces) corresponding to Riedel shears resulting from a left-lateral movement of the fault. This sense of displacement is also verified by the offset of the white lines to the right (within the white circle). Photo downloaded from USGS Photolibrary (http://www.usgs.gov).

to moderate seismicity may be tied with volcanic eruption, they are essentially separated processes, particularly for large earthquakes.

The southeastern boundary of the Caribbean plate and its present interaction with the South American plate is still a matter of controversy (e.g., Audemard and Audemard, 2002 and references therein). Yet there is general agreement that the plate boundary comprises several broad deformation zones where most relative plate motion is accommodated. The plate interaction also involves minor blocks, such as the North Andean, Maracaibo, and Panama blocks, whose affiliation and relationships to both plates have received several interpretations.

The subduction zone between the Cocos plate and Central America constitutes the southwestern boundary of the Caribbean plate, as well as the main seismic threat to the population established along the Central America isthmus. This is due to the shallow depth and seismic potential of the north-dipping subduction zone. Therefore, most damaging historic earthquakes correspond to this interplate seismicity, such as those that heavily damaged San Salvador city (El Salvador) in 1917 and Managua (Nicaragua) in 1972.

The Central America volcanic depression is a regional physiographic feature linked to the subduction polarity. It is highlighted by two large lakes, Managua and Nicaragua, and runs partly parallel to the active volcanic arc (Cowan et al., 2000). Although the tectonic nature of this depression is still a matter of debate, some authors have linked it to bounding faults with active seismicity (Dewey and Algermissen, 1974). The chain of active volcanoes is dissected by several steps, and it ends near the Costa Rica–Panama border in apparent relation to the flattening of the Cocos plate and the change of the subduction regime into a transform interaction between the Cocos and Caribbean plates (Taboada et al., 2000).

Earthquakes and active deformation in eastern Central America are concentrated along the Panama block boundaries. They relate to offshore deformed belts in many parts (North Panama and South Panama deformed belts) where plate interaction is accommodated mainly by shortening and strike-slip deformation. Other regions, such as its eastern boundary, also correspond to a zone of active but diffuse deformation, partly comprising the rain forest areas of northwestern Colombia. The Panama block seems to be colliding against the Colombian Andes, causing significant seismicity (Taboada et al., 2000).

2.2. South America

The Andean cordillera is considered to be the tectonic backbone of continental South America, and it has a long and complex history of terrain accretion along the former western boundary of Gondwana. It concentrated significant mountain-building processes during the Cenozoic, resulting in the highest non-collisional orogen worldwide (Ramos, 1999). This 8000-km-long chain has been traditionally divided into three main sectors: the Northern, the Central, and the Southern Andes (Fig. 2.1).

The most significant seismic hazard for South America in terms of recurrence and capability for producing very large earthquakes is the trench-related seismicity along the Pacific border (Fig. 2.1), which threatens the western Andean areas. This interplate seismicity is very shallow near the oceanic trench and coastal Pacific areas, and gets deeper below continental South America. The Mw 9.4 1960 Chilean earthquake, which is the largest earthquake ever recorded, was located at the shallow part of the subduction zone in south central Chile (Plafker and Savage 1970; Atwater et al., 1992). These shallow, trench-related earthquakes may also involve secondary effects that are sometimes even more dangerous than the earthquake itself, such as tsunamis and hillslope instability. This first-order tectonic feature caused the most important seismic crisis in South America. However, many other destructive earthquakes have been located within the South American plate, related in many cases to crustal features whose activity is connected to the Andean geodynamics. The seismic monitoring of these structures and their links with current seismicity is not as straightforward as at interplate margins.

The northernmost and southernmost ends of the Andes are currently dominated by strike-slip tectonics due to plate interaction, whereas the Andes comprised between 4° and 46° 30′ are considered to be the pure Andean-type orogen, where orogeny, magmatism, and earthquakes are basically driven by subduction (Ramos, 1999). The main geologic differences along the Andes are related to crustal nonhomogeneities linked to evolution of the terrain of western South America and to the geometry of the subducting Nazca plate (Barazanghi and Isacks, 1976, Ramos, 2008). The geometry of the subduction of the Nazca plate is the dominant factor that controls the characteristics of current seismicity and volcanism. Several latitudinal segments characterized by normal or subhorizontal subduction angles have been recognized based on interplate seismicity (Barazanghi and Isacks, 1976; Jordan et al., 1983; Ramos, 1999; Gutscher et al., 2000; Fig. 2.1).

At normal subduction segments, current deformation and crustal seismicity are concentrated within the Andean chain. Large earthquakes and active deformation at the foreland and other areas of the continental interior are rare. These segments exhibit active magmatism and a well-defined volcanic arc, hosting the most prominent Andean volcanoes.

Three segments where the Nazca plate subducts with subhorizontal angles have been recognized along the Andes (Fig. 2.1), in agreement with volcanic gaps of active magmatism. Another significant difference implies that active deformation processes, as well as crustal seismicity, are not constrained to the Andean orogen, but are also distributed within the foreland region as well (Jordan et al., 1983; Gutscher et al., 2000).

The tectonic setting of the Andes north of 4° 30′S is characterized by a complex interaction between the South American, Caribbean, and Nazca plates (Fig. 2.1), without a direct relation to active subduction zones (Audemard and Audemard, 2002). Earthquake sources are clustered along mountain chains, mainly in Venezuela and Colombia where major Quaternary deformation is concentrated.

This complex tectonic patchwork includes major features such as the northeast–southwest trending Boconó fault along the Venezuelan Mérida Andes and the east–west trending San Sebastián and El Pilar faults along the Caribbean coast of Venezuela (Schubert 1979, 1980, 1982, 1984; Soulas, 1986; Audemard et al., 2005), the Eastern Cordillera frontal fault zone in Colombia (Pennington, 1981), and the Dolores–Guayaquil megashear in Ecuador (Campbell, 1974). This belt of active deformation concentrates the highest known slip rates in continental South America (around 10 mm/yr) and is considered to be a diffuse boundary that detaches the Northern Andean Block from the remainder of South America (Pérez and Aggarwal, 1981; Schubert, 1982, 1984; Lavenu et al., 1995; Audemard et al., 2000, 2006; Taboada et al., 2000; Audemard and Audemard, 2002; Lavenu, 2006). These major deformation zones show a clear association with crustal seismicity and the location of historic damaging earthquakes.

The northern Andes comprise the Bucaramanga flat-slab. Although it exhibits strong differences with the other Andean flat-slab segments in terms of its geological past and current tectonic setting, all flat-lying subduction segments share a lack of active volcanism. This part of the Andean chain is also developed in close relation with a normal subduction segment of the Nazca plate in Colombia and northern Ecuador. Accordingly, it is characterized by active volcanism, namely, the Northern Volcanic zone. It is developed along the inter-Andean depression in Ecuador (i.e., Chimborazo, Cotopaxi, and Pichincha volcanoes) and along the Central and Western cordilleras in Colombia (i.e., the Nevado del Ruiz, Nevado del Huila, and Galeras volcanoes). Owing to eruption-related phenomena, they have produced many recent episodes of social concern and even catastrophes.

The long segment ascribed to the Central Andes (4°30′–46°30′S) is developed between the gulfs of Guayaquil and Penas. It is recognized as the typical Andean-type orogen, which exhibits a direct interaction between the upper South American plate and the subducting oceanic Nazca plate. It comprises two alternate segments of flat and normal subduction geometry.

The development of the Peruvian flat-slab segment (4°30′–14°00′S) has been related to the subduction of the Nazca aseismic ridge (Gutscher et al., 2000). The Cordillera Blanca, one of the Andes' highest areas and most breathtaking landscape, is located here, bounded by an active normal fault system (Schwartz, 1988; McNulty and Farber, 2002; Farber and Hancock, 2005). Shallow crustal seismicity characterizes the Eastern Cordillera and Subandean zone (Suárez et al., 1983; Dorbath et al., 1991). This region has witnessed two of the most prominent fault-related ruptures during historic earthquakes at the Quiches (M 7.25 in 1946) and Huaytapallana (M 6.2 in 1969) faults (Phillip and Mégard, 1977; Silgado, 1978). Although this latitudinal segment is also characterized by the lack of active volcanism, the active deformation and seismicity at the foreland areas are much less significant than in the other flat-slab segments (Ramos, 1999).

The normal subduction segment of the Central Andes, developed between 14°00′ and 27°00′S, is characterized by a widely distributed active volcanism (Central Volcanic zone) where the Altiplano and Puna plateaus stand out. They constitute andesitic-dacitic stratovolcanoes that erupted significant volumes of lavas and ignimbrites (Harmon and Rapela, 1991). This segment exhibits the widest section across the Andes (Fig. 2.1), where active deformation and potential seismogenic sources related to blind thrusting are concentrated at the Subandean zone (Dumont, 1996; Costa et al., 2006a).

The Pampean flat-slab (27°00′–33°30′S) documents one of the most widespread seismicity and active deformation at foreland areas (Jordan et al., 1983), being characterized by the block uplifts of the Sierras Pampeanas (Pampean Ranges).

According to GPS results, current deformation related to subduction is being accommodated at both margins of the Andean orogen (Kendrick et al., 1999). Therefore it has been proposed that the Andes itself behaves as a microplate at these latitudes (Brooks et al., 2003; Kendrick et al., 2003). Not very many active structures are reported at the western hillslope, although numerous Quaternary deformations have been described for the eastern Andean hillslope along the Precordillera piedmont (Costa et al., 2000a), including those related to historic destructive earthquakes in 1944 (Mw 7.0) and 1977 (Mw 7.4) (Mw: Moment magnitude). The shallowing of the subduction zone during the last eight Ma is considered to be the main reason for the eastward migration of magmatism and further cessation (Kay et al., 1991).

The seismic potential of the normal angle subduction zone of the south central Andes (33°30′–46°30′S) was verified by the large Chilean earthquake of 1960. Even 40 years after this event, GPS measurements indicate that the crust is still adjusting to this sudden strain release (Kendrick et al., 1999). Although generally clustered, crustal shallow seismicity does not show particular relation with active structures at the backarc.

The Southern Volcanic zone is developed overall in this latitudinal segment, as underlined by widespread volcanoes. Some of the volcanoes are active (i.e., the Lonquimay, Descabezado, Chaltén, and Hudson volcanoes), with contemporary episodes of ash fall.

The tectonic setting of the Southern Andes developed south of the gulf of Penas (46°30′S) up to the Beagle Channel area in Tierra del Fuego is dominated by the kinematic interaction among the South America, Scotia, and Antarctic plates. The strike-slip regime related to the interaction between the first two plates becomes dominant onshore in Tierra del Fuego, where the Magallanes-Fagnano fault constitutes a clear onshore transform boundary (Klepeis, 1994; Pelayo and Wiens, 1989). Two Ms 7.8 events took place within a few hours of each other in December 1949 and have been related to this fracture zone. Even if the total length of the seismic rupture remains unknown, it is considered to be one of the largest earthquake ruptures onshore in South America (Costa et al., 2006b).

3. Geomorphologic Analysis of Neotectonic Structures

The effect of tectonic features on landscape development and evolution has been widely recognized. However, the modification of tectonic-derived landforms by geomorphic processes under different morphoclimatic environments results in a wide variety of geomorphic signatures. To determine whether or not the resulting landforms are linked to active tectonic processes is not commonly a straightforward analysis. The essence of it involves recognizing the passive or active control of structures on landform assemblages.

For instance, scarps are the typical morphology related to faults, commonly represented by linear features. However, such morphologies can result from opposite processes such as Quaternary coseismic surface ruptures (purely tectonic), or they can evolve from ancient features (fractures, shear zones, and even lithologic anisotropies) enhanced by erosion. Alternatively, they can develop from many other situations not derived from structures that have undergone recent movements (purely morphodynamics). The successful discrimination between passive or active control by structures on landscape evolution then becomes a crucial issue.

A passive control is defined when the geomorphic signature of a tectonic landform is enhanced or overimposed by a geomorphic process, but not by the activity of the structure itself. On the contrary, an active control of structures on landscape implies the modification of a certain landform or landform assemblage due to a dynamic (continuous or periodic) tectonic process. Fault scarps, warping of alluvial surfaces, deflected drainages, and sudden changes in drainage patterns, among many reasons, could be a consequence of active control.

The geomorphic imprint of an active structure depends mainly on the dynamic interaction between the slip rate of a fault or the uplift rate of a fold and the erosion or sedimentation rate. Therefore, although the lack of diagnostic morphologies does not preclude the existence of shallow subsurface active structures with seismic capability, they provide the basic input in regional terrain analysis and semidetailed studies, favoring the selection of target areas for detailed paleoseismic studies.

4. Case Histories of Geomorphic Signature of Potential Seismogenic Sources

Through the following examples, we seek to illustrate the geomorphic expression of structures resulting from different tectonic regimes. Each example summarizes the evidence that led to the recognition of related Quaternary activity and consequently to more detailed *in situ* studies.

4.1. Geomorphic Signature of Quaternary Active Normal Faults: The Cordillera Blanca Fault System, Perú

The Cordillera Blanca is located in the Peruvian Andes and constitutes one of the most breathtaking Andean landscapes, with many peaks and glaciers above 6,000 m above sea level and deeply carved glacial valleys. The Cordillera Blanca fault system is the bounding structure of this mountain chain with a NNW trend and dominant normal-slip component of movement (Bonnot, 1988; Schwartz, 1988; McNulty and Farber, 2002; Farber and Hancock, 2005; Macharé et al., 2009. among many others), running approximately 200 km along its western hillslope (Fig. 2.3). The northern part of the fault system is expressed through a single fault trace dipping 35° to 45°W, whereas the southern section is characterized by several fault splays with a similar trend.

The Neogene vertical movements related to the fault activity and the Cordillera Blanca uplift are recorded in west-facing bedrock-cumulated scarps. The Quaternary and postglacial activity of the Cordillera Blanca can be recognized due to scarps affecting moraine deposits, whose estimated ages range from 11 to 14 ka (Farber and Hancock, 2005; Siame et al., 2006).

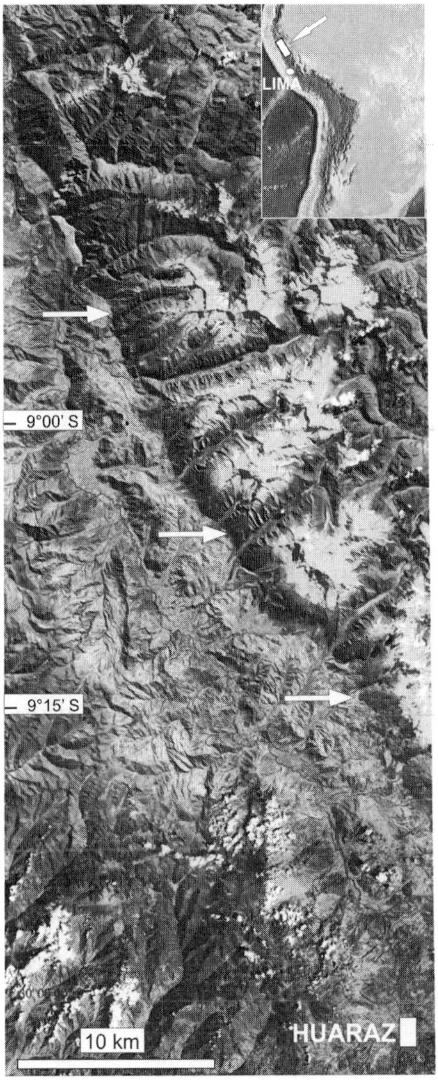

Figure 2.3 Satellite image of the Cordillera Blanca in Peru, where the snow-capped peaks stand out in light-gray -or similar-tones. The white arrows indicate the curvilinear trace of the Cordillera Blanca fault system.

Figure 2.4 shows multiple parallel scarps affecting unconsolidated moraine and hillslope deposits near the Cojup Creek. These linear features are directly related to the main fault trace, and a gravitational origin cannot be imputed. They are well preserved despite the significant slope angles, which suggest an active control due to repeated fault movements, together with the high resistance of the exposed bedrock to erosion. This fact accounts for the Late Pleistocene-Holocene activity of this fault and provides a rough estimation of its slip rate, ranging from 0.7 mm/yr

Figure 2.4 Oblique aerial photo (southeast-looking) of the western flank of the Cordillera Blanca at the Quebrada Cojup (middle-left). Main traces of the Cordillera Blanca fault system affecting moraine and hillslope deposits are indicated with arrows. Note the vertical offset of moraine arcs within the white circle. Photo courtesy of the Servicio Aerofotográfico Nacional, Peru.

(Bonnot, 1988) up to 3 mm/yr (Siame et al., 2006). Faulted moraine axis, as well as numerous slickenside striations, indicate that a normal faulting component of slip prevails, without a significant strike-slip component.

Geomorphic evidence of recent activity led to investigation of the related seismic record of the Cordillera Blanca fault at the Querococha Creek, whose postglacial activity is well depicted by the faulted edge of a moraine (Fig. 2.5). Schwartz (1988) found evidence for at least five earthquakes preserved in 14 kyr-old fluvial and glacio-lacustrine deposits, which yielded estimates of slip rates ranging from 0.86 to 1.36 mm/yr and derived recurrence intervals of 1500 years.

There are no reported historic damaging earthquakes related to this fault system (Silgado, 1978), and the zone is at present characterized by low to moderate crustal seismicity (Macharé et al., 2009). The active landscape imprints related to faulting and the prehistoric record suggest that the Cordillera Blanca fault system is a main potential seismic source. Because of the high hillslope angles, the whole area is also susceptible to earthquake-induced phenomena. A widely known natural disaster took place here in 1970 when massive avalanches (causing 5000 casualties) were triggered by a subduction earthquake. These avalanches started from huge ice detachments from the Nevado Huascarán, from altitudes higher than 6300 m, destroying the villages of Yungay and Ranrahirca, located 2000 m below.

4.2. Geomorphic Signature of Quaternary Reverse Faults: The La Rinconada Fault, Argentina

This reverse fault crops out at the eastern Precordillera foothills in arid western Argentina, located within the most active seismic corridor of the Pampean flat-slab backarc. Although most reverse and thrust faults exhibit a sinuous trace, the La Rinconada fault-related scarp stands out from the alluvial plain as west-facing

Figure 2.5 North-looking view taken from the Quebrada Querococha where the fault trace and the west-dipping attitude of the Cordillera Blanca fault are clearly visible. Note the vertical offset of the moraine arc at the skyline.

rectilinear esscarpments (Fig. 2.6). This fact is due to a high angle bedding-parallel fault surface interpreted as a consequence of flexural-slip folding (Costa et al., 1999, 2006a; Meigs et al., 2006). The La Rinconada fault has been identified as the seismic source of the Mw 6.8 earthquake that struck the region in 1952, although no evidence of coseismic surface deformation has been reported.

The fault scarp accounts for decametric displacement of older Quaternary strath terraces, even if the scarp amplitude is enhanced in many cases by subsequent streams controlled by this tectonic feature. Therefore it is difficult to estimate the overall slip because the alluvial surfaces on both fault walls do not correlate.

Gentle scarplets affecting young alluvial deposits (Fig. 2.7) have been identified due to the preferred vegetation lineament and subtle tone variation caused by the ponding of fine-grained material against the hanging-wall (to the east). They were thought to be related to low-angle propagating thrusts resulting in fold limb scarps (Costa, 2009a) rather than in rectilinear scarps. The fault-related stratigraphy investigated through trenches confirmed the low-angle thrusting of recent

Figure 2.6 Northeast-looking bird's-eye view of the clearly visible rectilinear trace of the La Rinconada fault, standing out from the piedmont alluvial plane of the Eastern Precordillera, south of San Juan city in western Argentina. The scarp is cored in Neogene sedimentary rock (with light colors in the photo).

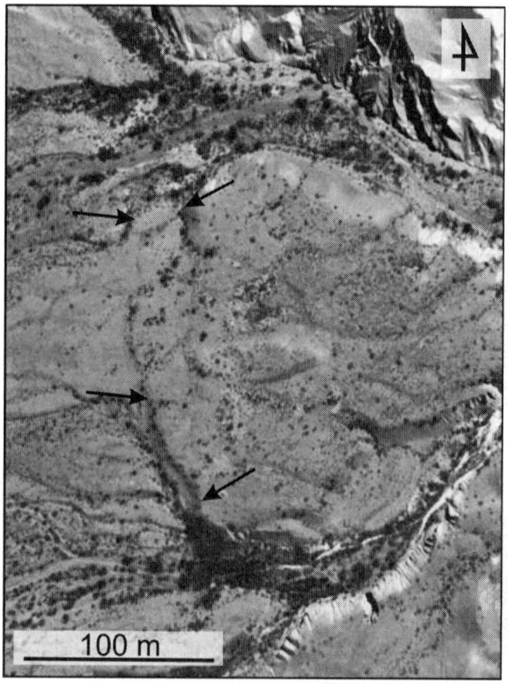

Figure 2.7 Gentle scarps in young alluvium (Holocene) (pointed out by black arrows). The fault propagation into unconsolidated and nonstratified deposits results in a lobate contour rather than a rectilinear trace. Image downloaded from Google Earth.

(Holocene?) fine-grained alluvium ponded due to recent fault movements. The upper propagating trace is flat-lying, giving rise to a monocline in the youngest uplifted alluvium rather than a clear surface faulting.

4.3. Geomorphic Signature of Quaternary Active Strike-slip Faults: The Boconó Fault, Venezuela

The Boconó fault system is a first-order structure that runs along the Mérida Andes in Venezuela, from the Caribbean coast down to the Venezuela–Colombia border. It concentrates the highest slip rates (up to 10 mm/yr) in northern South America and is probably the most studied and best known feature in terms of its neotectonic significance. Historical and instrumental seismicity also characterizes this region, and several significant earthquakes between the seventeenth and nineteenth centuries have been linked to this seismic source.

The Boconó fault has a sinuous trace that traverses the Mérida Andes for more than 400 km, flanked on both sides by low-angle thrusts (Audemard, 1999; Audemard et al., 2000; Audemard and Audemard, 2002). The fault usually runs in an axial position along this mountain chain at different altitudinal levels and morphoclimatic settings. One of the best exposed fault-related landforms as evidence of Quaternary activity is located near the village of Apartaderos (about 40 km northeast of the town of Mérida). This sector corresponds to a major drainage divide above 3000 m.a.s.l, where distal moraines formed during the Last Glacial Maximum (18 Kyr) have been laterally offset in plan view (Schubert, 1980b; Soulas, 1985; Audemard et al., 1999, Carrillo et al., 2006).

Figures 2.8 and 2.9 show the Los Zerpa offset moraine, which constitutes not only excellent kinematic evidence for the right-lateral regime of the Boconó fault,

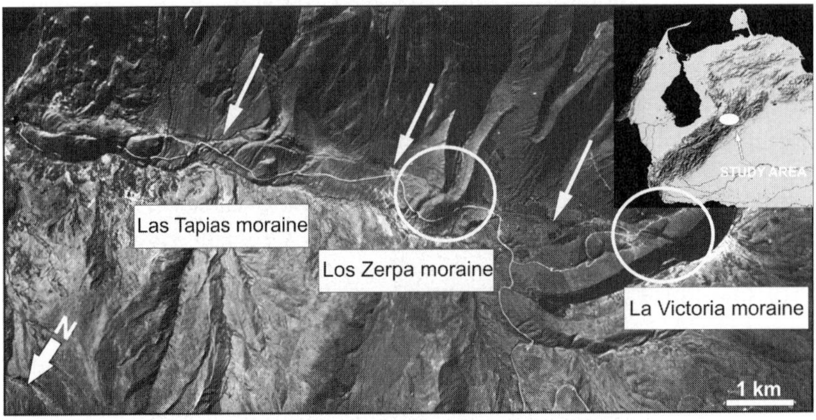

Figure 2.8 Aerial photo of LGM moraines affected by the Boconó fault at the Mérida Andes, accounting for Quaternary postglacial slip. The right-lateral sense of movement is indicated by the offset moraine crestlines within the white circles (Los Zerpa and La Victoria moraines). See detail in Figure 2.9. Aerial photo of mission 010455. Courtesy of Cartografía Nacional (currently IGVSB).

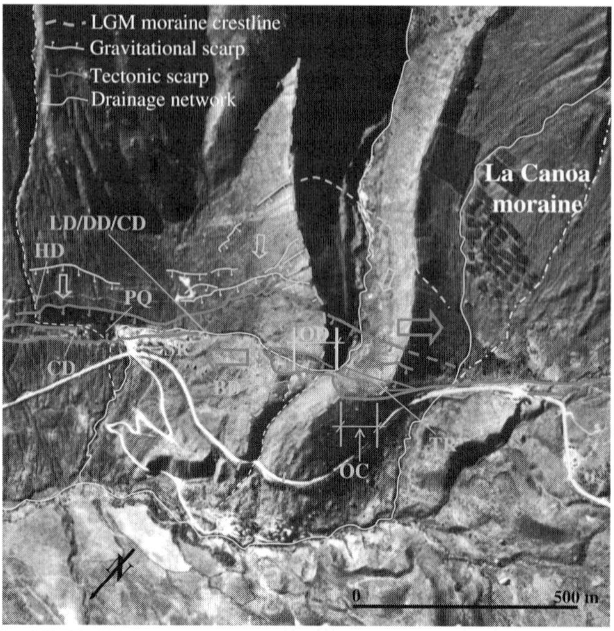

Legend: BD: Beheaded drainage; **CD**: Captured drainage; **DD**: Diverted drainage; **HD**: Hanging drainage; **LD**: Linear drainage; **OC**: Offset crestline; **OD**: Offset drainage; **PQ**: Ponded Quaternary deposits; **SR**: Shutter ridge; **TR**: Fault trench

Figure 2.9 Aerial photograph of the Los Zerpa moraine with the main trace of the Boconó fault outlined by fault-related landforms (translated from Audemard, 2009). Moraine crestlines are shown in yellow, while fault traces are highlighted in red. An active, only ~10-m deep, small pull-apart basin forms at the back of the frontal moraine, at a lazy releasing bend of the Boconó fault main active trace, which induces the gravitational collapse of the right-lateral moraine. OC and OD allow an estimation of the brittle dextral offset of the frontal moraine with respect to the lateral moraines. Also notice that the original late glacial drainage outlet is abandoned across the frontal moraine. The current drainage is deflected and captured to the east, giving the impression of a larger amount of dextral slip. Blow-up of aerial photo of mission 010455 (Courtesy of Cartografía Nacional; currently IGVSB).

but also a good constraint for the slip rate (in the order of 5 to 10 mm/yr) of this fault during the last 15 kyr (Schubert, 1980a; Soulas, 1985; Audemard, et al., 1999; Audemard and Audemard, 2002). The active fault trace is underlined here by a deformation zone, composed of north- and south-facing scarps bounding a depressed zone across the distal moraine deposits. Other morphologies related to prevailing strike-slip movements are also present, such as shutteridges, fault trench and related pounded alluvium, as well as secondary gravitational-induced scarplets (Audemard, 2009; Fig. 2.9).

The right-lateral displacement of the fault during the last 15 kyr resulted in the development of several diagnostic tectonic landforms highlighted in Figure 2.9. The tectonic offset of the frontal moraine has forced stream deflection and capture, with periods of alluvial ponding and upstream lake formation. The sedimentary record of lacustrine and fluvial deposits has been used to reconstruct the postglacial fault seismic history (Carrillo et al., 2006). It has also shed light on how the sliding of the right lateral moraine has been interplaying with tectonic fault slip, as recorded by the moraine-dammed paleolake at Los Zerpa (Carrillo et al., 2006).

A minimum post–15 kyr lateral offset of about 100 m can be measured at Los Zerpa moraine complex from at least two different criteria (OC and OD in Fig. 2.9). This value can be even larger if ductile deformation of the lateral moraine crestlines is taken into account; reaching values on the order of 10 mm/yr (Audemard et al., 1999). Similarly, a vertical throw of about 10 m can be estimated at the pull-apart basin affecting the frontal moraine (Audemard, 2009). This attests to the clearly dominant recent strike-slip component of motion of this fault, even inside transtensional jogs or bends (equal to ten times the vertical component).

4.4. Geomorphic Signature of Quaternary Active Folds: Montecito Anticline, Argentina

The Montecito anticline is located at the eastern foothills of the Southern Precordillera, Argentina, 50 km to the north of Mendoza city (1.5 million inhabitants), where the active Andean thrust front is underlined by current seismicity and historic destructive earthquakes. The arid climate and the scarce vegetation turn this Andean frontal section into a suitable area for landform preservation and outcrop exposure. The fold is located within a linkage zone between two oblique antithetic thrust systems, being interpreted as the surface expression of a west-verging fault-propagation thrust (Vergés et al., 2007).

The positive relief of the fold limbs reveals the doubly plunging fold trace, which clearly emerges from the surrounding alluvial plain (Fig. 2.10). The western fold limb stratigraphy exhibits a dynamic interaction between the fold growth and the Quaternary alluvial fan sedimentation recorded as onlap geometries (Costa et al., 2000b, Costa, 2009b). The anticline is cored with Late Tertiary continental beds (the light-colored unit in Fig. 2.10), whereas conglomerates and gravels of Late Pliocene and Quaternary age are exposed at their limbs.

The drainage network also suggests the Quaternary uplift of this structure. Some streams have been deflected, as indicated by parallel arc patterns at both fold periclinal closures. Other stream courses merged at the western outer limb and carved epigenic valleys across the structure where pseudo-meandric patterns dominate westward from the fold axis. The adjustment of the longitudinal profile of streams that run across the structure determined the development of alluvial fans, with their apex at the outer eastern flank of the anticline. Young terrace surfaces assigned to the Holocene are tilted against slope up to 12°W, accounting for the recent, and probable ongoing, activity of this structure. The geomorphologic analysis complements the stratigraphic evidence for Quaternary uplift of this fault-related fold and suggests that piedmont slope changes due to fold growth took place in a coeval relation with the surrounding alluvial plain development during the Quaternary. Although instrumental seismicity does not image this source for potential earthquakes at depth, we suspect that fold growth was accompanied by shallow prehistoric earthquakes.

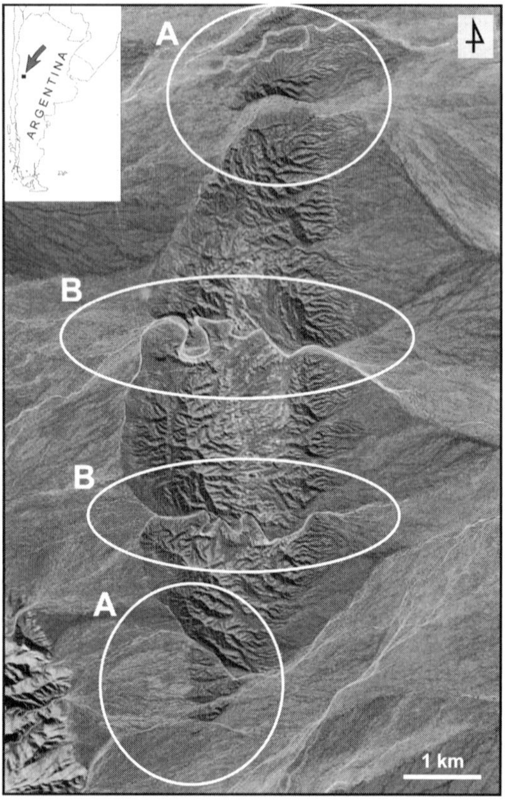

Figure 2.10 Aerial image downloaded from Google Earth of the Montecito anticline, in the Mendoza Precordillera piedmont (western Argentina). Note the parallel arc of the drainage at both periclinal closures and the water gaps related to the main two courses running across the structure.

5. Concluding Remarks

Because of their tectonic setting, many areas in Latin America and the Caribbean have been or could be seriously damaged by earthquakes. Considering the severe social and economic effects that these natural catastrophes can produce, identification and characterization of potential seismic sources with and without previous seismic records are mandatory for land-use planning and decision-making purposes.

Based on the concept that active geological structures during the Quaternary could be the source and epicenter of future seismic activity, many efforts during the last decades have focused on the identification and study of these features. Terrain analysis through aerial images has proved to be a necessary approach for recognizing potentially hazardous structures. This is particularly useful in areas lacking a suitable background of geologic/seismologic knowledge, as well as for structures without records of present seismicity.

Geomorphologic analysis of the active or passive control that tectonic features produce on landscape evolution is crucial for determining whether or not Quaternary movements have been experienced. This in turn leads to select targets prone to detailed paleoseismic studies whose main goal is to recognize evidence of prehistoric hazardous seismic crisis with associated surface deformation.

The examples selected in this chapter are case histories of structures resulting from different tectonic regimes under diverse morphoclimatic conditions. Their morphologic imprints allow the recognition of activity during the Quaternary, envisaged as a consequence of past earthquakes. These basic considerations help optimize the selection of suspected zones for field studies, which is crucial with regard to making the best use of time and resources when exploring large regions with basic data.

Almost 600 structures with proved or suspected activity during the Quaternary have already been recognized as a result of a recent international effort (Multi Andean Project-Geosciences for Andean Communities –http://can.geosemantica.net-). For most cases, their recognition was based on the geomorphic signature of these structures.

CHAPTER 3

The Changing Coastlines of South America

Federico I. Isla[a] *and* Enrique J. Schnack[b]

[a]*Consejo Nacional de Investigaciones Científicas y Tecnológicas – CONICET- and Centro de Geología de Costas, Facultad de Ciencias Exactas y Naturales, Universidad Nacional de Mar del Plata. Complejo Universitario, Funes 3350, 7600 Mar del Plata, Argentina. fisla@mdp.edu.ar*

[b]*Comisión de Investigaciones Científicas de la Provincia de Buenos Aires – CIC- and Laboratorio de Oceanografía Costera, Facultad de Ciencias Naturales y Museo, Universidad Nacional de La Plata, 64 y 120, 1900 La Plata, Argentina. eschnack@netverk.com.ar*

Contents

1. Introduction

In South America, as well as in most regions of the world, people prefer to live in coastal areas. Because of the diversity of coastal environments and of the physical factors resulting from the ocean–atmosphere interactions, the coast of South America is exposed to the effects of different processes. Tropical cyclones (hurricanes) occur in the Caribbean region, whereas extratropical storms are

Developments in Earth Surface Processes, Volume 13
ISSN 0928-2025, DOI 10.1016/S0928-2025(08)10003-7

typically produced in the Southwest Atlantic Ocean, from southern Argentina to southern Brazil. While wave action is dominated by west coast swell environments and ENSO effects from central Chile to Colombia, storm waves are predominant in southern Chile. East coast swell environments prevail along much of the east-facing coast of Brazil, and trade winds generate low-energy waves on the northeast-facing of South America from Brazil to Guyana (Fig. 3.1). In addition, the Pacific and Caribbean coasts are exposed to tsunamigenic events.

One of the most direct consequences of the land–sea interactions along the South American coast is their impact on shorelines, resulting in coastal erosion, loss of wetlands, and salt intrusion in coastal aquifers, among other effects. Even considering the global change scenarios, many of them forecasting a rise in sea level and an increase in the frequency and intensity of storms (IPCC, 2001), anthropogenic causes may be considered a major contributor to coastal change.

This review deals with natural processes affecting the coast of South America under different driving mechanisms operating over variable timescales, both episodic and secular. Because human activities are increasingly influencing coastal development, the effects of these processes are analyzed in the light of present and future scenarios.

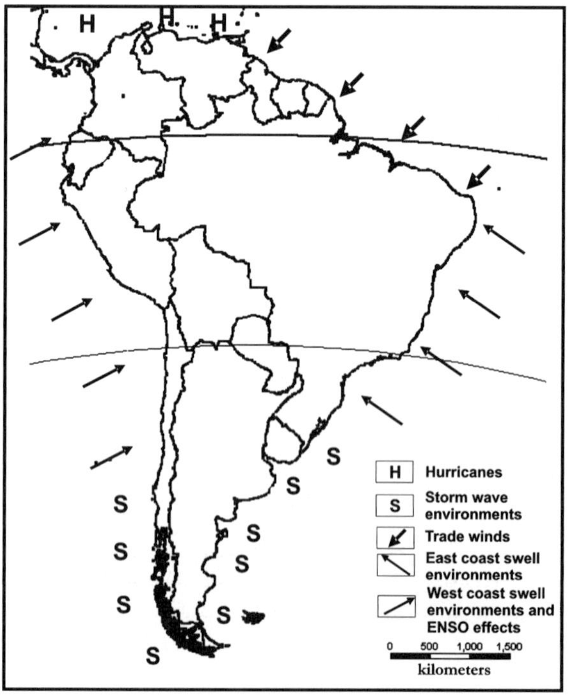

Figure 3.1 Major physical approaches on the South American coast. (*Source*: Modified from Davies, 1980).

2. Sea-Level Trends: Past, Present, and Future

In general terms, the Atlantic coastline is on a trailing-edge margin, and the Pacific coast is subject to tectonism in a similar way to that of the Caribbean coast (Fig. 3.2). Sea-level rise in the order of 1 to 2 mm/year is dominant in the whole continent as well as worldwide (Church et al., 2001; Munk, 2003; Alley et al., 2005). On the Pacific coast, only certain tectonic blocks, subject to earthquakes, seem to be uplifting: Tumaco (Colombia), Matarani (Peru), and Antofagasta (Chile). These uplifted areas are surely related to the oceanic ridges (Carnegie, Nazca, X) subducting below the South American plate. This fact makes it difficult to estimate sea-level trends.

Although the Pacific coastline of Colombia is subject to tsunamis and El Niño effects, a subsidence of 1.2 to 1.5 m in the last 500 years in the San Juan River delta is explained by coseismic effects (González and Correa, 2001). Within this tectonic domain, and considering a mesotidal regime, 62 barrier islands have evolved in a climate dominated by rainfalls and with dense vegetation dominated by mangrove swamps and rain forest (Martínez et al., 2000). These islands are very low (less than 2 m over the spring tide level), of short lengths (5.8–8.6 km), and dominated by beach ridges and washovers. They formed due to a high sand supply from rivers and a littoral transport from south to north (Martínez et al., 2000). On the other hand, on the Caribbean coast of Colombia, although the historical tide gauge of Cartagena sums a

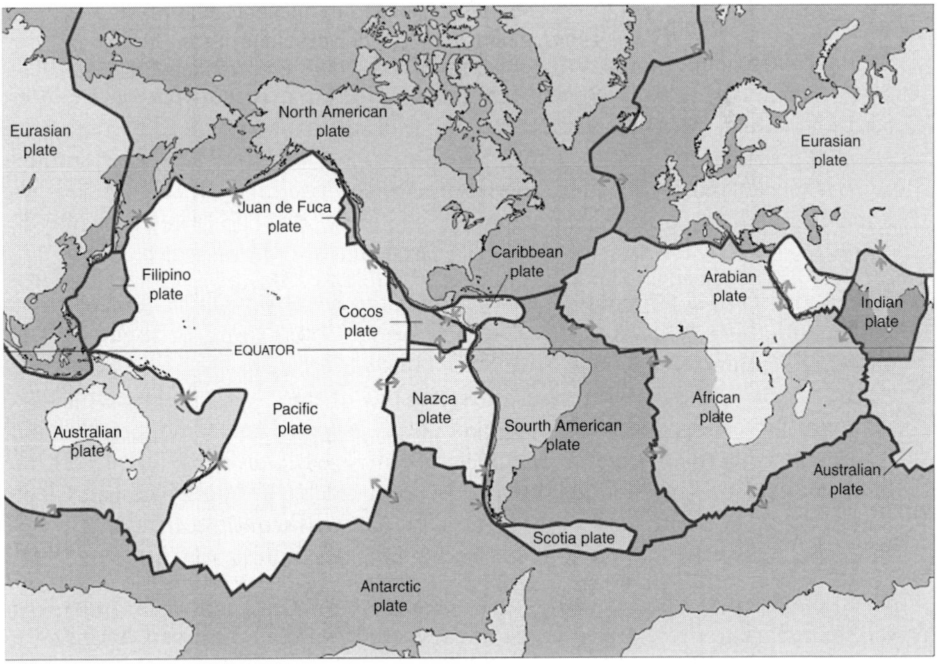

Figure 3.2 Tectonic setting of South America and adjacent regions. (*Source:* USGS, http://geology.er.usgs.gov).

sea-level rise of 22 cm (Correa et al., 2007), at the Gulf of Morrosquillo (Belt of the Sinú Delta) subsidence was estimated in 0.07 cm/yr (Correa and Vernette, 2004).

Sea-level trends along the seismic coast of northern Peru are not clearly explained. Tide gauges from Libertad and Talara indicate subsidence (Emery and Aubrey 1991). Beach-ridge plains related to river deltas can be assigned to the Holocene glacioeustatic sea-level fluctuation (Isla, 1989) or to coseismic uplifting trends (Patagonia). The beach ridge plains of the Chira and Piura rivers were conditioned to the availability of sediment related to former strong El Niño events (Ortlieb and Macharé, 1993). For the coast of Venezuela a sea-level rise of 2 mm/yr was estimated for the last 20 years (Almeida, 1995).

South America is a plate moving westward and away from the Mid-Atlantic Ridge. It is colliding mostly against the oceanic Nazca plate and its ridges (Malpelo, Carnegie, Nazca, X, Juan Fernandez, Selkirk, Mocha and Chile). To the north, the plate interacts with the Cocos and Caribbean plates. To the south, the South American plate also interacts with the Antarctic and Scotia plates. Within this tectonic setting, the trailing-edge continental shelf of Argentina has a minimum uplift of 8 to 9 cm/kyrs (Guilderson et al., 2000). There is evidence of the lowest sea level of −105 m about 15,000 years ago (Guilderson et al., 2000) and a sea level highstand during the Mid-Holocene (Isla, 1989; Ota and Paskoff, 1993; Martin et al., 2003; Martin et al., 2006a).

Over the last century, sea level rose globally about 1.0 to 2.0 mm/year, with water expansion from warming contributing 0.5 ± 0.2 mm (steric change) and the rest coming from the addition of water to the oceans (eustatic change) due mostly to the melting of land ice (Church et al., 2001). Different scenarios for sea-level rise have been proposed in relation to greenhouse concentrations. From different climate scenarios and model uncertainties, the loss of the Greenland Ice Sheet, plus the contributions from thermal expansion and a partial collapse of the West Antarctic Ice Sheet (WAIS) over the coming millennium, would lead to a sea-level increase of 1 m per century (Nicholls and Lowe, 2005). However, other models show that Greenland is the main contributor to the rise in sea level, whereas the East and West Antarctic Ice Sheets appear to be nearly balanced (Alley et al., 2005).

With regard to the tide-gauge records, three facts should be stressed:

1. Some tidal stations are located in estuaries. Although seasonal changes are not so evident in these stations (Emery and Aubrey, 1991), some of them may be subject to significant interannual ENSO effects.
2. South America is less subject than other regions of the Northern Hemisphere to glacioisostatic effects. Although some rebound is expected and some tectonic uplift has been estimated (Fuenzalida and Harambour, 1984; Gordillo et al., 1992), the highstand evidences of Tierra del Fuego were produced mainly by glacioeustatic effects.
3. The records in South America are usually scarce, discontinuous, and/or very short, being less than 100 years old (Pirazzoli, 1986; Emery and Aubrey, 1991).

Even considering these constraints, it should be noted that about 20 years have passed since the latter references, and a few records from the Southwest Atlantic can be considered a reliable reference for relatively stable areas on a subcontinental basis. Hourly sea-level measurements from the Mar del Plata (Argentina) tide gauge station for the period 1954–2002 have been used as the rough data to calculate a filtered series of absolute annual mean sea levels. The linear regression analysis of this series shows a trend of + 1.4 ± 0,01 mm/yr (Pousa et al., 2006; Fig. 3.3a). An even longer record for the Quequén Estuary tide gauge shows a 1.6 ± 0.2 mm/yr rise in sea level (Lanfredi et al. 1998). Moreover, an almost 100-year-old record for

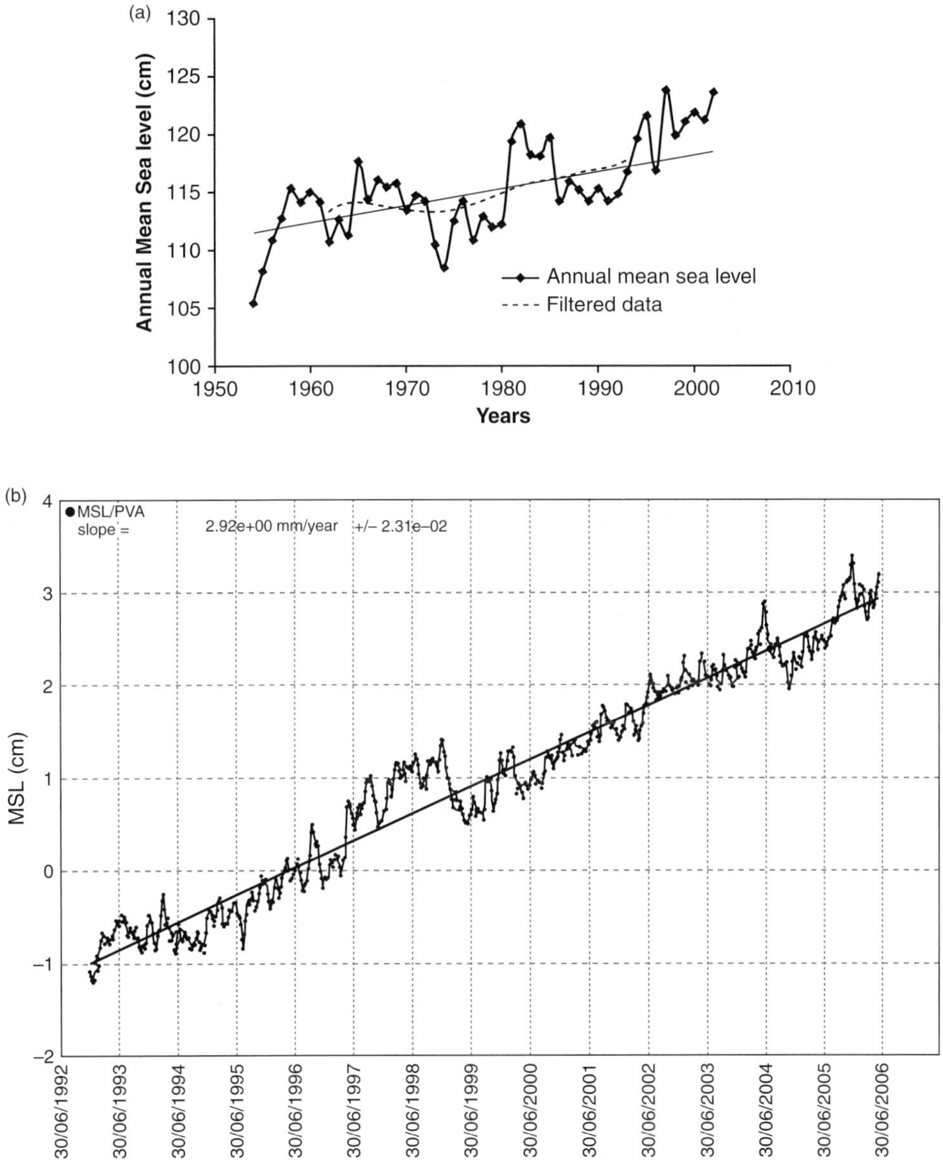

Figure 3.3 (a) Linear regression calculated from filtered data of absolute annual mean sea levels at Mar del Plata for the period 1954–2002 (after Pousa et al., 2006). (b) Global mean sea level since October 1992 as seen by the altimetry satellites. Seasonal variations have been removed. (*Source:* http://www.aviso.oceanobs.com/).

the Rio de la Plata at Buenos Aires shows a 1.6 ± m/yr rise (Lanfredi et al. 1998); these values are compatible with worldwide trends.

Because of the highly dense population living along coastlines or in the hinterland, sea-level rise as indicated by historical trends, or its acceleration between roughly 0.09 m/yr and 0.9 m/yr according to different scenarios (IPCC, 2001) would produce severe damage along the South American coast, particularly in low-lying areas. The more exposed areas are extended along the eastern and northern coasts of South America (Schnack, 1993; Brooks et al., 2006). Increases in coastal flood risks that are manifest in Orinoco, Amazon, Paraná-La Plata basins, as well as an exacerbation of erosion in beach systems, demand various responses to ameliorate the exposure of human populations to the hazards posed by the Sea-level rise (Brooks et al., 2006).

In conclusion, a future rise in sea level would result in an increase in coastal erosion and inundation, groundwater intrusion, migration, and loss of coastal wetlands (salt marshes, mangrove swamps), among other effects.

3. Storm Surges

A storm surge is a rise in sea level above normal tidal variations due to the action of strong winds blowing toward the land. When storm surges are the result of severe storms such as hurricanes, the low atmospheric pressure at the center of the depression can also raise the sea level. Thus storm surges are usually the result of two different physical processes, namely wind shear stress acting on the water surface and changes in the atmospheric pressure. Storm surges have been responsible for the extensive flooding of low-lying regions in many coastlines and have caused great loss of lives and property damage. Storm surges can be more severe if they coincide with a high tide or if they bracket several tidal cycles, particularly in the case of a perigean tide. Some researchers use the term *storm tide* to denote the sum of the astronomical tide and storm surge.

According to their origin, storms are classified as tropical and extratropical (Fig. 3.1). Tropical storms are generated in low latitudes, approximately between 5° and 25°, from where they move toward the coast in a somewhat unpredictable trajectory. Usually, they are of small extension, with strong pressure gradients and winds that, as in the case of hurricanes, can reach speeds over 240 km/h. These storms cause extreme flooding when they reach the coast, with water levels above 8 m in open coasts, or even higher in bays and estuaries. A coastal cliff retreat is also assumed to occur more rapidly with storms. On the coast between Aboletes and Turbo (Gulf of Urabá, Caribbean Colombia) common cliff retreats are of the order 0.5 to 2 m/yr, although maximum rates of 7 to 40 m/yr have been measured in Zapata, Damaquiel, and Punta Arboletes during the last 40 years (Correa and Vernette, 2004).

Extratropical storms, on the other hand, are generated at higher latitudes, between 25° and 60°, and cover much more extensive regions, in the order of hundreds of kilometers, around a low-pressure center not as clearly defined as in tropical storms, and with a slower displacement. In addition, the winds around the low-pressure center show a less symmetric scheme than in the case of tropical storms. Although low pressure and wind stress are the two main factors in any storm, wind stress is the primary factor in tropical storms. On the other hand, both low pressure and wind stress have similar importance in extratropical storms. These are typical of the North Sea, the northeast coast of the United States and Canada, and the southwest Atlantic Ocean.

3.1. Hurricanes

A hurricane is a severe tropical storm that forms in the North Atlantic Ocean, the Northeast Pacific Ocean, or the South Pacific Ocean east of 160° E. Hurricanes need warm tropical oceans, moisture, and light winds above them. The Saffir-Simpson hurricane scale, which has been used over the past 25 years, groups tropical storms (hurricanes) into five major categories as a function of wind speed alone (Dolan and Davis, 1994).

Hurricanes have great variation in frequency and intensity. From historical accounts, the coastal region with the greatest hurricane activity is South Florida (15% annual chance), while many areas in Latin America and the Caribbean have a 10% annual chance of experiencing a hurricane. Areas with smaller risk of impact (1 to 5% annual chance) are located in the southern part of the Caribbean, including northern Venezuela and northern Colombia (Pielke et al., 2003).

These intense storms originate in the Atlantic Ocean, off the northwest coast of Africa north of the equator (Fig. 3.4). They track from east to west and exhibit counterclockwise wind circulation around a center of low barometric pressure. As a result of this pattern, coasts facing east and north are the most exposed to impacts. Exceptionally, Hurricane Lenny (1999) formed within the region tracking from west to east.

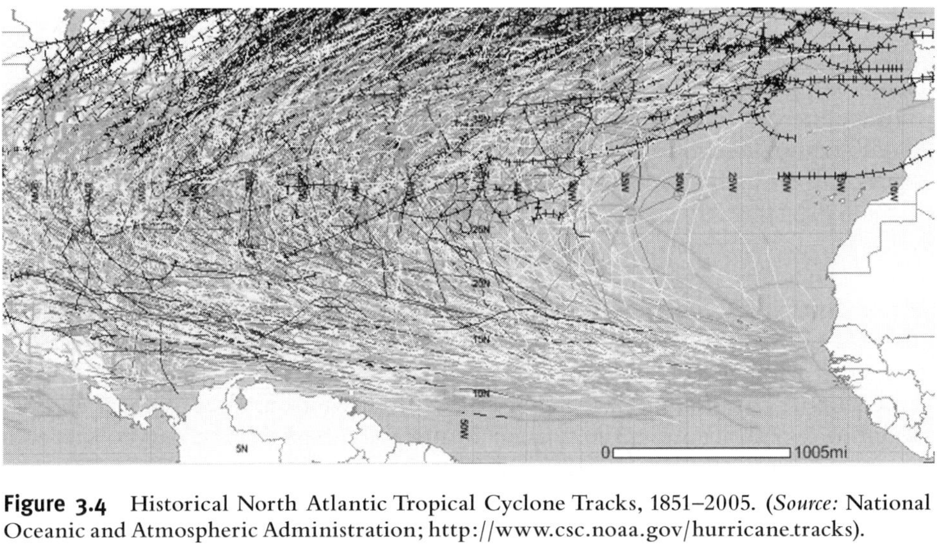

Figure 3.4 Historical North Atlantic Tropical Cyclone Tracks, 1851–2005. (*Source:* National Oceanic and Atmospheric Administration; http://www.csc.noaa.gov/hurricane.tracks).

The Atlantic Basin shows a very peaked season from August through October, with 78% of the tropical storm days, 87% of the minor (Saffir-Simpson Scale categories 1 and 2) hurricane days, and 96% of the major (Saffir-Simpson categories 3, 4, and 5) hurricane days occurring then. Maximum activity is in early to mid-September. Occasionally, a tropical cyclone may occur out of season—primarily in May or December.

Most of the hurricane tracks do not experience landfall on the northern South American coast but some of them affect the Caribbean coast of South America. Hurricane Joan, for example, impacted Venezuela and Colombia in October 1988, producing economic damage and human impact (Pielke et al., 2003). Hurricane winds generate and propagate deep-ocean swell that can cause substantial coastal change far from the storm center. The breaking waves from distant storm sources can transport sediments inland. These processes were recorded in the village of Tierrabomba, an island located to the south of Cartagena, Colombia. Sandy overwash deposits were deposited on the island by the effects of Hurricane Lenny, which was centered more than 500 km north of the Caribbean coast (Morton et al. 2006).

During El Niño events (ENSO warm phase), tropospheric vertical shear is increased, inhibiting tropical cyclone genesis and intensification, primarily by causing the 200 mb (12 km or 8 mi) westerly winds to be stronger. La Niña events (ENSO cold phase) enhance activity. Recently, Tang and Neelin (2004) also showed that changes to the moist static stability can also contribute to hurricane changes, with a drier, more stable environment prevailing during El Niño events.

Hurricanes affecting the coast of South America are basically within the Atlantic Basin, at low latitudes, and can be active on the northern coast of South America (Colombia, Venezuela). However, the first hurricane ever reported in the South Atlantic, Catarina, hit the southern coast of Brazil on March 28, 2004. This unprecedented event led some Brazilian meteorologists to deny that it was a hurricane at all; further analysis, however, has shown that it was (Pezza and Simmonds, 2005). In a detailed study of the storm, the authors describe its evolution from genesis on March 20, 2004 as an extratropical cyclone, through its strengthening to a category I hurricane before it drifted over land. This hurricane developed in an unusual combination of high sea-surface temperatures, low vertical wind shear, and strong mid-to-high latitude blocking (which interferes with normal east–west atmospheric flow). These conditions are functions of large-scale atmospheric circulation patterns in the region and could be related to climate change (Pezza and Simmonds, 2005).

3.2. Extratropical Storms

Many storm surges have been recorded along the Argentine coast simultaneously with the northward traveling tidal wave. The duration of these storm surges range from a few hours up to two or three days. They are basically produced by the combined action of an anticyclone located to the west of Argentina (semipermanent Pacific anticyclone) and a cyclone located over the Atlantic to the southeast of Argentina, the latter moving toward the east or northeast. Because of this situation, strong winds from the south or southwest and high water levels affect the whole Argentine coast, as well as the Rio de la Plata shores, Uruguay, and southern Brazil.

The most conspicuous and worse floods in the Rio de La Plata shores and the eastern, sandy coast of the Province of Buenos Aires are due to southeasters. These are very strong winds from the southeast, often caused by an anticyclone located over southern Argentina and the adjacent ocean. But for water to attain an extraordinarily high level, it is necessary, among other factors, that a depression be generated to the northern area of Buenos Aires, over Uruguay, and southern Brazil. The weather chart corresponding to the flood of April 15, 1940, induced the highest level in Buenos Aires (3.18 m over the predicted tidal level) since the beginning of records in 1905 (Fig. 3.5). Twenty five people were killed. Similar floods were experienced in 1958 and 1959, and more recently in 1989 and 1993.

Coastal plain flooding due to storm surges can be particularly destructive in areas where topographic gradients are extremely low, such as the Rio de La Plata shores, the Salado Basin, and around the Bahía Blanca estuarine complex. Every time this phenomenon has taken place many inhabitants have suffered

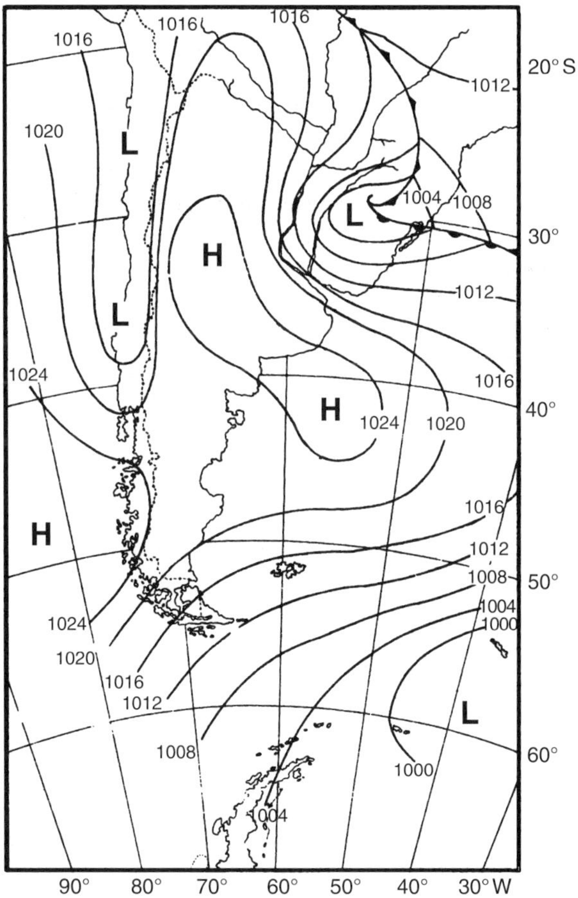

Figure 3.5 Synoptic chart of the severe southeasterly of April 15, 1940 in the southwest Atlantic Ocean. (*Source*: Balay, M.A., 1961).

Figure 3.6 Eroding coasts from Argentina. (a) cliff showing mass movements in Patagonia. (b) Barrier sand with severe erosion after a southeasterly, Santa Teresita (Buenos Aires).

severe property loss and other damage. Flooding is exacerbated by the Coriolis force piling up water as the storm surge travels along the Argentine coast, particularly within the Rio de La Plata.

Erosion processes on the sandy barriers of northern Argentina, Uruguay, and Brazil are typical. The coincidence of a storm surge with spring tides is thought to be of the utmost importance in coastal erosion. Storm surges (southeasters) are considered the most significant natural agent for coastal erosion on the eastern coast of Buenos Aires (Schnack et al. 1998; Fig. 3.6b). However, their effects are more severe in areas of heavy human intervention (beach sand mining, urbanization, coastal construction).

The Patagonian coastline, composed mostly of cliffs, is receding at a rate of about 0.5 m/year (Isla and Cortizo 2005). Many authors have mentioned variable rates of coastal retreat for the Buenos Aires coastline, ranging from a few meters/year to less than 1m/year (Schnack, 1985, Isla and Bértola, 2005). However, Patagonian cliff erosion reacts to different processes than those operating over the sandy barrier, heavily impacted by human activities (Figure 3.6). Beach erosion rates along the shoreline vary due to the temporal effects of the refraction of storm waves induced by linear shoals (Isla and Bértola, 2005). A similar situation seems to occur along the Uruguayan coast (Pivel et al., 2001). At Ensenada Algarrobo, Chile, a maximum coastline retreat of the order of 9.5 m/yr has been measured at Playa Tunquén (Martínez, 2001).

Logarithmic beaches from the north of Santa Catarina Island, southeastern Brazil, are subject to different erosion rates along their lengths. Between 1938 and 1994 Ingleses Beach, for example, was subjected to waves coming from the north and receded 0.5 m/yr at the northern end and 0.95 m/yr at the southern end. The Armaçao Beach, which is open to the eastern part of the island, retreats 0.1 m/yr at the northern part and 0.5 m/yr at the southern end (Abreu de Castilhos and Gré, 1997). In the State of Paraná, the pocket beach of Praia Brava in Caiobá is located between the capes of Caiobá and Matinhos. Between 1951 and 1969 this beach had a progradation of 42.1 m, it receded 1.8 m between 1969 and 1980, and the retreat increased to 14.9 m between 1980 and 1997 (Bessa and Angulo, 2001). The same area is subject to strong beach erosion caused by storms (Angulo et al., 2006; Fig. 3.7).

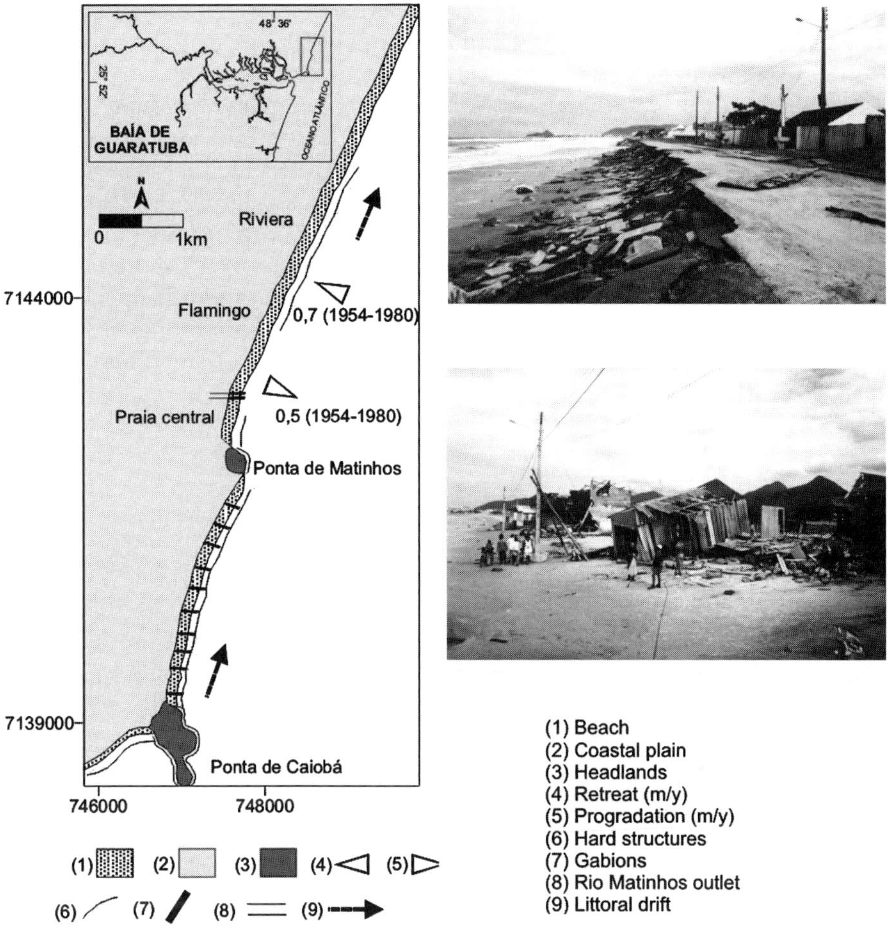

Figure 3.7 Coastline variations along a southern Parana stretch in the vicinities of Guaratuba Bay. Top right: destruction of the coastal road at Balneario Flamingo, after a surge in May 2001. Bottom right: Erosion at Matinhos Beach after the same storm. (*Source:* After Angulo et al., 2006b). Courtesy of R. Angulo.

In the State of Rio Grande do Sul, 81.3% of the beaches are subject to erosion, 11.5% are stable, and only 7.2% are accreting (Esteves et al., 2001). In Farol da Conceiçao (north of the Lagoa dos Patos inlet), coastal dunes were receding at a rate of 2.5 m/yr between 1975 and 1995 (Tomazelli et al., 1997). Further south, at Concheiros do Albardao (close to Hermenegildo), beach profiles were surveyed between 1991 and 1993. The coastline was approximately stable between 1991 and 1992; but in the interval May–June 1993, the shoreline retreated to 13–14 m per month (Klein and Calliari, 1997). Much of the erosion of the south of the state is caused by the ephemeral location of gullies (locally called *sangradouros*) that drain ponds and wetlands located between beach ridges. These *sangradouros* cause the erosion of frontal dunes and are increasing the action of rip currents. They are distributed between 0.5 to 1 per km, but they are much more densely spaced during the autumn (Figueiredo and Calliari, 2001).

Whereas hurricanes seem to be statistically more frequent, or stronger, in the last years, it is not yet known if extratropical storms in the southwestern Atlantic Ocean are more frequent or stronger. For the northwestern Atlantic Ocean, extratropical storms were more frequent on the North Carolina coast during the 1960s (Dolan and Hayden, 1981). In the Mid-Atlantic coast of USA, for example, a storm with waves of 3.4 m occurs every three months (in winter months they are more frequent); waves of 5 m occur every 3 years, and waves of 7 m happen every 25 years (Dolan et al., 1987). Statistical data are crucial to forecasting the threat of storms on the mid-latitude sandy barriers of South America (Argentina, Uruguay, and southern Brazil).

4. Tsunamis

Typical tsunamis consist of a series of high-energy, long–period waves of small steepness usually caused by ocean-floor displacements. Most tsunamis are produced by coseismic seafloor displacements; others are generated by underwater landslides; and a small number are the result of volcanic eruptions. Actually, tsunami generation involves complicated interactions among earthquakes, landslides, and sympathetic vibrations between the quake and the ocean above it. In general, scientists believe that it requires an earthquake of at least magnitude 7 to produce a tsunami. Calving of glaciers and the extremely rare meteorite or asteroid impact in the open ocean should also be considered among possible generation mechanisms. Of all 2250 tsunamigenic events historically known, only 223 (about 10%) resulted in human fatalities.

4.1. Transoceanic Tsunamis

Transoceanic tsunamis, capable of transmitting their energy far away from the source area, are quite rare events as compared to local and regional events. However, they are responsible for a considerable part of the damage and fatalities resulting from all tsunamis. The 11 transoceanic tsunamis that occurred in the world ocean during the last 250 years are responsible for 372,000 fatalities (Table 3.1), and 280,000 people were killed by the Indian Ocean tsunami of December 26, 2004. The signal of this tsunami was felt in several areas around the globe (Titov et al., 2005) and reached the east coast of South America at several localities within 20–24 hours. The highest waves on the Atlantic coast of South America oscillated from 0.15 m in Mar del Plata, Argentina, to 1.22 m at Imbituba Port, Brazil (see Table 3.2; Dragani et al. 2006). In the Pacific coast, at Callao (Peru) and Arica (Chile), wave amplitudes >50 cm were recorded, in agreement with the predominant direction of tsunami energy propagation (Titov et al., 2005).

Table 3.1 Historically Known Transoceanic Tsunamis That Have Occurred in the World Ocean during the Last 250 Years (*Source:* Tsunami Laboratory, Novosibirsk, Russia: http://tsun.sscc.ru)

Date and Place	Magnitude	Max Runup near the Source, m	Max Runup in the Far- Field, m	Fatalities
November 1, 1755 Lisbon	8.5	18	7.0	40000
November 7, 1837 Chile	8.5	8	6.0	many
August 13, 1868 Chile	9.1	18	10	3000
August 27, 1883 Krakatau		36	1.5	36000
February 3, 1923 Kamchatka	8.3	8	6.1	some
April 1, 1946 Aleutians	7.4	42	18	165
November 4, 1952 Kamchatka	9.0	18	9.1	>10000
March 9, 1957 Aleutians	9.1	15	10	none
May 22, 1960 Chile	9.5	18	12	1180
March 28, 1964, Alaska	9.2	68	6.0	123
December 26, 2005 Sumatra	9.3	34	9.1	280000

Table 3.2 Arrival Time (UTC), Travel Time (h), Maximum Wave Height (m), and Duration of Sea-level Activity (h). (Obtained from the sea-level records gathered at stations located in the southwestern Atlantic Ocean) (*Source*: After Dragani et al., 2006)

Station	Arrival time (UTC)	Travel time (h)	Max. Wave height (m)	Duration of sea level activity (hours)
Cananeia, Brazil[1]	Dec. 26, 20:45	19.77	0.20	54
Ubatuba, Brazil[1]	Dec. 26, 22:00	21.02	1.20	53
Port Stanley[2]	Dec. 26, 22:23	21.40	0.44	~48
Arraial do Cabo, Brazil (22°57′S, 42°01′W)[a]	Dec. 26 22:57	21.97	~0.90	~48
Mar del Plata	Dec. 27, 00:15	23.27	0.15	40
Santa Teresita	Dec. 27, 00:48	23.82	0.27	54
Imbituba Port (28°13′S, 48°39′W)[b]	Dec. 27, 03:20	26.35	1.22	~40
Puerto Belgrano	Dec. 27, 04:45	27.77	0.20	[c]

[a] Candella 2005 http://www.pmel.noaa.gov/tsunami/indo20041226/IEAPM.html
[b] Mello and Rocha 2005 http://www.pmel.noaa.gov/tsunami/indo20041226/tsunami-southernbrazil.pdf
[c] Truncated record–the tide gauge did not work after December 28, 2004, 02:00 UTC.
[1] Franca and de Mesquita, 2007.
[2] Woodworth, 2005.

The other 10 transoceanic tsunamis are responsible for 92,000 deaths, which accounts for only 13% of all tsunami-related fatalities. Three of them caused severe impacts on the coast of Chile, with a death toll of few thousand people altogether (Table 3.1). The characteristics of the three Chilean transoceanic tsunamis are described next.

November 7, 1837: Valdivia, Chile

An 8.5 Ms destructive earthquake hit the southern coast of Chile on November 7, 1837 with its epicenter near Valdivia, Corral, and Ancud. Waves reached 8 m at the nearest Chilean coast. Six-m waves were observed in Hilo, Hawaii, after almost 14 hours of propagation time (Walker, 1994). There is no quantitative data on the number of victims in Chile, but in Hawaii it caused 58 fatalities.

August 13, 1868 Arica, Chile

A destructive 9.1-Mw earthquake with its epicenter near Arica, northern Chile, resulted in 18-m tsunami waves that in 20–30 min after the quake hit the nearest Peruvian and Chilean coast. Data on resulting fatalities are fragmentary, but one can guess that at the nearby coast the tsunami took several thousand victims. Outside the source area, the largest waves (up to 10 m) were observed at the Chatham Islands at a distance of almost 10,000 km (DeLange and Healy, 1986). Along the east coast of New Zealand waves were 3 to 5 m high. These waves turned out to be the most severe far-field tsunami observed in New Zealand during the 160-year period of available observations. DeLange and Healy (1986) list this tsunami as having caused loss of life in New Zealand, but they do not give any numbers for fatalities. Five-meter waves reached Hawaii, causing 47 fatalities (O'Loughlin and Lander, 2003).

May 22, 1960 Chile

On May 22, 1960, a magnitude 9.5 Mw earthquake, the largest earthquake ever instrumentally recorded, occurred in southern Chile. The earthquake ravaged the vast area along nearly 1000-km of the Chilean coast. The main shock generated a destructive tsunami that hit the nearest coast with 8- to 10-m waves. The maximum waves, reaching 15 m in height, were observed along the 350-km section of the coast between Corral and Concepción. A ship of 3000 tons was washed onto the beach in Mocha Island (Saint Amand, 1961). In Lebu (Arauco Peninsula), a beach uplifted 1.5 m, but in the following five months it returned to its original level. The number of reported fatalities varies in different sources from 490 to 5700; 1000 fatalities is a reasonable assumption. In 15 hours, 8- to 10-m waves reached Hawaii (the reported maximum was a 12.1-m wave observed at Ahukini Point on Kauai Island) and caused 61 fatalities in Hilo, despite the advance warning and despite the warning sirens that sounded more than 3 hours before the first wave arrived. In 22 hours, the waves reached the east coast of Japan, still having heights of 5 to 6 m. More than 10,000 houses were destroyed and 122 people died (Tsunami Laboratory, Russia, 2006).

4.2. Regional Tsunamis

All historically known tsunami sources are located within the continental slope or the shelf at the average distance of 150 to 200 km off the coastline of South America. Tsunamis have been recorded in the Pacific coast, the Caribbean Sea, and the South Atlantic coast of South America.

4.3. Pacific Coast of South America

Due to seismotectonic features of the region, the sources of all tsunamigenic events are located within the area, with about a 20-minute propagation time limit. The sources of many South American tsunamis are located partly inland, so the first wave arrives almost simultaneously or shortly after the end of seismic shaking caused by the fault rupture (Gusiakov, 2005). In other instances, they are produced within the same basin, which in the case of the Pacific is large enough to permit tsunami waves traveling over long distances (Fig. 8).

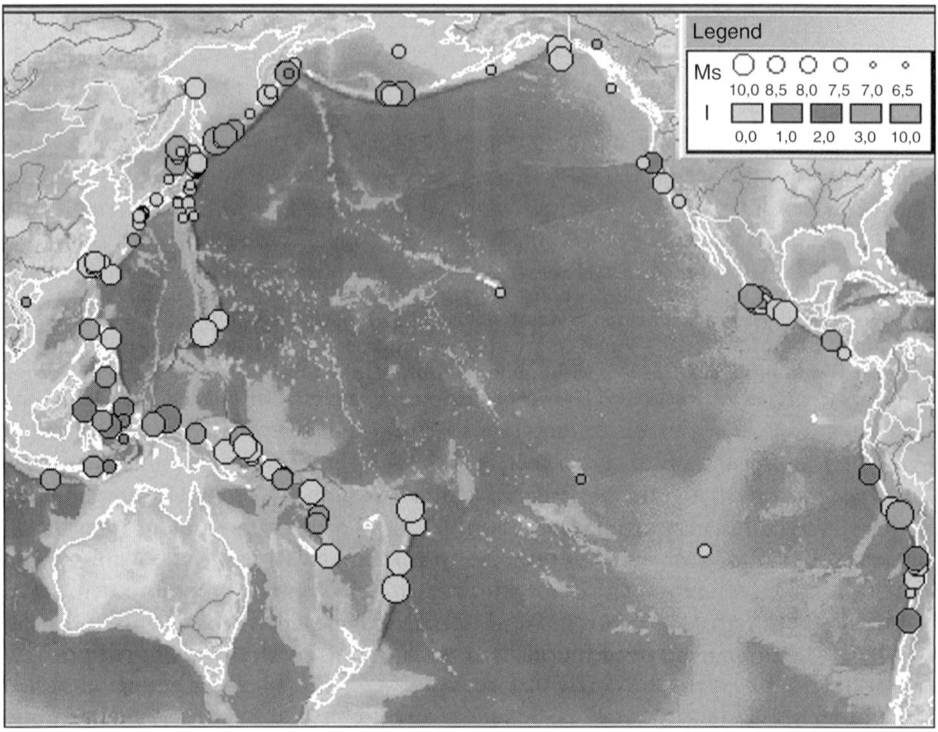

Figure 3.8 Recent tsunamis on the Pacific Ocean (1981–2003). Larger circles mean higher intensity. (*Source:* http://tsun.sscc.ru/recent.htm)

Following is a description of a few historical events.

On July 9, 1589, a tsunami affected the coasts of Lima. The level of the ocean rose 7 m, destroying properties in the 300 m close to the shoreline (22 victims in Lima). The tsunami was triggered by an earthquake (intensity VII) close to the coasts of Lima.

The coast of Pisco (Ica) was shocked (earthquake VI) on May 12, 1644. The coast was flooded and 70 inhabitants died. Another earthquake on June 17, 1678 (intensity VII), north of Lima, caused severe damage at El Callao Harbor; the sea receded before advancing. In October 20, 1687, another earthquake north of Lima (intensity IX) destroyed much of that city (200 inhabitants died).

A confirmed tsunami was recorded on November 26, 1705. Large waves hit the shoreline between Arequipa (Peru) to Arica (Chile), where the damages were more severe.

The worst tsunami affecting the coast of Peru was recorded at El Callao Harbor on October 28, 1746. Waves of 7 m caused the sinking of 19 ships (one of these was found 1.5 km to the interior). Diseases were estimated to be about 5000 to 7000. Other harbors affected were Chancay and Huacho.

Another tsunami affected El Callao on December 1, 1806. Waves higher than 6 m transported several ships to the shoreline, including an anchor of 1.5 tons.

A very detailed account of the earthquake that affected the Chilean localities of Valdivia, Concepción, and Talcahuano, among others, on February 20, 1835, was provided by Charles Darwin (1845):

February 20th.– This day has been memorable in the annals of Valdivia, for the most severe earthquake experienced by the oldest inhabitant. I happened to be on shore, and was lying down in the wood to rest myself. It came on suddenly, and lasted two minutes, but the time appeared much longer.... The tides were very curiously affected. The great shock took place at the time of low water; and an old woman who was on the beach told me that the water flowed very quickly, but not in great waves, to high-water mark, and then as quickly returned to its proper level; this was also evident by the line of wet sand.

March 4th.– We entered the harbour of Concepcion. While the ship was beating up to the anchorage, I landed on the island of Quiriquina. The mayor-domo of the estate quickly rode down to tell me the terrible news of the great earthquake of the 20th:– "That not a house in Concepcion or Talcahuano (the port) was standing; that seventy villages were destroyed; and that a great wave had almost washed away the ruins of Talcahuano." Of this latter statement I soon saw abundant proofs – the whole coast being strewed over with timber and furniture as if a thousand ships had been wrecked. Besides chairs, tables, book-shelves, etc., in great numbers, there were several roofs of cottages, which had been transported almost whole. The storehouses at Talcahuano had been burst open, and great bags of cotton, yerba, and other valuable merchandise were scattered on the shore. During my walk round the island, I observed that numerous fragments of rock, which, from the marine productions adhering to them, must recently have been lying in deep water, had been cast up high on the beach; one of these was six feet long, three broad, and two thick.... Shortly after the shock, a great wave was seen from the distance of three or four miles, approaching in the middle of the bay with a smooth outline; but along the shore it tore up cottages and trees, as it swept onwards with irresistible force. At the head of the bay it broke in a fearful line of white breakers, which rushed up to a height of 23 vertical feet above the highest spring-tides.

An earthquake originated in the coast of Chile on May 9, 1877, causing a tsunami that affected the shorelines of Japan, New Zealand, Samoa islands, California, and Hawaii. In the Eastern Pacific Ocean, high waves arrived from Pisco (Peru) to Antofagasta (Chile). The maximum wave measured was 23 m in Arica.

On January 10, 1878, the ocean flooded several coastal cities between Arequipa and Iquique. Maximum wave height was recorded at Tanna Island (12 m).

On January 31, 1906, an earthquake occurred in Ecuador at a depth of 25 km and with a magnitude of 8.6 (Mercalli). The area shocked was about 1200 km between Guayaquil and Medellín. Toward Bogotá the width of the earthquake was evaluated in about 350 km, covering an area of 300,000 km^2. Half an hour after the shock, a tsunami arrived at Tumaco (Colombia); the second wave arrived 20 minutes later, and the situation continued for about 4 hours. Fortunately, these waves arrived during low tide. Water levels of 2 to 5.9 m above the islands were estimated in the Colombia coastline (Herd et al., 1981). The low-lying coasts were more affected. Buildings settled close to the beach or at the estuarine areas of the rivers Santiago and Mataje were destroyed, and 1000 to 1500 inhabitants died. In Tola, 23 houses were destroyed. In Esmeraldas, the river flooded the village. In Bahía Caraquez, the water level rose 0.8 to 1.0 m in only 20 minutes. On the opposite hand, in Manta and Buenaventura, the sea level dropped 2 m.

On October 2, 1933, another earthquake (6.9 Richter magnitude) occurred offshore of Ecuador. Significant waves occurred at La Libertad, on the Santa Elena Peninsula. A submarine cable broke 25 km south of Salinas. Sea level dropped immediately to the land movement (10.30 A.M.), but after an hour the level rose to the high-tide mark. At mid-day the level dropped again to the low-tide level, and rose again at 14.00 hr. Fluctuations of about 2-2.5 m occurred in only 3.5 hr.

The earthquake that took place on April 1, 1946, on the Aleutian Islands affected the American coast from Colombia to Chile. Long-period waves were more significant in Iquique, where sea level rose 5 m. In Valparaíso, the ocean extended 100 m landward. Great alarm was caused, although there were no deaths.

On November 4, 1952, another tsunami coming from Kamchatka, Siberia, affected the Pacific coast from Ecuador to Chile; the most severe impacts were registered at the coastlines of Antofagasta (waves arrived at 8 A.M.) and Talcahuano (11 A.M.). In Chile, the sea level rose 3.7 m while the sea extended 500 m inland, but without victims. Tide gauges recorded 1.9 m in Libertad (Ecuador), and 2.0 in El Callao.

Another earthquake produced close to the border between Peru and Ecuador affected La Libertad coast on December 12, 1953. The magnitude of the event was about 7.3, and nondestructive oscillations were only 0.2 m in height.

On January 19, 1958, an earthquake (7.8 degrees Mercalli) occurred close to the border between Colombia and Ecuador; the areas affected were the coasts of Tumaco and Esmeraldas. Tsunami waves of 2.0 to 5.9 m caused 4 deaths.

The earthquake in Alaska (Kodiak, March 28, 1964) also affected the coasts of Peru and Chile; in El Callao a wave of 1.5 m was recorded.

On October 3, 1974, a seismic movement occurred at the sea in front of El Callao. The sea flooded some bays north of Lima (Chimú, Tortugas).

About three to four tsunami waves also occurred in Esmeraldas (Ecuador) on December 12, 1979. The so-called Tumaco earthquake was triggered at the frontier between Colombia and Ecuador. These movements took place 33 km deep. As the waves arrived during low tide, no significant damages were recorded (Herd et al., 1981) and induced a subsidence of 1.6 m (Correa and González, 2000).

On February 21, 1996, a seismic movement (6.9 Richter scale) originated 210 km to the southwest of Chimbote. It caused 15 deaths in Chimbote.

On November 12, 1996, an earthquake of 6.4 magnitude was recorded 93 km to the southwest of San Juan de Marcona (Perú) and at a depth of 46 km. A tsunami was triggered, causing property damage and human losses.

In the present century, a tsunami generated from an earthquake (6.9 magnitude on the Richter scale) occurred on the ocean close to Ocoña (June 23, 2001). Only three waves were remembered but their height was 8 m. Twenty three persons died and 69 disappeared.

Because of the risks threatening the Pacific coast of Colombia (earthquakes, tsunamis, local subsidence, flooding, and soil liquefaction), the decision was made to move the village of El Choncho to another location (Correa and Gonzalez, 2000).

4.4. Caribbean Sea

The Intra-Americas Sea (IAS) plate is characterized by subaerial and submarine active volcanoes, steep continental slopes, and frequent earthquakes. Some of the submarine earthquakes, eruptions, and subaerial or submarine slumps can generate tsunamis. In the central Lesser Antilles, several collapses have occurred in the last 10,000 to 20,000 years at volcanic sectors. The region has experienced tsunamis since at least the sixteenth century. The origin of these events has been both local and distant sources, and one or more events per century have been recorded: Venezuela, 1530; Jamaica, 1692; Martinique, 1755; St. Thomas, 1867; Puerto Rico, 1918; Dominican Republic, 1946. In addition, as mentioned earlier, the Lisbon earthquake of 1755 generated a tsunami with 6- to 7-m waves in the Lesser Antilles.

In the last 150 years, tsunamis have caused about 2000 victims in the IAS. The 1867 event in the U.S. Virgin Islands is a true precedent for the 1998 event that occurred in Papua New Guinea: superimposed earthquake epicenters, a great instantaneous tsunami, and densely populated coastal settlements. Historical events are shown in Figure 3.9, and Table 3.3 provides a list of tsunamis that affected the coast of Venezuela.

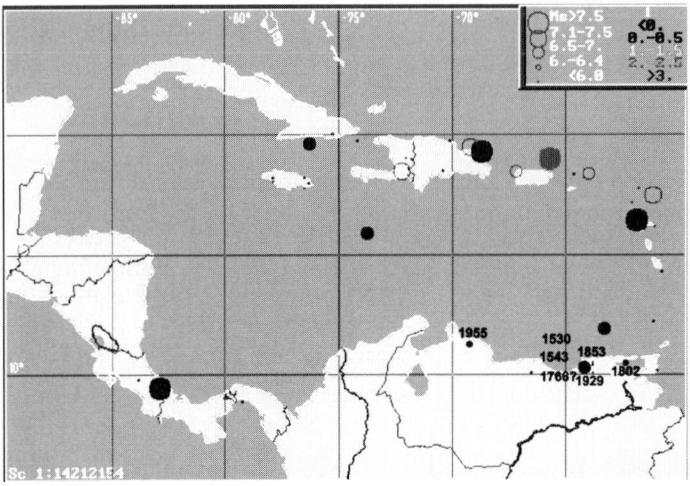

Figure 3.9 Historical tsunamigenic events in the Caribbean region from 1530 to 1991. There were 57 recorded events. However, this map does not present all events because source coordinates are lacking for 11 events. The dates are only shown for the South American coast where tsunamis with reliable information are indicated. The size of circle is proportional to the magnitude of the event. However, in the case of Cumaná five events are recorded with different magnitudes. (*Sources*: Lander and Whiteside, 1997; Schäffers, 2002; Tsunami Laboratory, Novosibirsk, Russia: http://tsun.sscc.ru).

Table 3.3 List of Caribbean Tsunamis Affecting the Coast of Venezuela (selected from Lander and Whiteside, 1997)

September 1, 1530	Paria Cumaná Cubagua	Sea rose 7.3 m and sank again near coast of Paria and at Cumaná and near Island of Cubagua. Ground opened, emitting black salt water and asphalt. Mountain at the side of the Gulf of Cariaco was cleft (earthquake). A fort and many houses destroyed, but not clear whether due to the wave, the earthquake, or both.
1543	Venezuela	Waves noted. City of Cumaná destroyed by earthquake?
October 21, 1766 [9:00 UT]	Cumaná, Venezuela	Very violent shocks raised Cumaná and caused the island of Orinoco to sink and disappear. In many places the water surface was disturbed. This is a possible tsunami report.
August 5, 1802	Orinoco River, Venezuela	Earthquakes at Cumaná caused the water of the Orinoco River to rise so high as to leave part of the bed dry. This could describe wave action near the mouth of the river, or bore action. The rudder of a vessel was broken.
July 15, 1853	Cumaná, Venezuela	A violent earthquake (MMI = X) in Cumaná was followed by a tsunami.
January 17, 1929	Cumaná, Venezuela	City was destroyed by an earthquake (Ms = 6.9), and a steamer off shore was endangered by a huge wave. The tidal wave following the earthquake caused much damage. Many sailboats were wrecked.
January 18, 1955	La Vela, Venezuela	A wave was reported; four ships were wrecked, and four waterfront buildings were damaged. An earthquake (Mb = 5.5) off the coast of Panamá is listed for this time.
September 20, 1968 [6 09 UT]		A report of a tsunami has not been verified. Hurricane Edna was passing north of Venezuela at this time, and an earthquake (Ms = 6.2) occurred near the coast of Venezuela.

4.5. Southwest Atlantic Ocean

No evidence is yet available for tsunami events in the southwest Atlantic Ocean. However, the southernmost portions of the ocean (Scotia Plate and Tierra del Fuego) are tectonically-active areas. In addition, glacial calving and submarine slumps can occur in the southern ocean. Hence exposure to tsunamis should not be disregarded in certain areas located in the southernmost region (Schnack and Pousa, 2004). On December 17, 1949, an earthquake affected the island of Tierra del Fuego. The epicenter was located in Dawson Island, southern Chile (7.75 Richter scale). At the Atlantic coast of the Grande Island (close to Rio Grande), the San Pablo lighthouse tilted 15 degrees (Isla and Bujalesky, 2004). In Punta Arenas (Chile), 3 deaths were reported as a consequence of the earthquake.

5. ENSO Effects

The ENSO (El Niño-Southern Oscillation) cycle is a fluctuation between anomalous warm (El Niño) and cold (La Niña) conditions in the tropical Pacific with a 2- to 7-year recurrence. ENSO is the strongest and most predictable natural variation of Earth́s climate on year-to-year time scales and has impacts on physical, geological, biological, and chemical processes in the oceans and atmosphere, and on terrestrial ecosystems. In addition, socioeconomic impacts are very important on a global scale (McPhaden et al., 2006). Changes in precipitation patterns, in response to El Niño warm the central and eastern Pacific and cause drought in Australia, Indonesia, and other areas in the region, whereas the islands of the central Pacific and the west coast of South America are frequently affected by heavy rains and flooding.

South America is particularly subject to interannual changes triggered mainly by the humidity and temperature transport from the western to the eastern Equatorial Pacific. They severely impact either the coasts of high slopes (Peru, Chile, Ecuador) or the low-lying coastal plains (Argentina, Brazil, Uruguay); but the main direct effects on the coastline occur in the Pacific region. Along this region, a positive phase (El Niño) usually results in heavy rainfall and the occurrence of mass movements on the slopes facing the coastal zone. Other effects are the flooding of rivers draining onto the shore, erosive processes, and other morphological changes at the shore and nearshore, as it was registered in Ecuador during the 1982–1983 event (Tutivén Ubilla, 1998). ENSO events have been historically recorded since 1541, but beach-ridge plains dated by the radiocarbon decay method to about 5000 years BP have been associated with heavy rains and high stages of the sea level. Therefore they were assigned to strong El Niño events (Ortlieb and Macharé, 1993).

A sea-level anomaly is also recorded during El Niño events, causing an increase in global mean sea level (Fig. 3.3b). The coastal and island populations of Ecuador were dramatically affected during the 1997–1998 El Niño season. Increases in sea level and wave action, due to Kelvin waves from storms in the northern Pacific, caused coastal erosion at the shoreline, destroying all structures near the beaches. In Punta Gorda, Esmeraldas, Ecuador, the average cliff retreat is between 1.5 and 2.3 m/yr, but it increased significantly during the 1982–1983 ENSO events (Santana et al., 2001).

The barrier island of El Choncho is related to the San Juan River delta (Western Colombia). The 1997 El Niño event was marked by was a rapid beach retreat, the opening of a new inlet, and flooding of small villages with destruction of coastal-protection structures (Morton et al. 2000). During this period, ocean waves were less than 1 m, and onshore wind speed was less than 2 km/hr. The occurrence of nonstorm washovers was therefore assigned to the regional increase of 20–30 cm in the sea level triggered by the thermal expansion of the Pacific Ocean (Morton et al., 2000).

6. Mud Volcanoes and Alt Diapirs

Along the Caribbean coast of western Colombia (Córdoba Department), mud volcanoes and salt diapirs are common not only in the coast but also on the continental shelf. Three mud volcanoes were identified between Punta San Juan and Punta Arboletes, and three areas from the surroundings on the shelf were thought to constitute other volcanoes (Correa et al., 2007). These natural bedfoms can be recognized by bubbles within the water column and may exude mud, gas, water, and sometimes even petroleum (Restrepo et al., 2007).

7. Anhropogenic Effects

Although shorelines in South America are affected in various ways by natural processes, human action has exacerbated erosion processes. One example is the beach erosion along the Buenos Aires coastline. In the past decades, many localities have been established in the beach areas bringing in an increasing number of seasonal and permanent residents; the problems that have resulted include unplanned urbanization, development of resort facilities, beach mining for building purposes, and salt intrusion into coastal aquifers. The most noticeable response has been the emplacement of hard structures to protect the beaches and the coastal facilities. But many of these structures have caused a diminished supply of sand to areas located northward (direction of the net littoral drift). One remarkable case of rapid shoreline retreat was observed at Mar Chiquita village, north of Mar del Plata (Argentina), where more than a 6 m/year retreat (Fig. 3.10) was registered from 1957 to 1979 (Schnack et al., 2004).

Figure 3.10 Erosion at Mar Chiquita Beach between 1957 and 1979. The retreat was controlled by defense works of local effect. The photograph shows the present situation. In other places along the Buenos Aires coastline, erosion rates are variable, depending on sand supply and degree of human intervention, whereas in other localities without occupation the coastal systems seem balanced or even accreting.

The shoreline of the Unare coastal lagoon (Venezuela) retreated at a rate of 7.5 m/year (150 m between 1961 and 1980; Pacheco and Suárez, 2004). The Unare River is flowing to the Caribbean Sea between the coastal lagoon of Unare and Piritu, supplying sediments to the longshore spits where the villages of El Hatillo and La Cerca were settled. Because this coast is not significantly impacted by physical effects (storms, hurricanes, earthquakes, sea-level rise), Pacheco and Suárez concluded that much of the erosion is caused by the sediment trapping by the 12 dams of the Unare River, whose accumulation was estimated in about 2,260,000 m^3/year (Pacheco and Suárez, 2004).

At the estuary of the Bio Bio River (Concepción, Chile), fluvial sand accumulates to form dunes at the northern margin. However, storm waves from the Pacific Ocean cause erode these dunes (Fig. 11).

Coastal erosion along the coastline between Ilheus and Vitória, eastern Brazil, is subject to strong trade winds from the northeast. Old inhabitants from Alcobaça pointed out that this erosion has increased in the last 30 years, when beaches receded 30 to 40 m. It was concluded that this erosive disequilibrium should be assigned to the peak in the deforestation activities (Addad and Martins-Neto, 2000). At the same time, in the city of Ilheus, construction of a jetty in 1971 led to the progradation of 340 m of the Praia do Porto (13 m/yr on the updrift side) and the erosion of 80 to 150 m at Praia do Norte (3-5.7 m/yr on the downdrift side; Apoluceno et al., 1997).

At the coastline of the Córdoba Department (Caribbean Colombia), much of the structures that were erected failed to prevent the rapid erosion process (Correa et al., 2007). The village of Necocli, at the inlet of the Urabá Gulf, located on a marine terrace formed during the Mid-Holocene sea-level highstand and attached to the hills, is an example of this risk of situation. A groin field was constructed because of the critical erosion (Fig. 3.12).

The preceding few examples explain some of the controversial influences humans have had on natural coastal systems. Human intervention has additional

Figure 3.11 Scarps of dunes at Hualpén (northern margin of the Bio Bio estuary) indicating seasonal storm effects on the beach.

Figure 3.12 Village of Necocli, Caribbean Colombia, from the air. Courtesy Aeroestudios Ltd. http://coastal.er.usgs.gov/coasts-colombia/.

consequences, resulting in the degradation of coastal salt marshes and mangrove swamps, or the exhaustion of water resources on dune fields. All of these interventions exacerbate the vulnerability of the coastal environments to natural hazards.

8. Concluding Remarks

Numerous speculations have been made about the future occurrence of episodic processes associated with global warming and the scenario of rapid sea-level rise. Tropical cyclones and extratropical storms, as well as, for example, ENSO-linked processes, may increase their frequency and intensity, and most likely their impact will affect larger coastal areas in South America.

1. It is widely accepted that global sea-level rise (between 1 and 2 mm/year) is causing impacts on coastal systems, whose effects can be seen in many sandy barriers around the world. An acceleration of sea-level rise, which could reach a global average between 0.1 m and 0.9 m during the twenty-first century, would certainly exacerbate erosion processes and produce serious constraints on coastal systems, particularly along the eastern and northern coasts of South America.
2. Interannual processes (ENSO) affecting the Pacific coast are also responsible for coastal changes in the eastern South American coast, and perhaps many effects of the teleconnection are still unknown.
3. Although independent from climate-driven mechanisms, the more unpredictable tsunamis may also pose an increasingly serious threat to coastal

populations in densely populated areas, both on the Pacific and Caribbean shores. An unlikely, but yet possible, coincidence of a storm surge, spring tide, and tsunami on a major coastal city would have devastating consequences for human lives and property. Such an event could occur in the Caribbean Sea, where coasts are exposed to tsunamis as well as tropical storms. Another case could be the Pacific shore if a tsunami occurred during a strong positive El Niño.

4. On a regional perspective, the vulnerability of coastlines to episodic processes will increase, particularly because coastal systems are already subject to human occupation and intervention. Statistical data on storm waves would permit better preparation for the climate changes affecting our coasts. Ecosystems, human lives, and property can be severely damaged; thus response strategies must consider the development of predictive tools, public education, alarm systems, and evacuation plans. Some of these procedures are already in place in many areas vulnerable to storm surges and tsunamis.

Acknowledgments

We wish to acknowledge the assistance of Armando Scalise and Jorge L. Pousa for their critical review of parts of the manuscript, as well as for providing useful information. R. Angulo, I. Correa, and C. Martínez provided some information about the coasts of Brazil, Colombia, and Chile, respectively. We also thank the reviewers for their useful suggestions.

CHAPTER 4

Disasters in Mexico and Central America: A Little Bit More than a Century of Natural Hazards

Irasema Alcántara-Ayala

Instituto de Geografía, Universidad Nacional Autónoma de México (UNAM), Circuito Exterior, Ciudad Universitaria, 04510 Coyoacán, México, D.F.

Contents

1. Introduction

During the course of the twentieth century, a significant number of scholars worked in the field of disasters (Barkun, 1974; Varley, 1991; Albalá-Bertand, 1993; Aysan, 1993; Blaikie et al., 1994; Tobin and Montz, 1997). The perspective of the early work, however, was mainly one that considered disasters to be a result of nature (Davis, 2002). However, many social scientists-geographers in particular-have argued that disasters involve interactions among social, economic, cultural, and political spheres of communities exposed to hazards (White, 1961, 1964, 1973; Kates, 1962; Burton and Kates, 1964; Burton et al., 1968; Cannon, 1993, 1994, 2003). Before Hurricane Katrina in 2005, it was common to assume that the effects of disasters had much greater impact in the so-called developing countries (Benson,

Developments in Earth Surface Processes, Volume 13
ISSN 0928-2025, DOI 10.1016/S0928-2025(08)10004-9

2003). However, the world witnessed how levels of vulnerability synchronized with the occurrence of natural hazards even in the most "developed" nations govern risk, and hence disasters. Although at the end of the last century much had been argued in terms of the catastrophic character of natural phenomena, at the present time scientific understanding points to the unnaturalness of disasters. Disasters are not natural (Maskrey, 1993a and 1993b). They grow up beyond nature, as they stream parallel to socioeconomic historical processes linked to development (Clarke and Munasinghe, 1995) and, most importantly, to the various aspects of vulnerability (Varley, 1991; Serrano, 2007)-in other words, to differential vulnerability (Winchester, 1992; Aysan, 1993; Cannon, 1993, 1994, 2003; Blaikie et al., 1994).

Because hazards are natural phenomena derived from the interaction among crustal and surface processes, in addition to climatic conditions, vulnerability can be regarded as the degree of susceptibility posed by populations due to their socio-economic, political, and cultural conditions (Alcántara-Ayala, 2002). Although vulnerability cannot be taken as a synonym of poverty, human groups with high levels of poverty are among the most vulnerable (Blaikie et al., 1994). At a global scale, disasters have quite dramatically affected different parts of the world, specifically in vulnerable regions such as Central America and Mexico (M&CA). Their impact can be easily understood as a synonym for a "vicious cycle" in which lack of development is a *cause* but at the same time also a *consequence* of those disastrous events. Therefore, achieving development is a difficult goal, while sustainable development is a challenge that very easily could become a dream.

According to the United Nations International Strategy for Disaster Reduction (UN/ISDR, 2004), more than 90% of the deaths related to disasters associated with natural hazards take place in developing countries. Their impact, as shown in statistics, suggests a global trend in which the number of disasters has increased, whereas the number of fatalities is decreasing, but the number of affected people and economic losses are increasing to a great extent. Latin America and especially Central America and Mexico are not the exception (Pettiford, 1995). Therefore, in order to have a better understanding of the impact of disasters in this region (Fig. 4.1), this chapter aims at providing general insights into the spatial and temporal dimensions of disaster occurrence in M&CA in the course of about a century. As a result of the available data derived from EM-DAT-the world emergency events database (www.emdat.be/)–the time framework set up for our anaylsis was 1902-2007, which is a period of 105 years. Although the objective of this chapter involves the evaluation of disasters in the region of interest, it should be clear that the analysis of a little bit more than a century of natural hazards implies not only time, but an understanding of the cultural landscape, where all these events have occurred.

According to the Emergency Events Database (EM-DAT), a disaster is a situation or an event that overwhelms local capacity, necessitating a request to a national or an international agency for external assistance. For a disaster to be included into the database, at least one of the following criteria must be met: (1) 10 or more people reported killed; (2) 100 people reported affected; (3) declaration of a state of emergency; and (4) a call for international assistance.

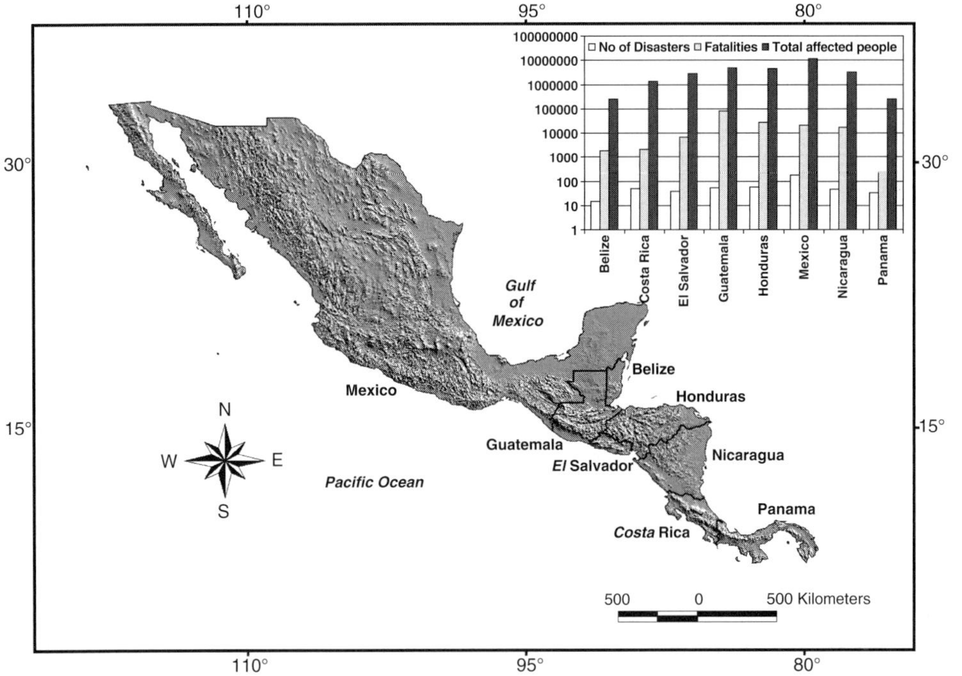

Figure 4.1 Overview of the impact of disasters associated with natural hazards in Mexico and Central America, 1902–2007. (*Source:* Based on OFDA/CRED database).

2. On the Nature of Mexico and Central America

Geomorphologically speaking, Mexico is joined with Central America throughout the Isthmus of Tehuantepec. However, given the historical nature of the country, the prevalence of regional identity and social heritage has harvested a cultural landscape that can be considered as firm or even more solid than the physiographical links. From south to north and east to west, surface processes and landforms resulting, such as coastal plains, dissected mountains, high plateaus, and major volcanoes, tropical natural scenarios for the occurrence of hurricanes, floods (Aragón-Durand, 2007), landslides, earthquakes, and volcanic eruptions are all rather common in the region.

A detailed evaluation of the physical and social atmospheres of Mexico and Central America is far beyond the objective and space of this chapter. Nonetheless, general geographic and demographic data are given in Table 4.1.

The area of Central America, 523,780 km^2, is roughly one-fourth that of Mexico (1,972,550 km^2). El Salvador is the smallest country; its area, 21,040 km^2, is the equivalent to just 4% of the total Central American surface. The population of Mexico, ~109,955,400 inhabitants, accounts for 72.70% of the total population in the region, whereas Belize makes up less than 1%. In terms of population density, El

Table 4.1 Baseline Data for Mexico and Central America

Country	Area (Km^2)	Population (2007)	Population density (hab/Km^2)	Unemployment (%)	Poverty (% population)	Indigence (% population)	GDP Current US$ (billions)
Belize	22,966	301,270	1.31	9.4	33.5	---	1.274
Costa Rica	51,100	4,195,914	82.11	4.6	20.3	8.2	26.24
El Salvador	21,040	7,066,403	335.86	6.2	48.9	22.1	20.37
Guatemala	108,890	13,002,206	119.41	3.2	60.2	30.9	33.69
Honduras	112,090	7,639,327	68.15	27.8	77.3	54.4	12.28
Mexico	1,972,550	109,955,400	55.74	3.7	39.4	12.6	893.4
Nicaragua	129,494	5,785,846	44.68	4.9	69.4	42.4	5.732
Panama	78,200	3,292,693	42.11	6.4	34.0	17.4	19.74
Total	***2,496,330***	***151,237,916***	---	---	---	---	***1,012.726***

Salvador has the highest (335.86 inhabitants/km^2), whereas as can be expected due to its surface, Belize has the lowest (1.31 people/km^2).

According to recent estimates for the year 2007 made by the United Nations Economic Commission for Latin America and the Caribbean (ECLAC), more than three quarters of the total population in Honduras live in poverty. In Nicaragua and Guatemala the percentage is higher than 50%, followed by El Salvador, where 48.9% of the inhabitants also are very poor. In Mexico, Panama, and Costa Rica the level of poverty is relatively lower, though still significant as it counts for 39.4, 34.0, and 20.3%, respectively. Moreover, among those living in poverty, a considerable proportion live under conditions of extreme poverty or indigence. The worst situation is found in Honduras, where more than 4 million people, the equivalent to 54.4% of its total population, are indigents. In the whole region under study, including Mexico, almost 27 million people live under conditions of extreme poverty (Table 4.1).

Patterns of social and economic development are different in the eight countries. Gross domestic product (GDP) is highest in Mexico (US$893.4 billion) and lowest in Belize (US$1.274 billion) and Nicaragua (USD 5.732 billions). Nevertheless, even if total regional GDP (USD 1,012.726 billions) is compared with the GDP of other nations such as Brazil with US$1.314 trillion or the United States with US$13.84 trillion, the economic overview of the region is fairly obvious. Unemployment in Central America is also significant, and although the highest rates are in Honduras, Costa Rica, and Panama, with 27.8, 9.4, and 6.4% of total population, respectively, the top levels of underemployment, 46.5 and 25%, are concentrated in Nicaragua and Mexico (Table 4.1).

As Mexico and Central America are situated in the Circum-Pacific Belt, also known as the Pacific Ring of Fire, where earthquakes and volcanic activity are particularly widespread, both subduction (Cocos-North American, and the Cocos-Caribbean plates) and transform boundary processes (Pacific-North American and the north and south Caribbean plate boundaries) control seismicity to a major extent (Fig. 4.2) and hence associated hazards (Tanner and Shedlock, 2004).

In terms of tectonic activity, different scientific efforts to produce global (Giardini et al., 1999; Shedlock and Tanner, 1999; Shedlock et al., 2001) and regional (Shepherd, 1993; Tanner, and Shepherd, 1997; Zúñiga et al., 1997; Shedlock, 1999) seismic hazards estimates have been undertaken in the last decades. Prediction pattern maps for Mexico (Lomnitz et al., 1999; Ordaz and Reyes, 1999), the Caribbean, and Central (Yong et al., 2002) and South America suggest that strong, damaging earthquake shaking is expected to take place in the flat-lying areas along or near the coasts or in valleys (Tanner and Shedlock, 2004). Furthermore, considering a 475-year period of recurrence, estimations made by Tanner and Shedlock (2004) indicated that great plate boundary interface earthquakes of $M_w > 8.0$ may occur at least once, and large earthquakes ($M_w > 7.0$) more than a few times. In a general sense, the largest seismic hazard values in Mexico and Central America take place in areas that have been, or are likely to be, the sites of the largest plate boundary earthquakes, whereas high-hazard values along the Pacific coasts of Mexico and Central America correspond to the subduction zone of the Cocos plate beneath the Caribbean (Tanner and Shedlock, 2004).

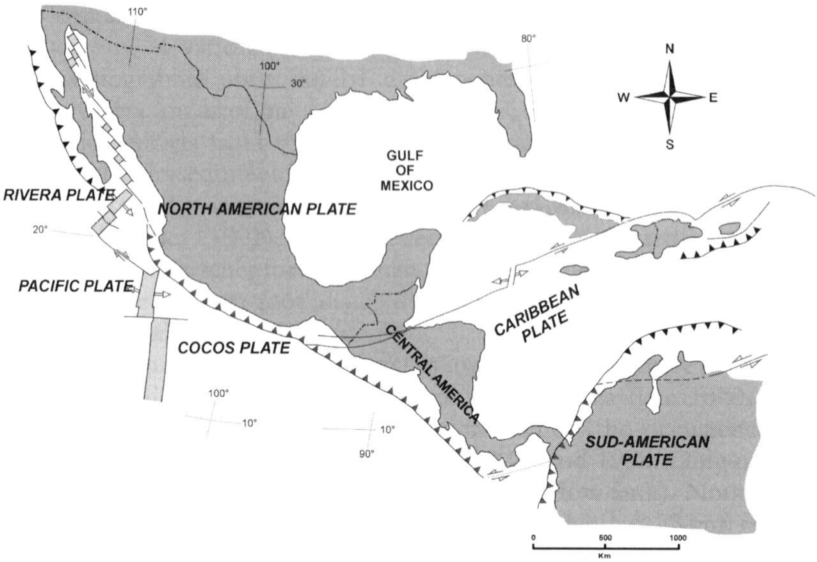

Figure 4.2 Tectonic setting of Mexico and Central America. (*Source*: de Cserna, 1990).

In Mexico, volcanic activity is concentrated mainly in the Trans-Mexican Volcanic Belt, about a 1000-km-long and 20- to 150-km-wide east–west strip that extends between the Gulf of Mexico and the Pacific Ocean (Mooser, 1972; Demant, 1978). This belt is composed of a large array of volcanic structures such as Late Tertiary and Quaternary stratovolcanoes, calderas, cinder cones, monogenetic fields, domes, and maars, mostly of the calc-alkaline type. In response to concerns about active volcanoes such as Popocatépetl, attention has been drawn not only to eruption dynamics and associated impacts dating back to pre-Hispanic times (Plunket, and Urunuela, 2006, 2008), but to the scientific and public reactions derived from volcanic crises (De la Cruz-Reyna and Tilling, 2008).

Historical and actual volcanic activity in Central America has taken place mainly on a volcanic front, formed by several narrow lines of active volcanic centers. The front follows a trend parallel to the strike of the subducting Cocos plate (Carr and Stoiber, 1990; Carr et al., 2003) and comprises about 40 volcanic centers, in addition to numerous emission centers (domes, cinder cones, maars, etc.). Ten major volcanoes and up to 300 smaller emission centers are concentrated in the Quaternary volcanic arc of Guatemala. In contrast, in Nicaragua, although the extension of the volcanic front is much smaller, there are 11 active centers. Along its 430-km volcanic chain, Costa Rica has approximately 120 emission centers. Tacaná, Santiaguito-Santa María, Cerro Quemado, Fuego, Pacaya, Santa Ana, Izalco, San Salvador, Ilopango, San Miguel, Cosigüina, San Cristobal, Telica, Cerro Negro, Momotombo, Masaya, Concepción, Rincón de la Vieja, Arenal, Paás, Irazú, Turrialba, and Barú are considered to be the most significant active volcanoes in Central America (Paniagua Pérez, 2002).

Linked to the occurrence of earthquakes in Central America are mass-movement processes, which are also a major hazard (Bommer and Rodríguez, 2002). Experiences have included the evaluation of historical and current landslides in Nicaragua (Devoli et al., 2007), Costa Rica (Alvarado et al., 2004), El Salvador (Blanco et al., 2002), and Guatemala (Coe et al., 2004), among others. Triggered not only by sesismicity, but induced by rainfall (Alcántara-Ayala, 2004a), landslides take place in scenarios where poverty, urban growth, lack of development especially in rural areas, land degradation and deforestation increase exposure as high levels of vulnerability are present. The number of events is enormous, but an unforgetable case that took place at Santa Tecla, at the capital of El Salvador (Jibson and Crone, 2001; Konagai et al., 2002, 2004; Evans and Bent, 2004) stands out. As a result of the earthquake of January 2001, a complex landslide known as "Las Colinas" swept away several houses, causing more than 500 casualties (Fig. 4.3).

Mexico is also very susceptible to landslides (Evans and Alcántara-Ayala, 2007). Major disasters associated with such geomorphic processes have taken place particularly in the last decade (Alcántara-Ayala, 2004a). According to EM-DAT, 239 people were killed in nine disasters associated with mass-movement processes during the period investigated. However, historical records analyzed by

Figure 4.3 In 2001, more than 500 people were killed in a disaster associated with an earthquake-triggered landslide in the capital of El Salvador.

Figure 4.4 Aerial view of the impact of floods in Tabasco, Mexico; 1,600,000 people were affected in 2007.

Alcántara-Ayala (2008) suggest that more than 3500 deaths associated with landslides have occurred from 1935 to 2006. The difference in numbers between these two sources is most likely due to three facts: (1) Landslide casualties are very difficult to quantify because they are frequently associated with floods and earthquakes; (2) the criteria established by the Centre for Research on the Epidemiology of Disasters (CRED) to be entered into the database may not always consider small events; and (3) information derived from particular historical records is more easily gathered at the national level.

Historical approaches (Garcia-Acosta, 2007) and modern processes studies have provided long-term perspectives for investigations of contemporary natural hazards. Estimations of recurrence intervals and evaluations of floods (Fig. 4.4) and droughts in Mexico (Mendoza et al., 2005, 2006), Central and South America, derived from palaeoclimate records suggest that expressions of regional climates are modulated by solar variability and, thus, by atmospheric circulation at a global scale (Schimmelmann et al., 2003). However, in an era in which disasters do not result exclusively from the dynamic nature of the Earth, but from risky constructions in vulnerable communities, it is crucial to understand the interactions among the exposed human groups and the fragile landscapes (Manuel-Navarrete et al., 2007).

3. Disasters in Central America and Mexico, 1902–2007

3.1. Spatial and Temporal Distribution of Disasters

According to the EM-DAT database (OFDA-CRED), 465 disasters associated with drought, earthquakes, floods, mass-movement processes, storms, and volcanic eruptions took place between 1902 and 2007. The consequences of those disasters included 160,296 deaths, 205,473 injured, 28,709,032 affected, and 2,933,862 homeless, which totaled more than 31 million affected people. Damages amounted to more than US$43,000 million (Figs. 4.5 and 4.6). Human losses resulting from all disasters were reported as highest in Guatemala, with 82,967 victims, whereas Mexico registered the largest amount of total population affected as it reached 12,402,464 people (Table 4.2).

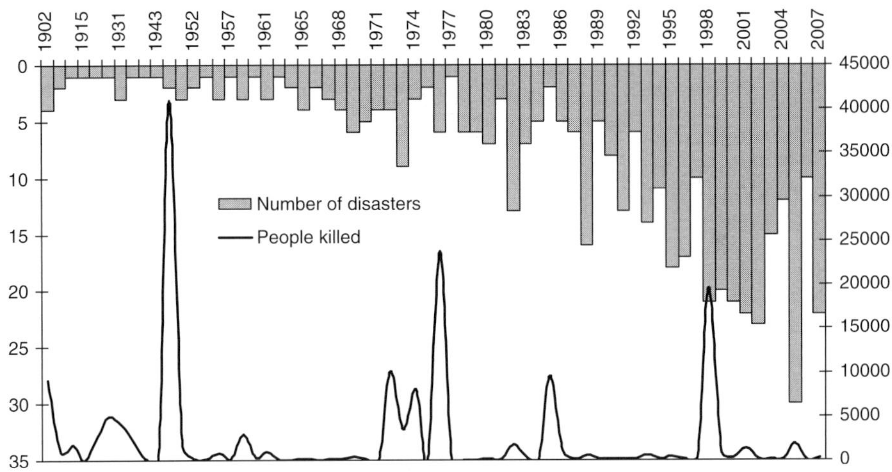

Figure 4.5 Number of disasters and people killed in Mexico and Central America, 1902–2007. (*Source*: Based on OFDA/CRED database).

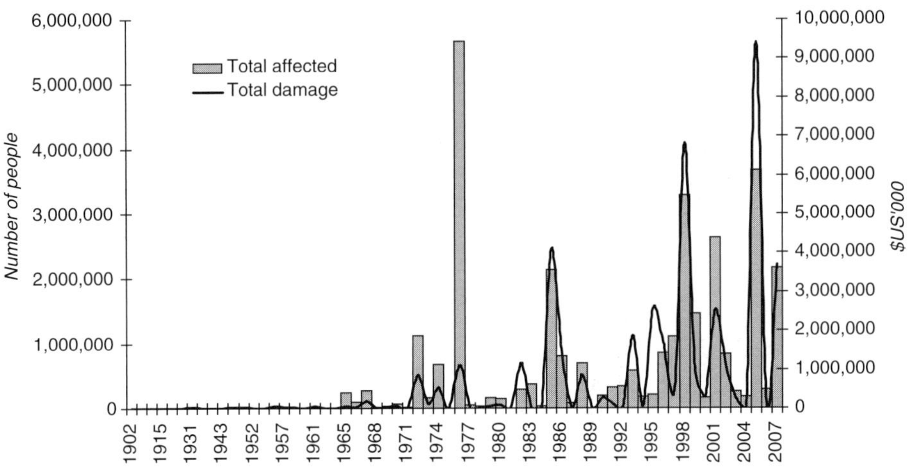

Figure 4.6 Total of people affected by disasters and economic damage in Mexico and Central America, 1902–2007. (*Source*: Based on OFDA/CRED database).

Table 4.2 The Aftermath of Disasters in Mexico and Central America, 1902–2007

Country	Num. of Disasters	People Killed	People Injured	People Affected	Homeless	Total Affected	Damage ($US'000)
Belize	15	1852	570	257600	0	258170	627557
Costa Rica	49	2149	8606	1347693	58916	1415215	931090
El Salvador	39	6812	47301	2894713	302500	3244514	4626810
Guatemala	54	82967	78028	4736301	1175815	5990144	2899550
Honduras	57	28114	12060	4581837	61004	4654901	4992479
Mexico	173	21191	36425	11434404	931635	12402464	26262210
Nicaragua	45	16984	21183	3210990	384192	3616365	2666952
Panama	33	227	1300	245494	19800	266594	102550
Total	***465***	***160,296***	***205,473***	***28,709,032***	***2,933,862***	***31,848,367***	***43,109,198***

(*Source*: Based on OFDA/CRED database)

Consequently, over a century, the occurrence of disaster in terms of space has been mainly concentrated in Mexico (173 disasters, the equivalent of 37.20% of the total), Honduras (57 disasters, the equivalent of 12.25%), and Guatemala (54 disasters, the equivalent of 11.61%), followed by Costa Rica (49 disasters, the equivalent of 10.53%), Nicaragua (45 disasters, the equivalent of 9.67%), and El Salvador (39 disasters, the equivalent of 8.38%), whereas a smaller number occurred in Panama (33, or 7.09%) and Belize (15 disasters, or 3.22%) (Table 4.2). Not surprisingly, the geographical distribution of disasters is highly correlated not only with the physical nature of this territory (mountain zones and flood plains), but with patterns of social underdevelopment.

With regard to the type of natural hazards associated with disasters for the same period, the frequency of floods and storms was the highest, with 157 and 146 events, respectively, whereas earthquakes totaled 79, volcanic eruptions 33, droughts 29, and mass-movement processes 21 (Fig. 4.7). It is important to stress,

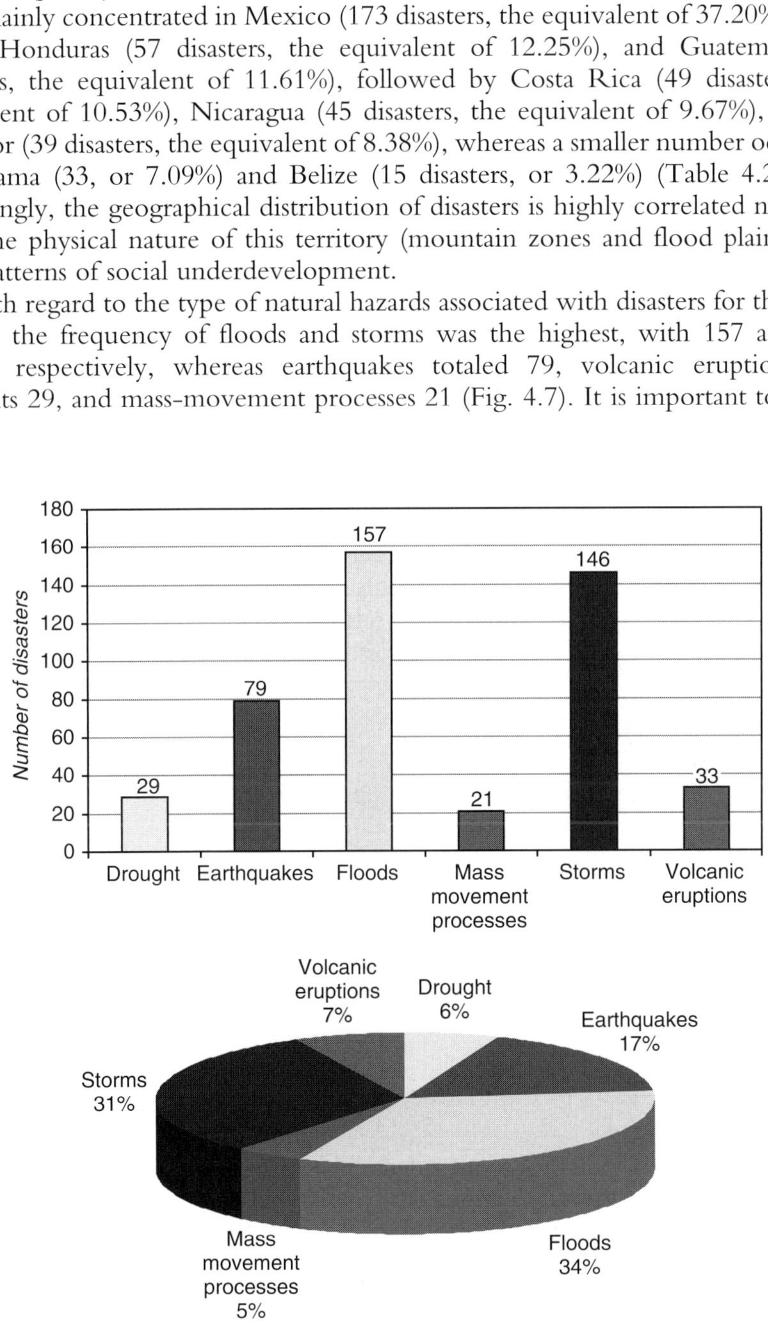

Figure 4.7 Number and percentage of disasters associated with natural hazards that occurred in the Period 1902–2007 in Mexico and Central America. (*Source*: Based on OFDA/CRED database).

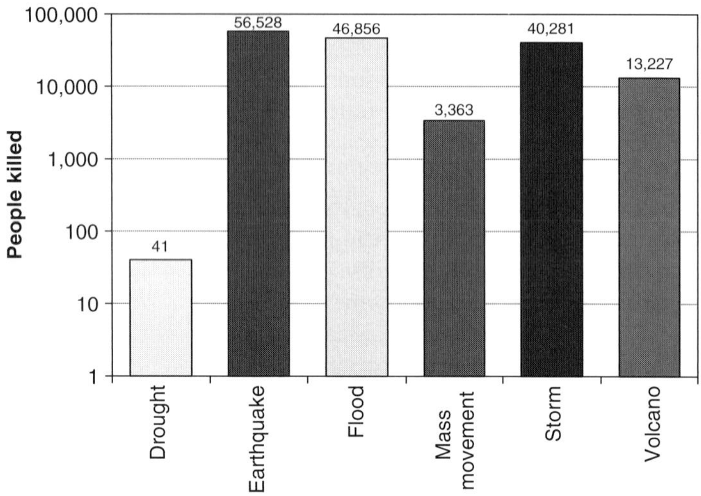

Figure 4.8 Number of people killed vs. natural hazards type during disasters in Mexico and Central America, 1902–2007. (*Source*: Based on OFDA/CRED database).

however, that given the nature of landslides, precise quantification is rather difficult; thus one should bear in mind that rainfall, earthquakes, and in a few cases even volcanic activity, act as triggering mechanisms.

The greatest number of fatalities in terms of hazard types is associated mainly with earthquakes, floods, and storms: they accounted for 143,665 victims, which equals 89.62% of total human losses. In addition, more than 13,000 people died in events associated with volcanic activity, while about 3400 people were killed due to landslides and 41 due to droughts (Fig. 4.8).

As can be observed in Figure 4.9, in the early years of the period analyzed, the highest number of victims came from disasters related to earthquakes and volcanic eruptions. Clearly, starting in the 1980s and 1990s, the number of disastrous events resulting from the occurrence of storms and floods amid vulnerable communities increased dramatically.

3.1.1. Major Disasters of the Century

In terms of people killed, the 10 major disasters that occurred in the region during the period under study (see Table 4.3) were concentrated in Guatemala (4), Honduras (3), Nicaragua (2), and. Mexico (1), respectively. The worst two episodes took place in Guatemala; the first involved 40,000 casualties resulting from a flood in 1949, whereas an earthquake in 1976 caused 23,000 victims. As 14,600 people died and 2,112,000 inhabitants were affected, the impact of Hurricane Mitch in 1998 was categorized as the worst disaster in Honduras and the third worst disaster on the top 10 list. In Nicaragua, during the 1972 earthquake, 10,000 people died, while, in Mexico, official numbers indicated 9500 victims during the 1985 seism. The total aftermath of the worst disasters was as high as 122, 232 deaths.

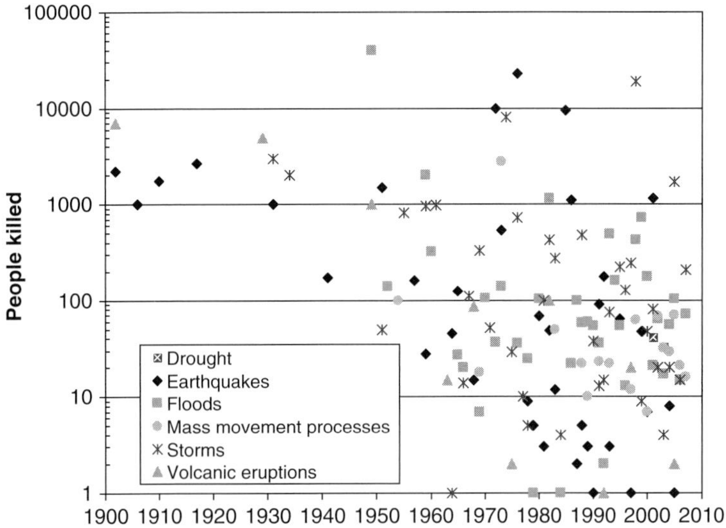

Figure 4.9 Temporal distribution of number of people killed in disasters associated with natural hazards in Mexico and Central America, 1902–2007. (*Source*: Based on OFDA/CRED database).

Table 4.3 Worst Disasters in Terms of Fatalities in Mexico and Central America, 1902–2007

Rank	Year	Country	Type	People killed
1	1949	Guatemala	Flood	40000
2	1976	Guatemala	Earthquake	23000
3	1998	Honduras	Tropical cyclone	14600
4	1972	Nicaragua	Earthquake	10000
5	1985	Mexico	Earthquake	9500
6	1974	Honduras	Tropical cyclone	8000
7	1902	Guatemala	Volcanic eruption	6000
8	1929	Guatemala	Volcanic eruption	5000
9	1998	Nicaragua	Tropical cyclone	3332
10	1973	Honduras	Landslide	2800
11	1917	Guatemala	Earthquake	2650
12	1902	Guatemala	Earthquake	2000
13	1959	Mexico	Flood	2000
14	1934	El Salvador	Tropical cyclone	2000
15	1910	Costa Rica	Earthquake	1750
16	2005	Guatemala	Tropical cyclone	1513
17	1931	Belize	Tropical cyclone	1500

(*Continued*)

Table 4.3 (*Continued*)

Rank	Year	Country	Type	People killed
18	1931	Honduras	Tropical cyclone	1500
19	1986	El Salvador	Earthquake	1100
20	1906	Nicaragua	Earthquake	1000
21	1931	Nicaragua	Earthquake	1000
22	1951	El Salvador	Earthquake	1000
23	1902	Guatemala	Volcanic eruption	1000
24	1949	Mexico	Volcanic eruption	1000
25	1959	Mexico	Tropical cyclone	960

(*Source*: Based on OFDA/CRED database)

The worst 25 disasters ranked in terms of total affected poeple are listed in Table 4.4. Ten are from Mexico, five from Nicaragua, four from Honduras, three from El Salvador, two from Guatemala and one from Costa Rica. Thirteen of the events involved tropical storms or cyclones; five were associated with earthquakes; and the other seven cases comprised three floods, three droughts, and one volcanic eruption. The population affected by these 25 events was as high as 23,974,890, from which 17,562,732 victims, the equivalent of 73.25%, were concentrated on the top 10.

Table 4.4 Worst Disasters in Terms of Total People Affected in Mexico and Central America, 1902–2007

Rank	Year	Type	Country	Total Affected People
1	1976	Earthquake	Guatemala	4,993,000
2	1985	Earthquake	Mexico	2,130,204
3	1998	Tropical cyclone	Honduras	2,112,000
4	2005	Tropical cyclone	Mexico	1,954,571
5	2007	General flood	Mexico	1,600,000
6	2001	Earthquake	El Salvador	1,334,529
7	2005	Tropical cyclone	Mexico	1,000,000
8	1998	Tropical cyclone	Nicaragua	868,228
9	1997	Tropical cyclone	Mexico	800,200
10	1986	Earthquake	El Salvador	770,000
11	1972	Earthquake	Nicaragua	720,000
12	1999	Storm surge/ coastal flood	Mexico	616,060

(*Continued*)

Table 4.4 (*Continued*)

Rank	Year	Type	Country	Total Affected People
13	1974	Tropical cyclone	Honduras	600,000
14	1999	Flood	Honduras	503,001
15	2002	Tropical cyclone	Mexico	500,030
16	1996	Tropical cyclone	Costa Rica	500,000
17	2005	Tropical cyclone	Guatemala	475,314
18	2001	Drought	El Salvador	400,000
19	1988	Tropical cyclone	Nicaragua	360,278
20	1992	Volcanic eruption	Nicaragua	300,075
21	1972	Drought	Honduras	300,000
22	1976	Storm	Mexico	300,000
23	1997	Drought	Nicaragua	290,000
24	1976	Tropical cyclone	Mexico	276,400
25	1967	Tropical cyclone	Mexico	271,000

(*Source:* Based on OFDA/CRED database)

3.1.2. Comparative Disaster Losses

Belize suffered the least number of disasters and was the second least affected, after Panama, in terms of people killed and total affected people. However, economic damages were lowest in Panama. Although Costa Rica was not among the most affected countries, more than 2000 people died during 49 disasters and about 1.5 million inhabitants were affected. In the case of El Salvador, 39 events caused approximately 7000 victims, more than 3 million of the total affected population, and US$4626 million. The highest percentage of victims were in Guatemala: 82,967 fatalities and almost 6 million total affected people as a result of 54 disasters. About 29,000 people were killed in Honduras in 57 events, in addition to 4.6 million affected inhabitants and damage of around US$5000 million. The highest number of disasters occurred in Mexico: as a result of 173 events, more than 21,000 people died and a total population of 12.4 million was affected; economic damage amounted to US$26,262 million. In Nicaragua, 45 disasters resulted in property damage totaling US$2666 million, approximately 17,000 people died, and 3.6 million people were affected.

3.1.3. *Exemplum diritas: Floods in Mexico, 1999*

At the regional level, in the provinces of Veracruz, Tabasco, Hidalgo, Michoacan, Jalisco, Oaxaca, Chiapas and Puebla, a storm and associated coastal floods took place between September 12 and October 29. Several days of intense precipitation derived from different hydrometeorological events caused 636 casualties, 60 people injured, 530,000 people affected, 86,000 homeless, a total affected population of 616,060, and US$451,300,000 of economic damages. Particularly in Tabasco the impact was considerable.

In Tabasco, 313,000 inhabitants, or 16.3% of the total population, were affected. The most severe impact was on 929 localities, which accounted for 48.4% of the province's total localities. At the beginning of August, antecedent rainfall caused river levels to rise to critical levels in Mazcalapa (14.35 masl) and Carrizal (7.88 masl). By September, the combination of tropical storms and cold fronts generated precipitation 12.5% higher than that registered in 1995 during Hurricanes Opal and Roxanne. Rainfall registered in both September and October were, respectively, 1.51 and 1.76 times higher than the monthly average. Floods resulted from three storms that took place on September 16, October 5, and October 23. The capital city of Villahermosa was mostly affected by the first two storms. A significant issue that emerged during this period arose from the fact that 14,800 million cubic meters of water were input to the dams in this region. As a consequence, 5000 million cubic meters had to be released from the Peñitas dam in Tabasco, and Carrizal and Grijalva rivers reached their maxium historic levels (Bitrán, 2000).

4. Hurricane Mitch: Beyond Political Boundaries

4.1. Understanding the Aftermath

Torrential rains, strong winds, floods, and landslides were associated with the occurrence of Mitch, a category 5 hurricane that took place in Central America at the end of October and the beginning of November 1998. Mitch's devastating regional impact, which was mainly concentrated in Costa Rica, El Salvador, Guatemala, Honduras, and Nicaragua, made it the worst disaster in Central America in the last hundred years. The devastation it caused was sadly a good example of how hazards and the occurrence of disasters respect no political boundaries; rather, they affect vulnerable communities of countries that may share similar socioeconomic, political, and cultural aspects constructed through historical development. The population directly affected by Mitch was almost 3.5 million-the equivalent of 10.9% of the area's total population, whereas the number of dead and missing reached about 20,000. In terms of human losses, Honduras and Nicaragua accounted for 74.59% and 21.83% of the total victims, respectively (Table 4.5).

An analysis of the rainfall that accompanied Mitch, reported in *Nature* (Hellin et al., 1999), indicated that despite being the most deadly hurricane in two centuries, precipitation was not extraordinary: rainfall totals and intensities measured at different time intervals (1, 2, 5, 10, 30, and 60 minutes) were smaller than values given by the updated maximum potential rainfall curve. At Cerro Guanacaure, located in southern Honduras, 698 mm of rain fell during a 99-hour period (between October 27 and 31). Even if Mitch was downgraded to a tropical storm on October 29, the most intense and prolongued rainfall took place between October 29 and 31, within 41 hours; this rainfall amounted 698 mm. Specifically, extreme rainfall intensities of 186 and 245 mm were recorded during six-hour

Table 4.5 Population Affected by the Disasters Derived from Hurricane Mitch in Central America

Item	Costa Rica	El Salvador	Guatemala	Honduras	Nicaragua	Total
1. Dead	4	240	268	5,657	3,045	9,214
2. Missing	3	19	121	8,058	970	9,171
3. Injured	—	—	280	12,275	287	12,842
4. In shelters	5,411	55,864	54,725	285,000	65,271	466,271
5. Total evacuated and direct victims	16,500	84,316	105,000	617,831	368,261	1,191,908
6. Population directly affected	20,000	346,910	730,000	1,500,000	867,752	3,464,662
7. Children under 5	10,400	180,393	379,600	780,000	451,231	1,801,624
8. Total population	3,270,700	6,075,536	11,645,900	6,203,188	4,453,583	31,648,907
9. %affected	0.6	5.7	6.3	24.2	19.5	10.9

(*Source:* ECLAC, 1999)

periods at the same time of the day (between 4:00 and 10:00 P.M., local time) on October 29 and 30. Considering such rainfall reports, Hellin and collaborators (1999) suggested that the severe damage that occurred in Honduras and Nicaragua was aggravated by the fact that the storm took place at the end of the rainy season on denuded and already soil saturated hillslopes. Such a perspective highlights how the physical phenomenon itself is part of a concatenation of hazards; in other words, it is the associated occurrence of multiple hazards (floods and landslides) that control risk and hence disaster occurrence.

Current marginalization commonly expressed as a high vulnerability degree played a significant role in Hurricane Mitch's aftermath since low-income communities were the most affected. Already low levels of living standards were worsened, personal effects and assets were lost, and infrastructure was severely damaged. According to estimations by the United Nations Children's Fund (UNICEF), children under 5 years of age were the most vulnerable group, making up 51.99% of the population that was directly affected by the disaster (Table 4.5).

According to the United Nations Economic Commission for Latin America and the Caribbean (ECLAC, 1999), Mitch's overall economic impact included 52% and 48%, respectively, of direct and indirect damages. Losses totaled $US6,018.3. The productive area was the most affected ($US3,906.9), followed by the infrastructure ($US1,245.5), the social sector ($US798.5), and the environment ($US67.4) (Table 4.6).

Table 4.6 Damage Caused by the Disasters Derived from Hurricane Mitch in Central America

Affected Sectors	Direct Damage	Indirect Damage Millions of USD	Total Damage
Social sectors	**551.8**	**246.6**	**798.5**
Housing	436.3	154.6	590.9
Health	53.8	78.9	132.7
Education	61.8	13.1	74.9
Infrastructure	**656.9**	**588.6**	**1,245.5**
Roads, bridges, and railways	528.1	541.5	1069.5
..*Energy*	28.6	30.1	58.7
..*Water and sewage systems*	74.6	16.8	91.4
..*Irrigation and drainage*	25.6	0.2	25.8
Productive sectors	1,824.1	2,082.8	3,906.9
Farming, fishing, and forestry	1,701.9	1,244.6	2,946.5
Manufacturing industry	32.8	575.2	608.0
..*Trade, restaurants, and hotels*	89.4	263.0	352.4
Environment	**67.4**	**0.0**	**67.4**
Total	***3,100.3***	***2,918.0***	***6,018.3***

(*Source:* ECLAC, 1999)

4.2. Casita Volcano: A Tragic Chapter within a Devastating Story

Even though Honduras was most affected by Mitch, undoubtedly, the most tragic single event occurred in the northwestern sector of Nicaragua. On October 30, 1998, more than 2000 people from the towns of El Porvenir and Rolando Rodríguez died as they were swept away by a mudflow. An intense precipitation episode of 1420 mm during the last four days of October, in addition to 564 mm of antecedent rainfall, triggered a complex mass-movement process (rockfall/avalanche) on the south flank of Casita volcano (Sheridan et al., 1999). Saturated former volcanic deposits and steep gradients produced a hyperconcentrated flow, which in less than 3 minutes devastated the area, killing almost all the population.

Casita volcano belongs to the structurally controlled San Cristobal Volcanic complex (Carreño, 1998). According to the first scientific report on the Casita volcano disaster (Sheridan et al., 1999), there were two source zones for the mass failure. They were located at 200 m SW and 100 m SE of the volcano summit. The main lithology comprised a brecciated dacite dome with high hydrothermal alteration. A NE-trending fault acted as a plane of weakness on which the main rupture took place as a result of a 20-m-thick, 60-m-high, and 150-m-long SE detachment on a gradient greater than 45°. The estimated rock fall volume was about 200,000 m^3; velocity was calculated in 15m/sec in the upper valley. A debris flow was the secondary process, and its source was situated 3 km above the two destroyed towns. The average depth of the flow was 3 m; it extended on an area of 25 km^2, over a 18- to 20-km distance (Fig. 4.10) (Sheridan et al., 1999).

The area of Casita volcano, like many others in the world (i.e., the province of Vargas in Venezuela), has been subjected to the occurrence of lahars since historical times. From the natural hazards perspective and particularly for mass-movement processes, it is crucial to analyze historical occurrences, in addition to morphological features expressed through the evolution of local landscapes, in order to understand the actual landforms dynamic. Under such a framework, conditions of potential future collapse in Casitas volcano have been analyzed by Van Wyk de Vries et al. (2000); they suggest that hydrothermal activity weakens the edifice, setting off flank spreading, changing the original shape, and steepening flank slopes. Therefore, not only Casita volcano, but also areas with similar scenarios require proper hazard and risk assessments (Salas, 2008), along with realistic measures to reduce vulnerability, thereby promoting and achieving disaster prevention.

5. Future Perspectives

According to the estimates produced by Intergovernmental Panel on Climate Change (IPCC) (2008a), in 2100, climate change will involve an increment of the global average surface air temperature by 1.1 to 6.4°C, as well as a significant rise in the sea level between 18 and 59 cm. Furthermore, it is very likely that hot extremes, heat waves, and heavy precipitation events will continue to become more frequent. More precipitation is expected at higher latitudes, and tropical

Figure 4.10 Complex mass movement process at Casita volcano, October 30, 1998; north-south extension is approximately 18–20 km. (*Source*: Open Sky Program, USA and Google Earth).

cyclone activity (Rosengaus-Moshinsky, 2006) is also very likely to become more intense, with larger peak wind speeds and heavier precipitation.

The scenario for Latin America, especially Mexico and Central America (IPCC, 2008b), indicates that those changes in precipitation patterns will affect the availability of water for human consumption, and agriculture. Desertification will tend to increase, thereby reducing the productivity of some crops and livestock, and the risk of flooding will increase considerably along the coasts due to the rise in sea level (Fig. 4.4). In this context, the International Strategy for Disaster Reduction (UN/ISRD, 2008) suggests that disaster risks will be affected by climate change in terms of the increase in climate hazards. In addition, the communities' vulnerabilities to natural hazards will be intensified by degradation of ecosystems, decreases in water and food availability, and changes in livelihoods. In addition, rapid unplanned urban growth will diminish the resilience of communities.

6. Discussion and Conclusions

6.1. Mexico and Central America: Chronicle of a Disaster Foretold

Year-to-year disaster impacts reflect a region's conditions of vulnerability. Undoubtedly, analysis of the consequences of a century of disasters in Mexico and Central America shows that we must consider that *there is a little bit more* than the simple occurrence of natural hazards over a hundred years. Clearly, the obstacles to preventing disaster are strongly linked with the lack of development in such countries; in this vein too, little attention is paid to reducing vulnerability. Disasters result from the "developing" conditions of vulnerable nations, and economic impact becomes a significant factor in a chain reaction that helps nourishing a "vicious cycle" in which lack of development is simultaneously the *cause* and *consequence* of those disastrous events. As a result, achieving development is a difficult goal, while sustainable development and disaster prevention are challenges that very easily could become a dream.

The analysis presented here in terms of the spatial and temporal occurrence of disasters in Mexico and Central America provides a useful benchmark against which vulnerability conditions should be compared. These results point to the need to downscale the understanding and particularly the vulnerability of disaster at the local level. In addition to understanding the importance of the interactions among hazards and vulnerable communities, it is crucial to consider resilience. Resilience is understood as the capacity to absorb a negative impact or faculty in order to recover from the effects of an external force or process, such as a natural hazard. Consequently, it is important to remember that, given their nature as constructed social processes, vulnerability and resilience equally share the specific dimensions of space and time.

One theme that emerges from many Latin American experiences is disaster scale (Velásquez and Rosales, 1999). Although the analyses of disaster impacts derived from the utilized database are unlikely to affect the region's perspective in the vulnerability aspect, they do raise questions about what would happen if small and medium (local) disaster events were considered in risk assessment and management. Taking a wider disaster scale view, we see that there are some doubts as to whether the impact of "large" disasters would be greater than the sum of all those "smaller" ones taking place in the region. These arguments would mark a clear break in the conception of magnitude and frequency developed by Wolman and Miller (1960), although the heart of this thought was conceived for forces in geomorphic processes. Human dimensions would have to be further explored along this line.

In examining not only all disasters that occurred during 10 decades, but the specific disaster impact of Hurricane Mitch in countries such as Costa Rica, El

Salvador, Guatemala, Honduras, and Nicaragua (Rhyner, 2006), we can point out several potential causal factors. First, it is mostly poor and indigent groups who are exposed to natural hazards because they usually settle down in areas that are geomorphologically susceptible to natural hazards (i.e., floodplains, steep hillslopes, volcanic flanks, etc.). As Alan Lavell (1994) pointed out, Central America is one of the most disaster-prone areas of the globe. Second, high levels of economic, social, cultural, and political marginality are poorly understood, and decision-making efforts are not being directed toward their appreciation and consideration. Third, scientific-pure and social-research and public education parties are not fully supported by government and private agencies, nor are they involved in risk management. Thus, natural hazards are not properly assessed, and efforts to reduce vulnerability are neglected. It would seem that prevention remains a hope, while reaction is still a reality.

The effects of disaster on human populations cannot be entirely appreciated through human, economic, and environmental losses. Social parameters related to individual impact cannot be easily quantified because human behavior is complex (Quarantelli, 1996). Diverse approaches to understanding hazards and disaster occurrence in different parts of the world have been undertaken in the last decades. However, only lately has comprehension of vulnerability and risk been addressed, considering both social and physical perspectives; particular attention has therefore been paid from a holistic view to achieve effective risk management (Cardona, 2001, 2004a; Cardona et al., 2003).

Most recent efforts to evaluate risk and vulnerability in Latin America have included the construction of different indicators such as Disaster Deficit (DDI), Local Disaster (LDI), and Prevalent Vulnerability indices (PVI) (Cardona, 2003b, 2004b, 2004c, 2005, 2006). The objective of those indicators involves the commitment to offering information on economic and social factors that affect risk and risk management to decision makers at national levels, so that they can identify risk and recommend adequate disaster risk management policies and actions. In addition, the use of such indices would allow comparison of profiles.

Above and beyond applying indicators, risk management should comprise a set of strategies, programs, and projects to coordinate reduction and mitigation actions designed to prevent, confront, and recover from disasters or devastating conditions. Supported by an understanding of the trilogy hazards-vulnerability-resilience (Fig. 4.11), as suggested in the Atomium Model (Alcántara-Ayala, 2009), risk management needs to be directed toward reducing vulnerability, lessening the social impact, guaranteeing the best possible security conditions, and decreasing material losses within a community-based strategy (Alcántara-Ayala, 2004b; Alcántara-Ayala et al., 2004) and a transdisciplinarity academic research framework.

It would be easy to forget that the first lessons about establishing civil protection agencies in the area of interest are rather new. More recognition should be given to the work developed by the National Centre for Disaster Prevention (CENAPRED) in Mexico, and the Coordinating Centre for the Prevention of Natural Disasters in Central America (CEPREDENAC), both of which have been dealing with disasters during the last two decades. In view of the fact that the causes of disasters can be easily traced to vulnerability factors, reducing vulnerability and

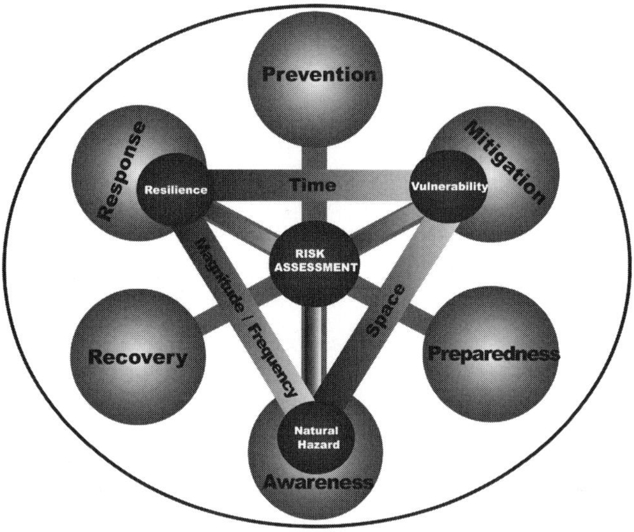

Figure 4.11 The Risk Management Atomium Model. (*Source*: Alcántara-Ayala, 2009).

enhancing resilience must be flagged as two of the key Latin American millennium goals, in addition to making it a global millennium goal devoted to disaster prevention (Alcántara-Ayala, 2008). The failure to acknowledge and address these issues will certainly be reflected sadly in the coming years if we continue to witness the effects of climatic change on unplanned settlements developed on fragile landscapes in both urban and rural areas.

Acknowledgments

Disaster information was kindly provided by the OFDA/CRED database (EM-DATA). Special thanks are due to Regina Below and Jean Michel Scheuren from Université Catholique de Louvain. Financial support provided by CONACyT in order to undertake landslides research in Mexico throughout the proyect 49844 is also gratefully acknowledged.

CHAPTER 5

Venezuela: The Construction of Vulnerability and Its Relation to the High Seismic Risk

Jaime Laffaille, Carlos Ferrer, *and* Klaudia Laffaille

(1) *Universidad de los Andes, Geophysical Laboratory and Foundation for the Prevention of Seismic Risk (FUNDAPRIS). Mérida, Venezuela. e-mail: jaime@ula.ve*
(2) *Universidad de los Andes, Geographical Institute, Mérida Venezuela. e-mail: carlosferrerve@yahoo.com*
(3) *Centro de Investigación en Gestión Integral de Riesgos, Mérida, Venezuela. e-mail: k.laffaille@gmail.com*

Contents

1. Introduction

When compared with countries such as Japan, Chile, and Peru, where great-magnitude earthquakes occur with relative frequency, Venezuela cannot be categorized as a "highly risky seismic country." Nonetheless, seismic activity has affected the country throughout its history. The most important reported events

Developments in Earth Surface Processes, Volume 13
ISSN 0928-2025, DOI 10.1016/S0928-2025(08)10005-0

have reached magnitudes close to 7.4, and several elements qualify this country among those at high seismic risk.

About 80% of Venezuela's inhabitants live in the northern part of the country, along a narrow belt of land between the Andes and the Caribbean Sea (Fig. 5.1). This area is related to the interaction zone of the Caribbean and South American plates that generate a stress field with a NNW–SSE maximum horizontal stress and an ENE–WSW minimum horizontal stress (strike-slip regime). The Bocono fault system or just Bocono's fault along the west of Venezuela is a main factor for risky seismicity, and the more important seismic events have been related to this system of faults. This interaction between the Caribbean and the South American plates is not a simple dextral transform fault; instead it is a broad active deformation zone resulting from a long-lasting oblique-collision process (Figs. 5.2). Nevertheless, a large portion of the right-lateral motion seems to take place along the dextral Boconó–San Sebastián–El Pilar faults, which are interconnected and have a historically long activity, while other, weaker earthquakes can be localized on minor associated faults.

The Bocono fault extends along more than 500 km from the Venezuela–Colombia border to the Caribbean Sea with a northeast general trend (Figs. 5.1 and 5.2). A rate of 2 cm/yr of plate motion at the offshore boundary has been recorded (Perez et al., 2001), half of which is accommodated by the San Sebastian fault (Audemard et al., 2000)

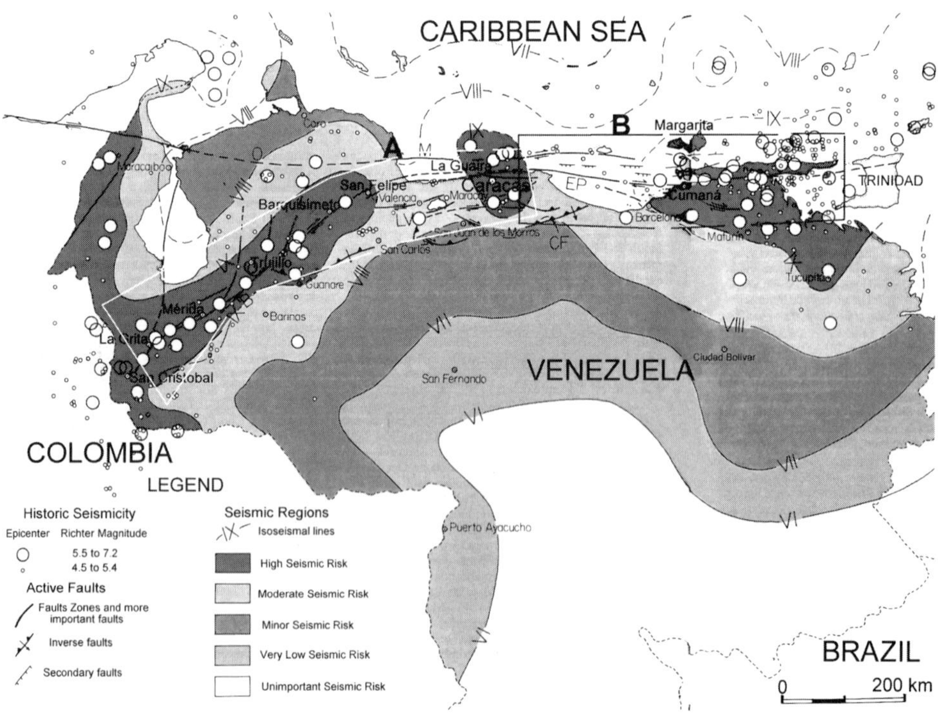

Figure 5.1 Main seismic regions of Venezuela. *Source:* Adapted from Schubert, 1984). A = area of Figure 5.2 and B = area of Figure 5.4.

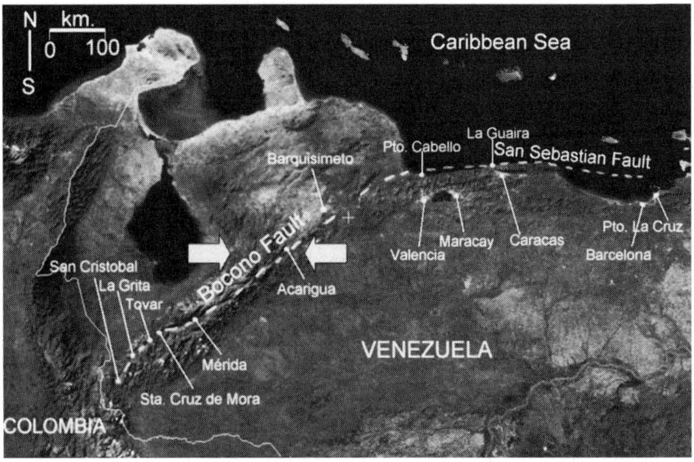

Figure 5.2 Main fault systems and cities in central-western Venezuela. The area is indicated as A in Figure 5.1.

(Fig. 5.2). The Bocono fault is morphologically expressed by a continuous straight alignment of longitudinal valleys, linear depressions, pull-apart basins, fault scarps, trenches, sag-ponds, linear ridges, and saddles that suggest that this major tectonic feature is active. The paleoearthquake reconstruction derived from different trenches suggests the occurrence of at least six to eight earthquakes in the past 9000 years in the Merida Cordillera, Central Andes (Audemard, 1997; Cowan et al., 1998), yielding a maximum average recurrence interval of about 1100–1500 years. Based on the northern strands and the average slip rate (2.6 mm/r), such an earthquake sequence should have accommodated about 23 m of slip since 9 ka, suggesting that the maximum slip per event ranges between 3 and 4 m. Results from a trench in La Grita (see description below) suggest a shorter return period of about 300 years. In this area the Holocene oblique-slip rate ranges of the fault system were estimated to range between 4.3 and 6.1 mm/yr (5.2+/−0.9 mm/yr) (Audemard et al, 1997).

Important cities such as Barquisimeto (~1,100,000 inhabitants), Acarigua (~300,000 inhabitants), Merida (~350,000 inhabitants), and San Cristobal (~300,000 inhabitants) are located in the seismic activity fringe of the Bocono system. Other cities, such as Caracas (~4,000,000 inhabitants, country capital), Maracay (~1,112,000 inhabitants), Valencia (a main industrial center of Venezuela of ~1,400.000 inhabitants), and La Guaira, are related to another big main system of active faults, known as the San Sebastian fault system that runs along a straight coastal line (Figs. 5.1, 5.2, and 5.3b).

The eastern region of the country presents a similar situation. Barcelona (480,000 inhabitants), Maturin (~500,000 inhabitants), and Cumana (~360,000 inhabitants) are geographically located in the most active seismic region of the country (Figs. 5.1 and 5.4). In this region the more active fault is the El Pilar system (Fig. 5.4); the most active zone of Venezuela, as deduced from the historical and instrumental data of Table 5.1 and its epicenters are plotted in Figure 5.1.

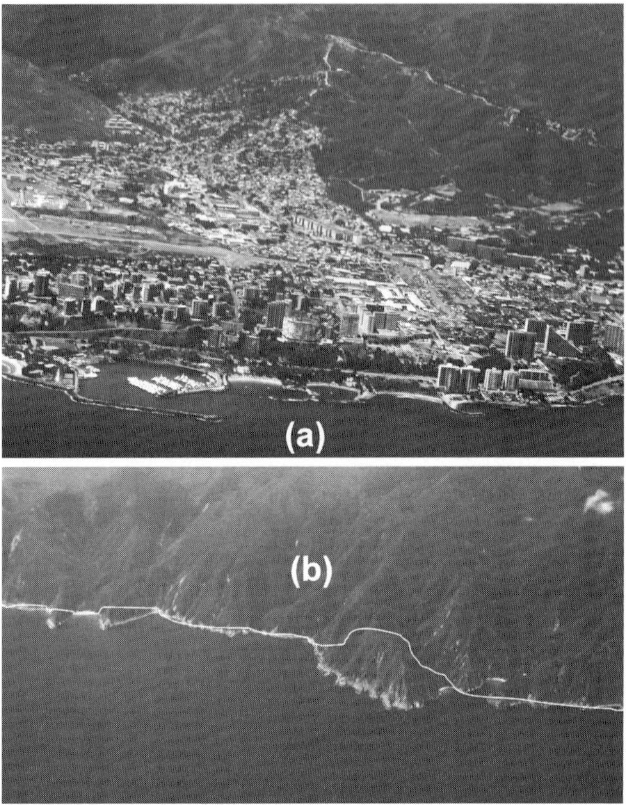

Figure 5.3 (a) City of La Guaira in the central coast of Venezuela. The city was built on alluvial fans that are crossed by the San Sebastian Fault. (b) Straight coastal line structurally controlled by the San Sebastian Fault.

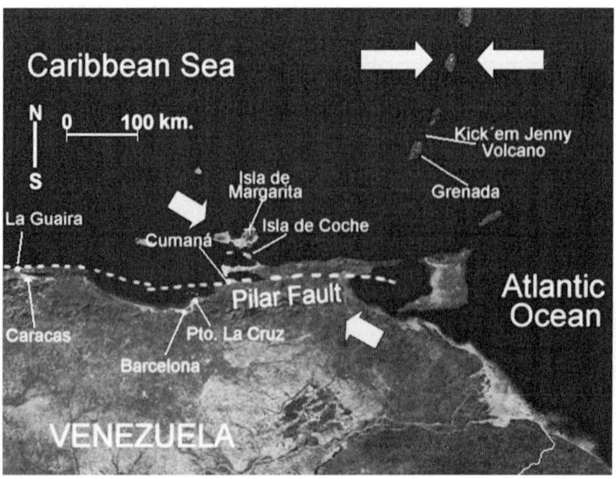

Figure 5.4 Main fault systems and cities in central-eastern region of Venezuela. Arrows indicate the relative convergence between the Caribbean and Atlantic plates.

Over 10 million people, taking in consideration only the main cities, live in the most active seismic zones of Venezuela. Such a concentration of cities and population in hazardous areas are a result of cultural, economical, political, and historical reasons. Because Venezuela has been and will still be affected by earthquakes, in this chapter we analyze some seismic events of the Venezuela history and what could be the seismic perspective in future.

2. Historical Record

The record of historical earthquakes and its main consequences are presented in Table 5.1, and selected earthquakes are described below.

2.1. The Earthquake of September 1, 1530: The Beginning of Venezuela's Seismic History

The first earthquake reported in Venezuela occurred on September 1, 1530, at 10 A.M. The epicenter was located at the eastern extreme of the faults system of Venezuela, and according to different documents (see for example, Centeno, 1940; Grases et al., 1999) it generated a tsunami that produced enormous damage in the region where the city of Cumana stands today (Figs. 5.4 and Fig. 5.5).

Figure 5.5 (a) Area affected by the earthquake and tsunami of 1530. (b) and (c) Present-day human occupation of risky areas affected by the tsunami of 1530.

Bartolome de las Casas (in Grasses et al., 1999) reports a description made by a captain named Castellón:

> *Suddenly the sea raised, and went beyond its ordinary limits in height four states (a state equals 1.96 meters), and got over the top of some trees by the mouth of the river (which is big and deep), and covered all the plains ... the ground started to shake terribly ... and then it shook several times that day. ... The ground cracked on several sides at the plains and the hills and from the cracks some salty, black-inked water, came out smelting like sulfur stone. A mountain range of the Gulf, called Cariaco ... cracked as much that it was split with a big gap in the middle of it. The houses of native people were destroyed, because they were made of wood and straw and some natives died ... the horror and fright caused more chaos.*

It seems very clear that Castellón was describing his experience with a tsunami, possibly generated by an earthquake.

It is not easy to estimate the magnitude of the event of 1530 because the Venezuelan coasts were practically uninhabited at that time. According to Alexander von Humboldt, who visited the area in 1799 (Humboldt, 1991), the old inhabitants of the region, where, as we have stated, today stands the city of Cumana, were probably safe from the constant sea floods by climbing up the surrounded hill where the old convent of San Francisco was located. Humboldt also comments that "it is a much generalized opinion in Cumana's coasts and Margarita Island, the Gulf of Cariaco owes its existence to a landslide with the ocean's bursting." By the time Humboldt visited Cumana, the city had about 15,000 inhabitants, and, according to this illustratious naturalist, the city would not grow much more because of the constant earthquakes and the nearby sea. Cumana had been affected by nine earthquakes of considerable magnitude (over 5.5 Richter magnitudes (Grases et al., 1999). Today this city has more than 350,000 inhabitants spread out over the whole region affected by the tsunami of 1530 (Fig. 5.5).

2.2. The Earthquake of 1610

The first important disastrous earthquake reported by the seismic history in the west of Venezuela occurred on February 3, 1610. It affected fundamentally the population of La Grita and the Mocoties River valley in the Andean zone (Fig. 5.1), and it was related to the activity of the Bocono fault. According to recent studies of the damage distribution and seismic intensities, the earthquake reached between 7.1 and 7.3 Richter magnitudes (Audemard, 1998, 2003). It killed more than 60 people and triggered a seismic landslide that affected the area that is now known as La Playa, in the Mocoties River valley (Ferrer and Laffaille, 1998).

Brother Pedro Simon (1987) tells us that the ground was shaking so strongly that it was almost impossible to walk. In this city, almost all houses built with mud (adobe) as well as the convent collapsed, rivers and streams became dry, and people thought that the water was being absorbed by the ground along cracks. Muddy water, however, was flowing abundantly the next day.

A hypothesis has been offered to explain what happened on February 3, 1610 (Ferrer and Laffaille, 1998; Lozada and Rodríguez, 2006) (Fig. 5.6). Because of the earthquake, there was a mass movement (landslide) in the Mariño mountain slopes (Fig. 5.6a). Then a landslide flowed along the canyon of a tributary of the Mocoties River (Fig. 5.6b), damming the main channel and generating a large lake that flooded the valley upstream during fourth months. Finally, the dam collapsed, the lake was drained, and a water wave flooded the valley where today is located Tovar city (Figs. 5.6 and 5.7).

2.3. The Big Earthquake of 1894

In 1894, in a place close to the area affected by the event of 1610, following the activities of the Bocono fault, a large earthquake occurred in the Venezuelan Andes destroying the city of Chiguara. Even today the landscape near

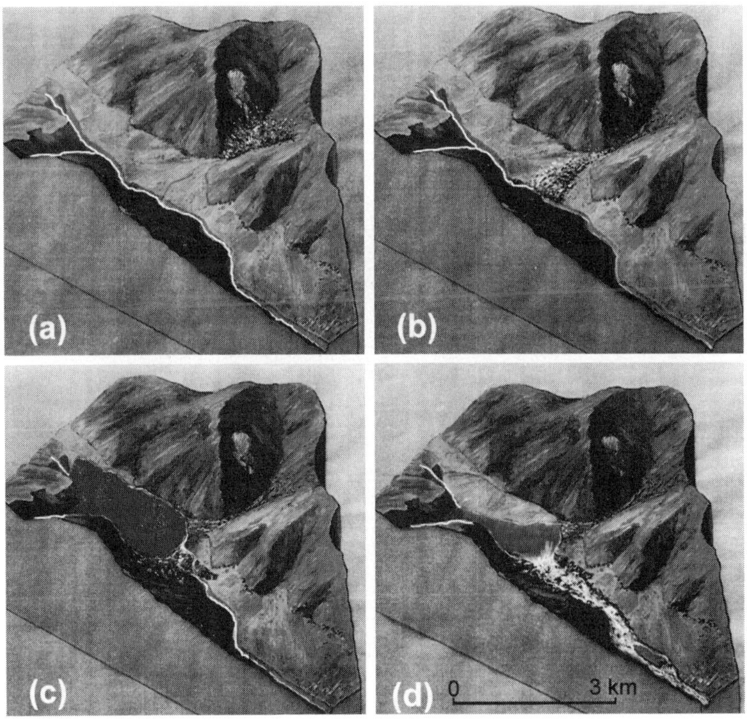

Figure 5.6 Model for the disaster of February 3, 1610. (a) The earthquake happens, and the mass movement begins (landslide) in Mariño mountain slopes; (b) the landslide moves with the water flow through a fluvial canyon tributary of the Mocoties River; (c) the landslide sediments are deposited on the Mocoties River Channel, generating a lake that floods the valley during four months; (d) the dam collapses generating a flood wave and the drainage of the lake.

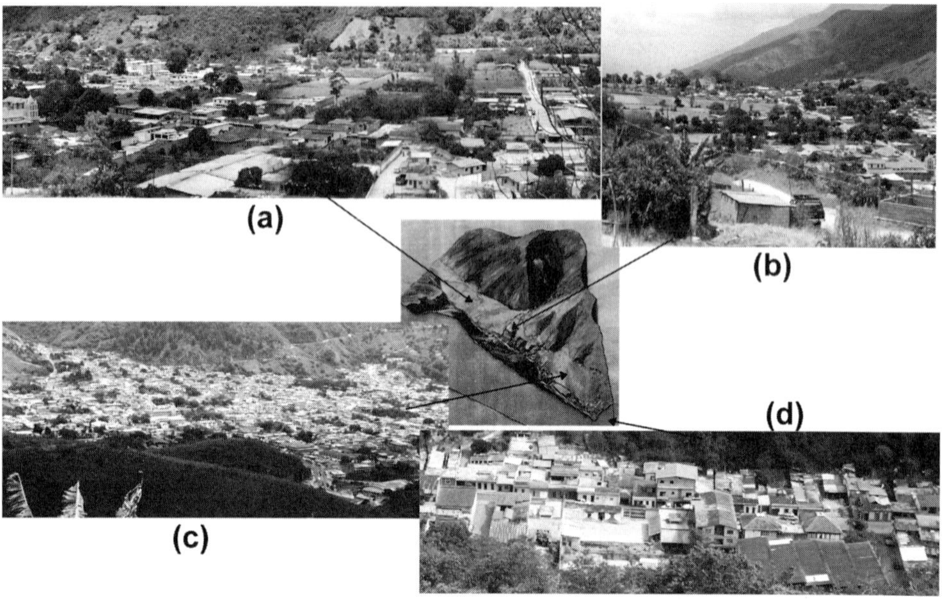

Figure 5.7 Current appearance of the area affected by the landslide, lake, and flood wave during the events of 1610.

Figure 5.8 (a) Chiguara City, built on a big landslide zone that shows clear signs of current activity (some scarps are indicated by dashed white lines). (b) and (c) Churches in the city of Merida that were partially destroyed by the earthquake of 1894.

Chiguara presents a large displaced mass and local scarps of the main landslide of 1894, which suffered partial reactivations trough time (Fig. 5.8a). Alamo (1894) commented that "all houses and buildings fell down, there were deep cracks on a big extension and the hills slid down."

The magnitude of the earthquake has been estimated to have been approximately a grade 7 on the Richter scale. It is ranked among the largest seismic events that ever occurred in Venezuela (Audemard, 2003). Centeno Grau (1940) calculated the damage area of this event to have spread over 7000 km^2, from Trujillo, Venezuela, where the tower of the San Francisco Church collapsed, to Pamplona (Colombia). It was especially intense in the region between Tovar and Mérida where churches and many buildings collapsed. The roofs of seven of the nine temples that existed in Merida city were partially destroyed (Rengifo and Laffaille, 2000) (Fig. 5.8b).

In Santa Cruz de Mora more than 100 people were killed (Febres, 1931), and Lagunillas, Chiguara, and Mesa Bolivar were totally destroyed. According to the testimony of Jose Ignacio Lares, an inhabitant of the region and an eyewitness to the 1894 earthquake, more than 70 tremors took place, and, during the next three days after the earthquake, a thick dust cloud rose by the Chama River covering an extension of more than 20 leagues (~112 km). Moreover, the river and torrents of waters flowed for over a month with mud, leaves, and forest debris (Rengifo and Laffaille, 2000).

Today the seismic scene would be notably more serious because substantially more people inhabit these vulnerable places than used to be the case. Houses and towns have not been designed to take into consideration the seismic condition and slope instability. This situation also concerns the public buildings where so many people concentrate during working hours, such as schools and hospitals. The Santa Cruz de Mora City Hospital, built beside a scarp of the Bocono fault, for example, obviously has been designed without any seismic-resistant engineering structure.

3. Earthquakes in Caracas

The city of Caracas has been severely affected by four seismic events: in 1641, 1812, 1900 and 1967 (Grases et al., 1999). Even though they cannot be classified as strong earthquakes (M > 7), two of them merit discussion in this context.

3.1. The Earthquake of 1812

The earthquake of March 26, 1812, is one of the most controversial seismic events in Venezuela's history because it affected the whole country, not only by doing direct physical harm to cities and populations, but by producing political, social, and economical effects. The earthquake also had destructive effects in Saint Vincent and Jamaica. The events started in Caracas on March 26 along the San Sebastian fault and affected four cities with landslides, rock falls, surface ruptures, and liquefaction. Three earthquakes and one volcanic eruption in Saint Vincent shocked the Caribbean area in 1812 (Altez Rogelio, 2008).

This event would be a special example of how the use and lack of information about a specific fact may be used for purposes that have nothing to do with the management of emergency situations or the disaster generated by such situations.

The earthquake happened during the War of the Independence of Venezuela and affected the pro-independence cities. In 1812, despite the war, Venezuela was making genuine progress as a nation, with a respected democratic government and an army capable of defending it. Optimism was evident, and the admission of Guyana and Maracaibo provinces into the federation was starting to be considered.

Then, on March 26, Holy Thursday, at about 4 or 5 in the afternoon, the disaster took place. It was a very hot afternoon, and the churches and streets were full of people who had come from nearby villages to participate in the rites and processions of the day. Suddenly, Caracas, a city of 32,000 inhabitants and the capital of Venezuela, became a bunch of ruins, and most of the churches were completely destroyed. La Guaira, Maiquetía, Merida, and San Felipe suffered the same fate, while Barquisimeto, Valencia, La Victoria, and other cities began to sway (Laffaille and Ferrer, 2003).

Religious fanaticism and superstition spread throughout the country. Taking political advantage of the tragedy, very influential people, supported by a group of priests, terrified the populace, claiming that the earthquake was God's punishment on the population for their support of the revolution.

Thus, in addition to inflicting immense material and human damages, the earthquake of 1812 contributed to the failure of Venezuela's first republic. There are extremely different versions about how many people were killed during the earthquake, with the numbers oscillating between 1000 and 30,000. The second number is certainly an exaggeration inasmuch as Caracas barely had over 30,000 inhabitants at that time. On the other hand, 1000 dead people is equally unbelievable when we consider that in the San Carlos barracks alone about 600 soldiers died. It is generally acknowledged that this earthquake was notably magnified by the historical conditions then prevailing. A very recent investigation (Altez, 2006) places this event in a more appropriate context of the seismic reality. According to Altez, the number of fatalities was likely close to 2000. The extension and intensity of the damages is less than what documents of that time indicate, being more related to the constructive characteristics of the buildings and site effects. This last aspect is supported by the damage distribution map made on a Caracas city plan in 1812 (Fig. 5.9a; Altez, 2006), which shows that the damages are distributed in a fringe that goes across the city in a north-northeast direction. According to the information presented by Yamazaki et al. (2005), the most important fringe of damages coincides with a discontinuity among zones of diverse sediment thicknesses (varying from 11–20 m to 31–40 m). This kind of perception was subjected to serious study following the Caracas earthquake of 1967.

3.2. The Caracas Earthquake of 1967

The Caracas earthquake of July 29, 1967, had an intensity of (M_W 6.6), the epicenter was located at 67.25W–11.0 N in the Caribbean Sea, and the focal depth was estimated to be 30.0 km. The city of Caracas was strongly affected by the earthquake during 35 seconds at 08:05 P.M.. More than 2000 people were injured, 245 were killed, and nearly 2640 buildings were damaged, resulting in economic costs of ~US$100 million. The quake actually consisted of four separate events that took place 25 to 30 km from Caracas, in front of the central coast, and

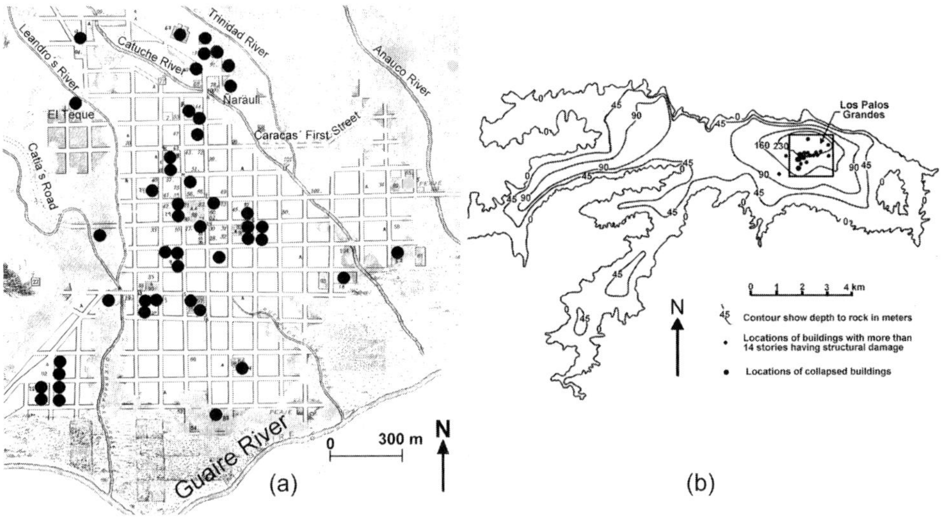

Figure 5.9 (a) Map of Caracas city in 1812, illustrating the sites where damage was reported (dark circles, Altez, 2006). (b) Main damage observed in Caracas during the 1967 earthquake. The contour lines that have been traced over the zone occupied by Caracas city in 1967 and the square correspond to the zone of Los Palos Grandes, where the most intense damage was observed.

were related to the fault system of San Sebastian (Suárez and Náblek, 1990) (Figs. 5.1 and 5.2). The San Sebastian fault system is thought to be dextral and transcurrent and is considered the main southern border of the Caribbean plate (Perez et al, 2001). This system is the main seismic source of the region and also generated the big earthquake of 1900.

The distribution of damages followed an unexpected pattern of zones with massive destruction and entire areas where the earthquake had very light effects. The most important damages were concentrated in the eastern zone of the city, particularly in a suburb called Los Palos Grandes (Fig. 5.9b), where different high buildings, designed and built according to current laws of design, suffered intense damages, including collapse. Similar buildings in other parts of the city were not affected as intensely, which indicated that the distribution of damage was related to terrain rather than to construction characteristics. It was concluded that the depth and characteristics of the underlying rock and regolith materials played a decisive role in the kind of effects that experienced in a particular place (Grases, 2002). Sediment thickness in excess of 300 m and related soil–structure interactions were identified as the principal factors in how the damage was distributed (Schmitz et al., 2005).

4. Tsunamis

As mentioned earlier, the Caribbean region is characterized by convergent, compressional, and collisional tectonic activity, which results in frequent occurrences of earthquakes and volcanic eruptions that can generate tsunamis through complex mechanisms. In the northeast of Venezuela, there is a less common subduction zone, where the convergence of two plates of ocean type (the Caribbean plate and the Atlantic plate) produces a volcanic insular arc (Lesser Antilles, Fig. 5.4). In the Lesser Antilles there are several volcanoes such as Sufriere Hills (Montserrat), Mount Pelée (Martinique), and Sufriere (San Vincent Island). The volcano closest to the Venezuelan coast is the submarine volcano known as Kick'Em Jenny (Fig. 5.4); it is located 8 km off of the northern coast of Granada Island and approximately 200 km from the coast of Venezuela (Fig. 5.4). With a 5000-m circular base, a 1300-m height above the seafloor, a pinnacle at 160 m below sea level, and vertical growth of nearly 4 m per year; the Kick'Em Jenny feature could be one of the main tsunami generators possible in the Caribbean Sea. Eruptions happened in 1939, 1943, 1953, 1965, 1966, 1972, 1977, 1988, and 1990. The most recent activity was reported on March 2003 (Smith and Sheperd, 1993). Although earthquakes can produce tsunamis, an alternate hypothesis has been proposed in the Venezuelan case. Older ideas have been refloated with new technologies and methods, and, for example, Beauperthuy (2006) proposed that sliding induced by earthquakes of big masses of alluvial sediments may be responsible for the local tsunamis that hit Cumana in 1530. As an explanation for this tsunami, Audemard (1999) proposed that a submarine broke off a segment of the El Pilar fault system.

Recently, traces of catastrophic floods have been identified, and maps of the potential tsunami hazard risk have been proposed. One analysis suggests that part of the Venezuela area can be impacted by tsunamis as well as the Lake of Maracaibo (Theilen-Willige, 2006). It is important to emphasize that some of these areas, such as Maracaibo, Puerto Cabello, Puerto La Cruz and Barcelona, are now highly populated (Fig. 5.10).

Figure 5.10 The tsunami-prone risky areas in northern Venezuela and the flooding directions (red arrow) derived by the traces of erosional features and abrasion. (Simplified from Theilen-Willige, 2006).

5. Urban Expansion and Types of Construction in Venezuela

Eighty-six percent of the Venezuelan population is urban, making it the seventh most urbanized country in the world. Since the earthquake of 1967, the population of Caracas has doubled to 5 million, growing at 3.1% per year, and the population density is 12,000 persons/km^2. The density is not uniform, however, and while the "formal" city averages 6000 persons/km^2, similar to the world average urban density, the barrios approach 25,000 persons/km^2.

Seismic-resistant design criteria for construction were neither updated nor adopted in Venezuela until the Caracas earthquake of 1967. Only in 1982 were the obligatory criteria for seismic constructions approved (Covenin Norm, 1756; Grases, 2002), but they were not implemented until some years later. The fact that a country possesses one or several norms that regulate how to design, calculate, and build constructions at seismically active zones, as well as laws that punish the violation of these norms, however, is not a guarantee of its effective application. These norms are not widely known either by the professionals who must use them or the people who really build constructions in the country (Laffaille, 2007). In the suburbs and slums of the main cities, the presence of houses called *ranchos* is very common. These houses are made of diverse materials, such as wood, cardboard, plastic or metallic rubbish, and zinc plates, and are not strongly constructed. Indeed, to worsen the situation; the slums usually are located in areas of low stability in steep slopes,

In rural zones, as well as in earlier zones or historic centers of cities, *tapia* (mud) and adobe, as well as *bahareque* (construction with logs, mud, and cane) (Fig. 11a and 11b), are still used. These systems perform very poorly or deficiently in the event of earthquakes because structures made with these systems vibrate with the same harmonic frequency as the ground on which they sit. This has been seen both in historic earthquakes and in very recent ones (for example, the earthquake in Ica, Peru, on August 15, 2007, where photos and press articles show the intense grade of destruction experienced by *tapias*) (see Chapter 6). Today the major cities of Venezuela have a number of modern buildings, but commonly the reinforced concrete that is used does not have a seismic-resistant design (Fig. 5.11c and 11d).

6. Microzonation Programs

In Venezuela, most of the detailed evaluations relating the damage to construction and terrain conditions were made after the earthquake of 1967, and seismic microzonation projects then started to be developed. Microzonation provides the basis for site-specific risk analysis, which can help mitigate earthquake damages. In general, seismic microzonation is the process of estimating the response of soil layers under earthquake excitations and thus the variation of earthquake characteristics on the ground surface. At this moment the government of

Figure 5.11 Types of construction of the working class in Venezuela: (a) Tapia; (b) Bahareque; (c) Brick of cement with elements of concrete; (d) Techniques and materials combined in indistinct ways.

Venezuela, through the Ministry of Science and Technology and the Venezuelan Foundation of Seismological Research, promotes and finances the microzonation of the major cities in this country. Such studies can be useful for planning and designing cities and their further development, but they do not resolve the existing situation. As shown in table 5.1, most of the houses and buildings do not satisfy Venezuela's earthquake-resistant design codes.

7. Discussion and Final Remarks

The high level of seismic risk that may be associated with Venezuela is mainly determined by one factor: the construction of vulnerability. This process continues in the present, even though the population and authorities both know that natural hazards can affect heavily populated sites. Natural hazards and vulnerability (social and physical) cannot be analyzed independently. Long periods between destructive earthquakes contribute to the society's false sensation of stability. It should be added that the native superstitious culture about earthquakes, plus the particular comprehension of their origins, has contributed to the apparent irrationality of risky settlement patterns in urban areas.

There are several different ways to understand the causes of a disastrous event. The first considers that the seismic forces have a magical, supernatural origin

Table 5.1 Vulnerability and Performance Level of Typical Venezuelan Buildings in Case of Intense Earthquakes. VI = less vulnerable, I = more vulnerable. ND = No Damage; TO = Total Operative; O = Operative; S = Surviving; NC = Near Collapsed; C = Collapsed; P = average behavior; x = likely behavior (simplified from Laffaille, 2007)

Vulnerability Level	Type	Performance					
		ND	TO	O	S	NC	C
IV	Venezuelan *ranchos*			x	P	x	
	Masonry						
I	River stones					x	P
II	Rock with mortar				x	P	x
I	Traditional mud bricks					x	P
IV	Reinforced mud bricks			x	P	x	
II	Mud wall				x	P	x
IV	*Bahareque* with tile roof			x	P	x	
V	*Bahareque* with roof light		x	x	P	x	
IV	Clay bricks mortar			x	P	x	
II	Clay or cement block				x	P	x
III	Clay or cement block with cement roof and reinforcement			x	x	P	x
IV	Clay or cement block with roof light and reinforcement			x	P	x	
	Reinforced Concrete						
V	Frames without resistant design		x	x	P	x	
VI	Frames with resistant design	x	P	x			
V	Walls without resistant design		x	x	P	x	
VI	Walls with resistant design	x	P	x			
	Steel Structure						
V	Frames without resistant design		x	x	P	x	
VI	Frames with resistant design	x	P	x			

whereby the attitude is that one cannot act against them; it is not possible to do anything concrete. All people can do is resign themselves to their fate.

The second perception is that if the "disaster" is related to the occurrence of a natural phenomenon such as an earthquake, a hurricane, the eruption of a volcano, or a landslide, then it is just a "natural disaster" and nothing can be done about. These two erroneous ways of thinking have predominated in Venezuelan society for a long time.

Clearly, these ways of approaching the seismic issue cannot generate effective governmental strategies of risk administration that will prevent or mitigate the disaster. After the dramatic consequences of the torrential rains and debris flows that affected mainly Vargas State's population (La Guaira, Venezuela's central littoral; see Chapters 1 and 3) in 1998, with thousands of deaths and economic losses in the millions, many official, private initiatives to implement a policy of risks have been promoted. Important efforts are being made to disseminate educational programs about disasters to society. However, it is not a simple task. On the one hand, the cities and their suburbs are growing much faster than can be handled by the institutions and organizations that set rules and ordinances of urban planning. On the other hand, economic and political conflicts of interest usually predominate over the town planning programs that take into consideration a rational inclusion of the risk factor. During the last century, petroleum started to be exploited, transforming the country and its society. The concentration of population in the big cities has been growing exponentially since that time. Meanwhile, the creation of vulnerability around urban places continues to grow, and Venezuela's seismic risk becomes greater every day.

CHAPTER 6

Natural Hazards and Human-Induced Disasters Triggered by Intense and Episodic Tropical Rains in the Venezuelan Mountains

Maximiliano Bezada[a]

[a]*Universidad Pedagógica Experimental Libertador-UPEL, Departamento de Ciencias de la Tierra, Caracas, Venezuela mbezada1919@cantv.net*

Contents

1. Introduction

Studies related to natural hazards, vulnerability, and risk of socio-natural catastrophes in Venezuela currently have drawn a lot of attention from state policies, leading to the allocation of resources through public institutions to implement and increase a culture of prevention and mitigation of socio-natural tragedies. These policies acquired greater priority after the catastrophic floods produced by the extraordinary rains that fell in December 1999 over the central Caribbean Mountains, which spread a great amount of debris floods and mudflows over the alluvial fans of the coastal zone, killing thousands of people. In fact, many natural hazards are present in the Venezuelan mountain areas, and the risk of a tragedy is directly proportional to the vulnerability stemming from the lack of urban and territorial planning programs.

Developments in Earth Surface Processes, Volume 13
ISSN 0928-2025, DOI 10.1016/S0928-2025(08)10006-2

In Venezuela, more than 50% of the population lives in the coastal-mountain geographic zone because most economic activities are concentrated in these areas. This dynamic has spurred the growth of urbanization in places where expansion is not advisable. For this reason, the population with the lowest economical resources has taken over the zones of maximum slope and the marginal areas of rivers and creeks. This has increased their vulnerability and their risk of being affected by a natural hazard, especially those hazards from geological and hydrometeorological origins (earthquakes, catastrophic floods due to extraordinary rains, and others) that have long affected these areas. This is how the likelihood of natural hazards producing human tragedy and catastrophe has increased sharply from 20 years ago and why it will explode in the upcoming years if an effective policy of risk mitigation is not applied. Presently, hundreds of thousands of people live in very vulnerable sites, and a potential catastrophe seems unavoidable, especially in the poorest neighborhoods located in intermountain areas of the main valleys and foothills of the northern Caribbean Mountains. Urban expansion in these areas since 1950 can be traced very well, as the result of population growth close to the center of main economical activities. Thus, we can observe the occupation of high-risk zones, such as alluvial plains, fans, cones, and areas beside rivers and creek channels. Also revealed is the corruption of some public officials who have granted permits to construction developers to build condominiums and houses in areas of extreme environmental sensitivity. During these processes, as Hernandez and Valbuena (2001) point out, the population's perception of risk, especially catastrophic floods, shows an important evolution. First, the fear of natural hazards predominates; second, the population adjusts its way of life to these hazards; and third, the general society overestimates the power of construction technology to build in risky areas. This chapter will present a brief revision of the natural and human factors of the main catastrophic floods and landslides reported on the Venezuelan mountains, and in particular we will look at urban growth and changes in uses of the land during the last 60 years.

2. Disorganized Urban Growth and Catastrophic Flood from Caribbean Mountains

On December 15 and 16, 1999, debris flows regionally affected the Venezuelan central coast, including several cities in the State of Vargas, which is located near Caracas (Fig. 6.1). The relief in this area is abrupt. The Sierra de Avila rise 2700 m. The narrow coastal area is no more than 0.5–4 km width. The rivers flow to the Caribbean Sea. The channels exhibit very steep longitudinal profiles that are broken at the footslope where a flat alluvial plain developed through alluvial fan deposition in contact with the coast. Debris flows came from all the basins of the northern face of the Venezuelan Caribbean range. This catastrophic event received global coverage because of the magnitude of loss of life and property.

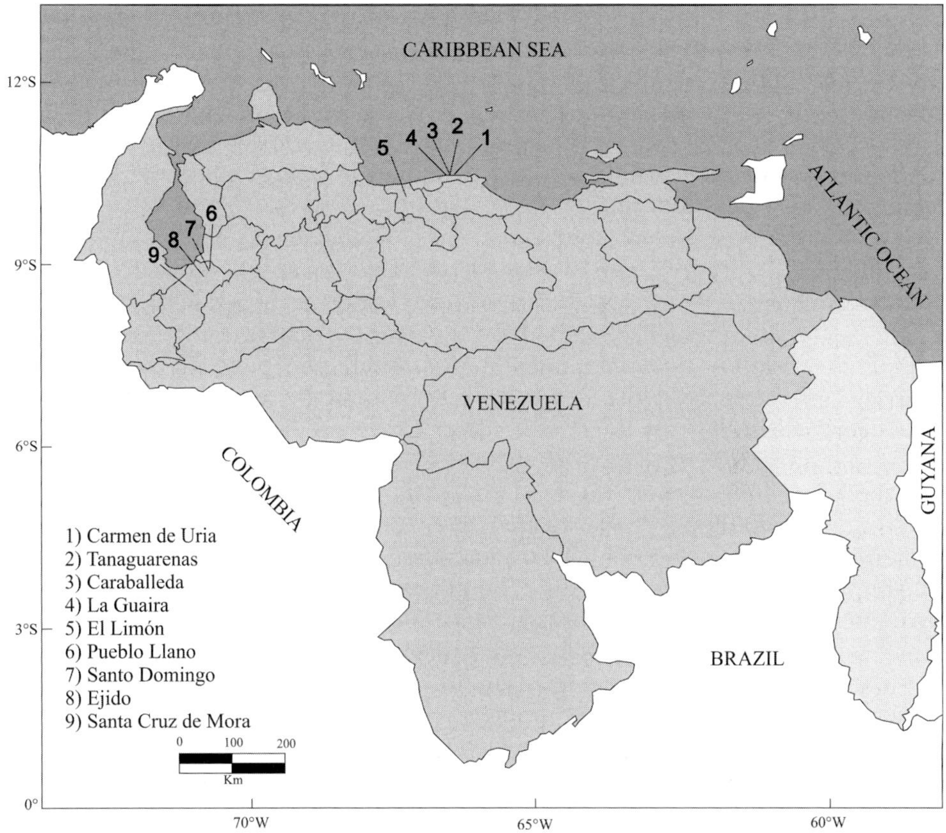

Figure 6.1 Location map. Main localities mentioned in the chapter.

This type of catastrophic event has been reported since colonial times. The oldest well-documented event, as reported by Humboldt (1816), was a massive regional flood that occurred in 1798. At that time, the only relevant town that existed in the Venezuelan central coast was La Guaira, which suffered great destruction and loss of many human lives. Many other such catastrophic events occurred subsequently, but we do not have good written records on any of them. In the last century, catastrophic floods were reported in 1900, 1938, and 1948 (Singer et al., 1983). Before 1999, the best documented event in Venezuela was captured on film and in press reports in 1951. This event was also a massive one, and it affected the entire central coastal zone. At that time, however, the main alluvial fans were not yet inhabited, and the material and human losses were only on a local scale.

All of these catastrophic floods were induced by extraordinary rains, which resulted from extratropical synoptic conditions associated with polar frontal systems. According to Goldbrunner (1961, 1976, 1984), these conditions generally affect the north of Venezuela during the months of November–February, with variable annual frequency and intensity. This atmospheric instability is reinforced by the Caribbean range along the north Venezuelan coast, which increases the amount and intensity of these extraordinary rains. As pointed out by Foghin (2001), however, these hydrometeorological events are unpredictable, and the prognosis of their return is not possible for practical purposes, since such synoptical situations do not have a linear behavior. At the same time, the population tends to forget catastrophic events until the next one comes along (Altez, 2005).

Besides the atmospheric conditions, the geological and geomorphic features act together to produce the massive debris flows over the alluvial fans. The Caribbean coastal range has as a northern limit the Venezuelan Caribbean coast, with steep slopes generated by east-west oriented faults. The coastal area is a rectilinear one with steep slopes, where the only flat areas are the alluvial fans themselves; these fans have been built by a series of Quaternary age debris flow deposits (Fig. 6.2). Ancient debris flow terraces that occur along the current water courses are located on terraces or fan terraces, which are the safest zones against the modern debris floods. The two best documented catastrophic floods that affected the Caribbean Venezuelan mountains occurred in 1951 and 1999, which according to Semetfav (2000) and Marnr (2000), had similar atmospheric conditions. Foghin (2001) compared the synoptic surface maps of February 1951 and December 1999 and showed their similarities (Fig. 6.3). These meteorological conditions are responsible for the great amount of precipitation that in just a few days surpassed the zone's earlier record for annual average precipitation (525 mm/y).

Figure 6.2 Central Venezuelan coast after the catastrophic flow of December 1999. The impact on the alluvial fans is notorious.

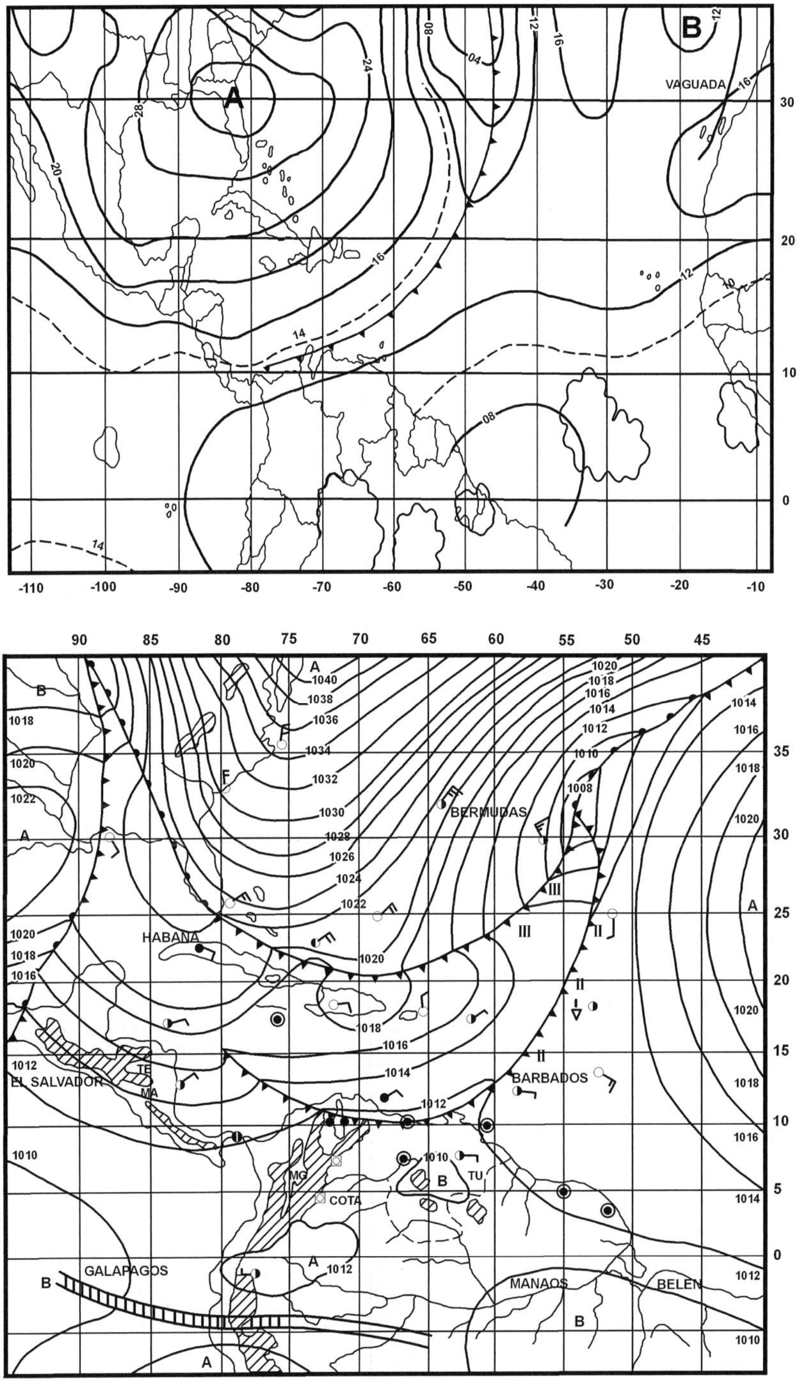

Figure 6.3 Maps of synoptic atmospheric conditions of the atmosphere during the catastrophic flows in the northern Venezuelan mountains in February 1951 and December 1999.

During the nineteenth century and until the mid-twentieth century, these alluvial fans were not used either as urban places or for agricultural purposes. They were occupied by dense forest cover that stabilized former alluvial and debris flow deposits. In this area, many secondary water courses can be identified in the aerial photos of 1936. The flat land between the fans was used as an agriculture zone, especially for sugar cane plantations. This land use ended after 1950 following the construction of a modern highway that connects Caracas with the coastal zone. After that time, use of this space underwent a drastic change. First, the plantation area and some small alluvial fans were set for future development, especially for houses, apartment buildings, golf courses, and recreational marinas. The debris flow of 1951 partially destroyed all the roads of this future development. In my view, this event, through the aerial photos that were taken, indicated the areas where no future development for houses should have been allowed. However, the developers overlooked the warning signs and continued their unsound urbanization policies.

Even worse, after 1958, with the beginning of democratic governments following the long period of military governments, Caracas and the neighboring land experienced a large demographic explosion as a result of the mass migration from the interior of the country to the main cities. This led to the development of the most dynamic part of the alluvial fans for building houses. Figures 6.4 to 6.5 show the urban history of Caraballeda and Tanaguarena towns developed on San Julian and Cerro Grande Rivers alluvial fans; two of the most affected areas by the catastrophic flow of December 1999. In the development of these areas, the courses of the main streams were altered and secondary channel were filled, the alluvial fans and low hills were also deforested and modified to permit building the urban infraestructure. The poorest people occupied the hills and the dangerous sites by the water courses. Human intervention increased the potential risk posed by the natural threats, maximizing the vulnerability of the new urbanized areas. Wrong policies established all the conditions for producing a catastrophe when new massive debris flows occurred. Hence, a catastrophic flood could have been predicted, creating a human tragedy once extratropical synoptic atmospheric conditions produced extraordinary rains; this in fact happened in December 1999, destroying large areas of the populated alluvial fans. The best example of bad urban practices and how human mistakes can create a very vulnerable area is represented by the development of Carmen de Uria town (Bezada 2000a and Bezada et al., 2000b), which it is explained step-by-step in the figures 6.6 and 6.7.

2.1. Causes of the Disaster of December 1999 at Vargas

The Vargas disaster happened in a place where the average annual precipitation record measured at sea level is 525 mm and the rainiest month is December, with an average of less than 60 mm. In December 1999, the extraordinary rains reached over 1200 mm, of which more than 700 mm fell on December 15 and 16.

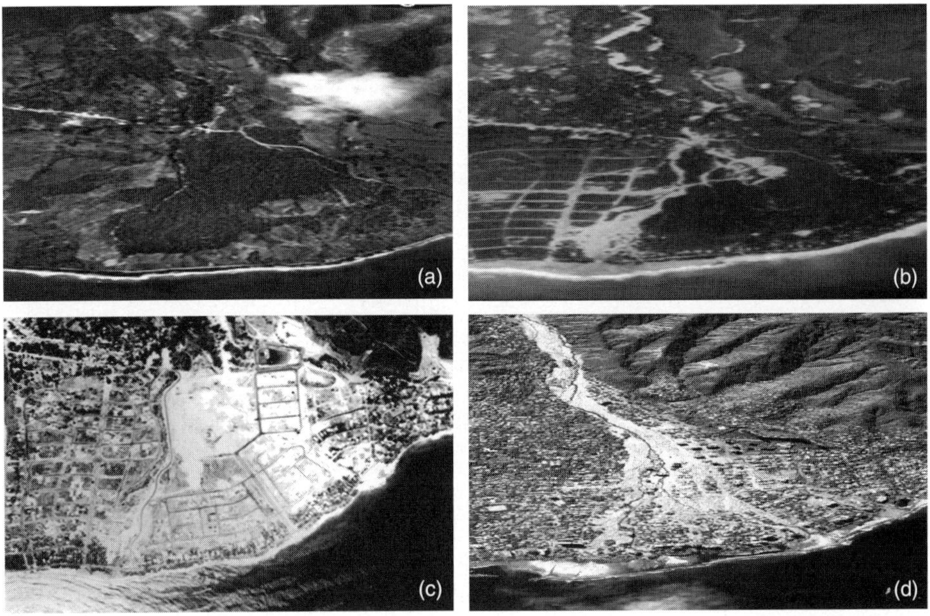

Figure 6.4 Retrospective analysis of aerial photos showing the change of the land use of San Julian River alluvial fans on the north Venezuelan coast. A. 1936: The alluvial fan is covered by an extensive forest that stabilized former debris flow deposits. The area is drained by several streams, with the main one on the left. Land to the right and left is occupied by sugar cane plantations. B. 1951: A big change in land use has occurred. The sugar cane plantations have disappeared, making way for roads for future urban development, which was partially destroyed by the catastrophic flow of 1951. In the center of the photo, we see the reactivation of secondary channels that produced land erosion and its deposition on several places on the coast. C. 1964: The forest over the alluvial fans has been destroyed, and the complete urbanization of the zone begins. D. 1999, after the catastrophic flood: The urbanized area is completely leveled by the debris flow.

Furthermore, the maximum precipitation (between 1200 and 1700 m.a.s.l.) over these two days, based on satellite images, was estimated to exceed 2000 mm (Semetfav, 1999; NOAA-NESDIS, 2000). These extraordinary rains generated large mass movements on the steep mountain slopes, resulting in catastrophic geomorphologic processes such as huge debris floods and mud floods over the foothills. The alluvial fans were reactivated, forming new lobes where the main population settlements are located. The debris flows and floods caused severe property destruction and killed approximately 5000 people (more than 20,000 from other estimations).

A massive amount of sediment flowed from 24 watersheds along 50 km of the coast during the storm and resulted in the destruction of deposits on alluvial fans and beaches (see fig.6.8). It has been estimated that 15 to 20 million cubic meters of sediment were deposited in the fan area (Larsen and Wieczorek, 2006). Sediment yield for the 1999 storm from the approximately 200 km^2 drainage area of watersheds upstream of the alluvial fans was as much as 100,000 m^3/km^2 (Larsen and

Figure 6.5 Retrospective aerial photos of the Cerro Grande River alluvial fan, where the town of Tanaguarenas is located, showing the change in land use since 1936. A. 1936: The center of the alluvial fan had an active river dynamic with several secondary channel streams. Both sides of the fan were used for sugar cane plantations. B. 1951: The agricultural use ceases and urban plans start. Roads were built over an area with very active fluvial dynamics. All the area was destroyed by the debris flow due to the extraordinary rains of 1951. C. 1999: Here we see the similarities of these catastrophic floods. The destroyed areas are the same as those in the past, with the difference this time being that the alluvial fans was completely populated.

Wieczorek, 2006). Large boulders (up to 12 m in diameter) were transported by the torrential flows, and the flows moved at an estimated velocity of 3.3 to 14.5 m/s (Larsen and Wieczorek, 2006).

More than 23,000 residences and apartment buildings were destroyed, and 65,000 were damaged in Vargas (Lopez et al., 2003). Roads, telephone, electricity, water, and sewage systems were severely disrupted. The total economic losses were estimated to be more than US$ 2 billion (Lopez et al., 2003).

A comparison of aerial photos from the last 63 years demonstrates that this catastrophe was induced more by human factors such as bad planning of urban zoning, without taking into account the natural dynamics of alluvial fans and the historical flood records. It is clear that the fans generate episodically, reactivating the alluvial lobes with a recurrence of several decades. Thus, former river channels were filled up; active channels were deviated from their main courses in order to use the

Figure 6.6 Aerial photos of the town of Carmen de Uria. A. 1936: The area was used mainly for agricultural activities. In the photo, we can see the hill that closed the valley in front of the sea and the Uria River running in the center of the image beside the Quaternary terrace. B. 1956. The area was totally transformed with bulldozer for building houses. C. 1994: The town of Carmen de Uria before the flooding. D. 1999: The town after the catastrophic flood of December 1999. The Uria River has now reoccupied its former course, and the entire town, constructed over the land fill, to the right of the Quaternary terrace, was completely destroyed. All the sediments coming from the upper basins plus the use for infill were removed, building up a deltaic plain downstream.

floodplains for urban developments. In other words, human action highly increased the vulnerability of the area, and this, rather than nature alone, was the main cause of the socio-natural catastrophe of December 1999.

3. The Torrential Avalanche of El Limon River

On the south side of the Caribbean mountains, catastrophic floods have been recorded from the Quaternary stratigraphy. The oldest known event is the torrential flood that buried pre-Columbian inhabitants in the Caracas area about 1000 years B.C.E. This area was discovered during the construction of the Parque Central complex in Caracas, the highest buildings in the city: during this construction, human remains were found embedded in debris-flow material (Singer, 1977).

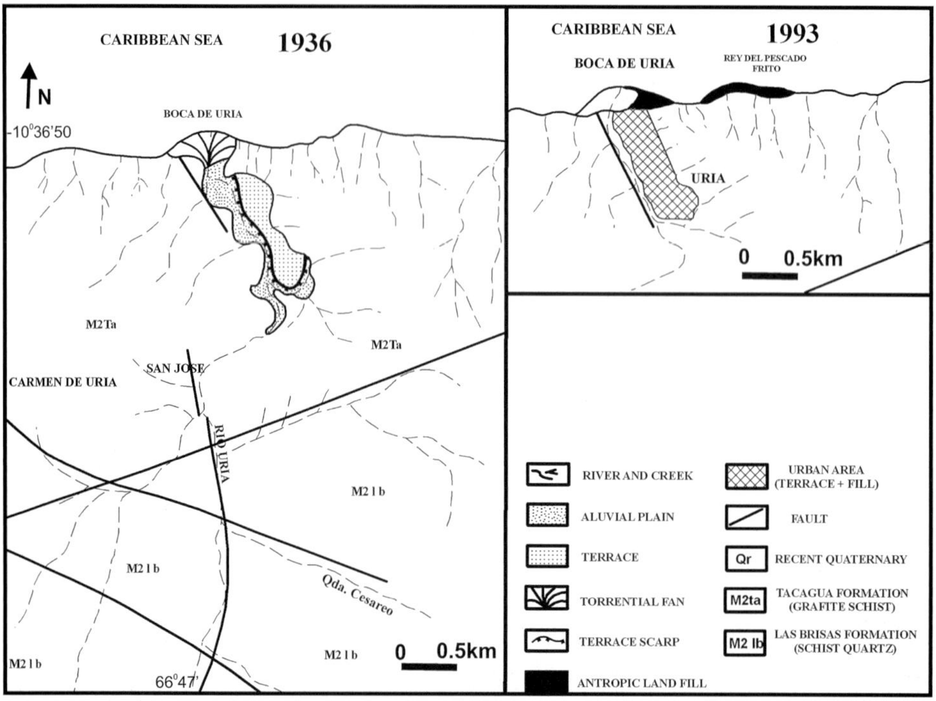

Figure 6.7 Maps shows the human transformation area where Carmen de Uria town was builded.

During the last century, the best recorded event on the southern face of the mountain range was the torrential avalanche that occurred in El Limon River, close to the city of Maracay, south of the Caribbean range, approximately 100 km west of Caracas (Fig. 6.1). On September 6, 1987, the area received 174 mm of rain in less than 5 hours. This event is associated with high precipitation due to a stationary tropical depression over the Caribbean Sea that generated locally intense rains producing torrential avalanches. A combination of physical factors such as geomorphologic characteristics, high slope, soil features, and outcrop of gneisses, which in cloudy forests develop a sandy weathering profile with a fragile structure that facilitate water saturation; all can induce mass movement-triggered debris flows. The residual superficial soils and subsuperficial weathering zone occur between 1600 and 2000 m.a.s.l. (Audemard et al., 1989). These geomorphologic/geologic factors, added to the intense rainfall, produced a catastrophic debris avalanche that killed more than 100 people and destroyed small towns located on the river's alluvial plain. More than 30,000 people were temporarily isolated by the event.

It is important to note that human intervention in the area has been one of the main factor influencing slope stability. Human-induced forest fires eliminated the leaf litter under the forest, allowing a major velocity of water absorption by the soil, which loosens its structure and produces mass flow, and is then transported by the streams mainly as a mudflow (Zinck, 1986a, 1986b).

Figure 6.8 Some details of the debris flow of the San Julian alluvial fan where several urbanized areas of Caraballeda were destroyed.

In the event that took place on El Limon River, the total amount of precipitation is unknown, as is true of other disasters produced by extraordinary rains in Venezuela. The magnitude of the rains in the high basin of El Limon River are not known in detail because the values used for the torrential avalanche evaluation were the ones registered in the pluviograph located in the administrative building of the Rancho Grande National Park, below the main area of mass movement. That station registered 174 mm in 4 ½ hours, with its highest intensity of 96 mm in 15 minutes (Audemard et al., 1989).

4. Catastrophic Floods in the Venezuelan Andes

The first catastrophic flow that is mentioned in the historical records of the Venezuelan Andes occurred during colonial times in 1767. This event was recorded in the colonial chronicles and was described as a destructive flood produced by the Mocoties River as a consequence of damming its channel by a seismically induced rock avalanche (Singer et al., 1983).

During the twentieth century, many floods in the Andes Mountains have been reported in chronicles and newspapers, but due to the low population density of the area until the mid-twentieth century, most of the towns were established in very stable and safe areas. Still, some of these events caused material destruction and loss of human lives. In order to draw possible evacuation routes, it is very important to know the areas that were affected and the path followed by the mud and debris flows outside the riverbed.

An example of a hazard map on the Andes is the research done by Hernández and Valbuena (2001). They constructed a map for the valleys of the Chama and Mocoties rivers along 142 km. This map showed that many towns were prone to hazards associated with torrential rains. Such is the case of the Ejido alluvial fans drained by the Montalban and Portuguesa rivers, which in 1933 and 1947 produced catastrophic floods that caused loss of human lives and destruction of agricultural lands. Information about this event was reported by Villamizar (2006a, 2006b), the official chronicler of Ejido, as well as local newspapers of the time. In November 1933, Ejido was flooded with mud and debris flows coming from the Portuguesa River. At that time the town had 5680 inhabitants, most of whom were engaged in agricultural activities. According to Villamizar's testimony, the oral tradition states that the amount of precipitation in that year was unusual. The local newspaper *Patria,* cited by Villamizar (2006b), reported that the flood destroyed the bridges, local market, sugar cane plantations, 24 houses, and many farm animals.

Ejido was flooded again on October 28, 1947. The main stream of the alluvial fans of the Montalban River experienced the most catastrophic flood as registered by newspapers and reported by Villamizar (2006a). This catastrophic flood caused 29 deaths, 20 missing persons, and important economic problems, due to the destruction of coffee and sugar cane plantations as well the death of cattle.

This catastrophic flood has been explained as resulting from a rock avalanche and mud landslide that damaged the Montalban River stream damming/blocking the river. When the detritus dam was broken, the flood event ensued. Jackson (2007), using cosmogenic isotopic data, estimated the ages of several rock avalanches, but they were older than these catastrophic floods. The detritus of older avalanches can be a source, however. The material loss was estimated at roughly US$ 1,200,000, which was a significant amount of money for the time.

Currently, the Canadian Geological Survey, in conjunction with the geological services of Andean countries, is developing an international cooperative program called the Multinational Andean Project: Geosciences for the Communities. In Venezuela, the National Institute of Geology and Mining chose the Montalban River Basin to develop its research because Ejido is heavily populated and because of the area's history of catastrophic floods. The project includes evacuation training and an educational program for the communities.

5. Recent Catastrophic Flow in the Venezuelan Andes

In the last five years, the number of catastrophic events in the Venezuelan Andes has increased. The main consequences have been increased land vulnerability produced by human occupation of the most fragile areas of the valleys. Among these more recent events has been the catastrophic flood that occurred in June 2003. The towns of Pueblo Llano and Santo Domingo, 100 km to the northeast of the city of Merida, were affected by torrential rains. This tragedy was caused mainly by a debris flow and landslides and killed 9 people, with 25 reported missing and 450 injured. This event was a warning of the high vulnerability and danger of the area. In February 2005, torrential rains once again produced tragedy on the west side of the country where 30,776 persons were affected. The most seriously impacted area was the valley of the Mocoties River, southwest of the city of Merida, with 41 deaths and 52 missing persons reported.

In the Merida area, the main geomorphologic processes are the landslides in the tributary creeks on the north side of the valley. The basins of the tributary creeks are underlain by very unstable material associated with the processes and products of weathering on schist and gneisses. This material, when collected by the tributary streams, produced mainly a mudflow, with big tree trunks that flooded over bridges and affected many streets and houses (Fig. 6.9). This mudflow material dammed the Mocoties River at a bridge close to the bus terminal. A temporary lake was created in front the bus station on the alluvial plain. The rupture of the damming sediments produced a wave of catastrophic flood over the alluvial plain where the bus terminal, the market, the soccer and baseball stadiums, and also many houses were irresponsibly built. All these constructions were partially destroyed, and most of the deaths occurred at the bus terminal and the surrounding area.

Hernandez and Valbuena (2001) have pointed out that the town of Santa Cruz de Mora is one of the critical areas that is at great risk in the event of torrential rains. If we analyze the growth of this town over 50 years, we can see the mistakes made in the expansion of this urban area. Figure 6.10A shows the town of Santa Cruz de Mora in 1950. When we compare the image of the same town in the year 2000 in Figure 10B, we can see that the town's expansion was built on the active alluvial plain and on the valleys of tributary streams. Local planning did not take into account the dynamic of the Mocoties River under conditions of torrential

Figure 6.9 A. Log dam at the Mocoties River and its tributaries, later producing a catastrophic wave when it collapsed. B. Mudflow coming from the removal of the weathering profile of schist and gneisses on the valley side of the city of Tovar.

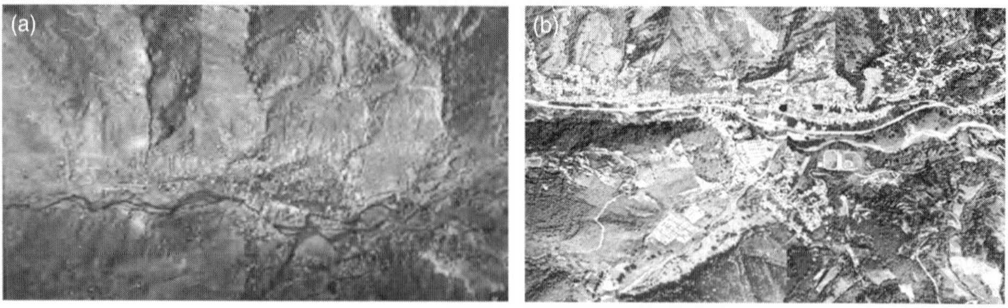

Figure 6.10 A. Santa Cruz de Mora in 1950. The town occupied the safest places above the Mocoties River and outside of the tributary valleys. B. Santa Cruz de Mora in 2000. The expansion of the city was built over the alluvial plain of Mocoties River, and also the newly populated area occupied the tributary streams of Mocoties River.

precipitation. The river simply reclaimed its territory but in the meantime produced tragedy and destruction. This was not a natural catastrophe; it was mainly a human-made tragedy. In Figure 6.11, we can see the areas affected after the hydrometeorological event. The catastrophe of Santa Cruz de Mora in 2005 occurred six years after the biggest disaster in Venezuela, at Vargas in 1999, and it was another sad example of the mistakes that should have been avoided.

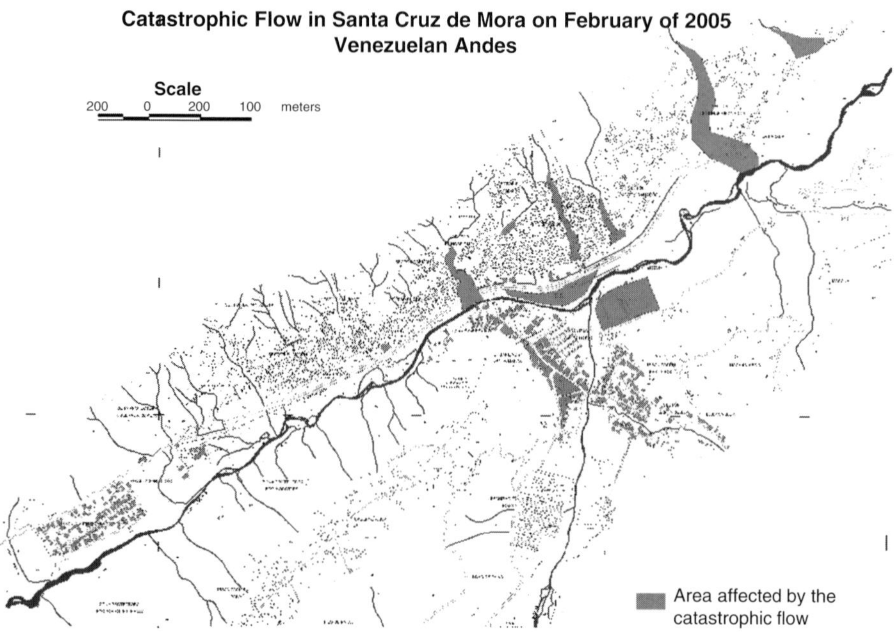

Figure 6.11 The dotted zone shows the area of Santa Cruz de Mora affected by a catastrophic flow in February 2005.

6. Final Remarks

All of the examples presented in this chapter show that the catastrophes were induced mainly by human error related to poverty and to the lack of governmental policies on territorial ordination, as well as to bad planning of new urban zoning in hazardous places. Urban expansion continues to be undertaken without consideration of the natural dynamics of mass movements, alluvial fans, and historical flood records. The continued development has increased the vulnerability of these communities. The same socio-natural catastrophes that have been affecting Venezuela in the past 10 years ago seem destined to be repeated. Currently, many other places in the Venezuelan mountains have the same vulnerability, especially the slump areas on the valley sides with steep slopes, riverbanks, and alluvial fans. The next tragedy will occur sometime in the near future, when meteorological conditions lead to large and heavy rains. How grave the next tragedy will be will depend on the corrective efforts that are made and on the application of new policies to protect the environment and human life.

CHAPTER 7

Convulsive Events, a Widespread Hazard in the Colombian Andes

Michel Hermelin[1] *and* Natalia Hoyos[2]

[1]*Departmento de Geología, Universidad EAFIT, Carrera 49-7 Sur 50, Medellín, Colombia.*
[2]*Laboratorio de Sistemas de Información Geográfica, Universidad EAFIT, Carrera 49-7 Sur 50, Medellín, Colombia.*

Contents

1. Introduction

Colombia's mainland stretches between 12°N and 4°S, with coasts on both the Caribbean Sea and the Pacific Ocean. Although the mountainous region occupies only about one-third of its territory, it is the most populated area since the Spanish conquest. Preference for this area was, among other factors, guided by its healthier climates, which meant less exposition to endemic diseases such as malaria. Most of the towns founded at that time, which later became regional capitals, were located at altitudes between 1000 and 2500 m above sea level (m.a.s.l). Furthermore, coffee plantations, which started in the mid-nineteenth century, fostered this tendency, as mountain coffee grows best from 1500 to 2200 m.a.s.l. Population increase (from 10 million inhabitants in 1950 to 41 million in 2004; DANE, 2005) has resulted in rapid urban growth and an almost complete destruction of the Andean forests.

[1] hermelin@eafit.edu.co, Phone 57+(4)2619500 ext. 331, Fax 57+(4)2664284. Corresponding author.
[2] nhoyosbo@eafit.edu.co, Phone 51+(4)2619500 ext. 893, Fax 57+(4)2664284

Developments in Earth Surface Processes, Volume 13
ISSN 0928-2025, DOI 10.1016/S0928-2025(08)10007-4

Climate is predominantly wet due to the influence of the intertropical convergence zone (ITCZ). The Andean region comprises three mountain ranges or cordilleras (Western, Central, and Eastern) and two interandean valleys–the Cauca valley between the Western and Central Cordilleras, and the Magdalena valley between the Central and Eastern Cordilleras (Fig. 7.1). Mountains above 4800 m.a.s.l are still capped by ice (Central and Eastern Cordilleras, Sierra Nevada de Santa Marta), while regions above 3500 m (*páramos*) display numerous remnants of Quaternary glaciations. Active tectonism and volcanism contribute to enhance natural hazards. Colombia benefits from a remarkable variety of climates, landscapes, and an outstanding biological diversity. However, it is also exposed to an impressive collection of natural hazards (Hermelin, 2005).

The term *convulsive* is used in this chapter in line with the definition given by Clifton (1988): "an extraordinarily energetic event of regional influence." Although Bates and Jackson (1987) define convulsive as a synonym of catastrophic, it seems preferable to use catastrophe in relation to a disastrous event, or even better not to use it at all (Gretener, 1984). Furthermore, contrary to Cliftońs opinion, convulsive events are perhaps much more common than it has been traditionally thought, particularly in wet tropical mountains. Without becoming "catastrophists" in a philosophical sense, Colombian geologists tend to be very cautious with respect to the classical uniformitarian view of progressive, imperceptible denudation of the Earth surface as being the main erosional process occurring in the tropical Andes. This perception is supported by the case studies presented in this chapter, showing that a major portion of the erosion and landscape change may be due to a few convulsive events.

From the innumerable natural destructive events that have struck Colombia during historic times, we have chosen seven as being representative: one of volcanic origin, three caused by earthquakes, and three triggered by rains (Table 7.1, Fig. 7.1). Although a recent paper (Schuster et al., 2002) on catastrophic landslides in South America gives short accounts of two of these events, they are analyzed from slightly different standpoints. Unfortunately, many of the events occurring in Colombia, though now duly reported by government authorities, are seldom the object of a careful evaluation. Some of them are described in local publications, but very few find their way to international publications. Our objective is to present a brief summary of the natural and anthropogenic conditions under which these events took place, followed by a short discussion on aspects related to their understanding and mitigation.

2. Case Studies

2.1. Armero—Chinchiná, 1985

There are numerous reports on the events that took place on November 13, 1985, which led to one of the deadliest volcanic-related disasters in recent history (e.g., Cárdenas, 2005; Mileti et al., 1991; Thouret et al., 1989; Mojica et al., 1985). Although a vast area was affected, we focus on the sites with the highest death toll, the cities of (former) Armero and Chinchiná.

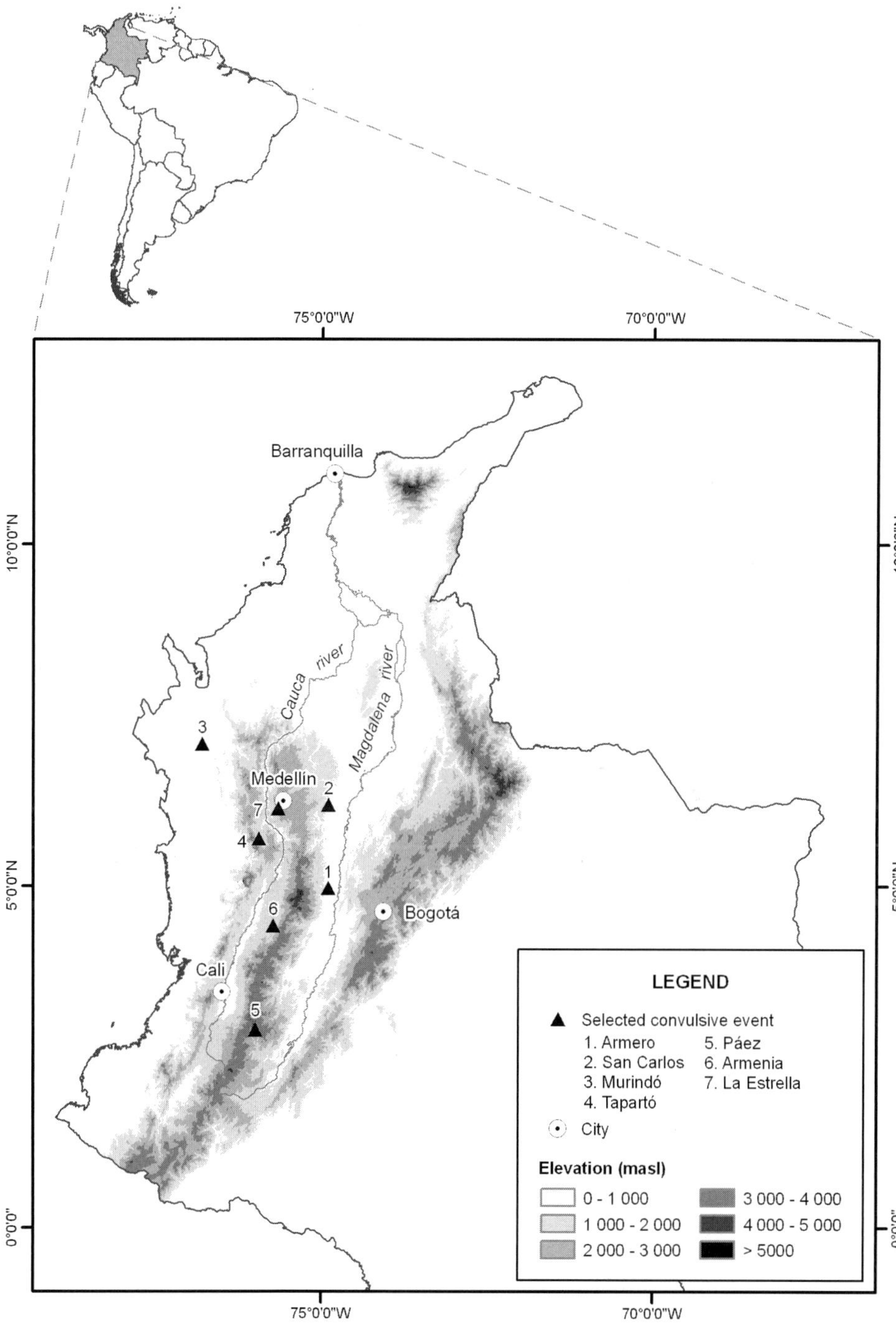

Figure 7.1 Location of selected convulsive events within the Colombian Andes. (Digital base data from USGS, 2004; CIAT & PNUMA, 1998.)

Table 7.1 Summary of Selected Convulsive Events in the Colombian Andes

Place	Date	Origin	Death toll	Estimated Economic Loss (US$ million)[a]
Armero – Chinchiná	Nov. 1985	Lahars (volcanic eruption)	23,000 deaths	200–300
San Carlos	Sept. 1990	Flash floods, landslides, and flooding (rain)	several tens of deaths	> 6
Murindó	Oct. 1992	Landslides and flooding (earthquake)	3 deaths 2–14 (fires)	17
Tapartó	Apr. 1993	Flash flood (rain)	59 deaths 70 missing	> 0.5
Río Páez	June 1994	Landslides, flash floods (earthquake)	1100 deaths	125
Armenia	Jan. 1999	Earthquake, landslides	1185 deaths	1857
La Estrella	May 2000	Landslides, flash floods (rain)	1 death1 missing	1

[a] Calculated at the time of the event.

The Nevado del Ruiz, an ice-capped volcano, is located in the Central Cordillera, approximately 140 km west of Bogotá (4°50.55′N, 75°16.20′W). It stands at 5321 m.a.s.l overlooking the country's two major interandean valleys, the Cauca to the west and the Magdalena to the east. This massive stratovolcano covers an area of 200 km^2; from its summit to approximately 4800 m.a.s.l; it is covered by glaciers with an area of 21 km^2 (Thouret et al., 1989; INGEOMINAS, 1985). The volcanic slopes are predominantly steep (20–30°), above 4000 m.a.s.l, and become gentler (around 10°) at lower elevations, spreading symmetrically to the floodplains of the Magdalena and Cauca rivers (Mileti et al., 1991). Currently, the volcano has one active crater, the Arenas, located near its northeastern flank (Thouret et al., 1989; INGEOMINAS, 1985). Several streams drain the volcano to the west and east, eventually flowing into the Cauca and Magdalena rivers, respectively.

Precipitation in the region follows a bimodal pattern, with wet conditions from mid-March through mid-June, and mid-September through mid-December. Maximum annual values are in the order of 3500 mm (1200–1900 m.a.s.l) on the western flank of the Cordillera and 2200 mm (1200–1600 m.a.s.l) on the eastern flank. Average temperature ranges from more than 24°C on the Magdalena valley to less than 0°C above 4800 m.a.s.l (Pérez, 1983). Major towns within a 60-km radius from the volcano are shown in Figure 7.2.

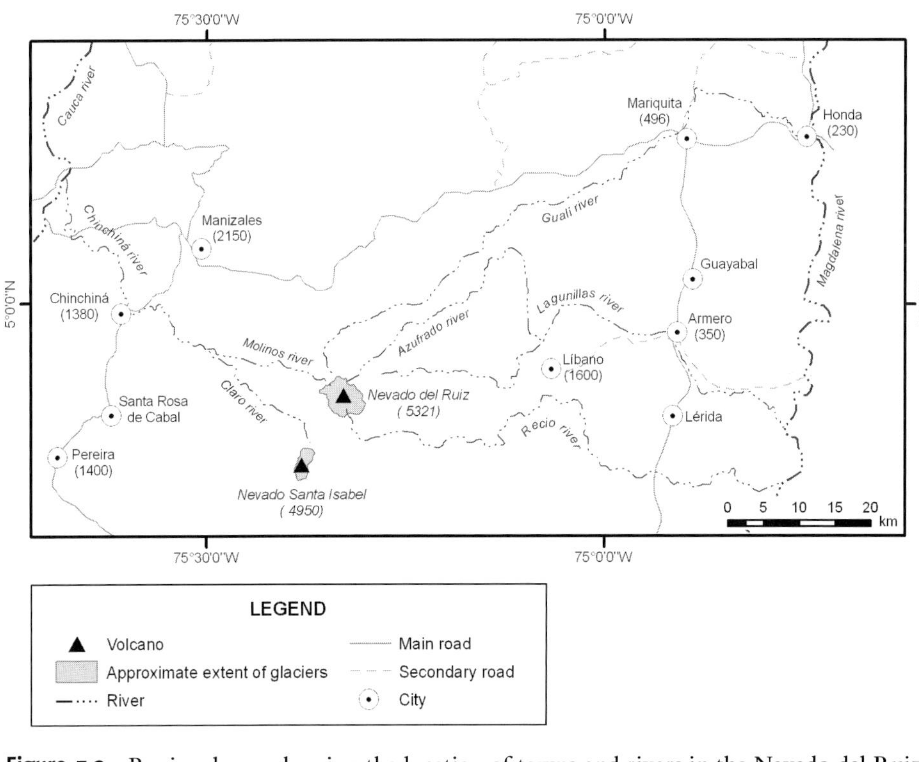

Figure 7.2 Regional map showing the location of towns and rivers in the Nevado del Ruiz vicinity. Numbers within parentheses indicate approximate elevation in meters. (Modified from IGAC, 1980.)

The Ruiz volcano's activity has been predominantly explosive since the late glacial (about 14,000 B.P.). Nine eruptions occurred between 13,800 B.P. and 1595 A.D., seven of which took place after 2200 B.P.; thus, an average return period of 250 to 500 years has been estimated (Herd, 1982). Historical documents report two major events, one in 1595 and another in 1845 (Espinosa, 2001). Both of these eruptions involved the formation of large mudflows; the 1595 lahars formed as a result of a volcanic eruption, while the 1845 mudflows were produced by a large mass movement on the glacier associated with a regional earthquake (Espinosa, 2001). The city of Armero was established around 1895, when the village of San Lorenzo (former name) was founded on the margins of the Lagunilla River, 1.5 km east of the cordilleran foothills (Mojica et al., 1985).

The volcanic activity associated with the November 1985 event started a year earlier with intermittent fumarolic and seismic activity near the Arenas crater. A team of national and international scientists evaluated the conditions of the volcano in early 1985 and stressed the importance of a proper monitoring system, which was installed by July 1985. Resultant data confirmed abnormally high levels of seismic activity, which increased by the end of August and culminated with a phreatic eruption on September 11 (Cárdenas, 2005; INGEOMINAS, 1985). A preliminary risk map was made public on October 7 by the Colombian Geological Service (INGEOMINAS). This map, which included the town of Armero, depicted areas affected by lava and pyroclastic flows, lateral blast, pyroclastic deposits, and lahars.

On November 13, a phreatic eruption took place at 3:05 p.m. (local time). Ash and small lapilli fell on several towns, including Armero (45 km east of the volcano summit). Around 9:09 p.m., two strong explosions took place, followed by a succession of pyroclastic flows and surges emitted from the vent. Around 9:30 p.m. a strong explosion occurred, resulting in the formation of a column of gases and pyroclastics that rose for more than 10 km above the volcano. While bombs, blocks, and lapilli fell within a few kilometers around the vent, ash spread to the east and northeast, as far as the Venezuelan border. Fumarolic activity and the accumulation of hot pyroclastic materials on the ice cap led to a large amount of ice being melted (around 10% of the ice cap equal to a volume of 0.06 km^3; Thouret et al., 1989). The mixture of water, materials from the eruption, and unconsolidated deposits near the ice cap resulted in the formation of mudflows that were channeled through streams draining the volcanic slopes. Their volume increased as they advanced down the river channels, due to the incorporation of materials from saturated fluvial and colluvial deposits (Cárdenas, 2005; Mileti et al., 1991).

Around 10:30–10:40 p.m., a lahar traveling down the Chinchiná River valley (west from the summit) swept through the town of Chinchiná. By 11:00–11:30 p.m., mudflows reached Armero in several pulses, covering an area of 3387 ± 10 ha (20 km distance in the W–E direction from the canyon mouth, 80–100 x 10^6 m^3 of material), and devastating 90% of the city (Fig. 7.3). Deposit depths up to 6 m were recorded at the local hospital but are estimated to have been on average 4–5 m. Temperatures may have reached 60–70°C based on medical reports on skin burns and survivor testimonies (Cárdenas, 2005; García, 1988; Mileti et al., 1991; Mojica et al., 1985). By 11:40 p.m., mudflows traveling down the Gualí River reached Mariquita (72 km east from the crater) and made their way

Figure 7.3 Aerial view of Armero a few days after the mudflows swept through the city.

to Honda, on the Magdalena River margins, by 1:00 a.m (November 14) (Mileti et al., 1991; García, 1988).

Mudflows reached depths of 20–40 m while in the river canyons, with average velocities of 18–60 km h^{-1} (Calvache, 2006; Mileti et al., 1991; Mojica et al., 1985). The total death toll was estimated at 23,000; 94% of the victims were Armero residents, while 6% were from Chinchiná (Presidencia de la República, 1986 in Cárdenas, 2005). In addition, over 200 000 people were directly or indirectly affected by this event. Direct economic losses were calculated to be $34,940 million pesos (US$ 211.8 million), equivalent to almost 1% of the country's GNP in 1984 (Cárdenas, 2005; Mileti et al., 1991; García, 1988).

2.2. San Carlos (Antioquia), 1990

San Carlos is a village located on the eastern slopes of the Central Cordillera, about 100 km east of Medellín, at an altitude of 1000 m.a.s.l. The village is drained by Río San Carlos, formed by several tributaries that originate on the dissected slopes of a Cretaceous granitic massif at elevations that reach 1800 m.a.s.l. The area was originally covered by humid tropical forest, with an annual precipitation of 5000 mm. During the last 150 years, the forest has been almost totally replaced by pasture, coffee plantations, temporary crops, and secondary forests. San Carlos River flows into the Punchiná dam, a major producer of hydroelectric power in the country.

On September 21, 1990, a 208-mm rainstorm fell in the upper catchment, between 8:00 p.m. and 1:00 a.m. (September 22). During the rest of the night, 23 more mm of rain fell in the area. The area affected by such a precipitation was

Figure 7.4 Large boulders transported by the San Carlos River torrential flow. The Calderas plant building (to the right) sustained considerable damage.

inferred from the distribution of more than 800 landslides and covered about 11 km^2. In addition to the landslides, much vertical and lateral channel bed scouring occurred. As a reference, the discharge of Quebrada Arenosa, one of the tributaries, which is normally less than 1 m^3 s^{-1}, reached 174 m^3 s^{-1}, while its transversal section near the Calderas hydroelectric plant, 5 × 3 m before the event, reached 30 x 20 m immediately after. Blocks with diameters up to 8 m were moved downstream, destroying most of the interior of the Calderas plant (Fig. 7.4).

Consequences of the event included several tens of deaths (20 reported by a local newspaper), 260 people evacuated, 27 houses destroyed and 30 more damaged, as well as several bridges and road sections demolished. Economic losses were estimated to be more than US$ 6 million at the time. Furthermore, much sediment reached the Punchiná reservoir, a key component of the country's hydroelectric system (Hermelin et al., 1992).

The recurrence of such an event, inferred by correlation of dated volcanic ash deposits, was estimated to be twice every 10,000 years. Previous geomorphic work at a subregional scale, however, did not provide any evidence of a torrential event of this magnitude (UN-ISA, 1984). One of the striking aspects of this multifaceted event was that land use had no significance in the occurrence of landslides; this fact was interpreted as a consequence of truly exceeding the resistance thresholds. In two hours, changes in this landscape probably surpassed the combined effect of average processes during thousands of years.

A final observation is that the rate of recovery was very rapid: three years after the event, it was almost impossible for an untrained observer to recognize the enormous changes that had affected the landscape. Vegetation growth had proceeded at such a rate that remaining evidences were almost imperceptible (Velásquez and Hermelin, 2005).

2.3. Murindó, 1992

The Murindó ("timber river" in the native Embera language) earthquakes, landslides, and flooding occurred in the Atrato valley in western Colombia. These events affected the alluvial valley itself, the western slopes of the Western Cordillera, and the Baudó Cordillera, located west of the Atrato River. The alluvial valley presents numerous swamps, which hinder surface communication. The sparsely distributed population is established on alluvial dikes.

The Western Cordillera in this area is formed by Cretaceous volcanic rocks and Tertiary sedimentary rocks, and presents steep slopes. The Baudó Cordillera is composed by Tertiary volcanic and sedimentary rocks. The region is extremely humid, with annual precipitation reaching 8000 mm yr^{-1}, and is almost completely covered by dense rain forest. Regional tectonic activity is controlled by the convergence of the Nazca and South American plates, at velocities of about 50 mm yr^{-1}, and several faults, with NS–NW trends, have been proposed (Velásquez, 2005).

The first seism occurred on October 17, 1992, with a magnitude of $Mw = 6.6$ and a depth of 14 km; its epicenter was located at 6.8450°N and 76.8060°W. A second one occurred the following day, with $Mw = 7.2$, a depth of 10 km, and was located 50 km north of the first one. In the meantime, several repeat events were felt, with magnitudes (Mw) around 5. The length of rupture was estimated at 120 km, and the events were considered by seismologists as a multiple event (Velásquez, 2005).

The effects of the earthquakes were multiple as well. In the area of Murindó, liquefaction was observed in saturated alluvial soils on river levees along a stretch of about 150 km, causing the formation of crevasses parallel to the streams. Tens of thousands of landslides occurred on relatively steep slopes, affecting mainly the western slopes of the Western Cordillera but also locations as far away as Apartadó (about 100 km NE of the epicenter). In nearby areas, denudation was pronounced, as vegetation and soils were completely stripped from the slopes (Murindó and Coredó rivers) (Fig. 7.5). The accumulation of trees in the lower part of the Atrato tributaries, particularly the Murindó River, formed dams called locally *palisadas* (palisades) (Fig. 7.6). These decreased the stream velocity, causing sedimentation and further flooding of alluvial terrains. As a result, the village of Murindó had to be evacuated. In many areas, forest vegetation died, being unable to withstand complete and permanent submersion.

There were only 3 victims from these phenomena due to the low population density in this area. A secondary effect was the sudden "eruption" of the Cacahual mud volcano, located about 130 km NE from the epicenter. Several houses were covered by mud, and others were burned by domestic gas from a home kitchen. Reported victims from house fires ranged from 2 to 14 (Velásquez, 2005; Martínez et al., 1994). Local building practices in the vicinity of the epicenter explain the absence of victims and the relatively small destruction. One-floor houses made of wood are well suited to resist seismic vibrations. Sizable damages, however, were reported in Murindó and surrounding towns, particularly on brick and concrete housing and

Figure 7.5 Extensive surface denudation due to mass movements associated with the Murindó earthquakes.

Figure 7.6 Temporary tree dams *(palisadas)* along the Murindó River.

infrastructure. In cities such as Medellín, located 150 km SW from the epicenter, 243 public buildings and 3400 houses were affected.

Economic losses for the Antioquia and Chocó provinces were US$ 17 million at the time. The village of Murindó was relocated on a levee of the Atrato River.

2.4. Tapartó, 1993

The Tapartó River drains the eastern slopes of the Western Cordillera for 27 km, until it flows into the San Juan River, a tributary of the Cauca River. Elevation in the region ranges from 600 m.a.s.l along the Cauca River margins to more than 4000 m.a.s.l at the cordillera's summit. Precipitation follows a bimodal pattern with wet conditions prevailing in April–May and September–October. Annual rainfall totals are in the order of 2065 mm at 1250 m.a.s.l (confluence of Tapartó and San Juan rivers) to around 4000 mm near the summits (Piedrahíta and Hermelin, 2005).

Regional geology and geomorphology include two major units (Piedrahíta and Hermelin, 2005; Calle and González, 1980). The upper watershed (above 1800 m.a.s.l along the Tapartó River) is characterized by steep slopes (locally greater than 100%) on Tertiary plutonic igneous rocks. Natural forests are the main land cover, while rock outcrops can be seen on the steepest slopes. The Tapartó River flows through this unit along its first 9 km, with an average gradient of 23.7%. The lower watershed is characterized by deeply incised valleys underlain by a sequence of Cretaceous sedimentary rocks. Quaternary deposits include slope, fluvial, and fluvio-torrential deposits. Major land uses include coffee and pasture for cattle. The Tapartó River flows within this unit from 9 km until its confluence with the San Juan River, with an average gradient of 6.4%.

Due to the lack of data, it is difficult to assess ground conditions prior to the 1993 event. It is clear, however, that there was no anthropogenic contribution to this event's onset, as the river headwaters are still covered by natural forests with no human intervention. It is also possible that by the time the event took place, soils were already at or near saturation, as April is one of the wettest months in the region.

On the night of April 25, 1993, a flash flood affected the lower 18 km of the Tapartó River. Since there are no rainfall gauging stations near its headwaters, it is not possible to quantify the amount of rainfall that originated the flash flood. Field evidence, however, pointed to a debris flow with high transport capacity. It originated in the upper watershed, where several landslides were observed in the days following the event. The generating mechanism was saturation of the thin soil layers that covered the steep slopes. Eventually, soil strength became negligible, and materials (vegetation cover, soil, rock fragments) moved down the slopes, mixing with water from the streams. While no field evidence was found, it is possible that the large amount of debris delivered to the river formed a temporary dam (Piedrahíta and Hermelin, 2005).

Debris flows advancing along the Tapartó River left no deposits in the upper watershed due to the steep slopes. On the other hand, two distinct processes were observed on the lower watershed: (1) channel incision was dominant from km 9 to 15, with deposition of boulders within the channel; this process was associated with a hyperconcentrated, turbulent flow, and (2) deposition on the floodplains from km 15 to 27. These deposits consisted of medium to coarse sand, suggesting a less concentrated, more rapid flow. Depth ranged from a few centimeters to 25–30 cm (Piedrahíta and Hermelin, 2005).

Figure 7.7 Destruction of a sugar mill by the torrential flow on the Tapartó River.

The affected area included past fluvial and torrential deposits adjacent to the river channel, with elevations up to 7.5 m above the average water level. Official records reported 59 fatality victims, 70 people missing, 200 homeless, complete to partial destruction of infrastructure (roads, bridges and schools), and more than $1300 million pesos (US$ 500,000) of additional damages (DesInventar, 2003) (Fig. 7.7).

The analysis of previous torrential events within this basin indicates that they are not restricted to recent years, as evidenced by (Piedrahíta and Hermelin, 2005):

- The presence of large boulders (up to 10 m diameter) of igneous rocks in the lower section of the Tapartó River, at least 10 km from their place of origin.
- Several layers of torrential deposits overlying buried soils within the floodplain and terrace deposits. A torrential deposit at the base of a sequence of 8 deposits yielded a date of 2210 ± 85 yr BP, providing a return period of 280 years.

There is a historic account of a similar event that took place in the early 1900s. Torrential floods also occurred in nearby watersheds with similar conditions in 1946 and 1991 (Hermelin, 1993).

2.5. Río Páez, 1994

The Páez River watershed is located in southwestern Colombia, on the eastern slopes of the Central Cordillera. The regional geological setting includes metamorphic, igneous, and sedimentary rocks from the Precambrian to the Tertiary, the oldest ones (metamorphic) forming the Cordilleran basement. Locally, there are Tertiary and Quaternary lava flows. Several geomorphic processes have modeled the landscape, as evidenced by glacial, fluvial, and slope deposits. There are numerous regional and local fault systems, with a generalized N–S and NNE trend (INGEOMINAS, 1994; París and Marín, 1979). Annual rainfall estimates are in the order of 2000 mm (IGAC, 1982).

On June 6, 1994, at 3:47 p.m. (local time), a 6.4 magnitude (Ms) earthquake occurred in this area, near the summit of the Central Cordillera. Its epicenter was at 2.87°N and 76.08°W, while its depth was estimated to be less than 10 km. The earthquake was attributed to crust movements along a high-dip, strike-slip fault with a NE trend (Falla de Moras) (INGEOMINAS, 1994). This event was followed by numerous aftershocks. By June 30 a total of 800 aftershocks, with magnitudes between 1.8 and 4.8, had been recorded by the National Seismological Network.

The earthquake triggered more than 3000 mass movements. Their occurrence was promoted by additional factors such as steep slopes, saturation of surface materials by heavy rain, and inadequate land use. Materials from the slopes entered the streams, forming debris flows that reached depths between 10 and 40 m, with maximum velocities of 50 to 60 km h^{-1}. Deposits with local depths up to 70 m were also reported (Wilches-Chaux, 2005; INGEOMINAS, 1994).

This seismic event affected 15 municipalities with an area of 10 000 km^2. The greatest destruction took place around the epicenter, in an area estimated to be 400 km^2. Intensity within this area was VIII. Destruction by debris flows, however, extended for approximately 54 km downstream from the epicenter, along the Páez River and several tributaries. The number of victims was estimated at 1100, most of them associated with debris flows. In addition, 8000 people were displaced from their homes. Damages included complete destruction of houses (1650) and communal buildings (15), partial failure of houses (3200), and infrastructure damage (bridges, water supply systems, and roads). Economic losses were estimated at $124,000 million pesos (US$ 125.5 million) (Wilches-Chaux, 2005; DesInventar, 2003; INGEOMINAS, 1994).

This region has been affected by significant seismic events since historic times. Espinosa (1993, in INGEOMINAS, 1994) reported 10 events with intensities above VI from 1566 to 1983, four of which partially destroyed the state's capital of Popayán, approximately 80 km southwest from the Páez epicenter. A return period of 50 to 60 years has been estimated for earthquakes with magnitude (Ms) greater than 5.0, while a return period of 150 years has been estimated for earthquakes with magnitude greater than 7.0 (INGEOMINAS, 1986).

2.6. Armenia, 1999

One of Colombia's most representative regions to foreigners is the central coffee-growing region (Eje Cafetero) located on the western slopes of the Central Andean Cordillera (4°20′N–5°05′N and 75°30′W–75°55′W). Three regional capitals, Manizales, Pereira, and Armenia, serve as population, economic, and cultural centers. Their combined population in 1999 was estimated to be around 1 million (DNP, 1997). Elevation ranges from 1000 to 2500 m.a.s.l. Annual precipitation follows a bimodal pattern, with wet conditions in March–May and September–November. Annual rainfall presents strong local variations but generally increases from 2000 mm at lower elevations (1250 m.a.s.l) to around 2400 mm near 2000 m.a.s.l. Average temperature at

1400–1600 m.a.s.l is 19–20°C, with an average lapse rate of 0.83–0.87°C per 100 m change in elevation (CENICAFÉ, 1989).

Two physiographic units can be differentiated: (1) the steep slopes of the Central Cordillera, made of metamorphic, igneous, and sedimentary rocks covered by pyroclastic deposits from the Tertiary-Quaternary, and (2) hills on Pliocene-Pleistocene volcanic-sedimentary deposits formed by glacial, alluvial, torrential, and volcanic processes. Quaternary deposits include pyroclastic and alluvial deposits (González and Núñez, 1991). Regional tectonics are dominated by the Romeral fault system, which stretches for more than 800 km from Ecuador, through Colombia, bordering the western foothills of the Central Cordillera with a NS general trend (González and Núñez, 1991).

Documentation of historical earthquakes in this region is scarce or nonexistent, owing partially to the late arrival of settlers into the area (late 1800s). Historic records and international seismological networks report the occurrence of three large (magnitude Mb > 5.3) earthquakes in 1973 (2) and 1988. Only one of these events had a similar origin as the 1999 earthquake (shallow fault). Nonetheless, the regional seismological network had registered shallow (depth < 33 km) earthquakes during the period 1987–1999, with magnitudes 2.4 < Ml < 4.0 (Rosales and Meyer, 2005).

On January 25, 1999, at 1:19 p.m. (local time), a 6.2 magnitude (Mw) earthquake occurred with its epicenter at 4.38°N and 75.64°W, about 18 km south of Armenia (approximately 175 km west of Bogotá). Its depth was 10–34 km according to various sources. It was generated by movement along a high-dip, strike-slip fault with a N10°E trend. A major aftershock (Mb = 5.8) occurred on the same day, at 5:40 p.m. Minor aftershocks were recorded for up to three years (Rosales and Meyer, 2005; Cardona, 1999). The earthquake triggered several geomorphic processes, including mass movements, mud and debris flows, surface cracks, and subsidence. Slope instability was increased by the unusually high rainfall of December 1999 through February 2000 associated with La Niña. These processes affected an area of 1500 km^2 (Rosales and Meyer, 2005; Cardona, 1999).

Twenty-eight municipalities suffered direct damages, with the greatest losses occurring near Armenia (Figs. 7.8 and 7.9). The death toll reached 1185 (78% from Armenia), 160,397 people were left homeless, 400,141 were directly affected, and 1,534,500 were indirectly affected. Infrastructure was severely damaged, particularly schools (27% damaged beyond recovery) and hospitals. Total losses, which included damages to houses, infrastructure, and the productive sector, amounted $2,786,000 million pesos (US$ 1857 million). Seventy-three percent were related to housing damages. Total losses represented 2.2% of the national GDP for 1998 (CEPAL, 1999).

The extensive construction damage was related to (Rosales and Meyer, 2005; Cardona, 1999; Rodriguez, 1999):

- The nature of the earthquake (shallow): Damages in Armenia showed a NE–SW trend, coinciding with the fault's trend. In addition, accelerogram records indicated a strong vertical component.

Figure 7.8 Widespread destruction at Armenia during the January 25, 1999 earthquake (Photo by R. Rochel, Universidad EAFIT.)

- Earthquake effects were amplified by several factors: Deep volcanic ash deposits and soils (up to 30 m thick in Armenia), artificial landfills, topographic effects.
- Poor quality of building materials and construction techniques. According to the national seismic hazard assessment (AIS et al., 1998), the region is within the high seismic hazard zone. Many buildings however, did not follow the specific construction guidelines as either: (1) they had been built prior to the enactment of the National Earthquake Resistant Construction code (1984), or (2) developers and locals did not know or ignored these guidelines. Critical institutions for emergency response, such as hospitals, police, and fire departments, lost personnel as their buildings partially collapsed.

Figure 7.9 Collapsed building at a street corner in Armenia (Photo by R. Rochel, Universidad EAFIT.)

2.7. La Estrella, 2000

Although the event at La Estrella was of relatively low magnitude, its consequences on the affected community were very heavy; it is for this reason that it is included in this chapter.

The Aburrá valley, within the Central Andean Cordillera, extends from 5°58.73′N to 6°31.04′N and 75°13.40′W to 75°43.42′W. It covers an area of 1152 km^2 between 1300 and 2800 m.a.s.l. This valley is home to the country's second largest metropolitan area, with 3.3 million inhabitants. The area of interest (5 km^2) is located within its southern section, in the municipalities of La Estrella and Sabaneta, with elevations ranging from 1650 m.a.s.l to 2250 m.a.s.l. The area is drained by several streams that flow from west to east into the Medellín River, the most important one being La Bermejala and its tributary La Llorona. Precipitation follows a bimodal pattern with wet conditions prevailing in March–May and October–November. Average annual rainfall is 2300 mm yr^{-1}. Major land uses include pasture and shrub with some small urban areas. Forest plantations are important toward the watershed divide (Cadavid and Hermelin, 2005). Geomorphology includes steep slopes (>100%) on igneous and metamorphic rocks, hills (12–45% slope) developed mostly on slope deposits, and alluvial deposits (floodplains, terraces, and torrential deposits) (Cadavid and Hermelin, 2005).

There are several geomorphic evidences of previous events similar to the May 2000 event. These include relict mass movements, mud and debris flow deposits, and torrential flows up to 4 m deep along the major streams. Three similar events were registered by local newspapers and governmental institutions (1982, 1999, and 2004), all of them related to heavy rainfall events leading to mass movements and flooding (Cadavid and Hermelin, 2005).

On the night of May 29, 2000, a torrential rain fell on the study area for approximately 7 hours. Three rain gauges within 10 km recorded rainfall totals of 40 to 62 mm, the highest figure being recorded at the closest station (1.5 km to the SE). Only one of these stations (7 km to the south) recorded rainfall intensity data (11.5 mm h^{-1}). The event was localized, as other gauges within 7 km recorded less than 6.6 mm (Cadavid and Hermelin, 2005). Local inhabitants reported that rain was most intense around 2-4 a.m. (May 30). As a result, 264 mass movements occurred with the following characteristics (Cadavid and Hermelin, 2005):

- Mudflows: Accounted for 61% of the total number of mass movements and occurred mainly on steep slopes underlain by volcanic-sedimentary rocks.
- Complex slides: Represented 26% of the total number of movements. Mostly on steep slopes developed on relict slope deposits.
- Slumps: Represented 14% of the total and occurred mostly by reactivation of older mass movements. Steep slopes (>49%) underlain by slope deposits and landfills were the most affected.
- Torrential flows: Formed by materials from mass movements along streams, due to stream undercutting. These flows moved boulders with diameters up to 2 m.

Heavy rain acted as the triggering factor, leading to generalized undercutting along stream banks followed by the re-activation of mass movements on the slopes. These materials were delivered to the streams, increasing their erosive power. Additional factors controlling the occurrence and spatial distribution of movements were slope and type of surface material. Land use did not seem to play an important role (Cadavid and Hermelin, 2005).

The event mainly affected the municipality of La Estrella, with an estimated population of 49,902 in 2000 (DNP, 1997). As a consequence of this event, 1 person died and 1 person was reported as missing. Direct economic losses amounted to $2450 million pesos (US$ 1 million), equivalent to 20% of La Estrella municipality's annual (year 2000) budget (Cadavid et al., 2004). Indirect losses related to road destruction, interruption of water supply, and displacement of numerous families were not accounted for.

3. Discussion and Conclusions

This overview can be considered representative of only natural disasters related to complex natural phenomena in Colombia. It should be remembered, however, that the country is also exposed to tsunamis, hurricanes, coastal erosion, and mud volcanism.

Natural disasters are more common than was previously thought. Their prevention, analysis, and relief efforts need to consider several factors, as illustrated by the case studies presented:

- High-energy events are the least frequent, and historical reports and statistical data are scarce. As a result, there is often no collective memory of similar events (e.g., Armero). This is worsened by the rapid vegetation recovery that takes place in some of these areas (e.g., San Carlos, Tapartó). Only during the last decades have governmental institutions kept a fairly consistent record. More reliable data are now available due to the improvements of the national organization for disaster prevision and relief.
- The increase of population has had several adverse consequences:
  - Indirect influence on events such as floods, flash floods, and landslides through deforestation, land-use change, and so on.
  - More people, properties, and infrastructure are progressively exposed to natural hazards.
  - The occupancy of dangerous areas in urban zones by emigrants coming from the countryside contributes to the increase in vulnerability (e.g., as at La Estrella).
- Results from technical studies are seldom included in regulations and disaster prevention programs (e.g., Armero, La Estrella). One fortunate case was the development of the 1984 National Earthquake Resistant Construction code, an outcome of the 1983 Popayán earthquake (southwest Colombia). The Armenia event, however, is a good example of the catastrophic consequences that arise from poor regulation enforcement.
- Finally, the influence of global change may increase events of climatic origin (Hermelin, 2005).

With respect to prediction, an enormous effort is being made to keep active volcanic centers under close observation and to increase the national seismic network and keep it in good condition. Furthermore, local meteorological and hydrological networks are being installed in large cities (Medellín, Bogotá), since national networks do not have the spatial density required for satisfactory prediction analyses given the mountainous conditions of the country. This initiative, which should go together with detailed geomorphological mapping, should be followed in the entire territory.

More efforts are needed to understand the processes themselves, through direct observation, laboratory tests, and modeling. The development of new methods to identify and to characterize natural disasters must be fostered, as for instance the excellent initiative of DesInventar, proposed by OSSO-Univalle (Observatorio Sismológico del Suroccidente-Universidad del Valle). A more detailed knowledge of the Colombian landscape and its processes, the social and economic aspects of risk prevention, should also be encouraged. Only a combined effort in these fields will allow inhabitants to coexist with such a demanding natural environment as the Colombian territory.

CHAPTER 8

Geomorphology of Natural Hazards and Human-induced Disasters in Ecuador

Fausto O. Sarmiento

Department of Geography, The University of Georgia, Athens, GA. 30602. USA

Contents

1. Introduction

The active geology of Ecuador is often cited as the natural outcome of the young orology that fuses volcanic, tectonic, and plutonic episodes and their propensity for disastrous manifestations to the urban and rural population in the Andes. Often associated with these catastrophic events, the incidence of a capricious topography enhances conditions prone to massive impacts that have tripled in the last couple of years (Serbín, 2007). Despite the loss of lives and other calamities, no direct link between the explosive eruptions of the many volcanoes of the country can be coupled with the ecological risk that nearby communities internalize to live in such a mountainous territory. People know of the danger but remain rooted in the territory appropriated through equally capricious historic developments. On one hand, there is an environmental determinism that highlights the uncertainty re locating the epicenter of earthquakes, which have altered the course of history and the fate of entire towns unable to escape nature's wrath. On the other hand, traditional planning with a possibilistic approach reveals the likelihood of survival for the many people leaving in

Corresponding author: *Fausto O. Sarmiento, fsarmien@uga.edu*

Developments in Earth Surface Processes, Volume 13
ISSN 0928-2025, DOI 10.1016/S0928-2025(08)10008-6

the shadow of the volcanoes. Yet, none of the urban centers have been planned as a response to the location of geological faults or to seismic vulnerability. Finally, the periodicity of major climate events, whether with circadial occurrences—such as the freezing, gelid nights in the tropical highlands; or with circannual phenomena—such as the torrential rains in April and May; or with the decadal occurrence of mega storms—such as those associated with the El Niño-Southern Oscillation (ENSO) that wreaks havoc in coastal communities, is never taken seriously, even though the weather services have correctly forecasted them. In most cases the incoming events are not anticipated, but they are feared (Sarmiento, 1988) because they can randomly occur at different scales: from minor perturbations to major disturbances or larger disasters, even catastrophic events.

As an indication of the detrimental effect of small-magnitude disasters in Ecuador, during the 1990s, some 1686 people died as a direct result of those events, distributed among landslides (509), epidemics (342), flooding (338), drought (114), fire (67), explosions (32), seismic activity (32), marine tides (7), and anthropogenic causes (245). Among those events created by people are listed traffic accidents, airplane accidents, civil unrest, gas tank explosions, and intentional fires. Of course, major disasters have also occurred in the past when entire populations were wiped out, as happened in the Riobamba earthquake (1797), the Latacunga earthquake (1645), or the Ambato earthquake (1949). Overall, for the risk analysis toolbox developed to improve regional conditions of preparedness for disasters, the Development Program of the United Nations (UNPD 2005) suggested a Disaster Risk Index (DRI) of 59.0 for casualties from four hazards (CRED), including earthquake, drought, flood, and tropical cyclone. This DRI is relatively high and puts Ecuador in the top five South American countries needing to develop risk-management programs and education/awareness campaigns to minimize the effects of disaster, particularly in poor, rural areas.

According to Hugo Yepez (2002), director of the Geophysical Institute of the Polytechnic University in Quito, there is a need to completely reconceptualize disaster. To change the deterministic paradigm for the disaster-prone region of mountainous Ecuador, it is important to acknowledge a dearth of information about the intricate relation of geomorphologic features and disasters that has been prompted by human interventions (Alexander, 1997). It is clear that in Andean cultural landscapes, the human imprint has easily impacted the disappearing forest cover, increasing soil erosion, channeling rivers, spilling nutrients, watering into formerly drier lands, and building structures that have affected the slope lands and controlled the rate of processes or the intensity of flows, according to the geopolitics of place instead of the laws of gravity alone (Allan et al., 1988). With the changing demographics of mountain regions and increased pressure on the use of resources to satisfy the demand of growing cities in the lowlands, the assessment of rural highland Ecuador's vulnerability to disaster involves matters of geopolitics and social dynamics that are transforming the community-driven, subsistence-based economy into a more global, urbanized, soil-hungry, input-friendly, and industrially based economy (Knapp, 1991; Gade, 1999; Blanchard-Boehm, 2004), which brings the vulnerability of the traditional Andean system and its ecological risk assessment to a higher level of detail.

Amidst the choices for avoiding disasters in culturally rich sites conserving biodiversity, the Andean region shows potential to become an exemplar of different options that will permit the protected landscape approach to lead to less ecological risk or the increased adaptability of mountain communities exposed to global change (Bruce, 1994; Sarmiento, 2006). However, the trend described with the aid of general circulation models of global climate change has received minimum attention from the social sciences in the Americas, particularly in the Andes. There physical and biological drivers are seen as purely geomorphic features with no connection to a human agency. Little research has yet been done on the human dimension of global climate change (MRI 2006), particularly when the natural and cultural boundaries intersect. This makes it much more appropriate to emphasize human impacts on the geomorphology of the Andes by analyzing resilience, adaptation for resource allocation, and human drivers of landscape change in the context of globalization and sustainable development.

2. Ecuador and Anthropogenic Disturbances

The mosaic of Ecuador's geoforms reflects ecological gradients along elevation clines associated with its equatorial locale and active orogeny working in an apparent equilibrium to confirm the view of a unified, regional identity. Meteorological conditions favor a cluster of intensity episodes, with seasonality options occurring in four main natural regions: (1) the Oriente (Amazon Basin) or *cisandean* domain to the east; (2) the Sierra (Highland plateau) or *interandean* domain in the Andes Mountains; (3) the Costa (Coastal lowlands) or *transandean* domain; and (4) the Galapagos Islands or the *oceanic* domain (Fig. 8.1). The active backbone of the Andes, with topographic and climatic extremes, receives fertility inputs of volcanic origin. The lowlands at each of the two outer slopes receive the intense and constant input of alluvial origin. In some cases there is a feedback in the piedmont associated with erosion, deposition, and other geological dynamics, which is pointed out as a natural mechanism for soil diversity in the tropandean landscape. Thus, soil condition also varies because of the differential age, exposure to weathering, type of parent material, and intensity of anthropogenic disturbances, creating microsites and catenae of enhanced diversity within the region. Table 8.1 shows the events that have triggered disasters in equatorial mountain environments.

The human imprint on the landscape is noticeable even in areas where the extent of forest cover now obscures the evidence of ancient human occupation, starting with the Anthropocene in South America, between the Pleistocene and Holocene. I have argued elsewhere (Sarmiento, 2002, 2005) that even the name Andes itself hints at the cultural essence of the mountain environment by using the archaic Castellan shorthand to describe the terraces (or *andenes*) widespread in the region of first contact with the Europeans in their conquest of *Tawantinsuyo*, the *Inka* empire. The entirety of the built mountain chain (or *Cordillera de los andenes*) became known for these stoned or drywall terracing (or *andenerías*), which allowed for agriculture and forestry production in spite of steep topography. Sites such as *Ciudad Perdida* in Colombia,

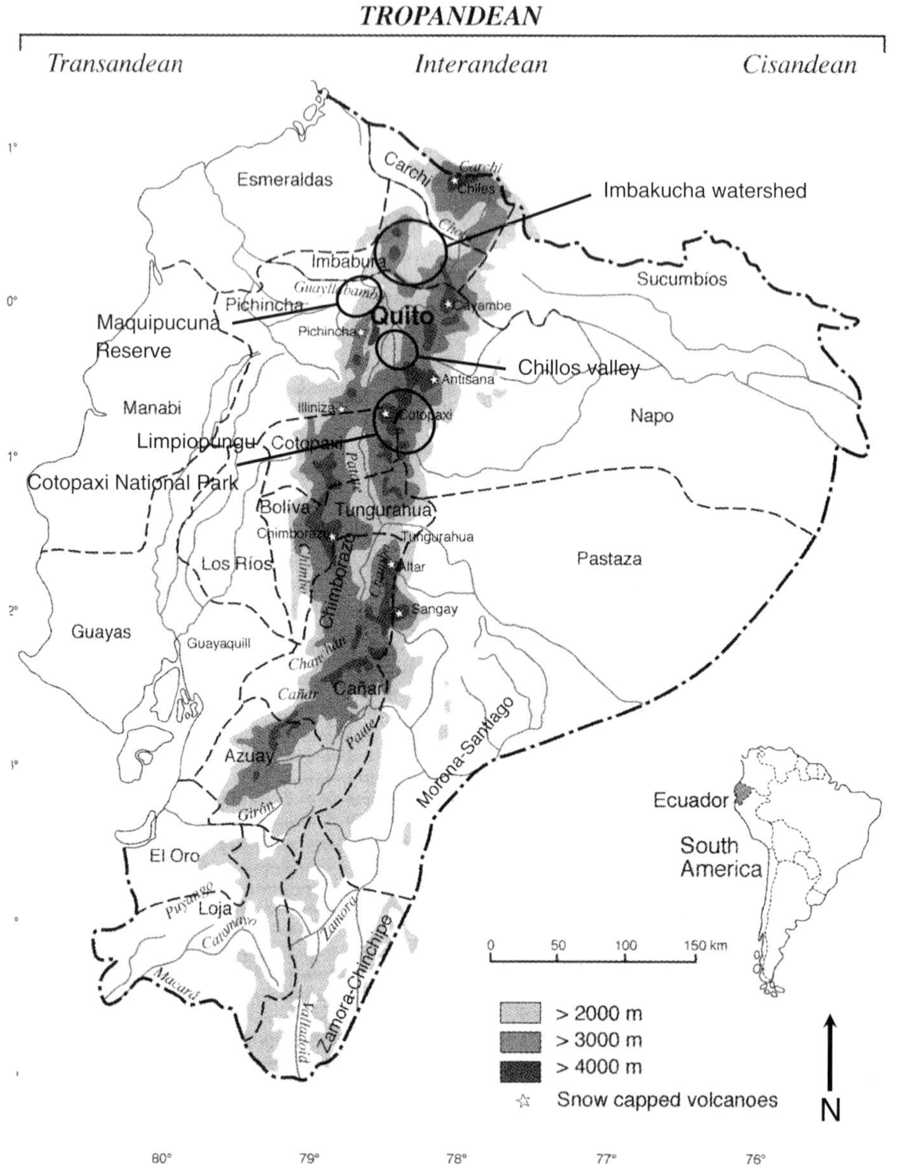

Figure 8.1 The map of Ecuador with the different tropandean domains. The transandean valleys are shown opened to the Pacific coastal plain, the interandean valleys are shown in the mountain basins, and the cisandean valleys are shown opened to the Amazon plains. Several field sites where the study was conducted are listed.

Mullituru in Ecuador, *Pajatén* and *Machu Picchu* or the great extent of the *Colca* canyon in Peru, *Tarijal* in Bolivia, or *Bilche* in Argentina are examples of the once widespread constructed mountain landscapes in the staggered display. These heavily settled agrosystems flourished in challenging conditions of climate and topography, since

Table 8.1 Disasters Ranked by the Mortality Outcome. The disasters are listed by the initiating agent, together with the feasibility of preparedness as exhibited by risk-assessment instruments available at the national level and emergency response plans at the local level, and the location affected, discriminated between the lowland and highland. The list follows the United States Office of Foreign Disaster Assistance (OFDA) for the worldwide classification of agents initiating disastrous events

Initiating Agent	Mortality	Preparedness	Lowland	Highland
Avalanches	low	no	0	1
Civil strife	middle	no	1	1
Drought and famine	middle	yes	0	1
Displaced persons	high	no	0	1
Earthquake	high	no	0	1
Fire	middle	no	1	1
Flood	middle	yes	1	0
Heat wave	high	no	1	0
Landslide	high	no	0	1
Storms	low	no	1	1
Technological accident	middle	yes	1	1
Tropical cyclone	low	no	1	0
Volcanic eruption	low	yes	0	1

geomorphic processes of the young geology are still evident in the orogeny and associated flooding episodes, constant tremors, occasional earthquakes, rockslides, and volcanic eruptions of record-breaking magnitude that impacted those sites.

A first attempt to explain human-induced disasters in the equatorial Andes is seen in the case of the mega-fauna extinction. In the *Inga* brook, near the volcano *Ilaló* in the Pichincha Province, a discovery of a mastodon, with the obsidian arrowhead still encrusted in the petrified skull, illustrated a clear human influence in reducing the pressure of mega herbivores. Charcoal found in the *Inga* site with other Clovis artifacts suggests the initiation of fire and the use of burning as a factor in anthropogenic changes to the mountain landscape as early as 12,000 B.P. Hence, fire, agriculture, and hunting have acted upon the mountain landscapes of Ecuador since antiquity, often becoming driving forces in the current appearance of highland grasslands (Grubb, 1970; Keating, 2007), the tropandean forest belt (Sarmiento, 1993), and the overall tropical mountain dynamics (Troll, 1968; Brush, 1976; Ellenberg, 1979; Zimmerer & Young, 1998; Sarmiento, 2003).

Table 8.2 lists different mountain systems that are found in contemporary Ecuador. Throughout the area, the most notable feature is the changed land use and land cover of the current configuration, due to forest conversion into pasture. The highland grassland is created/maintained by the repetitive use of fire as mowing mechanism, weeding the sprouts and saplings that hence do not regenerate into arboreal dimensions. This is the process known as paramization, a clear indicator of the degree of hemerobiotic state (read culturally dependent) of this mountainous region (Sarmiento, 2005). Just recently in environmental discourses,

Table 8.2 Summary Table for the Top 20 Highest Mountains of Ecuador, indicating the risk associated with volcanic activity and seismic potential. A column listing the presence of human drivers (also known as anthropogenic impacts) is presented

Mountain *	Elevation	Domain	Volcanism	Impacts
Chimburazu	6310 m	Transandean	Ancient	desertification
Kutucpachi	5897 m	Interandean	Active	paramization
Kayampi	5790 m	Cisandean	Ancient	paramization
Antisana	5758 m	Cisandean	Active	paramization
Kapac Urku	5319 m	Cisandean	Ancient	desertification
Illiniza (South)	5263 m	Transandean	Dormant	paramization
Sangay	5230 m	Cisandean	Most Active	paramization
Illiniza (north)	5116 m	Transandean	Dormant	paramization
Tungurawa	5023 m	Cisandean	Most Active	paramization
Kariwayrazu	5018 m	Transandean	Dormant	paramization
Kutacachi	4944 m	Transandean	Dormant	paramization
Sinchulawa	4873 m	Interandean	Ancient	paramization
Corazón	4782 m	Transandean	Ancient	paramization
Pichincha	4776 m	Transandean	Most Active	paramization
Chiles	4723 m	Transandean	Dormant	paramization
Rumi Ñawi	4722 m	Interandean	Ancient	paramization
Imbabura	4621 m	Interandean	Dormant	paramization
Cerro Hermoso	4571 m	Cisandean	Dormant	paramization
Ata Kazu	4455 m	Transandean	Dormant	paramization
Pasochoa	4199 m	Interandean	Ancient	paramization
Sumaco	3780 m	Cisandean	Dormant	minimal
Reventador	3562 m	Cisandean	Most Active	minimal

* Vernacular Kichwa is preferred as a descriptor of the mountains listed; thus the phonetic alphabet is preferred over the Castilian version.

herds of cows have been termed the most dangerous influence on the planet: the bucolic, pastoral environment long associated with grazing livestock masks the deleterious additions they make to greenhouse gases and and their role in deforestation. In the highlands of Ecuador, it is easy to understand why, after paramization, several disasters can be considered "natural" (e.g., see Berz, 2004), although they are really induced by management (or the lack thereof).

2.1. Geomorphology and Disasters

As Hewitt (1997) has pointed out, mountain lands pose exceptional risk for human activities and some unique dangers. Recently, Ecuador has experienced the reactivation of volcanic activities in or nearby areas of heavy human occupation, such as Quito, the capital city; at the base of the Pichincha volcano; and at Baños, the idyllic tourist town at the base of the Tungurahua volcano. A visit to Roman Catholic churches in Baños and El Quinche near Quito helps us to understand how the syncretic attributes of faith in the Virgin Mary (or Mother Earth or *Pachamama*) as protector of eruptions and earthquakes, as these figures are depicted in colonial paintings, remind mestizos and indigenous inhabitants about the supernatural power of disasters. These are typical manifestations of ecological risk, where neither the periodicity can be figured out nor the reactivation can be easily triggered. Compounding the geoecological risk, neither the uncertainty of the occurrence can be diminished nor the duration of the event can be forecasted (UNDP, 2006). Furthermore, the explanations of why this happens and where it happens cannot be easily communicated to the public at large. Figure 8.2 shows the impact of the explosive cloud generated in both volcanoes and the serendipitous event of ashfall to the lowest human density areas. No major disaster-related injuries were reported after the Pichincha eruption, other than obvious problems with breathing ash-laden air in the capital city. But the potential for an enormous disaster is evident, with unexpected social and economical effects on local populations and on the country as a whole (Gavidia, 1991). Table 8.3 shows the aggregated values of the top 10 natural disasters in Ecuador, where it is clearly shown that the most damaging one (cf. lost of life) is the earthquake. The most common disasters affecting populations, however, are drought and flooding, and the most damaging ones (cf. economic impact) are earthquakes, landslides, and flooding (or a combination of these three). In order to be included in the database of OFDA/CRED (2007) and to be labeled disasters, natural events have to comply with four requirements: (1) 10 or more people are reported killed, (2) 100 people are reported affected, (3) the government calls for international assistance, and (4) the appropriate national authority makes an official declaration of a state of emergency, which in Ecuador is the National Directorate of Civil Defense, an organization that works directly under the president of the Republic.

2.2. Andean Farmscapes and Human Impact

There are no data that help correlate the huge deforestation of highland Andean forests associated with the expansion of the agricultural frontier with landslide or flooding episodes in downstream communities. Empirical observations, however, unequivocally show that they are closely linked. It is clear that the siltation of Daniel Palacios dam, in Azuay Province, is directly linked to the increased erosion of the Paute River watershed (Harden, 2004). It is also clear that the coloration of rivers flowing through the interandean domain has continued to be darkened by the load of sediment they carry, as happened with the events following two consecutive earthquakes in 1987 near the Reventador volcano. These earthquakes produced mass-wasting devastation downhill for the pushed vegetation that were then converted in weak earth dams, and following a season of heavy rains, the dams subsequently burst open (Schuster et al., 1996; Chatelain et al., 1999). Despite the aggravation of isolated communities living in distant Amazonian villages, they were not considered environmental refugees, as this is a rather new phenomenon linked to global climate change (Bates, 2002) and to metropolitan areas

Figure 8.2 Volcanic eruptions of (a) the Pichincha volcano and (b) the Tungurahua volcano have thus far spared heavy human settlements in the city of Quito and the town of Baños. It can be seen that fortunately the fumes of the dangerous clouds have been pushed away from major human settlements, affecting mostly farm animals and wildlife.

that affect thousands at once (Thouret, 1999). Rural farmscapes have not yet been designated as "refuges" for displaced populations. Distant rural communities do not often make headlines. Yet, they are exposed to geoecological risk and suffer the consequence of disastrous events in disproportionate numbers.

As noted elsewhere for developing countries (El Masri and Tipple, 2002) disaster mitigation leading to sustainability in Ecuador is intrinsically linked to the development of mega-infrastructure or other type of human agency. Not just dams and reservoirs, but other facilities built in mountain areas have also affected

Figure 8.2 (Continued)

geoforms and soil properties. Major urban expansion on the fringes of the capital city pushes shanties to higher elevations without basic services, reaching the critical contour levels of the greenbelt, but breaching protected forests. Political maneuvering makes provision of potable water or sewage infrastructure one of the most sought-after compensations for the social support of squatters that thus extend the urban frontier in amoeboid fashion, which exacerbates geoecological risk. The pseudopodia of urban growth add to the complexity of risk management, as they multiply the risk by increasing the edge-to-core ratio of municipal responsibility, thereby having to manage multiple, disorganized, dislocated, and fragile structures. Landslides in the waterlogged, naked slopes provoked by the urban-frontier expansion have almost always punished those living in the margin with loss of lives and what little property they had; this is a good example of the paradigm that disasters and poverty levels are correlated and intertwined.

The electrification of rural areas has also had great impact, with the roads and installations of workshops and factories polluting faraway places. The so-called penetration roads have further dug into more distant and steeper slope lands, increasing the risk for rockslides and landslides in the sites where the road has been laid, often cutting the resistance of the angle to precipitous declines. As a result, the area becomes prone to landslides that wash away the road and, in some cases, the traffic on such a road. Studies of landslide regeneration (e.g., Myster and Sarmiento, 1994) have shown that the origin of landslides can be correlated with the existence of a footpath, a mountain trail, or a full-fledged road that was built ignoring the unique circumstances of mountains (so-called mountain specificities) not only in the physical, vertical constraint but also in the sociocultural arena. It is important to note that ancient road systems, such as the great *Inka* road or *Kapac Ñan*, fed by the smaller roads or *Inka Ñan*, which in turn were fed by even smaller footpaths, whether *Chaqui Ñan* or *Kuluncu*, utilized mainly the ridgeline of mountain terrain. In contrast, colonial and neocolonial

Table 8.3 Summary Table for the Top 10 Natural Disasters Recorded in Ecuador, with information taken from EM-DAT: The OFDA/CRED International Disaster Database at the Catholic University of Louvain, Brussels, Belgium. The CRED/EM-DAT team is currently working on enhancing the economic damage figures, so these statistics should be treated with caution

Disaster Type	Date Recorded	People Killed	People Affected	Economic Damage (US$)
Earthquake	Aug. 5, 1949	6,000	100,000	20,000
Earthquake	Mar. 5, 1987	5,000	150,000	1,500,000
Epidemic	May 1969	400		
Epidemic	June 13, 1991	343		
Flood	Nov. 1982	307	700,000	
Landslides	May 9, 1993	250		
Flood	Oct. 1997	218		271,000
Earthquake	May 14, 1942	200		
Epidemic	Jan. 1992	200		
Landslide	Mar. 28, 1993	200		500,000
Drought	Mar. 1964		600,000	
Volcanic	Aug. 14, 2006		300,013	150,000
Flood	Mar. 24, 1992		205,000	20,000
Flood	Aug. 4, 1983		200,000	
Flood	Apr .8, 1970		140,500	
Volcanic	Nov. 3, 2002		128,150	
Epidemic	Mar. 2000		100,000	
Flood	June 1989			15,000
Flood	Mar. 6, 2002			13,000
Volcanic	Aug. 14, 2001			10,975

roads cut the slopes to position the road parallel to the river flow, whether at midelevation embankments or directly onto the floodplain at the valley bottom, making them easy targets for slope stabilization failures and colluvial and alluvial wash flows. Thus technology for road construction and the machinery associated with it often increase the probability of disaster and multiply the geoecological risk.

2.3. Roads and Increased Fatality

Modern road building has brought a lot of destruction to mountain forests; not only because it allowed settlers to colonize upper areas, but also because of the massive destruction inflicted in the process of constructing the road itself (Solberg et al., 2003). This destruction occurs mainly when the cut of the talus is followed by the dumping of rubbish to the river bottom, or when the vertical cut does not have appropriate staggering support or terracing that allows for gullies and culverts that become focal points for risk. Because of the lack of resources and poor planning, no tunnels or stilt columns suspending bridges among brooks can be found in Ecuadorian highways or even secondary roads, unlike the environmentally friendly construction seen in the Swiss Alps, the Italian Autostrada del Sole, the German Autoban, or many other tunneling highways in the developed world. Often, not even low-tech solutions, such as gabions, retention walls, contention fences, or vegetated mats, are built-in with the new road. They are familiar after-the-fact "improvements" on already affected highways. Hence, an increased fatality rate is associated with "technical" risks, such as driving in mountain roads without barriers, illumination, traffic signals, security ramps, speed traps, breaking landing paths, or flexible lanes for ascending traffic. An example of this situation is the world-famous "Road of Death" in the Yungas of Bolivia. Many "roads of death" are similarly found throughout the equatorial Andes, particularly in the mountain passes of the Quijos, Pastaza, Paute, and Zamora rivers toward the Amazon lowlands, and in the Mira, Guayllabamba, Toachi, Chambo, Chanchán and Catamayo rivers toward the coastal plains. Often shrouded by the fog of the cloud forest terrain that engulfs the mountain passes, these "roads of dead" are decorated with small, white crosses dotting the many turns where fatalities occurred.

The rugged topography of the Andean incline was conquered in the early 1900s with the railway connecting Quito with Guayaquil. When coming down from the town of Alausí, following a steep gorge named the "devil's nose," the railroad zigzags through the pass as the most parsimonious of effort-limited travel route. This design, which at that time was deemed the most important transportation engineering work, now serves only as a tourist attraction. A plan to build a railroad to the Amazon region crossing the divide from Salcedo toward Puerto Napo never materialized. At present, no railroad is operating between the coastal plain and the highlands, owing to the recurrent destruction of the area because of flooding, tremor-triggered landslides, mudslides, and rockslides. Despite having several tunnels instead of deep cuts in gorges around the area of Lita and Estación Carchi, the northern railroad from Ibarra to San Lorenzo has also been stopped. Most people from the towns along the railroad have emigrated to the cities, and few return to the quasi-mystical ghost towns along the tracks. Not only the lack of maintenance is a problem there, but the critical habitats crossed by the rail line had also been altered, mainly to open agricultural frontier facilitated by a brand-new highway that parallels the railroad. The once pacific travel to the sugar-exporting port in Esmeraldas Province is now a major danger due to paramilitary intrusion from Colombia and continuous terrorist threats.

Terrorism is an important factor when dealing with the human impact on geomorphology, particularly at three levels. First, when the threat of violence exists, most activities in the areas affected are performed with resolve, often rushing construction or forest clearings, thereby increasing the likelihood of erosion and pollution. Second, the culture of violence and fear requires different patterns of settlement, changing the traditional stilt-shacks by the river edge to a basement-fortress in the center of a large radius cleared in the *terra firma* forest, away from the river, with all the mass movement that the new architecture entails. Third, when acts of terror have occurred, such as in explosions on bridges and on the oil pipeline, severe episodes of pollution and destruction have been registered when crude oil escaped toward the Quijos River and polluted several kilometers of pristine white waters.

The trans-ecuadorian pipeline (OTE: Oleoducto Trans Ecuatoriano) was built in the 1970s in record time, following the boom of the oil exploration. It was meant to facilitate the export of crude from the Amazon region collected in the town of Nueva Loja or Lago Agrio, passing through the Andes Mountains to the terminal harbor of Balao in Esmeraldas where cargo ships and tankers fill up and refill the medium-density outgoing petroleum. With the hassle of the deadline, the pipeline was built aboveground, making it a much easier target for terroristic acts. With that lesson, the new oil pipeline available for heavy oil (OCP: Oleoducto de Crudos Pesados) was entirely built underground and is designed to anticipate risk and monitor possible problem areas. When the first pipeline was built across the country, a vast amount of land was damaged, mainly for the penetration road that allowed colonization of the recently open frontier. Deforestation soon followed, ostensibly to provide a demonstrable productive patch of pasture or a marketable harvest of agricultural goods. The major expansion of the agricultural frontier in the Amazon region of Ecuador, with the ensuing deforestation and degradation of the mountain terrain, as well as the vast areas of the Andean foothills, is one of the major forces that has altered the landscape in the last 40 years. No other localized, geoecological event registered in the past has been able to inflict such damage and transformation of the land cover and land use to the extent that deforestation in the Ecuadorian Amazon has.

In fact, the human induction of disaster by deprotecting forested slope lands is a major geomorphologic condition that speeds the erosion and weathering of exposed soils and rocks, and produces an unnecessary toll on rural communities. Figure 8.3 shows how the human agency has radically changed the mountain landscape, which cannot be readily translated to disaster statistics. Table 8.4 presents a summarized

(a)

Figure 8.3 Denudation of the Andean slopes represents a major human impact in the Ecuadorian Andes. Often, after deforestation and burning of the original biota, the affected lands are prone to catastrophic events as seen in (a) the Guamani pass. Forest to grassland change in highland Ecuador is a major theme as it ignites not only controversy from traditional conservationists that believe the Páramo to be a natural ecosystem, but also multiplies the elements of risk and the casualties of the lack of environmental planning in the mountains, as seen in (b) the headwaters of the Toachi River.

(b)

Figure 8.3 (Continued)

version of disasters in Ecuador from 1904 to 2006, emphasizing that the risk of floods is the major potential disaster in the tropandean landscapes of Ecuador.

2.4. El Niño Periodic Disastrous Floods

The impacts of El Niño-Southern Oscillation (ENSO) events, particularly in the transandean domain, exemplify this conundrum associated with natural hazards. Taken as a true decadal cycle resulting from the warming of the surface waters of the Pacific Ocean that flow closer to the shore around Christmas (hence the name El Niño referring to the coming of "baby Jesus" in December), the hazards associated with the torrential rains of ENSO are exacerbated by careless agricultural practices upslope. Descriptions of the flooding associated with this phenomenon from the European explorers of the sixteenth century relate to the magnitude of water alone, especially around the city of Babahoyo where navigable waterways were usable to carry cargo toward the Guayaquil harbor. Most of the area associated with the mouth of the Guayas River is an alluvial terrace, which is subject to inundation during the rainy season and which carved new mangrove islands at every flood in the delta. Further from the shore, legendary flooding of the Bula-Bula River provided synchrony to a fertile agricultural enterprise in the Andean foothills. At present, descriptors of catastrophic episodes there are often linked with the images of chocolate-water full of debris, destroying not only the infrastructure of bridges, highways, and railroad, but also the comfort of a pleasant livelihood. Development is halted with ENSO. Some observe that there is a 50-year backward trend occurring after every El Niño and that the country will never be able to free itself from the grasp of this disaster.

Finally, some human impacts on the geomorphology are rather subtle but can still make important changes in the insular domain. The Galapagos Islands, thought

Table 8.4 Adapted Summary Table of the Natural Disasters in Ecuador, during a century of recorded major events (1904–2006), with special reference to the average value per event (p.e.) listed in the database of the OFDA/CRED International Disaster Database, housed at the Catholic University of Louvain, Brussels, Belgium. It is likely that the human agency exacerbates the conditions of risk conducive to disasters, particularly in torrential rains/droughts, epidemics, landslides and wildfires. Indeed, anthropogenic landscapes are thus more vulnerable than "natural" states

Disaster Type	Ave	Human Agency	Events	Killed	Injured	Homeless	Total Affected
Drought		Medium	2	0	0	0	634,000
	p.e.			0	0	0	317,000
Earthquake		None	16	11,336	486	214,867	398,303
	p.e.			709	30	13,429	24,894
Epidemic *		High	11	1,000	0	0	159,689
	p.e.			91	0	0	14,517
Flood		High	22	889	259	115,436	1,577,277
	p.e.			40	12	5,247	71,694
Landslide		High	12	1,099	120	180	81,456
	p.e.			92	10	15	6,788
Volcano		None	10	6	13	7,200	546,883
	p.e.			1	1	720	54,688
Wildfire		High	2	0	0	0	800
	p.e.			0	0	0	400

* Includes meningitis, equine encephalitis, typhoid, cholera, dengue fever, plague, malaria, and unknown.

to be one of the best examples of wilderness in the planet, indeed reflects the pressure placed by previous anthropogenic changes associated with plant and animal invasions. For instance, on San Cristobal Island, it is easy to find orange trees, not as orchards, but as forests of oranges growing wild through seeds spat out by thirsty tourists. The same happens with many other introduced plants, such as the cascarilla tree, the national tree of Ecuador (*Cinchona succirubra*) which is threatened in the cloud forests of southern Ecuador, its original locality; but, becomes a pest due to its invasibility to Galapagos and successful spread on Santa Cruz Island. Animal-produced impacts on the land cover are also affecting the landscape in major ways, mainly through the presence of feral dogs, cats, and even donkeys, all of which plague Isabela Island. On most islands of the archipelago, it is the action of feral goats that hints at the subtle impact of human drivers in changing today's landscape.

3. Conclusion

Between the narratives of environmental determinism and possibilism, understanding the human impact on the geomorphology of Ecuador produces an interesting duality. The capricious rugged terrain with elevations, such as Mount Chimburazu, which is more than 6300 m above sea level, imposes a sense of wonder and explains the fatalistic approach of many mountain villagers who cling to their holy Virgin Mary for protection and compassion in times of geoecological risk. However, popular belief tells the tale of Saint Marianita de Jesus who prophesized that Ecuador would be destroyed not by earthquakes, eruptions, or flooding, but by corrupt, incompetent governments. This dictum reverberates in the unconsciousness of many Ecuadorians who reify some notions of indigenous people; they consider mountains their guardians or *Apus*, and they respect them. The modern world in the Andes exhibits an inherent lack of respect in its quest for global markets. Thus far, governments are being blamed for having Ecuador stocked between the vulnerability of farmscapes exposed to disaster and the need to develop better adaptability to the geoecological risk. Thus far, Saint Marianita de Jesus remains uncontested.

The risk is hitherto augmented by lax social-response mechanisms or disaster management, including risk transfer, wherein politically oriented development planning—or its lack—does not account for vulnerability to natural, let alone human-, induced disasters. As Kahn (2005) has pointed out, rich and poor nations alike share responsibility for reducing the death toll and human suffering caused by the increased frequency of natural disasters. Cataclysmic events are still mystical.

CHAPTER 9

Natural Hazards in Peru: Causation and Vulnerability

Kenneth R. Young[1] *and* Blanca León[1,2]

[1] *Department of Geography and the Environment, University of Texas at Austin, Austin TX 78712, U.S.A.*
[2] *Museo de Historia Natural, Av. Arenales 1256, Apartado 14-0434, Lima-14, Peru*

Contents

1. Introduction

At any given moment, sudden changes due to Earth system processes may rearrange the planet's surface, with consequences for both natural systems and people's welfare and livelihoods. Over long time spans, the physical dynamism of the Earth is intrinsically linked to the evolution of life and the formation of landscapes. Abrupt changes can promote landform degradation, formation, and transformation. Within this dynamism, the development paths of human societies have often resulted in increased exposure to conditions leading to the loss of life and infrastructure. People must evaluate their individual and collective vulnerabilities, in part based on knowledge of the natural frequencies and magnitudes of hazards.

Corresponding author: [1] *Kenneth R. Young kryoung@austin.utexas.edu, Phone: (1)+512-471-5116, FAX: 512-471-5049* [2]*blanca.leon@mail.utexas.edu,*

Developments in Earth Surface Processes, Volume 13
ISSN 0928-2025, DOI 10.1016/S0928-2025(08)10009-8

Peru is located on the western edge of South America, one of the most tectonically active locations in the world (Cobbing, 1978; Veblen et al., 2007). Its location on the Pacific volcanic rim, the presence of the relatively young Andean Mountains on active plates and faults, and the occurrence of an expansive Andes-to-Amazon interface provide a series of natural hazards from seismic events, and also from others that are seasonal or periodic such as floods and mass movements (Fig. 9.1). These features can isolate settlements, making transportation and accessibility difficult (Fig. 9.2). Climatically, Peru's western side is dominated by the influence of the Pacific Ocean, making it xeric to semixeric, with trade winds from the east often blocked by the Andes. Deep valleys that drain mostly toward the Amazon Basin connect the high Andean plateaus and peaks to the more humid Amazonian side. Eastern Peru is dominated by humid tropical lowlands, but the steep eastern Andean slopes also provide a long but narrow corridor of cloud forests on rugged mountain relief.

In this chapter, we provide an overview of natural hazards in Peru, especially in regards to distribution, vulnerability, and risk management (see also Bernex, 2002). The importance of natural disasters has been recognized in general as a social cost, which is a factor in development planning (Alcántara-Ayala, 2002; Arnold et al., 2006). However, exposure to risk is variable through time and from place to place, while vulnerability also varies spatially in relation to the ability of people to predict, prepare for, and recover from natural hazards.

2. Natural Hazards and Human Vulnerabilities

2.1. Seismic Events

2.1.1. Earthquakes, Tsunamis

Seismic events in Peru are caused by the collision of two plates, the Nazca and the South American (Demets et al., 1990). Epicenters of most of the seismic events affecting the Peruvian coast and western Andean slopes occur along the intersection of those plates (Degg and Chester, 2005), which runs generally N-S, becoming closer to the coastal margin south of 11°S. In the case of earthquakes on the eastern Andean flanks, they are caused by isolated faults, and in the southern Andes some events are linked to volcanism.

Peru is affected by shallow, intermediate, and deep earthquakes. Tavera and Buforn (2001) examined the distribution of 19 recent earthquakes and tied the spatial patterns to likely causation processes. The shallow quakes were oriented along generally E-W trending axes, commonly associated with reverse faults, whereas the intermediate-sized earthquakes were aligned in relation to the Peru–Chile trench. Deep-foci earthquakes were distributed E-W on the border with Brazil and N-S on the border with Bolivia. These and other data allowed those authors to relate Peruvian tectonics to flat subduction of the Pacific plate in northern and central Peru, and to normal subduction in southern Peru. These same basic differences have consequences for the distribution of volcanoes, as will be discussed later in this chapter.

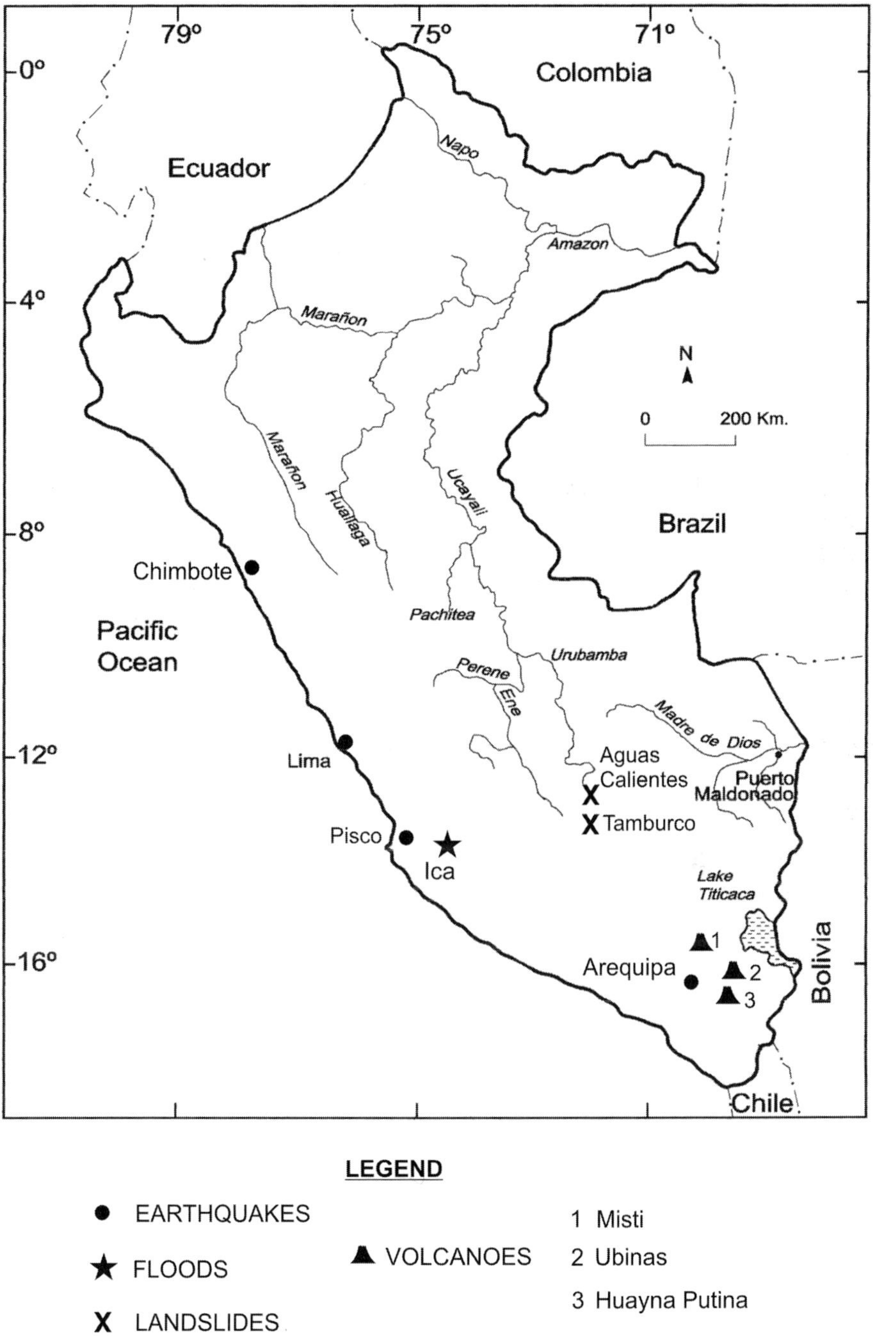

Figure 9.1 Main localities related to disasters mentioned in the chapter.

Figure 9.2 East view of Huascarán icecap,and road to Olympian Pass in Ancash cutting through steep rocky slopes.

Other parts of Peru experience earthquakes due to faults rather than plate movements. These earthquakes can take people by surprise as happened in September 2005 in the Moyobamba–Alto Mayo area of northern Peru (Table 9.1).

Probably the most hazardous areas for seismic activities occur along the coast south of 8°S. The three major seismic events during the last six years have caused millions of dollars in direct economic losses due to infrastructure destruction, not accounting for postrecovery costs.

Table 9.1 Recent Major Earthquakes in Peru

Seismic OccurrenceDate	Epicenter Location	Human Losses	Magnitude (M_w)
August 15, 2007	Near Chincha Alta, Ica-Lima	514	8.4
September 26, 2005	75 km NE of Moyabamba, Alto Mayo basin	5	7.5
June 23, 2001	175 km SSE of Puquio, NW Arequipa	138	8.4

Data from USGS (http://earthquake.usgs.gov/eqcenter/eqarchives/year/byyear.php).

On June 23, 2001 there was an 8.4 moment magnitude (M_w) movement in south central Peru (Table 9.1). Apparently, only two other quakes in historical times (1604 and 1868) exceeded this magnitude (Giovanni et al., 2002). An area of 120 by 300 km was also exposed to some 220 aftershocks (Tavera et al., 2006). Robinson et al. (2006) showed that a very large, contiguous area, some 6000 km^2 in size, blocked the further propagation of waves by the presence of faults related to a fracture zone of the subducting plate. These authors were able to map a remaining seismic gap, corresponding to a part of southern Peru and northern Chile with the hypothetical potential for an M_w 9 earthquake in the future, along with its associated hazards.

Byers (2004) compared infrastructure damage of this particular earthquake to others worldwide. Only part of the destruction was due to the direct effects, with others resulting from ground settlement, rock falls, soil and rockslides, and cracks in supporting elements. Audemard et al. (2005) described the extensive soil liquefaction that occurred throughout the area affected by the 2001 earthquake, although the effects of the quake were spatially restricted to certain sandy soils in major river valleys and the coastal plains. A strange occurrence of a "grass flow" caused by this earthquake in the high Andes of northern Chile was observed by Naranjo and Clavero (2005). In effect, they reported a slurry flow originating from an Andean wetland due to liquefaction. This event caused rock, soil, and plant materials to flow downhill at up to 50 km/hour for a distance of more than 14 km. These authors point out that previously these relatively flat highland features were not recognized as a potential source of hazards for people living downslope.

Keefer and Moseley (2004) provided an even larger context to consider the damage associated with the 2001 earthquake. They enumerated and described a host of surface and landform features altered by the quake, including rock falls and several thousand landslides, ground cracks, and collapsed subsurface feeder pipes and aboveground rills that connect to drainage channels. Each landslide delivered several hundred cubic meters of colluvium to the ravines and dry channels of this hyperarid region. In June and July 2002 an unusual rain event led to tremendous flash floods and the catastrophic reworking of surface materials loosened and dislocated by the earthquake a year previously. The authors demonstrate that larger seismic events (such as the 1604 M_w 8.7 earthquake) and more rainfall associated with especially strong El Niño events (such as in 1607 and 1608) could cause regionally significant landform effects, which they describe as a "shattered landscape."

Alva-Hurtado (1985) recorded other cases of soil liquefaction in Peru, showing that most were associated with strong seismic events and were primarily located along the coast. An M_w 8.4 earthquake in August 2007 included two separate movements (Sladen et al., 2007) that also produced liquefaction and ground rupture (Tavera et al., 2007).

Raised marine cliffs occur along the length of the coastline. Bourgois et al. (2007) pointed out that they result from major earthquakes in the past. They examined in detail the ties among geological features, tectonic forces, and geomorphic processes in northern Peru occurring not only on land but in adjacent marine zones. There are north-to-south differences along the coast owing to heterogeneity in plate subduction, but there are also variable expressions of subsidence processes offshore and subduction erosion in deeper waters. Over long time spans, eustacy thus would interact with (1) tectonic uplift, (2) landform processes including wave-caused erosion, and (3) marine processes, such as avalanches, all of which in turn are providing materials that may be later subducted under the South American plate.

Other hazards caused by seismic activities are tsunamis. In Peru, the occurrence of tsunamis has been documented for over 400 years, although these events apparently caused the destruction of human settlements as early as 880–890 B.C.E. (Bird, 1987). At time spans involving humans and their exposure to tectonic risks, tsunamis are of great concern in coastal Peru. They may be caused by earthquakes on land or offshore, and they may originate in Peru or at times in other places, with waves traveling across the Pacific Ocean.

Tsunamigenic earthquakes in Peru of M_w 8.52 and 8.64 would have recurrence periods of 50 and 100 years, respectively, according to calculations by Kulikov et al. (2005). In turn, these events would produce tsunami wave heights of 11.2 and 13.7 m, respectively. Major earthquakes in 1586, 1724, and 1746 had 24–28 m tsunami runups. Houses and infrastructure near sea level can be exposed to direct damage from the wave and from flooding afterward. The 2001 earthquake produced a tsunami that traveled and killed 23 people in other parts of the Pacific Basin (e.g., Goring, 2002).

Kulikov et al. (2005) showed that all major earthquakes in Peru are strong enough to be considered tsunamigenic. However, there are also large place-to-place variations in runup that have to do with local landscape-scale features such as bathymetry, offshore profiles, and coastal topography. A 6-m-high tsunami wave, which resulted from the 2001 earthquake, occurred 20 minutes after the quake. Unfortunately, however, this was 10 minutes after the sea margin first receded, leading several hundred people to their death as they gathered fish off the dry floor of the ocean (Kuroiwa, 2002).

On February 21, 1996, an earthquake (M_w 7.5) near Chimbote (9.6°S), north of Lima (Fig. 9.1), set off a tsunami that killed 12 people (Bourgeois et al., 1999). The epicenter was 130 km offshore, so ground movements felt on land were minimal, and most of the hazard to people was expressed through wave action. The rupture process leading to the quake happened near the marine trench. Ihmlé et al. (1998) suggest that this class of earthquake may produce a tsunami and fatalities in places otherwise considered to pose low risk for seismicity.

Okal et al. (2006) evaluated tsunami risk for the coastal area near Pisco (13.7°S), motivated in part by the need to plan for the construction of liquefied natural gas facilities designed to receive gas sent over the Andes in a pipeline connecting to the Camisea gas fields. Seismic modeling was used to predict a recurrence interval of about 53 years for a tsunami capable of causing substantial damage, and of 140 years for a tsunami causing catastrophic destruction. The most recent earthquake of this magnitude was in the Pisco area on August 15, 2007 (M_w 8.4), while the previous one of that magnitude was in 1868, 139 years earlier.

Various researchers have used the spatiality of seismic events in Peru to produce hazard maps (Dorbath et al., 1990). Sharma and Candia-Gallegos (1992) utilized the history of earthquakes over 90 years to subdivide the country into five zones with boundaries accurate to about 20 km plus or minus. Their analysis combines and mixes attributes of recurrence, attenuation, magnitude, and source area, so it cannot be used for specific local planning, but it does help with organizing national efforts. Degg and Chester (2005) provided an updated analysis and summary, also putting the social and institutional responses in Peru into context. The map of Degg and colleagues (Degg et al., 1999) was created using a grid of 40 x 40 km and offers much more useful resolution of seismic hazards.

Tectonics constitutes an important background for understanding geomorphic and geological processes in other contexts and places. Dumont (1993) pointed out that some of the lakes in the Peruvian Amazon owe their shape and location not to current lentic or lotic processes, but instead to neotectonics and subsidence. Hanuš et al. (2000) show that mining districts in the Andes are located in seismically active fracture zones, where the ores have accumulated over the millennia. The four time epochs from the Miocene to the Eocene that coincide with metallogenic activity also can be associated with four time periods of especially active plate subduction, thus connecting macrotectonic processes to local economic outcomes.

2.1.2. Mass Movements (Landslides, Avalanches)

Landslides occur on steep slopes. Most of them are caused or triggered by other natural disasters, such as earthquakes and heavy rainfalls (e.g., Keefer and Moseley, 2004). Large landslides linked to faults have been documented in several places in central and southern Peru. For example, near the Patacancha-Tamburco fault in the central highlands (Carlotto et al., 2006), on February 18, 1997, in Tamburco, Apurimac, 220 people died due to the collapse of a slope (Fig. 9.1).

Many mass movements are triggered by earthquakes (Keefer, 2002). However, the Mantaro landslide of April 25, 1974, one of the largest landslides known in the world, occurred without seismic activity (Berrocal et al., 1978). Instead, it was caused by a sudden loss of cohesion of unconsolidated sediments and soil, probably caused by a rise in the water table. About 400 people died, a village was destroyed, and perhaps 1300 x 10 m^3 of rock and soil was moved in less than four minutes. The natural dam formed in the Mantaro River was 3.8 km long, 2.5 km wide, and eventually backed up water 32 km. Forty-two days after the landslide, the Mantaro River overtopped the dam and began to flow once more.

Over time, mass movements and erosion create and maintain slope profiles. Seismically active regions release a predictably large amount of colluvium due to landslides triggered by earthquakes (Keefer, 1994). Dykes and Welford (2007) proposed that long-term landform development in wet tropical mountains is driven by (1) chemical incision, (2) uplift, and (3) formation of steep lower slopes that exceed bedrock failure thresholds, creating large landslides and the characteristic steep topography. However, by far the most numerous landslides observed in that site in a humid mountain area were due to road cuts, as is characteristic in tropical montane zones (Young, 1994).

A debris flow on April 10, 2004 killed 11 people and destroyed a portion of the town of Aguas Calientes (Fig. 9.1), downhill from Machu Picchu in a very humid area on the eastern slopes of the southern Peruvian Andes. Although caused by steep topography and high rainfall, the fact is that these events are on the increase (see Carreño and Kalafatovich, 2006), which in turn suggests that land use and landscape degradation by people are also implicated as causal factors. The town of Aguas Calientes has tripled its population in 25 years, putting more people in high-risk zones. In addition, mitigation efforts by local people and government authorities that consisted of rechanneling part of the Alcamayo River appear to be in vain, given that debris flows would not be stopped. Brooks et al. (2005) described a 2.5-km stone wall built by local people before European contact to contain debris flows in northern Peru. Concerns that slope instability would threaten the archaeological site of Machu Picchu itself were allayed by research carried out by Vilímek et al. (2007), although they stress that people living in Aguas Calientes or other places that are on debris fans remain in danger during and after large rainfall events.

Many people in Peru also live downslope from moraine-dammed proglacial lakes. Hubbard et al. (2005) studied a large rock avalanche that had crashed into Safuna Alta Lake in Ancash, producing a wave more than 100 m in height, which then overtopped the moraine dam. Enough remained of the moraine so that the dam did not fail and the lake's water was not catastrophically released. These authors show that the safety margin for producing a large rotational failure of this natural dam was almost exceeded, and it would have failed under slightly more dramatic conditions. Ongoing warming in the high Andes due to global environmental change is expected to increase risks of glacial-lake outburst floods (Vilímek et al., 2005a), especially given the high permeability of moraine material and the additional possibility of increased likelihood of further rock falls, icefalls, or landslides (Vilímek et al., 2005b). Carey (2005) states that about 30,000 people have died in Peru in the last 70 years due to disasters associated with glaciers. He believes that a disjunction between what scientists and government officials say and what local people believe will further increase the vulnerability of people living nearby or among high Andean peaks, who must cope in some manner (Fig. 9.3).

Figure 9.3 Local man examining a massive landslide at 2800 m, which had formed a landslide dam in the river below on the eastern Andean slopes in San Martin.

Over long time periods, changes in climate as mediated through ENSO (El Niño-Southern Oscillation) and ocean temperatures will in turn cause changes in the frequency and geomorphic importance of mass movements (Trauth et al., 2000). On time spans much closer to direct human perceptions, mass movements affect road construction and maintenance costs (Plessis-Fraissard, 2007). The loose materials built into tailings near most mines in the Andes are also prone to instability and the risk of failure (Garga and de la Torre, 2002).

2.2. Volcanism

Volcanoes in Peru are clustered in the southwestern part of the country in the central Andes, comprising a complex of active and dormant volcanoes dating back from the Pleistocene/Neogene boundary (Thouret et al., 2001, 2005). Degg and colleagues (Degg et al., 1999; Degg and Chester, 2005) have mapped those areas in Peru that are exposed to volcanic hazards.

The compound stratovolcano Misti sits adjacent to the city of Arequipa in southern Peru, putting potentially 800,000 people in danger (Fig. 9.1). Over the last 50,000 years, ash falls have occurred every 500 to 1500 years and pumice falls every 2000 to 4000 years. Thouret and colleagues (Thouret et al., 2001, 2005) also estimated that more than 200,000 people in the city are living in places susceptible to pyroclastic flows and lahars. An eruption 2050 to 2300 years ago sent pumice flying more than 25 km away, with cooled pyroclastic flow deposits now to be found under lahar-caused deposits. Ironically, the historic downtown of Arequipa is built of "sillar," blocks of rock made of volcanic ash. The 2001 earthquake particularly damaged these structures, knocking down, for example, the cathedral's towers (Kuroiwa, 2002).

In 1600, the explosive eruption of Huaynaputina volcano 75 km from Arequipa (Fig. 9.1) caused global climate change (de Silva and Zielinski, 1998), as seen in sulfites in dated Greenland ice cores, reduced tree ring widths in North America, dust in the sky over Europe, views of a reddish sun in China, and the coldest summer in 1601 in more than 600 years. The tephra falls and pyroclastic flows of that eruption covered an area 40 by 70 km, and tephra reached more than 300 km from the explosion (Thouret et al., 1999). Subsidence of the dome and eruption from 20 km deep down in the magma reservoir kept the volcano from forming a caldera (Lavallée et al., 2006).

Certain kinds of mineral deposits containing iron oxide, copper, and gold are associated with the volcanic arc of southern Peru and northern Chile (Sillitoe, 2003). However, most hazard concerns are focused on the disadvantages of being near the volcanoes, particularly for the inhabitants of Arequipa (Degg and Chester, 2005). Volcanism in southern Peru also threatens water resources, as occurred in April 2006, when the Ubinas volcano (Fig. 9.1), the most active volcano in southern Peru (Thouret et al., 2005), threatened Las Salinas Lake, a body of water that provides irrigation for the agriculture of that part of Peru. In terms of vulnerability, given the lack of access to economic resources and social capital, Degg and Chester (2005) remarked that women and the elderly are the most vulnerable groups. Local land-use planning is ineffective, and people often must pragmatically choose increased risk from natural hazards over other sacrifices in where and how they live.

2.3. Rain and Floods

Floods account for the third most important disaster between 2003 and 2007, according to the government agency INDECI, the Instituto Nacional de Defensa Civil (http://sinadeci.indeci.gob.pe/Sinpad/Estadistica/Frame_Esta_C2.asp), although floods were considered the most important natural hazard type between 1993 and 2002 (INDECI, 2004). Every year there are floods in Peru's Amazon lowlands. Unless they are extreme in height or duration, their effects on daily life and livelihoods are not drastic because local people expect them. People in those areas often have land-use strategies that accommodate the risks and benefits associated with those floods (Chibnik, 1994).

However, much of Peru is subject to much less predictable ENSO variations, causing years of floods and droughts to occur in the 3- to 8-year cycle of ENSO, and even more irregularly when large El Niño events occur, as they did in 1983, 1992, and 1998. For example, northwestern Peru's climate is arid to semiarid, but during a strong El Niño year that area receives several times its typical annual rainfall. There are flash floods in the desert and associated damage to houses and infrastructure. An extended land–sea breeze system brings Pacific Ocean moisture over land, while trade winds carry in additional moisture from the Amazon (Bendix, 2000). Another consequence can be increased sediments and contamination in rivers with headwaters affected by mines (Tarras-Wahlberg and Lane, 2003). Romero et al. (2007) has shown that rainfall associated with El Niño varies along an altitudinal gradient in a Pacific River basin in Cajamarca in northern Peru.

In January 1998 Ica, a city on the hyperarid coast of central Peru, was surprisingly hit by floods caused by El Niño (Warner and Oré, 2006) (Fig. 9.1). There were 70 people left dead, and 25,000 houses were damaged or destroyed. As is typical with ENSO, places normally dry are wetter than normal, while the reverse ("oscillation") is the case elsewhere. There was a severe drought in Ayacucho in the same month on the other side of the Andes, creating a water crisis for Andean urban populations, especially in the rapidly growing city of Huamanga.

Many surface features of the dry coastal areas of Peru result from fluvial events that occur infrequently. A rainfall event in February 2001 in southern Peru was particularly unexpected, given that it was not a strong El Niño year. Many ephemeral streams began to flow, with overbank flooding and mudflows (Houston, 2006). Less surprising to local people were the rain and floods in early 1998 that came during an El Niño year. Modeling by Magilligan and Goldstein (2001) showed this event to have been a 50- to 100-year flood in terms of recurrence intervals. In the years following the flood, the channel narrows and lateral bars develop, processes that Manners et al. (2007) said have occurred during many strong ENSO events of the Holocene. These authors believe that floods such as these would alter floodplain agriculture and even cultural development.

Longer term records put the severe 1982–1983 and 1997–1998 El Niño years into perspective, in these cases based on marine sediment cores. Rein (2007) showed that coastal floods and the resulting sediment deposition in marine environments for those El Niño years were the largest recorded in 1000 years. However, floods of these magnitudes occurred more frequently 2000 to 4000 years ago, while the largest floods recorded occurred in the early Holocene. Using the same techniques, Rein et al. (2004) localized a drought period with no strong flooding during the 800 to 1250 A.D. period. Rein et al. (2005) showed that ENSO first became strong 17,000 years ago, with strong El Niño events occurring every 60 to

80 years. Many streams in the Atacama desert were last perennial in the late Pleistocene (Nester et al., 2007).

During the El Niño event of 1997–1998, drought conditions affected the Amazon Basin, causing tree mortality, although the degree of influence was quite variable spatially across the basin (e.g., Laurance et al., 2001). However, drought appears to have recently become a much more unpredictable event in the wetter, Amazonian side of Peru. A severe drought not associated with El Niño affected part of the Department of San Martin in February 2007 during the peak of the rainy season (*El Comercio*, February 19, 2007).

2.4. Other Hazardous Events

The current Peruvian governmental agency for natural hazards, INDECI, recognizes several additional disaster types. Six are anthropogenic in nature, such as social danger (terrorism, delinquency), and others are broadly environmental: global disasters/global warming, ozone layer reduction, deforestation, fire, pollution, chemical spills and dangerous substances.

Forest fires are caused mostly by people during seasonally dry weather. These fires affect both human-built and natural environments. According to INDECI records, fewer than 1% of fires are caused by electric storms, and these have been recorded only for high Andean sites. Amazonian forest fires are usually linked to the clearing or expansion of fields for crop agriculture for which burning is commonly used.

Among climate-related hazards, INDECI (2004) recognizes at least four other disasters types related to climatic conditions. Tornado-like winds (which INDECI calls *vendavales*) occurring between 2003 and 2007 were listed as the country's most important natural disaster in terms of number of events, with over 2974 cases registered. Along the Peruvian coast the presence of winter winds called *paracas* are associated with sand blowing that disrupts transportation and causes some structural damage in Ica and Lima (Fig. 9.1). In the Peruvian Amazon, strong winds occur not only with moving rain fronts but also as a result of air masses coming from the south and east (Foster and Terborgh, 1998). Other events listed by INDECI related to climate are damaging precipitation such as snow or hail; hail is a frequent hazard to crops in the high Andes.

In addition, high Andean sites are exposed to very low temperatures when air masses come from the south in the Southern Hemisphere winter. In the last four years, the sudden occurrence of low temperatures (called *heladas*) is the fourth most important natural disaster recognized by INDECI (http://sinadeci.indeci.gob.pe/Sinpad/Estadistica/Frame_Esta_C2.asp). Changes related to climate change, such as a tendency for drier conditions, or glacier retreat (Vuille et al., 2003) are recognized by the Peruvian strategic plan (INDECI, 2004), but mostly in relation to their impact on coastal urban areas, although impacts are also important in the highlands (Young and Lipton, 2006).

In general, the amount and type of sediment transport are thought to affect landforms and to result in a particular topography given enough time (Schneider et al., 2007). The recent availability of digital elevation models (DEMs) permits the more efficient measurement of morphometric properties and the testing of landscape evolution models. High channelized and hillslope sediment transport and erosion should lead to higher relief mountains and numerous branching valleys. These are typical of many places in the northern and eastern Andes with humid climates.

Human impacts alter water flows, triggering increased mass movements and increasing soil erosion through land-use practices (Inbar and Llerena, 2000; Harden, 2006; Fig. 9.4). Other impacts include dams and mining (Latrubesse et al., 2005). Commonly overlooked in rural development schemes is the hydropower potential available in mountains with high relief (Madueño Luján, 2003).

2.5. Living with Danger

We personally experienced the August 15, 2007 earthquake (M_w 8.4) on the fourth floor of an apartment complex in Lima. First there was an eerie silence, and then the entire building shook and swung for over 3 minutes. We also witnessed the nearly complete lack of coordination and communication among authorities in response to infrastructure loss and tsunami warnings, which were bungled for Lima. Many lives, more than 500, were lost in the Pisco area and in adjacent Andean sites (Fig. 9.1). Little appeared to have been learned regarding either mechanisms to alert local people or plans to coordinate rescue efforts immediately after the earthquake.

The August 15, 2007 earthquake brought before the public in Peru a series of fresh debates regarding organizational structures and political responses to natural disasters. Effective leadership in times of crisis has been discussed in terms of

Figure 9.4 Traditional agricultural fields on steep, eroding slopes in Ancash.

narcissistic leaders, who either benefit from or block needed responses during and after disasters (King, 2007). Unfortunately, this appears to have been the case during the recent events, as was also the case during the 1998 El Niño floods in central Peru. The El Niño floods were discussed by Warner and Oré (2006), who pointed out that despite a strong local participatory system there was a lack of continuity and appropriate responses at the regional and national levels.

More than half of Peru's population is poor (see the INEI: Instituto Nacional de Estadística e Informática. Website: http://www.inei.gob.pe, under "Información Socio demográfica"), which means that they usually occupy buildings that are not properly constructed or reinforced. The seismic events of August 15, 2007 also generated a discussion about the need for an assessment of proper housing materials. The proposal to reinforce adobe housing (Giesecke, 1996) appears to be sound, despite the public statements often made that more expensive materials should be used. In addition, profound but mostly untreated mental health issues arose from the quake and the sad aftermath (Fraser, 2007).

Peru's present-day population is mostly urban. About 57% of the population lives along the coastal desert region, where the largest cities are located, including the capital Lima. The coastal region in Peru is probably the area most prone to hazards related to earthquakes, tsunamis, volcanoes, landslides, and floods. The official institution in charge of coordination, planning, and control of disasters is INDECI, although in practice the agency is often overwhelmed or overlooked. INDECI's current prevention plan dates from August 2004 and makes a distinction between natural and human-made disasters.

Historical perspectives, including geological and geomorphic studies, help us understand long-term effects and to put floods and volcanic eruptions into context. Keefer and Moseley (2004) have shown how a recent colossally seismic event in southern Peru, combined with the occurrence of an El Niño event, could have happened in the seventeenth century. Past differences in the frequency of El Niño can affect cultural development (e.g., Sandweiss et al., 2001). The importance of a historical perspective is also demonstrated with the case of climate change where the lack of accurately recorded data can limit understanding of causalities (Vuille et al., 2003). But history by itself is not enough: despite the fact that data exist on the magnitude, trajectory, and characteristics of past volcanic events near Arequipa, there is still no urban planning in that and other growing cities (Thouret et al., 2001; Polk et al., 2005). Mapping risk exposure is incomplete without a broader understanding of human vulnerabilities (Alcántara-Ayala, 2002).

As people continue to move from rural to urban areas (Paerregaard, 1997), some kinds of risks and vulnerabilities increase. Growing Andean cities push up onto steep slopes (Fig. 9.5), typically placing the poorest people in zones prone to mass movements. In Peru, the largest cities are on the coastal plains, with 8 million people in Lima and over another million in Arequipa and Trujillo. Ironically, some of the most seriously threatened real estate of Lima (Degg and Chester, 2005) is where relatively wealthy inhabitants live, often in apartment buildings perched on the margin of the sea cliffs (Fig. 9.6) overlooking the

Figure 9.5 Cusco, one of the largest cities in the Peruvian Andes, where urban expansion covers adjacent hillslopes.

Figure 9.6 A tall building sits on top of Quaternary colluvium along the coast of the city of Lima.

ocean. A third of the population of Lima is poor, with no running water, high exposure to environmental contamination, and few economic alternatives. There are also social disjunctions, separating poor from rich. More than that, the goals and risk management orientations of rural and urban dwellers differ profoundly from the pronunciations of disaster experts or spokespersons (Carey, 2005; McEntire and Fuller, 2002). More knowledge is important, but also crucial is better understanding.

CHAPTER 10

Geomorphology of Natural Hazards and Human-induced Disasters in Bolivia

Edgardo M. Latrubesse[a], Paul A. Baker[b], *and* Jaime Argollo[c]

[a]*Centro de Investigaciones Geológicas-CIG and LATYR- CONICET, Universidad Nacional de La Plata, Calle 1 N^0 644, (1900) La Plata, Argentina*
[b]*Duke University, Division of Earth and Ocean Sciences, Durham, NC 27708, USA*
[c]*Universidad Mayor de San Andrés, Instituto de Geología y Medio Ambiente, La Paz, Bolivia*

Contents

1. Introduction

Bolivia is a geographically and culturally diverse nation with an area of 1,098,580 km^2 and a population of approximately 9.5 million. Approximately 63% of the population is urban, and 69% of this urban population resides in three major cities: La Paz (including El Alto), Cochabamba, and Santa Cruz. The country is divided into seven main geologic-physiographic provinces, from west to east: the Cordillera Occidental of the Andes (Western Cordillera), the Altiplano, the Cordillera Oriental of the Andes (Eastern Cordillera), the Subandean belt, the Chaco-Beni plain, the Pando block, and the Brazilian Shield (Fig. 10.1).

Corresponding author: *Edgardo M. Latrubesse, Centro de Investigaciones Geológicas-CIG and LATYR- CONICET, Universidad Nacional de La Plata, Calle 1 N^0 644, (1900) La Plata, Argentina, latrubesse@yahoo.com.br*

Developments in Earth Surface Processes, Volume 13

ISSN 0928-2025, DOI 10.1016/S0928-2025(08)10010-4

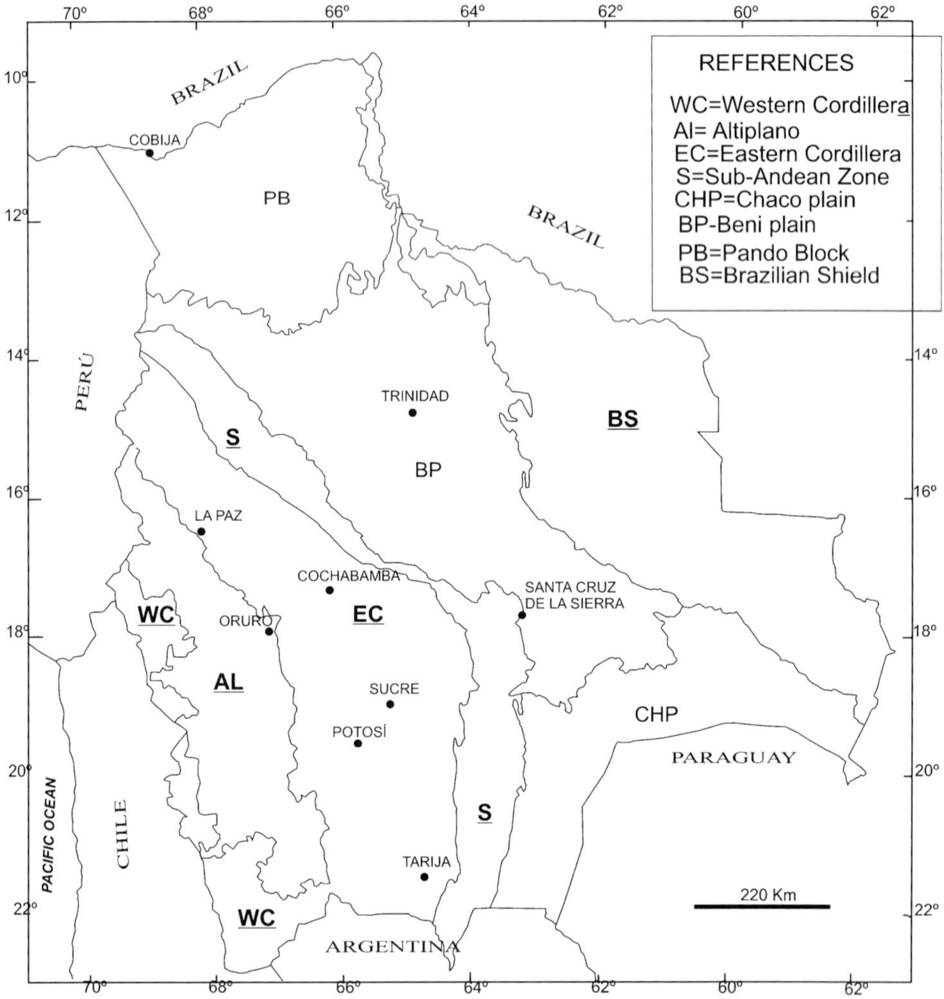

Figure 10.1 Main morphotectonic units of Bolivia and cities.

The Cordillera Occidental is the active volcanic arc formed by the eastward subduction of the Nazca seafloor plate underneath the South American continent. Thus much of this mountain chain is composed of Neogene-age magmatic and volcanic rocks. The range contains at least 18 "potentially active" volcanoes along the frontier between Chile and Bolivia (de Silva and Francis, 1991). The highest mountain in the Cordillera Occidental of Bolivia is Nevado Sajama (it is also the highest mountain in Bolivia, 6542 m above sea level, m.a.s.l.). It is not categorized as "potentially active," and its 25,000-year-old icecap (Thompson et al., 1998) suggests that it has not had a major eruption for at least the past 25,000 years. Sajama and its neighboring volcanic peaks, Nevados Payachata and Quimsachata, are (as of November 1984) the only Bolivian peaks in this range covered by active glaciers (Jordan, 1999).

The Altiplano is a wide endorheic basin with an area of about 110,000 km^2. It is elongated in a roughly north-south direction, parallel to the bounding ranges of the Cordilleras Occidental and Oriental. The average elevation of the Altiplano is about 3700 m.a.s.l. Despite its great elevation, for the most part the Altiplano is a relatively flat landscape with a surficial cover of Quaternary lacustrine and alluvial deposits in lower elevations between 3653 and 4000 m.a.s.l. The plain is interrupted by isolated ranges and hills with elevations from 4000 to 5300 m.a.s.l. The Altiplano is characterized by the presence of large salt pans or *salares* such as Salar de Uyuni and Salar de Coipasa and large lakes such as the salty, shallow Lago Poopo and deep, fresh Lago Titicaca. Salar de Uyuni is one of the flattest surfaces on Earth, Earth's largest salt flat (ca. 10,000 km^2), the site of several generations of large paleolakes (e.g., Fritz et al., 2004), and at 3653 m.a.sl. is the lowest point on the Bolivian Altiplano. The Altiplano was probably formed (meaning that it became a topographically separate sedimentary basin) in the late Eocene (e.g., McQuarrie et al., 2005) and was rapidly uplifted from near sea level around 11 million years ago (Ma) to near its present elevation by 6.7 Ma (Ghosh et al., 2006). The important cities of La Paz/El Alto and Oruro are located on the Altiplano. Mining, government, tourism, and commerce are all important sectors of the economy; small-scale traditional agriculture remains a major livelihood and economic resource.

The Cordillera Oriental has a complex geological origin and character. It comprises part of the Andean fold and thrust belt and contains a variety of plutonic, volcanic, metamorphic, and sedimentary rocks with ages ranging from Proterozoic to Quaternary. High peaks of the Cordillera Oriental include Illampu (6421 m.a.s.l.), Illimani (6402 m.a.s.l.), and Huayna Potosi (6088 m.a.s.l.), among others. As of 1984, 16 peaks in the Cordillera Oriental were glaciated, with a total glaciated area of about 550 km^2(Jordan, 1999). All of these glaciers are retreating rapidly (Francou et al., 2005). As 30 to 40% of the potable water of La Paz/El Alto is derived from glacial meltwater of the Cordillera Real (the highest range in the Cordillera Oriental), the retreat and eventual loss of the glaciers will represent a large economic loss (Vergara et al., 2007). The cities of Potosi, Sucre, and Cochabamba are located in this range.

An east-west elevational cross section of the Cordillera Oriental is asymmetric: the western flank with the Altiplano has local relief up to about 2500 m, while the eastern flank drops abruptly more than 5000 m from the highest peaks to the eastern lowlands. The Subandean zone is the easternmost, active frontal fold-and-thrust complex located between the Cordillera Oriental and the eastern lowlands. The majority of the drainage systems flowing to the Amazon Basin (comprising 720,000 km^2) or to the La Plata Basin (comprising 229,500 km^2) have their headwaters in the Cordillera Oriental and Subandean zones. The most important Bolivian tributaries to the Amazon include the Mamore, Madre de Dios, and Beni, Itenez. The main Bolivian river draining to the La Plata Basin is the Rio Pilcomayo.

The Chaco-Beni plain lies east of the Subandean zone. This region is part of the Andean foreland basin: its sediment cover overlies the Precambrian Brazil Shield. The plain covers about half the area of Bolivia, and its drainage is either northward toward Amazonia or southward toward the La Plata Basin. This region contains Bolivia's most productive agricultural and pastoral lands as well as large natural gas reserves. The cities of Santa Cruz and Trinidad are located in this zone.

Bolivia lies entirely inside the tropical region; thus all of the country is characterized by a wet season ranging from October to April and a drier season from May to September. Most of the wet-season precipitation derives from the South American summer monsoon (Zhao and Lao, 1998). In the north the Altiplano is wetter and warmer (500 to 1000 mm/y; mean annual temperature ~12° C), becoming progressively more arid and colder to the south (<100 mm/y; mean annual temperature ~2° C). The eastern lowlands also have a north-south climatic gradient with semiarid to arid conditions (~350 mm/y) prevailing in the Chaco close to the Bolivia–Paraguay border and much wetter conditions (>2000 mm/y) northward in the Amazon region.

The diversity of the topographic relief in Bolivia is a factor in the diversity of its regional climatic conditions. For example, local relief produces orographic lifting in the Andean and Subandean eastern flank, where the Yungas forests occur; local convective rains in the Altiplano and Chaco; and rain shadow effects in inter-Andean valleys and on the Altiplano. Steep slopes also produce significant local variations of solar exposure, a factor controlling locations of glaciers and arable lands. The eastern flank of the Cordillera Oriental of Bolivia is characterized by temperatures that decrease with increasing elevation (lapse rate about 5°C/km) and precipitation that peaks between about 1200 and 1800 m.a.s.l. Summertime rain rates are highest (>5000 mm/y) over the Chapare region in the foothills of the Cordillera Oriental (Killeen et al., 2007).

Cold air incursions, called *surazos* in Bolivia and Peru and *friagems* in Brazil, begin as cold-core anticyclones in the South Pacific (Garreaud, 1999). These cold surges penetrate northward from the Argentinean plains to the Chaco and Beni plains as far as the Amazonian region. They can produce abrupt drops of temperature of as much as 10°C (Ronchail, 1999). *Surazos* can also reach up onto the Altiplano and the Andes. In the Cordillera Occidental, the arrival of *surazos* is accompanied by winter snow in the higher peaks such as Nevado Sajama.

The largest source of interannual variation of precipitation in much of tropical South America is related to the El-Niño-Southern Oscillation (ENSO). In general, El Niño events are registered as dry years on the Altiplano and in the Cordillera Oriental (e.g., Garreaud and Aceituno, 2001), but some El Niño events bring wetter-than-normal conditions south of the Beni plain toward the Chaco. La Niña events, on the other hand, tend to produce wetter-than-normal conditions across much of Bolivia. There is some evidence that regional precipitation amounts also vary on longer (decadal to multidecadal) timescales (e.g., Hastenrath et al., 2004) and that this variation is partly controlled by sea-surface temperatures in the tropical Atlantic (e.g., Liu, 2008).

2. Volcanoes

Modern volcanic activity in Bolivia is limited to relatively isolated regions of the Cordillera Occidental. Large explosive eruptions have been recorded in the Pliocene and Pleistocene, forming pyroclastic and lava plateaus in the Cordillera Occidental. The most recent volcanic activity seems to belong to Volcan Parinacota

(18° 10′S and 69° 09′W), which was active ~8000 B.P. (Clavero et al., 2002). Da Silva and Francis (1991) judged its hazard to be mainly associated with mudflows and debris flows that could threaten the nearby Arica-La Paz Highway and some small communities located nearby. Considering the low volcanic activity in recent times and their mostly isolated locations in areas with very low population density, volcanic hazards and vulnerabilities on the Bolivian side of the international border are considered to be quite low.

3. Earthquakes

Because of the eastward-increasing depth of the Nazca lithospheric plate as it subducts under the Pacific coast of South America, earthquakes in Bolivia do not often attain magnitudes greater than 6 on the Richter scale. The predicted values of peak ground accelerations (a key indicator of susceptibility of buildings to damaging during earthquake ground movement) due to seismicity also decrease eastward. Maps and descriptions of historical seismic activity in the country are available online through the San Calixto Observatory website (http://www.observatorio-sancalixto.org/). Here we summarize from that website some of the most significant earthquakes to have impacted Bolivia in historic times.

Most of the historically damaging earthquakes in Bolivia have occurred in the Cordillera Oriental. In 1650 an earthquake destroyed the dome of the cathedral in Chiquisaca. A quake damaged adobe buildings in Santa Cruz in 1871, in San Antonio (now Villa Tunari) in 1871, and in Yacuiba in 1887 and 1899 (also causing injuries). In 1909 an earthquake in Sipe Sipe destroyed several adobe buildings and caused 15 deaths. In 1925, 1958, 1976, and 1998, earthquakes occurred in the region of Aiquile. The last of these, on May 22, 1998, was also the most destructive. This magnitude 6.6 earthquake was at that time the largest shallow-focus temblor in Bolivia in the past 50 years (Funning et al., 2005), and it caused damage over a region of about 100 km diameter, including the villages of Totora, Aiquile, and Mizque. More than 100 people were killed. In Cochabamba city, earthquakes in 1942, 1943, 1959, and 1972 caused some damage to buildings. A strong earthquake destroyed about half of the capital city of Sucre on March 27, 1948, killing 3 persons and injuring several more. In 1957 an earthquake in Postrervalle (southwest of Santa Cruz) destroyed several adobe buildings. In 1947 an earthquake with magnitude of 6.4 caused widespread damage in Consata (La Paz Department). A second, weaker quake was recorded in Consata in 1956.

On June 9, 1994, the largest deep-focus earthquake on Earth occurred 636 km beneath the Beni lowlands near Rurrenabaque. It is believed that the earthquake rupture occurred within the downgoing Nazca plate, deep under the South American continental lithosphere (Silver et al., 1995). The earthquake was felt in almost the entire country. In Cobija it caused cracks in walls and made tall buildings sway in La Paz. The earthquake is thought to be responsible for five deaths in the mountains of Peru and for minor damage to many buildings in Brazil and Peru. The quake was felt in many parts of South America and North America, as distant as

Toronto, Canada (reported in "Significant Earthquakes of the World, 1994," U.S. Geological Survey).

Because of its sparser population and smaller area, there has been less economic and human loss due to seismic activity in the Cordillera Occidental of Bolivia. In 1995 a magnitude 5.3 temblor destroyed most of the adobe buildings of the town of Cumujo (just northwest of Salar de Coipasa). A second quake of magnitude 4.6 revisited the region in 2001 and caused minor damage to adobe buildings in the town of Coipasa. A much stronger earthquake of magnitude 6.9 occurred in Potosi Department along the frontier between Bolivia and Chile on November 17, 2005. The quake caused power outages in Tocopilla, Chile.

Minaya and co-workers (2005) considered possible damage to the city of La Paz from large earthquakes with epicenters in southern Peru or northern Chile. They concluded that the probability of an earthquake of magnitude 9 (Mw) in those parts of Peru or Chile was high and that such a remote earthquake might result in an intensity of V on the Modified Mercalli scale (felt by nearly everyone), but that there was no chance that such earthquakes could generate intensities of VIII or higher (causing damage even to well-built buildings) in La Paz or elsewhere in Bolivia.

4. Floods and Droughts in Bolivia

In economic terms, floods and droughts are the most costly natural disasters in Bolivia. Floods are induced by anomalously high precipitation and runoff, whereas droughts are caused by anomalously low rates of precipitation. Of course, the hydrologic and economic impact of both flood and drought can be exacerbated by direct human activities such as population growth in affected regions, deforestation, the covering of regions of natural recharge by construction of impermeable barriers (roads, buildings, and parking lots), and the construction of housing developments in regions prone to landslides.

In Bolivia, the interannual variability of precipitation, beyond the range of normal wet-season/dry-season variability, is often ascribed to ENSO variability. Indeed, during El Niño events of the past few decades (Fig. 10.2), the Altiplano and Andes of Bolivia, as well as western parts of the lowlands, are often drier than normal, especially during the DJF season (Fig. 10.3, 1950–2002). Farther east, precipitation during El Niño events often is above normal. However, during past La Niña events, DJF precipitation throughout Bolivia, on average, has not been significantly different than normal (Fig. 10.4, 1950–2002).

Whereas anomalous precipitation in Bolivia can sometimes be associated with ENSO conditions, ENSO is clearly not the only cause of anomalous precipitation. For example, in both of the past two years (DJF, 2006–2007 and DJF, 2007–2008) there has been anomalously high precipitation (Figs. 10.5 and 10.6) and devastating flooding in much of lowland Bolivia. Newspaper accounts (Keene, 2007; Claure, 2008) quote meteorologists attributing both cases to El Niño conditions. In fact, as described previously, El Niños are most often associated with *dry* conditions in

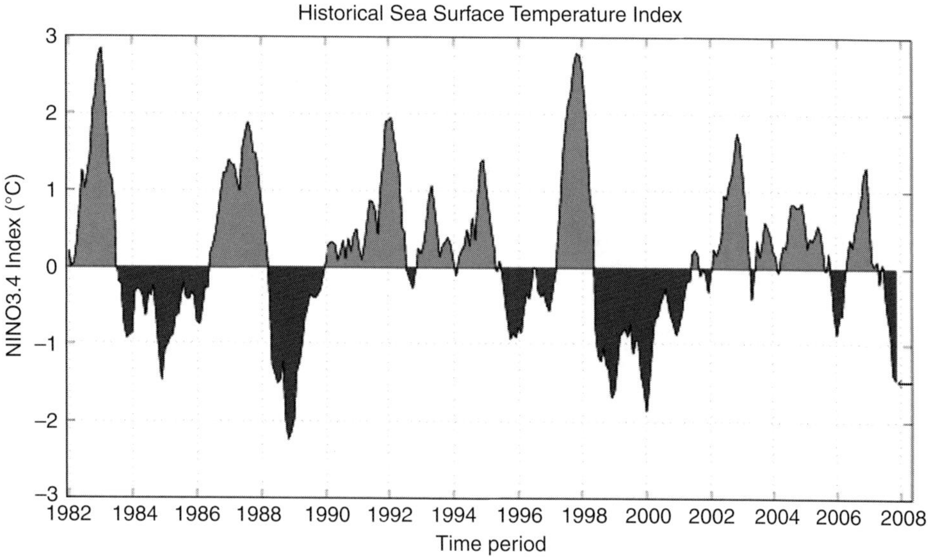

Figure 10.2 Niño 3.4-based index of El Niño (Red) and La Niña (Blue). From http://iri.columbia.edu/climate/ENSO/currentinfo/archive/200712/ENSO.Quick.Look

Bolivia. Moreover, although DJF, 2006–2007, was indeed a weak El Niño (Fig. 10.2), the period of DJF, 2007–2008, was actually a weak La Niña. Thus, either phase of ENSO *can* be associated with flooding conditions in lowland Bolivia; in fact, it seems likely that ENSO was not responsible for flooding in either year.

No matter the distal cause of anomalously high or low precipitation, the result can be catastrophic. The 1982–1983 El Niño is a case in point. During that event, floods affected 700,000 people in lowland Bolivia, and drought affected 1,600,000 in the Bolivian highlands. Economic losses totaled US$ 837 million, and 40 deaths were reported. Drought affected the Altiplano and Andes, with low discharges recorded in the Desaguadero, Iauca, Mauri, and Marquez rivers, as well as some minor rivers, such as Yanapollera, which provides water to the town of Uyuni. In Potosi rainfall during 1983 was half of the average, producing the worst drought in 30 years (at that time). Water supply for the city was insufficient. In the lowlands, strong rainfall, 350 mm, fell on Santa Cruz Department in March, nearly three times the monthly average. On March 17 the total precipitation recorded was 120 mm. The heavy rains caused extensive flooding of more than 150,000 km^2, strong erosion in the upland areas, and sediment accumulation in valleys and low-lying regions affecting the public works, housing, and basic services. It is estimated that 9500 houses were affected by floods in Santa Cruz Department and 5000 more in rural areas. The floods of the Piraí River produced damage estimated at US$ 37 million, and 900 people were reported missing (CEPAL, 1984).

Between 1990 and 1992, nearly 2 million people were affected in Bolivia by floods and droughts. Flooding during 1991–1992 affected more than 40,000

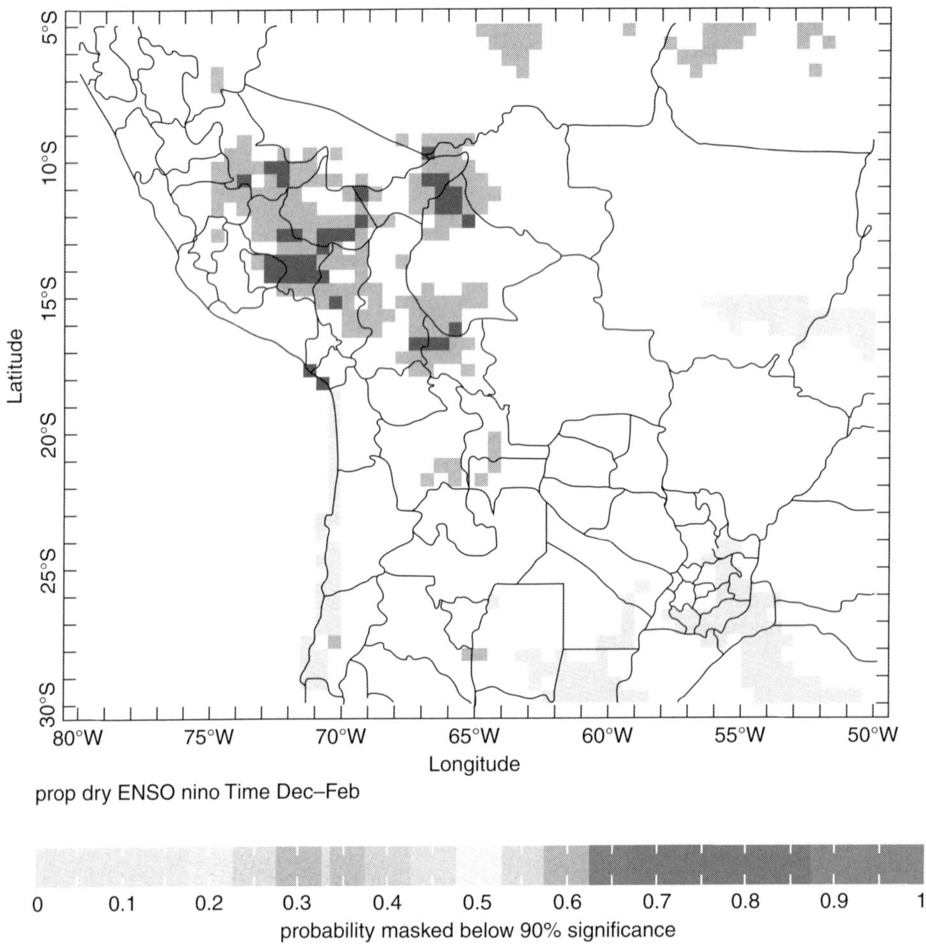

Figure 10.3 DJF precipitation in much of Bolivia is significantly lower than normal during El Niño events. http://iridl.ldeo.columbia.edu/SOURCES/.IRI/.Analyses/.ENSO-RP/.ver1950-2002/.0p5deg

inhabitants from 160 communities in northeastern Bolivia, causing agricultural and ranching losses greater than US$ 16 million.

The very strong El Niño of 1997–1998 had significant impacts in Bolivia. Drought affected much of the Altiplano and the south of the country, and flooding was widespread in the eastern lowlands. Approximately 77,000 inhabitants were affected by flooding in eight of the nine Bolivian departments (the only exception was Chuquisaca) (PAHO, 2000). The incidence of malaria increased as a result of the flooding. In the highlands, dry conditions contributed to a 5-m recession of the Chacaltaya glacier. In total, the results of flood and drought caused economic losses estimated at US$ 527 million (53% due to drought, 47% to flood).

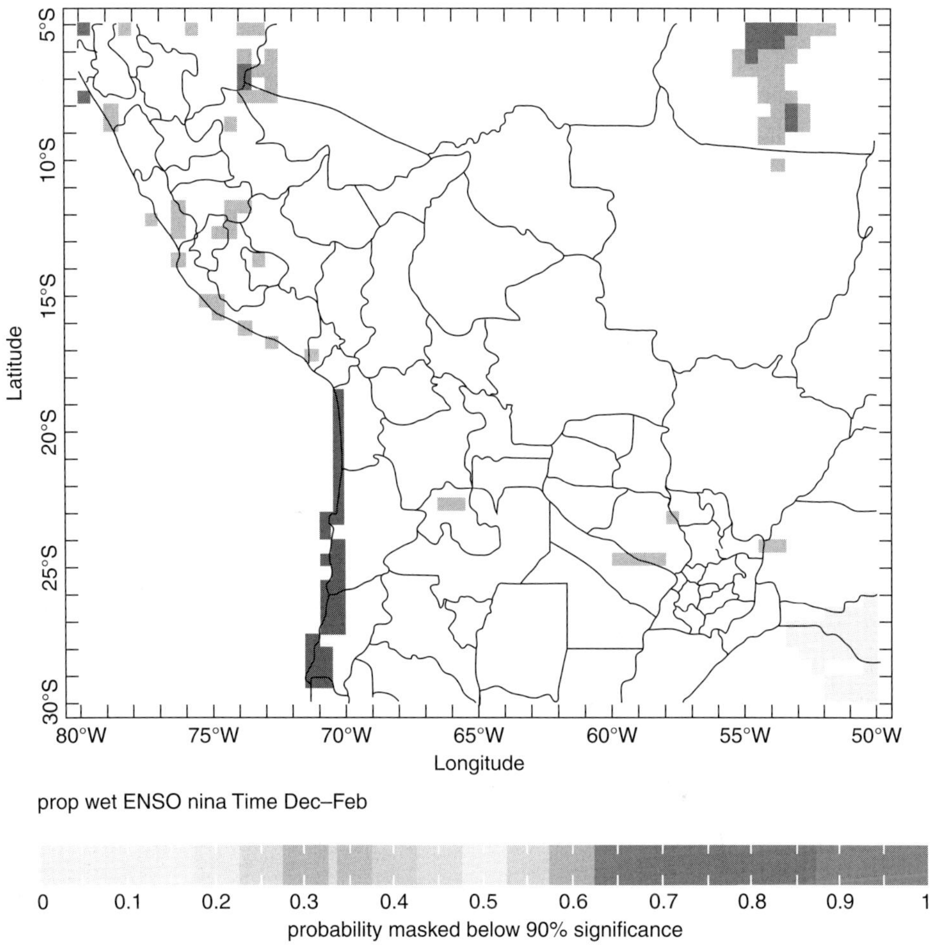

Figure 10.4. DJF precipitation in Bolivia is not significantly greater than normal during La Niña events. http://iridl.ldeo.columbia.edu/SOURCES/.IRI/.Analyses/.ENSO-RP/.ver1950-2002/.0p5deg/

The floods of 2006–2007 were dramatic. An intense rainy season started in January 2006 and continued for about 112 days. An area of about 800,000 km^2 was flooded. Flooding was most severe along the Mamore River. The Beni capital city of Trinidad was inundated, crops were destroyed, and 22,000 cattle drowned in the Beni. According to the Dartmouth Flood Observatory (http://www.dartmouth.edu/~floods/index.html), 63,000 persons were displaced from their homes and 41 deaths were reported. The total damages throughout Bolivia were estimated at US$ 90 million. The flood was described as the worst flood of the past 30 years. One year later it was surpassed.

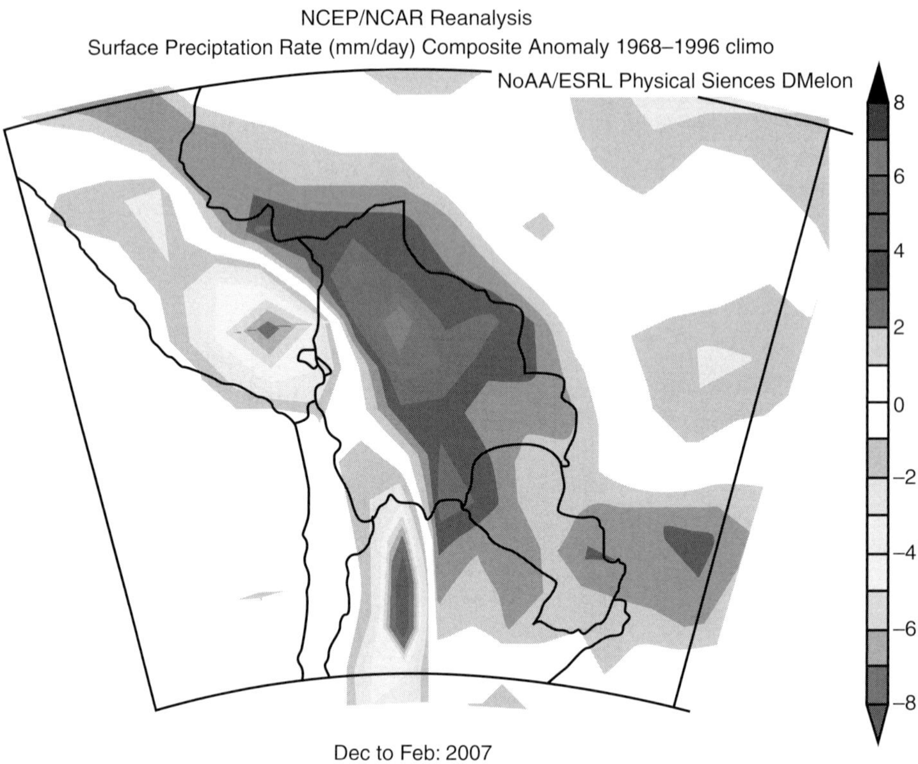

Figure 10.5 DJF 2007 was also unusually wet in the lowlands and dry in the highlands. http://www.cdc.noaa.gov/cgi-bin/Composites/printpage.pl

During the last month of 2007 and the first three months of 2008 (as of March 25, 2008), the eastern lowlands again underwent record flooding. Flooding extended over an area of about 870,000 km^2. Again, Trinidad was the city that suffered the most. For much of the region, it was the third consecutive year of serious flooding. An estimated 240,000 persons were displaced from their homes and 73 died. At this time, economic losses are estimated at US$ 200 million (Dartmouth Flood Observatory).

Water and sanitation systems were deleteriously affected by flooding (United Nations, 2008). The incidence of several diseases increased concurrently with the flooding: acute diarrheal diseases, acute respiratory diseases, dengue, leptospirosis, hemorrhagic fever, yellow fever, and malaria. Rates of malnutrition and psychological disorders also rose. Serious damage affected roads and bridges, schools, and houses. Food resources were seriously damaged with loss of many cattle and crops. The total impacts of this latest and largest flood have not yet been fully accessed at the time of this review.

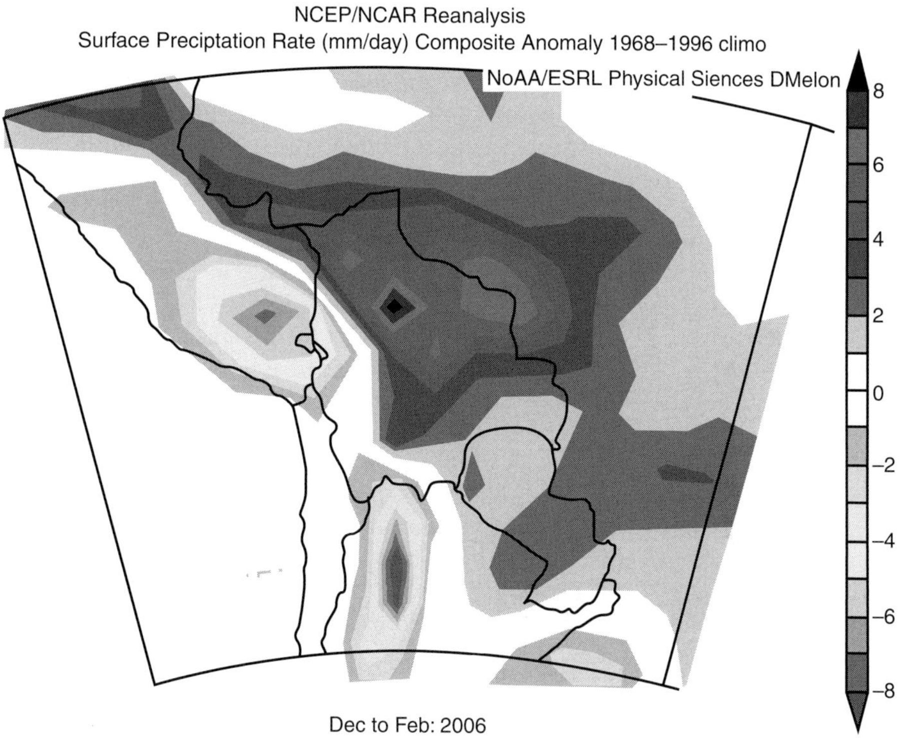

Figure 10.6 DJF 2008 was anomalously wet in the lowlands and dry in the highlands of Bolivia (mm/day).

5. Landslides

Throughout many parts of Bolivia, landslides cause a large loss of life and property. We describe two case studies, the landslide in Chima town and the landslides in La Paz city.

La Paz is the largest city of Bolivia with an estimated (2005) total population of 1,640,000 (840,000 for La Paz and 800,000 for El Alto; data from Instituto Nacional de Estadítistica, Bolivia). An estimated 60% of the urban population of La Paz proper is housed in self-built settlements (O'Hare and Rivas, 2005).

Landslides are a particularly important hazard in La Paz. The steep slopes and unconsolidated bedrock, the seasonality of precipitation and fluctuating water level, construction (often informal) on the unstable slopes, and deficient urban planning, all contribute to the city's vulnerability (O'Hare and Rivas, 2005). Mass movement is typically triggered by ground saturation following intense summer convective precipitation events. The unusual and spectacular geomorphological situation of the city is another contributory factor. The city was founded in 1548 on the site of an indigenous community named Chuquiago, along the steep valleys of the

Choqueyapa and Orkojahuira rivers. City growth spread up (to 4000 m.a.s.l.) and down (to 3200 m.a.s.l.) the La Paz valley, a sedimentary basin filled with unconsolidated, continental Cenozoic deposits. At the head of the valleys, the steep-sided walls of the Altiplano form huge natural amphitheatres that are now largely covered by informal settlements. The latter are concentrated in the highest elevations and on the steepest slopes of the city, those most vulnerable to large landslides; but floods, landslides, and other mass movements, can affect almost all parts of the city, including the affluent barrios of the Zona Sur.

O'Hare and Rivas (2005) discuss two particular events in detail. A landslide following heavy rainfall produced a damaging landslide in the Cotahuma Barrio of western La Paz on April 6, 1996. This slide killed 27 people and destroyed 80 homes. Several days of heavy rain in March 2003 caused a series of mass movement events at several locations around the city. Hardest hit was the barrio of Llojeta where a landslide destroyed the homes of 75 families.

The peculiar setting of La Paz also subjects the city to flash flooding following periods of intense hail or rain. Such rainfall events are brought about by unusually heavy convective activity that can be produced during any summer wet season. Such a short, but intense, midafternoon storm brought torrential rain and hail to La Paz on February 19, 2002. Drainage along small tributaries funneled into the Choqueyapu River. Hail and debris helped to block the drainage water system of the downtown, and the water overflowed along some streets, including the busy main thoroughfare (El Prado). The event caused 77 deaths, and one damage estimate exceeded US$ 60 million (Enever, 2002). In March 2002, the Municipal Government of La Paz (GML) created the Permanent Committee of Attention and Management of Disasters and in June 2002, with the support of the UNPD, set up programs for flood management in La Paz city.

One of the more infamous landslides was that of Chima, a town located in the Department of La Paz. "Chima," originally Chima Jaucata, in the Aymara language means "castigated place." The population of Chima is 2614, and its main economic activity is gold mining. The gold occurs in Miocene conglomerates and modern alluvial deposits. The area is known as the Gold District of Tipuani, the richest and most productive gold mining region of Bolivia, one that has produced 995 tons of gold during its history.

Detailed studies of the disaster were undertaken by an international team of the CYTED Network XIII, and our discussion is culled from several publications produced by that group (Orche 2003a, 2003b; Orche et al, 2003, among others). The area of Tipuani-Guanay is considered one of the most hazardous mining areas of Bolivia, having suffered repeated avalanches and floods. In 1952 a landslide killed 400 inhabitants; in 1971 a landslide killed 20 inhabitants; another in 1991 killed 20 more. In 2001 the area suffered from flooding. On March 31, 2003, a landslide killed 169 persons and injured 11. Floods again afflicted the town on December 18, 2003, destroying several houses.

The landslide of March 31, 2003 is illustrative. At the time of the landslide, the town of Chima had a population of about 2000 and contained 621 homes (Orche et al., 2003). The main mining activity by the Cooperativa Aurifera Chima Limitada was concentrated on Cerro Pucaloma and produced a

Figure 10.7. The town of Chima after the landslide disaster of March 31, 2003. (Photo by E. Orche).

300-m-high, steep, unstable, and talus-covered face. Intense rainfall during March infiltrated the artificial fissures on this face and exacerbated the unstable conditions. From March 26 to 28, 2003, a group of experts from CYTED network XIII visited Chima and, noting the risky situation of the town in relation to the mining activity, alerted residents and local authorities. The CYTED technicians further pointed out that overload produced by water saturation of the adjacent mountain could cause its talus slope to fail. Only three days later, on March 31, a rotational slide, transformed downslope into a debris flow, advanced rapidly to the Rio Tipuani. The total volume of the debris fan was estimated at 400,000 m^3. A lobe of this flow swept away a portion of the city killing 169 inhabitants (including those missing), affecting another 690 (35% of the total population) and destroying 149 houses (one-quarter of the town) (Fig.10.7). Economic losses were estimated at US$ 1.2 million. As a consequence of this landslide, the Unit of Risk Management was created in La Paz Department with the support of UNPD.

6. Conclusions

Bolivia is subjected to a wide variety of natural hazards. The damaging effects of these natural hazards are often exacerbated by human decisions and activities. The Bolivian government has only limited resources to plan for and to mitigate against disaster.

The most serious hazards facing the nation are hydrometeorological events, particularly floods, droughts, and landslides. In many of the cases that we have mentioned, economic losses and loss of life could have been avoided or, at least, minimized with better planning and selective investment of resources into hazard mitigation.

Acknowledgment

The authors thank especially Professor E. Orche for providing valuable information on the Chima disaster.

CHAPTER 11

Soil Erosion in Brazil from Coffee to the Present-day Soy Bean Production

Selma Simões de Castro[a] *and* José Pereira de Queiroz Neto[b]

[a]*Universidade Federal de Goias*
[b]*Universidade de Sao Paulo, Department of Geography, AV. Prof. Lineu Prestes 338, Butanta, 05508-900 - Sao Paulo, SP - Brasil*

Contents

1. Introduction

Until the first half of the twentieth century, coffee was the main crop and the major factor in Brazil's national and international economy. However, during the second half of that century, soy production in several states, along with cattle-raising, became a major factor influencing the expansion of agriculture and the incorporation of new areas into the territory. This agricultural expansion was also responsible for a marked increase in erosional processes.

Erosion is a set of processes connected to denudation of the land (Hole, 1968) and is responsible for relief modeling and sedimentary deposits (Tricart, 1966). The concept of *accelerated erosion*, related to release of particles from the soil used for

Corresponding author: *Selma Simões de Castro, Universidade Federal de Goiás-IESA, Campus II, (74001-970) Goiania, GO, Brazil, selma@iesa.ufg.br*

Developments in Earth Surface Processes, Volume 13
ISSN 0928-2025, DOI 10.1016/S0928-2025(08)10011-6

agriculture and cattle-raising, and transportation and deposition by rainwater, is already well established in the field of agronomy.

In Brazil, processes of accelerated erosion have been classified as:

1. Erosion caused by diffuse or laminar runoff, also called sheet erosion.
2. Erosion caused by concentrated runoff, which some authors differentiate as:
   (a) rill erosion, which can be shallow, meaning that they can be undone by soil preparation for agriculture, or deep; the latter is also known as gully erosion;
   (b) erosion gullies, or *boçorocas*, when the rills become extremely deep, reaching all the way down to the water table (Rio Grande do Sul, 1985; Bertoni and Lombardo Neto, 1985, Lepsch, 2001).

Most of the research on erosion has focused on laminar erosion, using the Universal Soil Loss Equation (USLE) proposed by Wischmeier and Smith (1978). This classic theoretical model served as the basis for evaluating potential soil loss caused by erosion. The equation is as follows:

$$A = RKLSCP$$

where:

R = rain erosivity index
K = soil erodibility (soil loss measured in a standard unit with 22.13 m L and 9% S)
L = slope length, soil loss relative to the standard unit
S = slope gradient, soil loss relative to the standard unit
C = use and management, soil loss in a cultivated area, with a given management system, relative to fallow land
P = conservationist practices, soil loss in cultivated areas using these practices, relative to cultivated areas along the line of the most pronounced declivity

Bertoni and Lombardi (1985) and Ponçano and Christofoletti (1987) stated that the USLE serves two purposes:

1. To assess soil loss for different conditions of soil use.
   To estimate tolerable soil loss for different types of crops.

After it was proposed, this equation was revised many times and the letter R was added to it. It is now known as RUSLE. It considers, among other specifications:

- The inclusion of databases such as an erosivity map, world data on erodibility or soil identification, in which the application of the USLE erodibility nomograph was not applicable.
- Greater flexibility regarding the length of the slope.
- The incorporation of variations derived from the characteristics of the canopy, soil cover, relief roughness, incorporation of residues, and other protection practices.

These changes were made in response to the problems encountered in its application in tropical regions, such as in Brazil. The largest data bank on the results of the application of both the USLE and the RUSLE can be found in the Instituto Agronômico do Paraná (IAPAR), which has been developing these studies since

1974 (Anjos; Van Raij, 2004). They point out that the mean annual soil loss in that state reached 22.7 tons/ha for different types of soil, soil preparation, and crops. They also ranged from 0.8 tons/ha for no-till planting for the soy-wheat rotation in oxisol up to 105 tons/ha for conventional grain crops in red and yellow Ultisols. By means of the USLE, several authors have also presented estimates for soil loss caused by erosion for some Brazilian states, such as São Paulo and Paraná (Castagnoli, 1966; Bellinazy et al., 1981; Bragagnoli, 1994).

It is important, however, to highlight the fact that one of the problems in applying this equation has not been solved in either its country of origin or other countries. The USLE equation does not take into consideration two main geomorphologic processes: erosion caused by concentrated runoff (rills and erosion gullies), resulting in linear erosion, and mass movements (Leprun, 1981; Queiroz Neto, 1978; Oliveira, 1990, Oliveira and Queiroz Neto, 1994), which cause soil creep, landslides, and similar movements.

With regard to linear erosion specifically, although some older studies deal with these phenomena in Brazil, only the most recent studies focus on the processes and mechanisms involved in this kind of erosion (IPT, 1986; Oliveira, 1994; Oliveira and Queiroz Neto, 1995; Salomão, 1994, 1999; Salomão and Queiroz Neto, 1995; Marinho et al.; 2006, among others). The position, direction, and intensity of water flows, as well as the hydraulic gradient and the possibility or not reaching the water table, among other aspects, are evaluated in these studies. Although there is an universal equation for evaluating the potential of linear erosion, several studies have been conducted in Brazil on susceptibility, risk, and modeling but monitoring of erosion is not common, In the case of gullies development, the studies take into consideration phenomena such as *piping* and mass movements (Guerrard (1992), Imeson and Kwaad (1980), Kirkby (1965), Pilcher (1953), Prandini (1974)). More recently results by Salomão (1994, 1999) and Marinho et al. (2006) were presented on the Cerrado region.

Active erosion processes causing surface runoff of rain water in Brazil have been described by Bertoni and Lombardi Neto (1985), DAEE/IPT (1989), Faria (1996), and Lepsch (1983; 2002), among others. Quantitative data on erosion derive mainly from agronomy, and they are obtained from standardized experimental plots subjected to different management and crop types (Marques et al.; 1961; Bertoni and Lombardi Neto, 1985). These plots help determine soil and water loss for different types of soil and declivity, under natural and artificial showers (rain simulators), for the purpose of establishing agricultural practices that can help minimize/control the effects of erosion (Bertoni and Lombradi Neto, 1985). These data serve as the basis for estimating the intensity of erosional processes.

Research carried out on the current dynamic of the slopes includes recording different erosional processes, especially those caused by concentrated runoff of rain water. Research was carried out through the interpretation of remote sensor data (aerial photography and satellite images, mainly from Landsat) accompanied by field observations for validation and recording in topographic maps, or more recently, georeferencing with GPS. These works aim at determining the degree of interference of factors, which are responsible for the intensity and position of the erosional processes and its features (DAEE/IPT, 1989; Santana, 1995), however they rarely present quantitative data. Even if they do not quantify the material moved by the erosion, they are extremely important for evaluating the presence and intensity of these processes in both time and space.

Collective or mass movements are prevalent in the Planalto Atlântico (Atlantic Plateau) of São Paulo (Queiroz Neto, 1978), and their mechanisms are studied in geomorphology (e.g., in Furian, 1995; Furian et al., , 2003). Their incidence in that part of the State of São Paulo was confirmed by the erosion map for the state (Gouveia et al., 1997).

Finally, as Oliveira (1994), Oliveira and Queiroz Neto (1997), and Campagnoli (2005), for example, observed, records of sediment production, traditionally dealt with in hydraulic engineering, is an indirect way to evaluate the intensity of erosional processes. Even though these records are still very sketchy in Brazil, they allowed Campagnoli (2005) to develop a map of the country representing areas that have the potential to produce sediments.

These elements serve as the basis for an examination of erosion processes connected to the cultivation of coffee and soy. Principles and definitions of accelerated erosion are presented while dealing with the incidence of these processes, induced by the practice of agriculture in Brazil. These crops are symbolic because they are examples of the agricultural expansion that has taken place during the last two centuries in Brazil. During that time, the general economic model did not change significantly and has been supported by single crops destined to the international market. In fact, they do not show any significant changes either from the historical standpoint of incorporation of additional land and the production of new spaces resulting in an organization of the territory, or from the effects on the environment.

Coffee as a permanent crop was introduced in Brazil by slave labor and remained a manual process for two centuries, until the second half of the twentieth century. Today only part of the work is mechanized, but the soil does not suffer strong disturbance. In contrast to coffee, the planting of soy is a heavily mechanized seasonal activity and has involved soil correction and fertilization from the first time it was planted, the second half of the twentieth century. Some areas used irrigated systems, which yielded up to two harvests a year, multiplying the mechanical work and its influence on the environment. No-till planting or a straw cover for soy is a relatively recent management system and produces beneficial effects with regard to controlling erosion.

This chapter highlights the erosive effects of these two crops, as well as the climate conditions in which they were planted. Even though these crops have suffered and still suffer from erosion caused by surface runoff of rain water, their effects are very different because of their management systems, time in the field, and other features.

First, a brief summary of the natural conditions of the Brazilian territory where these crops developed will be presented and will include an examination of relief, vegetation, climate, and soils. This analysis will be done by defining morphoclimatic domains, as initially proposed by Ab'Saber (1973), although more recent nomenclatures have adapted these domains. Next, an outline of the implementation and development of the coffee and soy crops in Brazil will be presented on the basis of historical records and demographic data, focusing on pioneering fronts and modern practices. Finally, a case study is presented on accelerated erosion triggered by modern agricultural practices in the upper Araguaia River in midwestern Brazil.

2. The Interrelationship of Coffee, Soy, and Erosion in Brazil: Geographic Context and Background

2.1. The Morphoclimatic Domains of Brazil

Ab'Saber (1973) first defined Brazil's "large morphoclimatic and climatological/ botanical units" (Fig. 11.1). These units can be used to better understand the agricultural capabilities and development of the country. The Brazilian territory

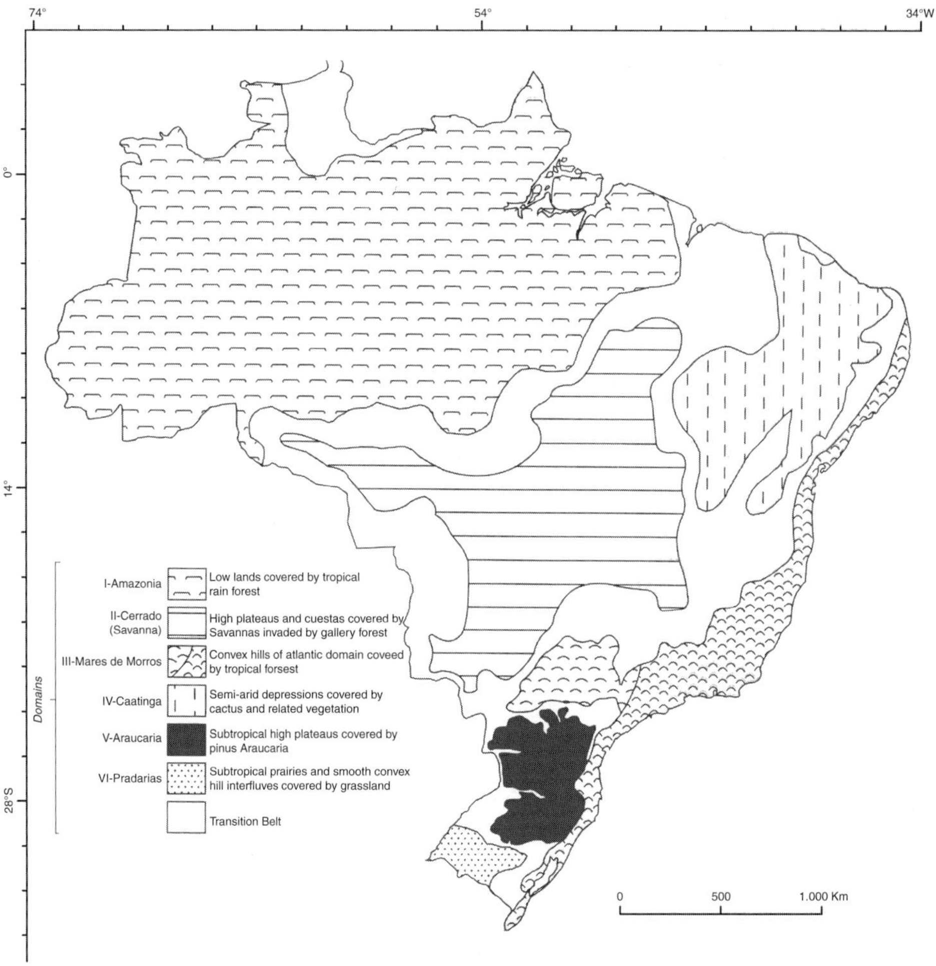

Figure 11.1 Map of Brazil, showing locations of study sites and the morphoclimatic domains identified by Ab'Saber (1973, 1977).

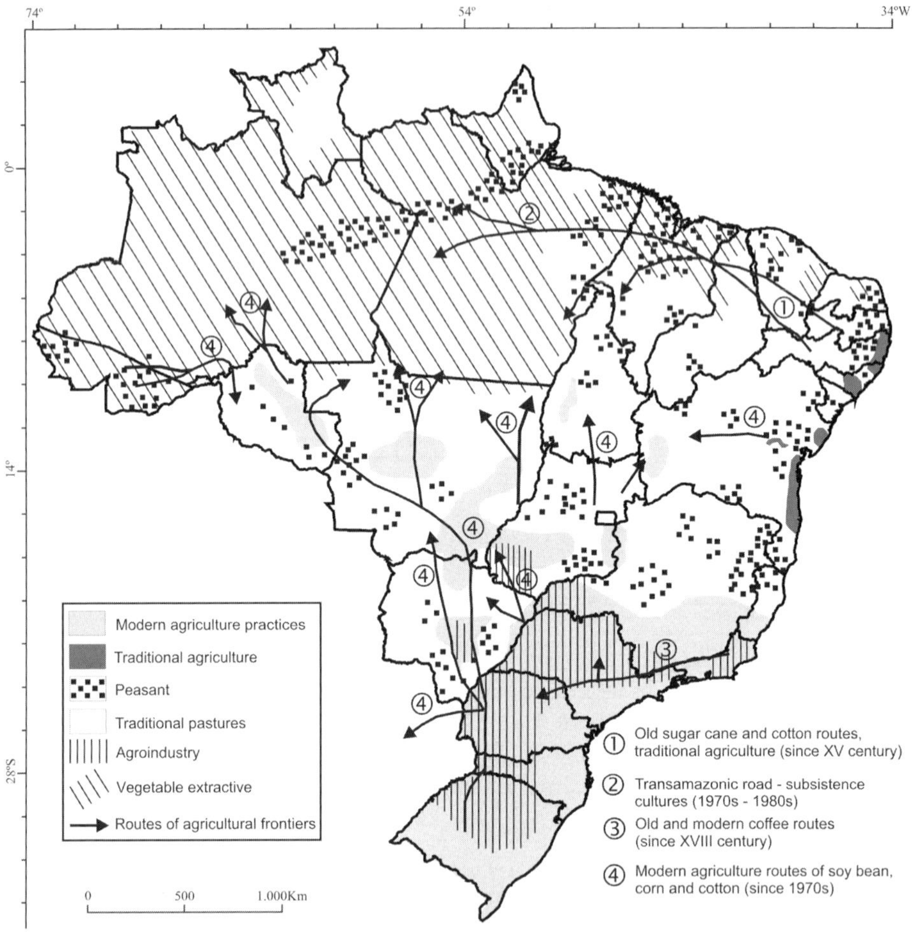

Figure 11.2 Pioneer fronts and different agricultural and pasture activities.

was occupied over a long period of time by agricultural frontiers, as presented in Figure 11.2. Such fronts were responsible for occupying and structuring the land, and the result is the current social and spatial configuration.

2.2. Coffee:Brief Considerations

Coffee first expanded throughout the coastal region within the Mar de Morros (polyconvex or hilly relief) Domain of the Tropical Atlantic, Brazil. Expansion continued toward the west, occupying the State of São Paulo and the north of Paraná, all the way to the limit of the Savanna Domain in Central Brazil. Currently, the expansion of this crop has included the State of Minas Gerais, still within the Mar de Morros Domain). The first coffee plants coming from Belém, in the State of Pará, the north region, Brazil (Amazon), arrived in the surrounding area of Rio de Janeiroin the southeast of the country, toward the end of the eighteenth century.

At the beginning of the nineteenth century, coffee plantations reached the limits of the Mar de Morros Domain, expanding toward the State of Espírito Santo (east), Minas Gerais, and São Paulo (southeast). Throughout the nineteenth century and during most of the twentieth century, the crop in the São Paulo area found favorable conditions that caused it to expand more (Monbeig, 1952).

Since it was introduced, coffee has always been managed manually, first with slave labor, which began in 1870, then, as of 1889, with free workers (mostly Italian immigrants), after slavery was abolished. It is important to point out, however, that even today, most of the agricultural work is still done by hand.

The evolution of coffee as a crop was followed by an increasing in production, population growing and the destruction of the forested areas of the State of São Paulo.

- In 1836, coffee production was limited to only one hundred fifty 60-kg bags (9000 kg), forested areas occupied around 80% of the territory (20,450 ha), and the population of Brazil was around 5 million inhabitants (6% in São Paulo);
- In 1870, coffee production reached 3,764,000 bags, and the population had grown to 10 million inhabitants; deforestation in São Paulo destroyed almost 20% of the primitive forests;
- In 1900, production reached 11,367,371 bags, the Brazilian population had grown to 18 million inhabitants, with 2,282,000 living in São Paulo (13%), and deforestation had affected more than 30% of the primitive forests.

Between 1800 and 1900, the Brazilian population grew 260%, while in São Paulo the growth was 760%. Actually, the population growth that took place between 1870 and 1900 coincided with the arrival of immigrants (900,000 people between 1887 and 1900), raising the growth rate in 6%.

Deforestation in the State of São Paulo continued until around 1970, by which time only 10% of the primitive forest remained. This deforestation activity followed the expansion of the coffee plantations until the first half of the twentieth century, but one must remember that an expansion in the area used for cattle-raising and for planting sugar cane and other crops were concomitant factors. The most significant expansion of sugar cane plantations occurred after the 1970s as a result of government incentives for fostering the production of fuel alcohol.

During the 1920s, the northwest of Paraná was used to plant coffee, and in little more than 50 years the crop completely occupied that region, making the state the largest coffee producer in the country. However, because of the frosts that affected plantations throughout the 1970s, the crop progressively lost space and reached its lowest level during the first years of the 1990s. Coffee would then be replaced largely by soy and wheat. The biggest problems for this crop in the south of the country were the low temperatures because of the frost and some phytosanitary problems. Thus, coffee started to look for new spaces, with more amenable temperatures and free from soil-conservation problems. The new region found was to the south of the State of Minas Gerais, which quickly became the largest producer. At the same time, a significant increase in planted areas occurred in Espírito Santo, while Paraná dropped to fourth place in terms of Brazilian production (Table 11.1).

Table 11.1 Coffee Production in Brazil by State, 1995 to 2000 (in tons)

States/Years	1995	1996	1997	1998	1999	2000
Paraná	250	800	1,500	1,600	1,400	1,500
São Paulo	1,800	3,000	3,500	3,125	3,300	3,200
Minas Gerais	9,200	15,000	16,000	16,500	17,500	18,000
Espírito Santo	3,100	5,300	6,500	6,000	6.200	6,300
Others	2,500	3,000	3,500	3,200	3,000	3,200
Total for Brazil	16,850	27,100	31,000	30,425	31,400	32,200

Source: Pessoa, A., 2006.

The movement of this crop took place within the limits imposed by the Mar de Morros Domain, in the Brazilian southeast. Therefore, migration to the State of Minas Gerais largely represented a rediscovery of environmental conditions—climate, relief, and soil—that prevailed during the first expansionist wave in the nineteenth century. The presence of coffee in the State of Espírito Santo and its prevalence from the nineteenth century to the present time can also be classified within the environmental conditions of the Brazilian southeast.

In his notes, Monbeig (1952) had pointed out that the expansion of the pioneering front had been a driving element, but the occupation of the space took place through what can be called a multicrop system, in which the most important crops were rice, corn, beans, cotton, sugar cane, and cassava, among others, in such a way that this production played an important role in the local, regional, and national economy. Monbeig's notes also indicate coffee's role in the erosion process in southeastern Brazil, as a result of the destruction of the original forests and related hydropedological imbalances, as will be seen below.

2.3. Soy: Brief Comments

Soy was introduced in Brazil toward the end of the nineteenth century, in the State of Rio Grande do Sul (south of the country), which is located within the Pradarias (Prairie) Domain. Morphoclimatic conditions in this area were completely different from those where coffee was planted. Expansion, however, began during the second half of the century, first in the southern states and later, closer to the end of the century, in the internal tropical plateaus of the Savanna Domain, principally in Middle West region.

The history of soy, brought from the United States toward the beginning of the twentieth century as a remedy, is completely different from the history of coffee. Soy was implemented in Rio Grande do Sul mostly owing to fiscal incentives given to wheat. Soy entered as part of a rotation crop. From the beginning, this crop used technical management systems, from sowing to harvest, requiring intense and heavy mechanization and strong fertilizers and biocides. As a result of the international scenario, soy expanded, first to the State of Paraná (Southern), toward the mid-1970s, replacing coffee.

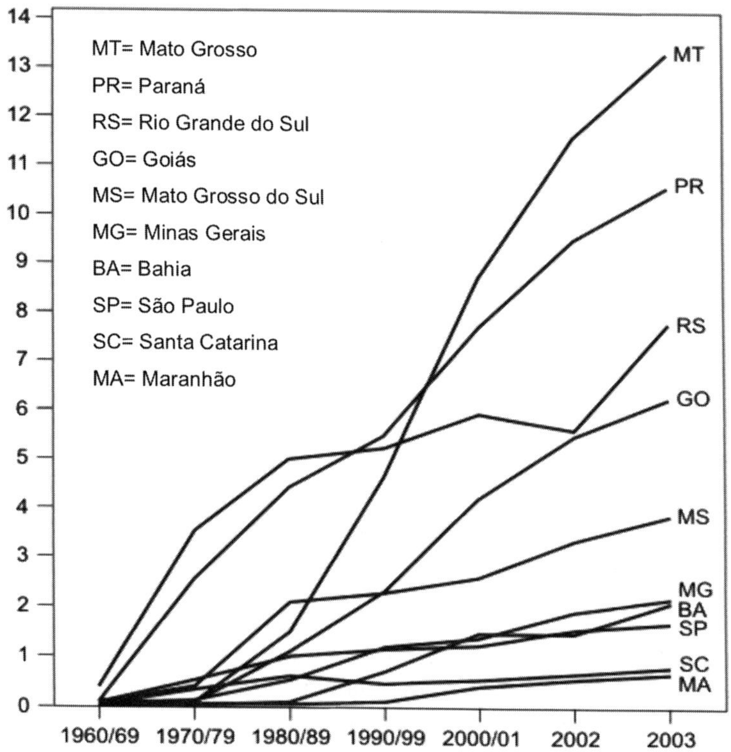

Figure 11.3 Soy production in different Brazilian states from 1960 to 2003.

Figure 11.3 shows that around that time production became significant in every state where soy was planted, reaching around 40 million tons in the year 2000, transforming Brazil as the second world producer of soy. In the beginning, this expansion did not influence the growth in the Brazilian population. However, after this crop expanded toward the central part of the country, entering the Savanna Domain, important internal migrations took place. The yearly statistical reports published by IBGE (the Brazilian Institute of Geography and Statistics) show that the population growth for that region was higher than the overall growth for the country. At the same time, there was a marked decrease in the demographic growth rates in the southern states, especially in Rio Grande do Sul, indicating that a significant percentage of that population had moved to central Brazil.

With regard to cultivated areas, productivity and production data on the soy bean in Brazil from 1960 to 2003 (see Table 11.2) shows an enormous increase in crop area from 0. 4 to 18 million ha, related to an increase in total production from 0.5 to 50 million tons and productivity from around 1100 to 2764 kg/ha. The most important period of increase was 1999–2000; the increase in productivity during that the time was very important, more than double.

Table 11.2 Soy Bean Production in Brazil from 1960 to 2003

Years	Cultivate area (Million ha)	Productivity Kg/ha	Production (Million tons)
1960–1969	0.4	1089	0.5
1970–1979	4.9	–	7.3
1980–1989	9.5	1721	16.6
1990–1999	11.3	–	24.0
2000–2001	13.6	2557	34.8
2002	16.3	–	41.9
2003	18.0	2764	50.0

Source: Embrapa CNPSO, 2007 (www.cnpso.embrapa.br).

Interestingly, the State of São Paulo was not planted with soy. Instead, this crop went directly to the Savanna Domain in central Brazil, advancing toward the Amazon (north) and northeastern areas. This was the case because when soy started to expand as a crop, during the 1970s, all available land was taken by sugar cane so that producers could benefit from the government's tax incentives to foster the production of fuel alcohol (the Pro-Alcohol program).

In the Savanna Domain, soy could be grown in favorable conditions: deep soils with adequate physical and hydrological characteristics, which could be corrected and fertilized; gentle slopes favoring intensive mechanization; and an appropriate climate, in which the dry season begins after harvest (Fig. 11.3, States of Mato Grosso—MT and Goiás—GO). On some properties, irrigation makes it possible to have two harvests a year, which multiplies the influence of this crop over soil degradation and erosion. This new policy is generally called Modern Agriculture.

The States of Goiás, Mato Grosso, and Mato Grosso do Sul have the most important soy yields. They were the main beneficiaries of a governmental policy called Polocentro, which was part of the II National Development Plan (1975–1979). The goal of the policy was to promote development of the savanna by incorporating the region into the national productive system. Important tax incentives and heavy investments in infrastructure, including roads and electricity, served to ensure the success of the National Development Plan. Figure 11.4 shows the area encompassed by the Polocentro.

Figure 11.5 shows the agricultural and pasture fronts of the States of Goiás. Cattle-raising activities were almost exclusive until 1930, but since that time coffee began to be planted in the south of Goiás, as a result of the expansion from Minas Gerais into an area of Oxisols (Red Latosols) that did not suffer the consequences of frosts. The occupation of the southwest of Goiás with the so-called Modern Agricultural Practices, however, took place just in the 1970s, when the Polocentro governmental program was implemented and soy crop and farmers from the south arrived.

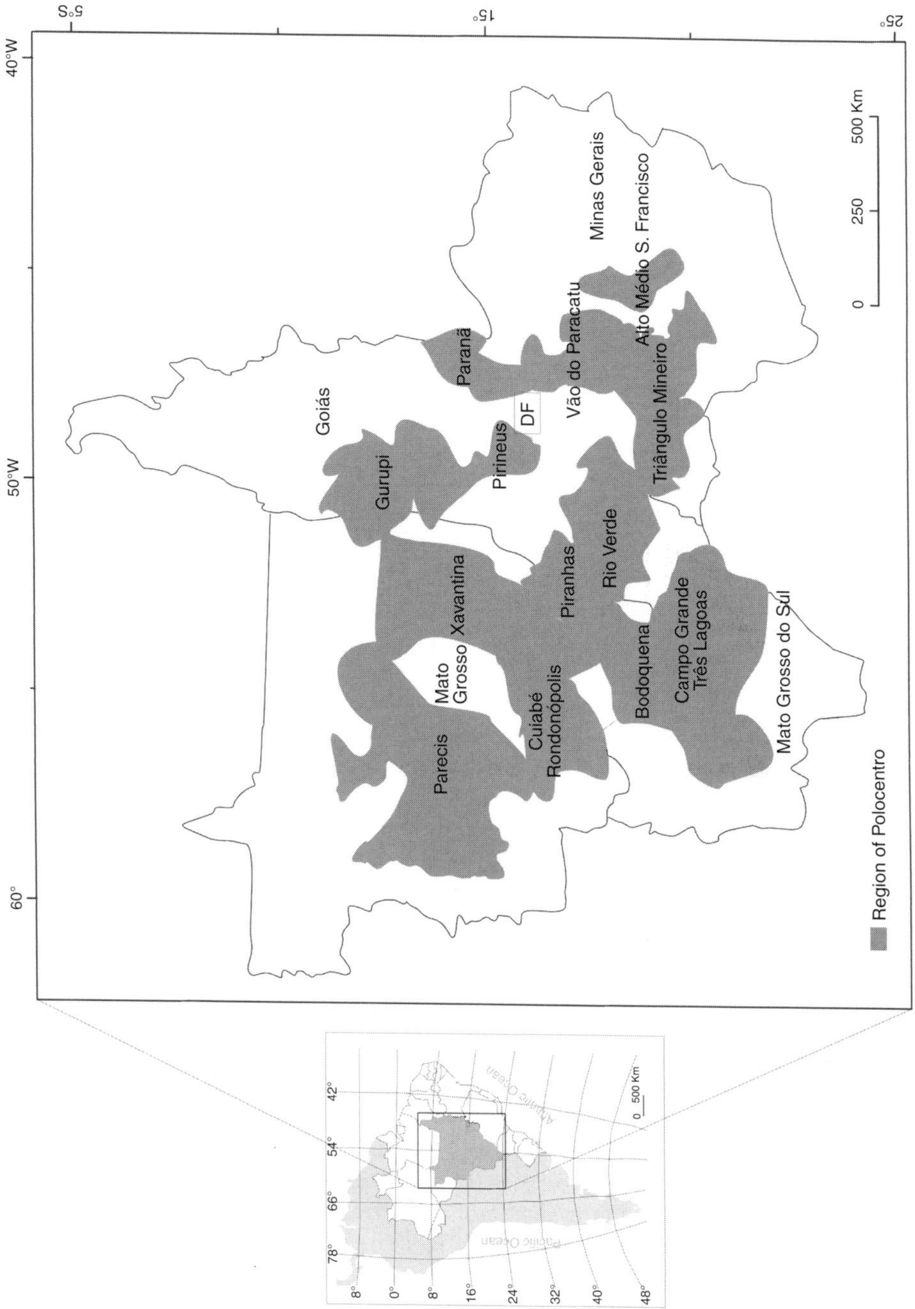

Figure 11.4 Area encompassed by the Polocentro.

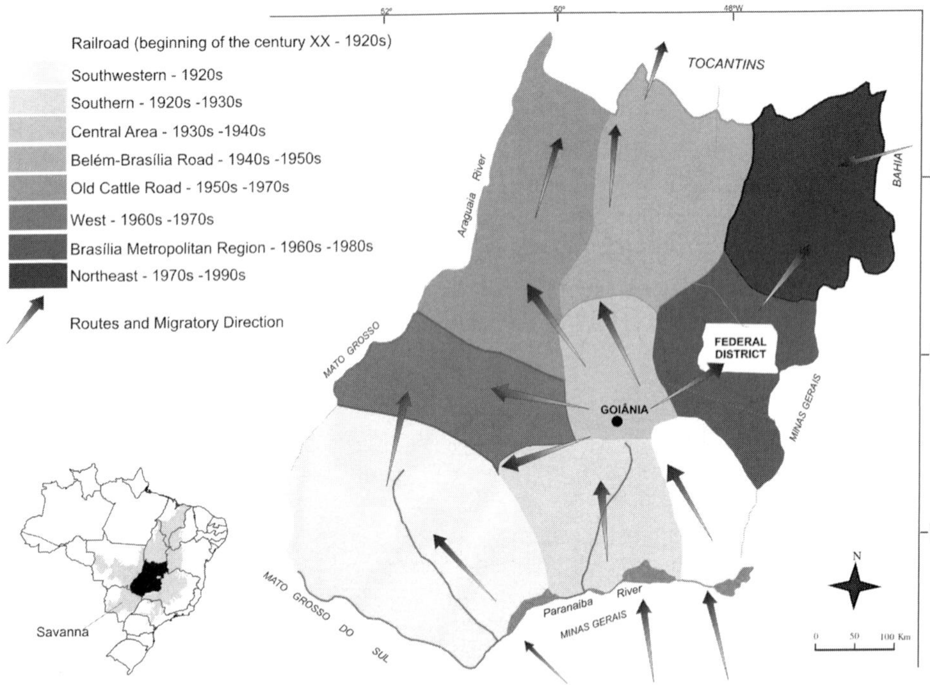

Figure 11.5 Routes of agricultural, pastural, and migratory directions by decades in Goiás State, Brazil.

3. Erosion and Agriculture

The data in this section regarding factors C (cultivation and management) and K (soil erodibility, as per USLE) come from research carried out by the Instituto Agronômico de Campinas and are valid for the State of São Paulo (Bellinazzi et al., 1981; Bertoni and Lombardi, 1985). This research served as the basis for Bellinazzi et al. (1981) to present and evaluate soil loss caused by erosion on the order of 215 million tons/year in São Paulo (Table 11.3).

The data presented in Table 11.4 show that different types of soil loss caused by accelerated erosion are associated with coffee and soy. Whereas coffee is a permanent crop and requires very little mechanization, soy is an annual crop and must be renewed. It also requires intense and continuous mechanization while it is being developed, all the way to harvest time. These are determining features that help define the intensity of erosive processes (C value on the USLE) (Table 11.5):

Another important factor in evaluating the relative intensity of erosion caused by these different crops is soil erodibility (factor K). The soil survey carried out in Brazil shows that the most common types of soils are the deeply weathered and ancient Oxisols and the strongly leached and acidic former forest soils with relatively low native fertility known as Ultisols. The first soil survey in the State of São Paulo, conducted by the Soil Commission in 1960, already pointed to the greater erodibility

Table 11.3 Soil Loss Caused by Erosion in the State of São Paulo, for Different Types of Soil Use (adapted from Bellinazzi et al., 1981)

Types of soil use	Area occupied		Land loss		Land loss
	Km^2	%	tons/year	%	tons/ha
Annual crops	3,257	13	68,288	31.7	20.9
Temporary crops	1,487	6	20,030	9.3	13.3
Permanent crops	1,615	6	1,453	0.7	0.9
Pastureland	10,405	42	4,162	1.9	0.4
Forest/reforestation	3,227	13	51	0.02	0.01
Others	4,867	20	121,681	56.4	25.0
Totals	24,860	100	215,666	100	8.7

Table 11.4 Soil Loss (tons per hectare per year) Caused by Coffee and Soy (adapted from Bellinazzi et al., 1981 and Bertoni and Lombardi, 1985)

Type of Culture	Cycle	Land loss tons/ha/year	Percentage of loss by erosion in the State of São Paulo	
			A%	B%
Soy	Annual crop	20.1	12.2	5.3
Coffee	Permanent crop	0.9	0.9	0.4

A = related to losses caused by all vegetable covers; B = related to total loss in the State of São Paulo.

Table 11.5 Erodibility (factor K) of Oxisols and Ultisols in the State of São Paulo (adapted from Bertoni and Lombardi Neto, 1985)

Type of Soil	Erosion limits by horizon (factor K)	
	Surface	Subsurface
Oxisols	0.11 to 0.22	0.03 to 0.22
Ultisols	0.18 to 0.54	0.11 to 0.45

of Ultisols; which was later confirmed by several authors (Queiroz Neto, 1978; São Paulo, 1989; Salomão, 1994). On the basis of data obtained on experimental plots, Bertoni and Lombardi Neto (1985) presented erodibility values for these two types of soils in the State of São Paulo (factor K), showing that Ultisols are more likely to erode (Table 11.5). Carvalho (1994) pointed out that two units of Ultisols over the Caiuá sandstone (Cretaceous), in the northwest of Paraná State, are strongly to very strongly susceptible to erosion—for example, Quartzipsamments soils (no developed sandy soils), in contrast to Oxisols, most of which are moderately or weakly susceptible. Oliveira and Castro (2005) recently confirmed this susceptibility, including the high risk of gully processes developing in these soils in the southwest of Goiás State in Savanna Domain, which was verified by Xavier, Castro, and Barbalho (2005) and Marinho, Castro, and Campos (2006). However, it was primarily in these soils that soy was planted, in a highly technified management system, thereby inducing erosion.

Couto et al. (1990) studied the role of land use and mining activity in the erosion and production of sediments in the São Lourenço River in Mato Grosso State. Soy was planted in the higher altitudes of the plateaus, where the gentle relief is covered by Oxisols, but the crop was strongly mechanized and the presence of Ultisols induced a higher incidence of laminar erosion. Ribeiro et al. (1997) indicated that several latosol units occurred in the upper surfaces of the plateaus of the Middle Paranaíba River Basin, predominantly used to plant soy and corn, which also require intense technological management systems.

These observations indicate that from the State of Paraná in the south through central Brazil, Oxisols are more often used for planting soy than are Ultisols. If, on one hand, Oxisols are less susceptible to erosion, on the other hand, soy, an annual crop requiring large-scale use of heavy machinery, from sowing to harvest, generates serious consequences. The first, and perhaps the most important consequence, is the appearance of a deep compacted layer that causes decreased infiltration of rain water, thereby increasing surface runoff and the risk of more intense erosions.

Estimates of soil loss caused by erosion, calculated with the USLE, can only evaluate the erosion caused by diffuse surface runoff and do not take into consideration the erosion caused by concentrated runoff (Leprun, 1981; Oliveira, 1994). For some researchers, rills and erosion gullies would be more efficient for the transportation of soil by erosion than diffuse surface runoffs and, therefore, would produce more sediment (Oliveira, 1990, 1994; Santana, 1991; Oliveira and Queiroz Neto, 1994; Santana and Queiroz Neto, 1995).

Observations on the incidence of rills and erosion gullies have been carried out by means of aerial photography, Landsat and RADAR, and field controls. The Orientações (Guidelines) project, which aimed at combating erosion in the State of São Paulo (DAEE/IPT, 1986, 1988, 1990, 1992), detected the presence of more than 9000 large erosion gullies in the Paraná Sedimentary Basin. From this total, 80% are found especially on Ultisols deriving from the Bauru Group sandstone (Upper Cretaceous of the Paraná Sedimentary Basin), and only 5% are related to urban areas. Although the amount of displaced soil has not been calculated, its importance is attested by the volume of sedimentation found on valley floors and reservoirs (Oliveira, 1994; Oliveira and Queiroz Neto, 1994). Erosional gullies developed through the valley floors, exposing layers containing human artifacts (Oliveira, 1990, 1994, Oliveira and Queiroz Neto, 1994, Oliveira et al., 1992, 2001). Two meters of sediment were deposited shortly after a disturbance of the original forested landscape along the valleys. The massive deposition occurred during the beginning of the deforestation and decreased progressively as the valleys became more equilibrated systems.

Santana (1991) and Santana and Queiroz Neto (1995) used aerial photography to study the evolution of the erosion caused by concentrated runoff in the 22-year period 1962–1984. They were able to verify that during that period rills and erosion gullies developed very slowly or not at all, except when a new type of soil use (a new crop or new management techniques) was introduced. In that case, aerial photos showed that concentrated runoff became intensified. On the other hand, small rills were erased, so to speak, by soil preparation, especially in the case of annual crops.

These findings show that erosion and sediment production are extremely important during the first phases of land settlement and normally involve deforestation. A marked decrease was observed as agriculture stabilized, but they became more intense when a new crop or a new type of management was introduced. These findings made it possible to build a hypothetical model of erosion evolution in the Western Paulista Plateau to show these facts (Fig. 11.6).

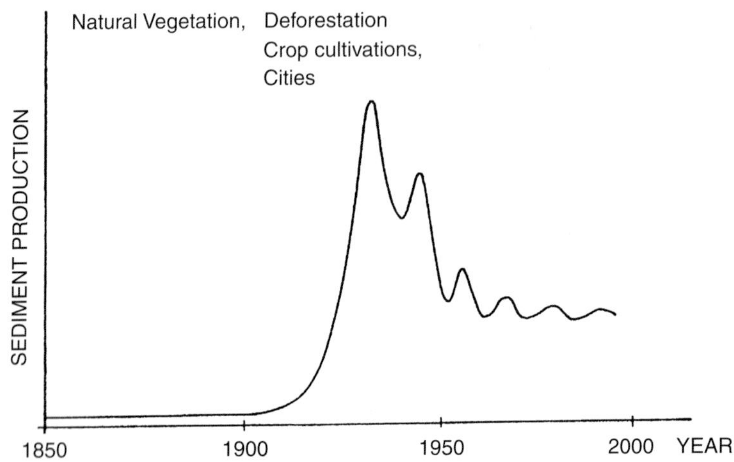

Figure 11.6 Hypothetical model for sediment production in the Western Plateau of São Paulo.

Campagnoli (2005) produced a map of areas with sediment production in Brazil. This map provided a comprehensive view of the erosive potential within the Brazilian territory and clearly showed that this potential is related to the cultivated areas (Figure 11.7). Areas with a high potential for sediment production generally correspond to those that are currently used for grain production, especially soy—areas to the west of the gaucho countryside, Mato Grosso do Sul and Mato Grosso, and the Central Plateau from Goiás to Tocantins.

The State of São Paulo can be included in the category of low to moderate sediment production, and areas with high to very high sediment production surround the Brazilian Amazon. These two observations deserve additional comments: as already mentioned, soy was not planted in the State of São Paulo, except as a secondary crop. In addition, coffee plantations were replaced by sugar cane, which causes very little erosion, or by pastureland, which protects the land more than coffee does (temporary crop and pastureland, respectively, Table 11.4).

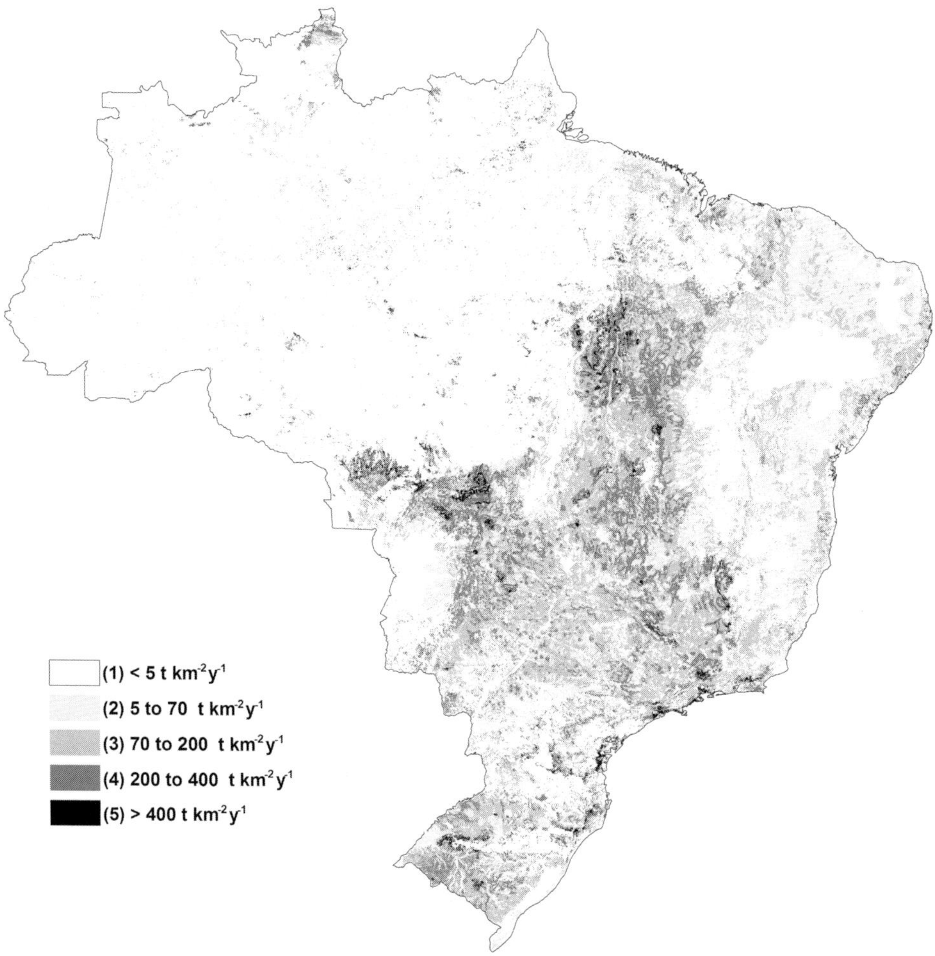

Figure 11.7 Map of actual sediment production areas in Brazil.

Finally, in the Atlantic Plateau of southeastern Brazil (Mar de Morros Domains), signs of processes connected to mass or collective movements are commonly seen, as a 1962 photographic survey of the State of São Paulo shows (Queiroz Neto, 1978). While studying slopes affected by mass movement in the Serra do Mar (Sea Ridge), Furian (1995) suggested the importance of pedological cover as a conditioning factor of water circulation on the soil. This is responsible for triggering those processes, which take place during intense and prolonged rainfall in short periods of time. The polyconvex relief of the hills, characteristic of the Mares de Morros (hilly) Domain, and the geological substrate formed by igneous and metamorphic rocks help determine the presence of these processes (Furian et al., 2003). Coelho Netto (2003), presenting a synthesis of several works realized in this domain, considers the gully and the mass movement process to be like a headwaters and slope evolution conditioned by geological lineaments that promote preferential hydrological routes. However, Coelho Netto (2003) analyze the successive erosion-deposition cycles that affect this domain since 10,000 years ago, but does not concentrate on more recent processes, related to the intensive deforestation produced when the introduction of the coffee culture during the last 200 years.

Another work carried out by IPT has made an important contribution to knowledge of collective soil movements in urban zones, including the means of prevention (Cunha, 1989). The erosion map of the State of São Paulo (Gouveia et al., 1997) registered the frequency of these collective movements, which are accompanied by a moderate to average incidence of small and larger rills and gullies in the Atlantic Plateau. This situation contrasts with the Paraná Sedimentary Basin, over sedimentary rocks, where a high incidence of small rills, larger rills and erosion gullies was noted.

Out of the Atlantic Plateau, located in the Brazilian Midwest in the headwaters of the Araguaia River, as Castro (2005) and Marinho, Castro, and Campos (2006) have reminded us, is another recent example of the relationship between agriculture and erosion, especially since the 1970s when areas with inadequate soils were incorporated by soy growers. Erosion is unquestionably the biggest environmental problem. This area is characterized by a high residual tabular plateau, which is a more elevated, flat, and divisor surface corresponding to the *Chapada* (Plateau), rising more than 875 m high, supported principally by aeolian sandstones (Botucatu Formation) from the Jura-Triassic, intercalated by basalt rocks (Eo-Cretaceous from Serra Geral Formation), and Cretaceous, (Bauru Group) covered by clayish soils related to the thick red Oxisols (Drago, 1983; Mamede et al., 1983; New et al., 1983). This area is dominated by intensive agricultural activity favoring grain crops in general and large rural properties. These properties rarely make provisions for legal reserves, and they also do not make allowances for a permanent preservation zone along their boundaries, measuring at least 100 m, in accordance with environmental legislation.

The smoothed plateau is surrounded by scarped and dissected borders that are approximately 50 m high. Related to the talus and rocky footslope, Cambisols and Litholic Neosols (Lithosols) developed over the sandstone formations. Commonly, waterfalls and mass movements occur where the area is deforested. This pattern was triggered in the 1980s when the area was destined to become extensive pastureland.

At the footslope area, a slightly dissected surface, or pediment surface, extends from 825 m to 675 m.a.s.l. The slopes generally oscillate between 8 and 12% on the lower concave segments, and this area is the headwater of the first- and third-order streams.

This last surface is dominated by thick, typically dystrophic to sandy alic Quartzipsamments from the aeolian sandstones of the Botucatu Formation. Hydromorphic Quartzipsamments prevail on the lower segments of the slopes with more pronounced declivity, as well as in the broad concave area of the headwaters of the top and middle thirds of the large watershed areas. In valley floors with river basins and floodplains we see the predominance of Typic Haplaquox, which is normally dystrophic, as well as Histic epipedon (NEW, 1983; Boulet, 2001; Marinho and Castro, 2003; Marinho et al.; 2006). Gullies are predominant in Hydromorphic Quartzipsamment soils only in the context of a detailed scale, which affect the slopes reaching Quartzipsamment soils. After discharging the sediments on the valley floor, they can bury the saturated soils and even change the course of the Araguaia River (Castro et al., 2004). Both kind of soils are generally occupied by pastureland, and their original cover is normally degraded.

The soil use in this region shows that from 1965 to 2003, the deforestation rate for the upper basin of the Araguaia River was above 60%. Table 11.6 summarizes the recent situation (1999), as calculated by a geoprocessing technique over a Landsat TM5 image and validated in the field.

Water susceptibility to linear erosion (Xavier et al., 2005, Salomão 1994, 1999), and laminar erosion, developed on the basis of an adapted USLE (Oliveira and Castro, 2005) demonstrates the moderate to high degree of erosion susceptibility that is generally attributed to the fine sandy nature and low degree of cohesion of these soils, the long gentle slopes (700-3500m), with small differences in elevation (≈ 50 m). It was noted that the susceptibility was more pronounced in and around the headwaters and valley floors (Fig. 11.8A, B).

Table 11.6 Land Use and Occupation Classes and Erosion Occurrences on Upper Araguaia River Basin. After Barbalho et al., 2001

Vegetation	Area		linear erosion sites	
	km^2	(%)	No.	(%)
Riparian forest	120.00	7.91	0	0.00
Wet land	32.99	2.18	0	0.00
Dense savanna vegetation	268.78	17.72	3	3.30
Sparse savanna vegetation	248.62	16.39	3	3.30
Agriculture	509.63	33.60	33	36.36
Pastureland	334.43	22.05	52	57.14
City and others	2.26	0.15		0.00
Total	**1,516.71**	**100.00**	**91**	**100.00**

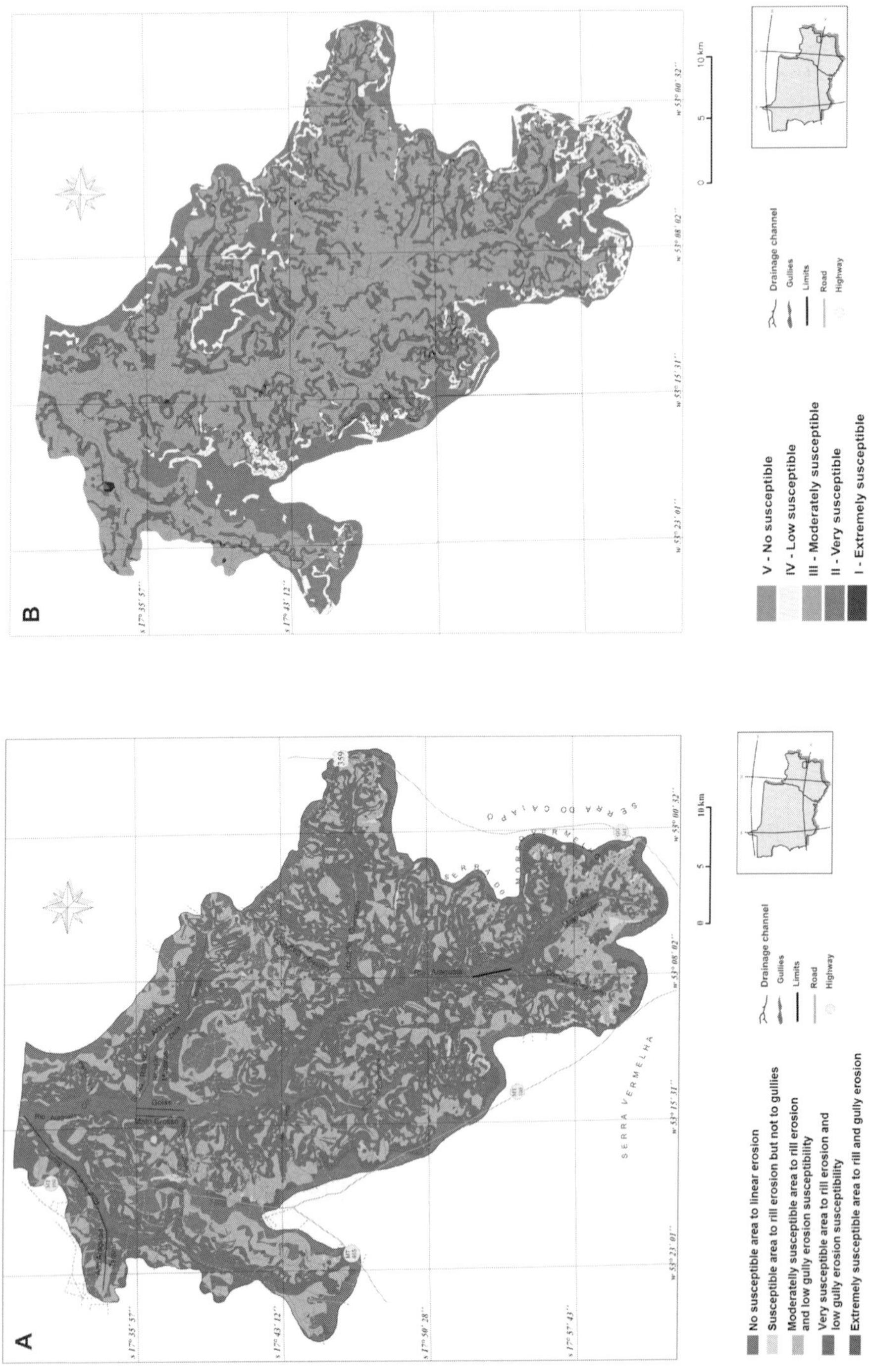

Figure 11.8A-B Susceptibility of rill-gully (A) and laminar-sheet-wash (B) erosion.

When one analyzes the degree of discrepancy (or discordance) between potential and current use of the land, one can observe that gullies are concentrated, especially in areas with the highest and maximum discrepancy. Thus, one can establish an empirical and unequivocal relationship with the use given to the land and the agricultural and cattle-raising management imposed on it (Castro, 2005; Marinho et al., 2006). The degrees of risk for linear and laminar erosion are presented in Figures 11.9A and B, respectively, and were developed on the basis of the methodologies proposed by Salomão (1999). Erosion

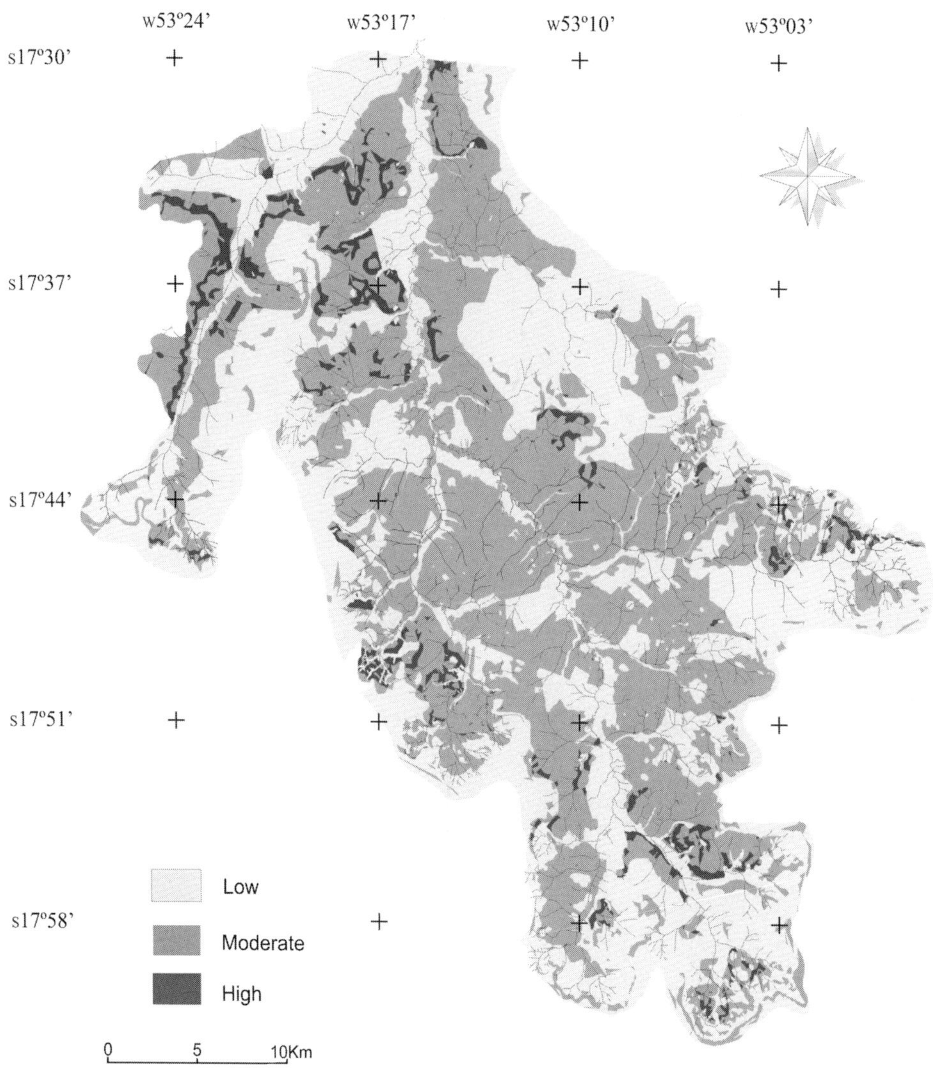

Figure 11.9 Risk of rill-gully (A) and laminar-sheet-wash (B) erosion.

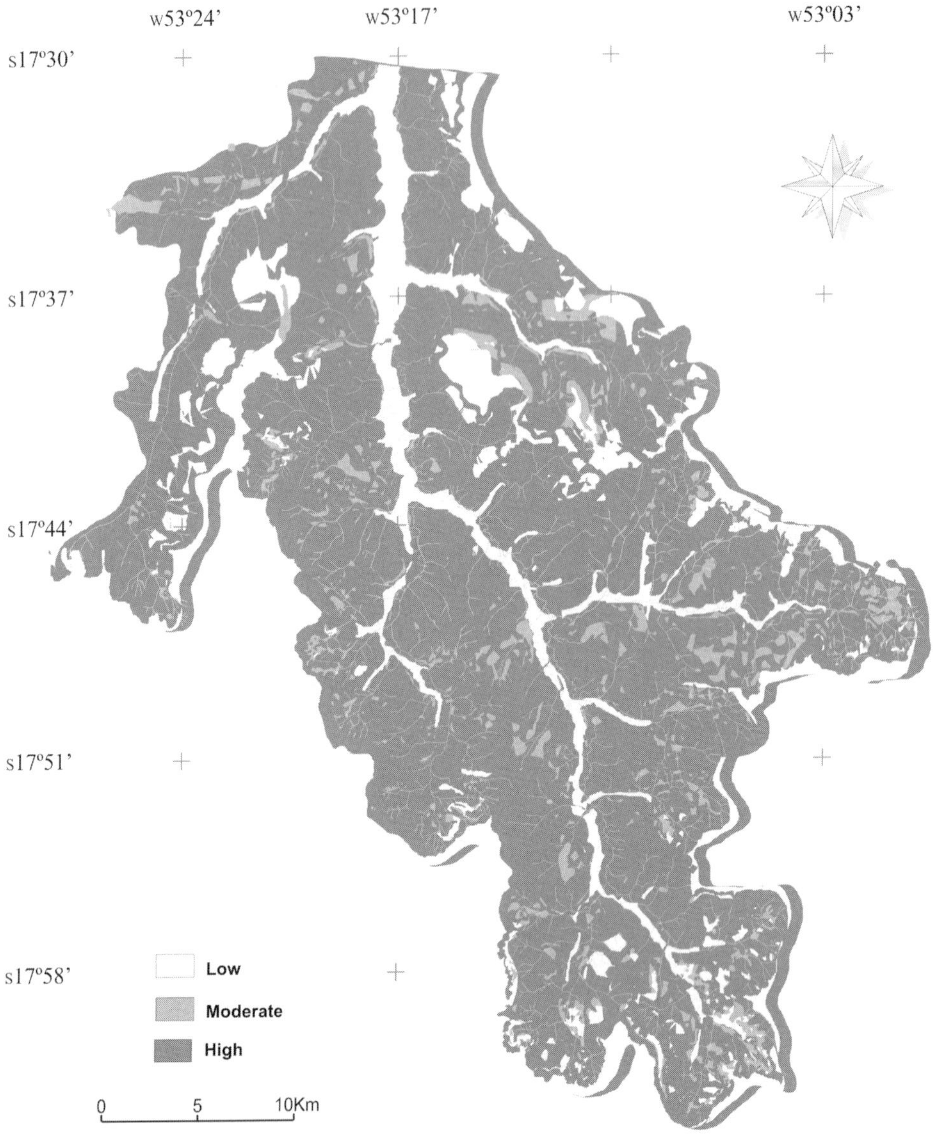

Figure 11.9 (Continued)

sites are concentrated in high-risk areas for linear erosion, with rills, and in low-risk areas, with gullies. However, these instances are mostly of gullies, indicating that many may have begun as rills and developed into gullies. There is also a certain spatial correspondence between susceptibility to linear and laminar erosion, although laminar erosion is more restricted in terms of area (four sets).

In short, erosion sites may be related to the evolution process of headwaters that are undergoing an erosive dissection process at the plateau edges. Erosion has been intensifying as a response to intensive and inadequate occupation of the land. The land use showed a disregard for environmental legislation, which beginning in the 1970s already required that the headwaters and river banks, as well as terrains with marked declivity, needed to be considered fragile permanent preservation areas. They have not been.

4. Development Models for the Gully Erosion Process in Brazil

Surface water flows are frequently associated with the formation of smaller and larger rills. Subsurface flows, however, are associated with piping phenomena, which are capable of generating soil subsidence by continual removal of material generally located in the soil-rock contact point. They can even cause internal washouts, which are a condition for the formation and evolution of erosion gullies (Guerrard, 1992; IPT, 1986; Salomão, 1994, 1999). In this sense some currently models can be presented.

First, in the Mar de Morros Domain (Atlantic Plateau), several studies have shown the structural piping control by lineaments of Precambrian rocks in the headwaters of first- and second-order streams deforested 200 years ago and occupied by coffee plantations and pastures, promoting strong and successive sedimentation in the floodplains (Coelho Netto, 2004). This process is dominant in deforested headwaters, but sometimes it can start on the slopes without connection to the channels or connected on the basis of the slopes or finally mixed, complex.

Second, in the area previously occupied by coffee, which is currently occupied by pastureland and other crops, and even in urban or periurban areas (expansion zones). Gullies generate on gentle medium and long slopes, with Oxisol–Ultisol–Gleysol, in the west of the State of São Paulo and over sandstones of the Bauru Group (Upper Cretaceous) of the Paraná Sedimentary Basin. The linear erosion studied by Salomão (1994, 1999) showed the presence of surface and subsurface flows (suspended and underground) related to the different soils along the slopes converging from the upper third of the slopes, and responsible for the development of erosion gullies (Fig. 11.10).

In the Brazilian Midwest, under the Savanna Domain, two cases can illustrate the gully process: where coffee had been planted, currently occupied by pastureland, in the south of the Brazilian Midwest, old paths used by the ox carts gave rise to numerous rills along these paths and dirt roads. These rills degraded and became full-fledged gullies. In that same region, the convex relief (hilly) with ruptures associated with shallower soils, such as concretionary Cambisols (Inceptisols or Brown Forest soils), which are found halfway up the slightly convex slopes immediately below thick Oxisols, favor a rapid progression of rills into gullies (Souza and Castro, 2000). The concentration of subsurface lateral flows

HORIZONS	TEXTURES AND STRUCTURES	HYDRAULIC CHARACTERISTICS			
		K	POROSITY		
			PT	P > 30 / PT	P < 30 / PT
1	SANDY WITH LAMELLAE (A)	117	35	0,5	0,5
2	SANDY (A)	82	38	0,5	0,5
3	SANDY CLAYISH WITH POROUS STRUCTURE (BI)	95	36	0,5	0,5
4	SANDY CLAYISH WITH POROUS STRUCTURE (BL)	89	37	0,5	0,5
5	CLAY WITH BLOCK STRUCTURE (BT)	8	36	0,2	0,8
6	CLAY (C)	2	36	0,1	0,9

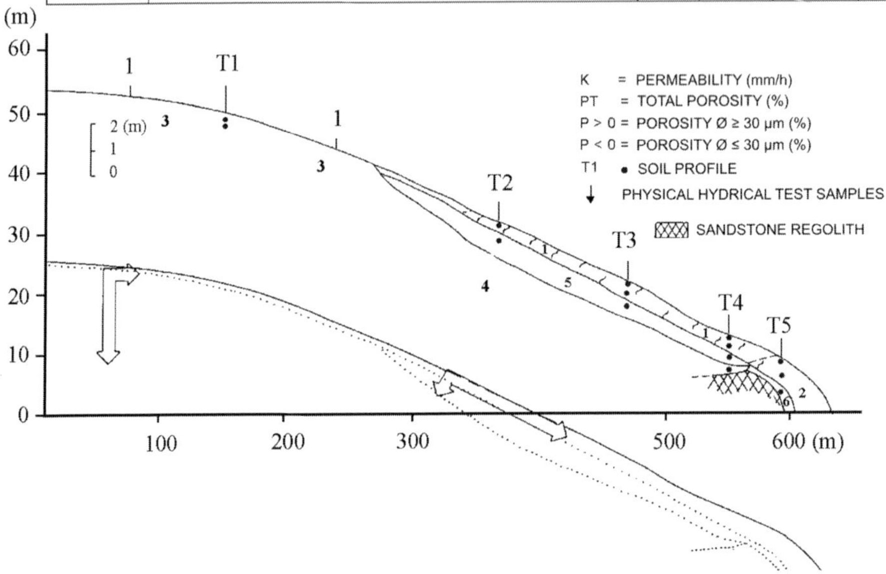

Figure 11.10 The slope, the toposequence axis and its soil profiles and the underground water flow.

above them where *piping* caused the removal of fines (clay and fine sands) also generated a sinking of the topography near the top of the watershed and formed broad zones that contributed to formation of a gully. This gullying also favors the concentration of lateral surface flows, linear erosion, and mass movement. Furthermore, the process is associated with episodes of heavy rains ($\geq 50\,\mathrm{mm\ h^{-1}}$) in the beginning of the rainy season and can be also connected to the preexistence of a path.

Still, in the savanna region, principally in the Upper Araguaia River Basin occupied by pastureland and a region that had previously been planted with soy after intense and fast deforestation, a study by Marinho et al. (2006) of a toposequence comprised of Quartzipsamments–Hydromorphic Quartzipsamments–Gleysols, implemented a piezometric station with a three-year monitoring activity. The study showed that the water table was deep (around 30 m) on

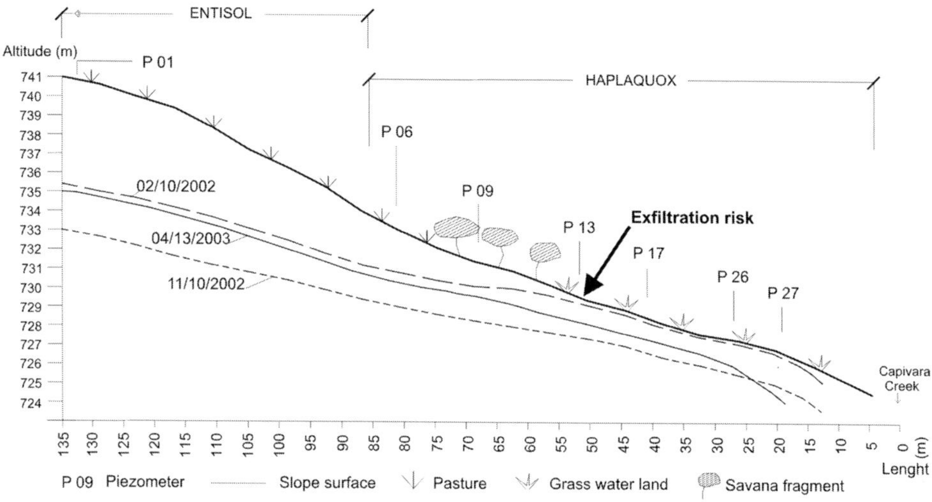

Figure 11.11 The water table level in a representative topographic section, Upper Araguaia River Basin.

the largest part of the slope until the lower third where, through the last 60 m until the stream, the hydraulic gradient was increased. Water could even exfiltrate at the end of the rainy season (February), when the flow joined the surface flow, as is shown on Figure 11.11. Both caused the appearance of gully erosion on the valley floor, which evolved headward on the slope. So this model of gully process is based on convergence of flows associated with the ascension of the water-table level. It is runoff that is related to the heavy rains and facilitated by concentrated flux over cattle paths or small roads cutting the topographic levels.

The fact that in all these cases no erosion gullies were noted when the slopes were covered with the original vegetation leads one to believe that the linear erosion originated in the hydropedological and geomorphological imbalance caused by deforestation, the absence of conservationist practices, and the choice of inappropriate sites for these crops. Therefore, the problem does not exactly lie in the coffee or soy bean, but in soil management practices that frequently have no conservational measures or in the region's lack of potential to support the crops.

However, coffee seems to have caused less water erosion than soy and that could be explained by the fact that it is a permanent, long-standing crop, which is not the case for soy, which includes mechanically intensive seasonal management. The erosion processes are located in the headstreams and in the borders of the high plateau or hillslopes, in a position susceptible to structural control by the rock substratum or strong lateral pedological changes (texture and structure) that induce subsuperficial and superficial convergence of lateral flows (Fig. 11.12).

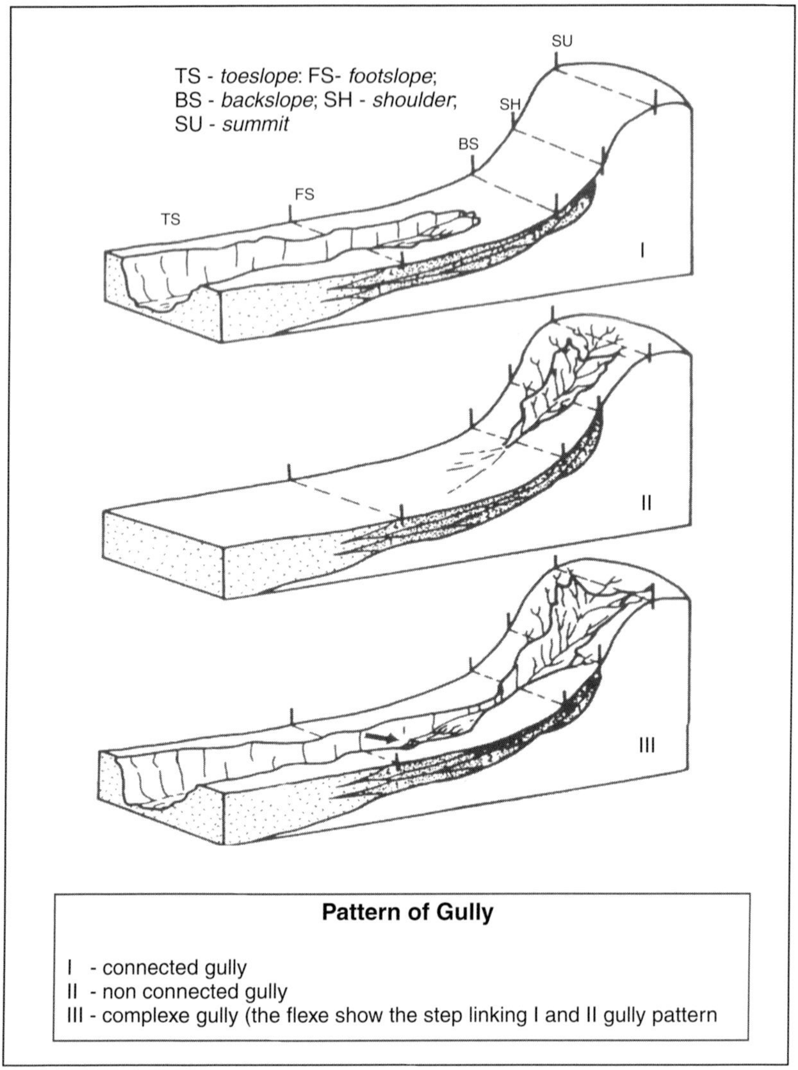

Figure 11.12 Pattern of gullies. Flexes refers to the break in slope.

5. Conclusion

Coffee seems to have had less impact on water erosion than soy crops. That finding can be explained by the morphoclimatic domains that they occupied, since these two crops require different types of management. One is a permanent and long-standing crop (coffee) that causes little interference with the soil, whereas the other (soy) intensely interferes with the soil every year, because of crop rotation, which is required to control fertility.

It is also important to remember that the start of the gullying process can be controlled by preferential lineaments or differential soil catenaes along the slopes. Both factors can control the development of subsurface piping, and can also be initiated by concentrated flows related to the cattle paths, roads, as well as the absence or inadequacy of conservational practices in soil management.

In this sense this chapter shows that the relationship between use, management, and water erosion increases exponentially because crops tend to incorporate areas that are not adequate to this activity. This is particularly true of fragile, sandy soils with a low degree of cohesion and a high susceptibility or strong change in texture and structure in soil cover on the river floodplains, which facilitates the initiation of piping. That is the essence of the problem—the hydrological unbalance initiated in the risk zones. So, it seems that the absence or inadequacy of management planning, especially in these zones, is a major concern.

CHAPTER 12

Landslides and Disasters in Southeastern and Southern Brazil

Ana Luiza Coelho-Netto[a,*], André de Souza Avelar[a], *and* Willy Alvarenga Lacerda[b]

[a]*GEOHECO/Geo-Hydroecology Laboratory – Geography Department, Federal University of Rio de Janeiro – UFRJ, Brazil*
[b]*Civil Engineering Program – Geotechnical Department Federal University of Rio de Janeiro – UFRJ, Brazil*

Contents

1. Introduction

Disasters constitute events, either natural or provoked by humans, that have an adverse effect on a vulnerable ecosystem, causing human, material, and environmental damages, and consequently economic, ecological, and social loss. The intensity of a disaster depends on the interaction between the magnitude of the adverse event and the system vulnerability, and is quantified as a function of damage. Generally, disasters are of mixed nature, natural and man-made. The present text focuses on disasters derived from extreme rainfall and related to the occurrence of mass movements, or landslides, in watersheds under different land

* **Corresponding author**: *E-mail address: ananetto@acd.ufrj.br, Address: Geography Department, IGEO-UFRJ, Av. Athos da Silveira Ramos, 274, Ilha do Fundão, CEP 21941-590, Cidade Universitária, Rio de Janeiro, RJ, Brazil Tel.: (0 55 21) 2590-1880*

Developments in Earth Surface Processes, Volume 13
ISSN 0928-2025, DOI 10.1016/S0928-2025(08)10012-8

uses. Today, the intensity of disasters is potentially increased by global climatic changes, linked to accelerated and disordered growth of large urban-industrial centers and to the concentration of poorer populations in high-risk areas. On the other hand, the degradation of forests and, therefore, the loss of hydrological and mechanical functions, make the slopes vulnerable and create the potential for risk-phenomena occurrences that can propagate in headwaters up to the main channel trunk or even up to reserve terminals, especially in the case of coastal basins that drain directly to lagoon systems or to the sea.

This chapter focuses on disasters that occurred in the southern and southeastern regions of Brazil, in close association with mountainous physiographic domains (Fig. 12.1). These phenomena can attain a catastrophic dimension during intense rainfall periods, particularly in densely inhabited regions with slums concentrated in risk areas. The catastrophic dimension could be estimated from the capacity for resilience of natural ecosystems or the capacity for recuperation and functional rehabilitation of degraded land and the associated social and economic systems. These extreme landslide events produce an excess of sediment yield that converges into the channel network that drains the watersheds, causing immediate mass sedimentation of the drainage system in the lower plain or coastal flatlands (so-called *baixadas*). Consequently, this rapid sedimentation tends to favor large floods that may

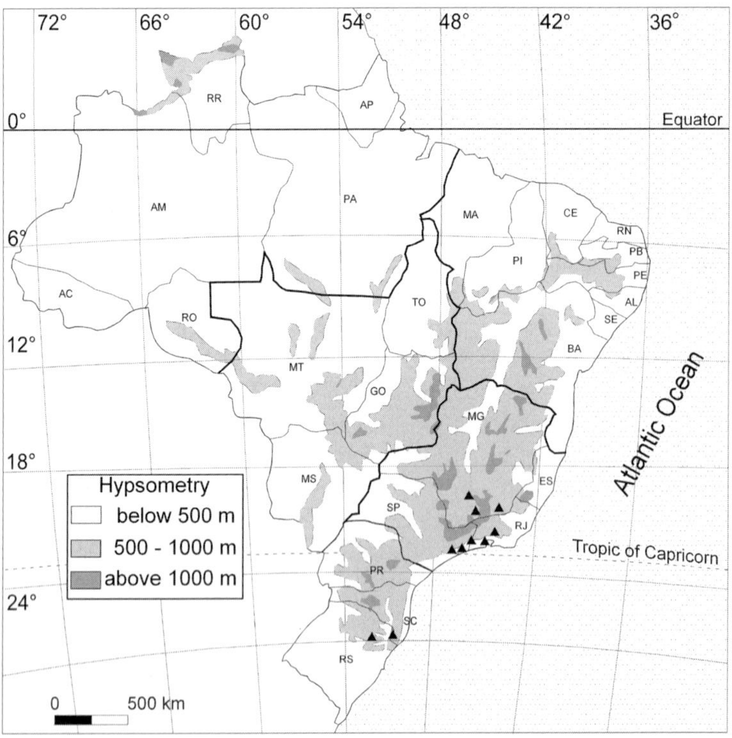

Figure 12.1 Location map of areas where major landslide events occurred in southeastern and southern Brazil.

also cause serious damages and loss in inhabited urban and industrial plains or even in areas of forest, agricultural, or pasture production areas. Landslides can even catalyze the breakup of roads, ducts, and tailings dams; these last two cases may cause the spreading and toxic contamination of soil and water.

Landslides causing disasters represent one of the principal problems related to slope occupation in Brazil. The majority of these phenomena occur in the mountainous steep lands of the *serras* and coastal ranges along the Brazilian Atlantic region. These disasters are more common in these domains because, since Portuguese colonization, the rising and expansion of the country's major cities have been concentrated on this region. Historically, urban colonial nuclei started along the coastal lowlands and Quaternary fluvial plains, expanding over smooth hills and rapidly during the twentieth century as urban-industrial centers toward the mountainous coastal massifs and *serra* region. As an example, Figure 12.2 illustrates the spread of the metropolitan area of Rio de Janeiro city around and onto the steep slopes of the coastal massifs: Tijuca massif on the right side, Pedra Branca massif on the left side, and Gericinó massif in the northern area.

Coastal massifs and mountainous escarpments that drain to the Atlantic Ocean can reach altitudes above 1200 m, while inland the mountainous escarpments can reach an altitude up to 2000 m and relief amplitude of more than 700 m. On lower hills, relief amplitude ranges between 200 and less than 100 m, while gently inclined and wider valley bottoms prevail. The origin of mountainous relief along the coastal

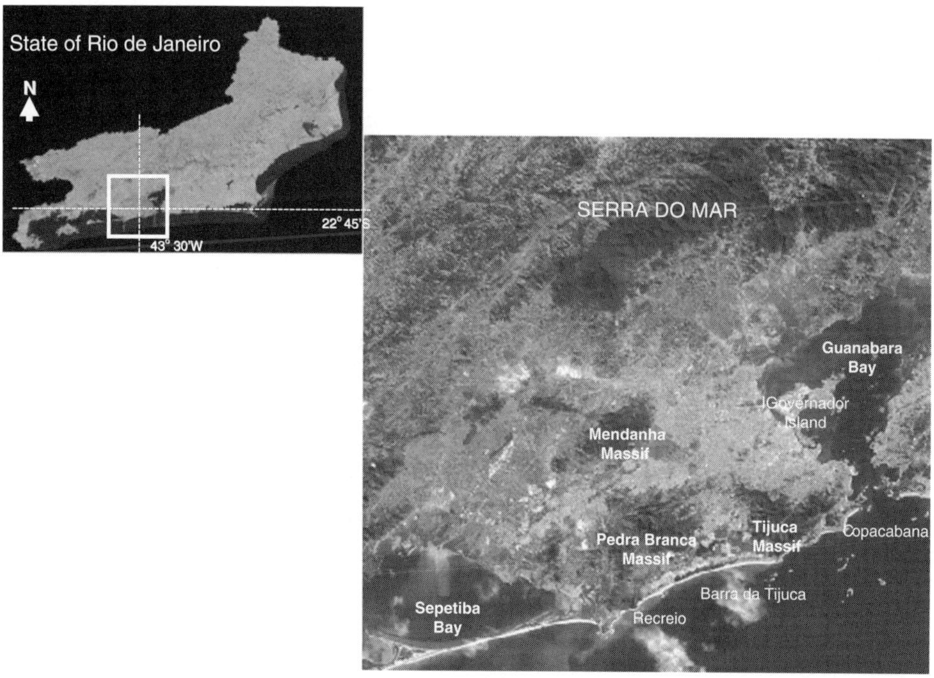

Figure 12.2 The mountainous compartments of Rio de Janeiro municipality and surrounding metropolitan region (see above the area location at Rio de Janeiro state). One may observe a portion of the Serra do Mar and the coastal massifs surrounded by the densely urbanized and flat lowlands called *baixadas*. Forest remnants are still found in these compartments.

strip is related to geomorphological processes unleashed by tectonic movements responsible for the opening of the Atlantic Ocean starting in the mid-Mesozoic Era.

Overall, it should be emphasized that in the southeast region of Brazil, more than tectonic actions, the actual configuration of relief is associated with lithologies and structures originating from the Brazilian thermotectonic event, configuring an extended fold belt dated between the Middle and Upper (Precambrain) Proterozoic Era. In these regions, metamorphic rocks of high grade occur that are characteristic of the presence of gneisses, charnockites, schists, amphibolites, and phyllites associated with plutonic igneous rocks, principally granites, but also syenites and gabbros. Due to the marked occurrence of metamorphic foliation, folding, thrust fault, and fractures in these rocks, there is clear structural control of geomorphological work, resulting in the lithostructural orientation of drainage channels, valleys, and divisors. On the other hand, in the southern region, the *sierras* and escarpments are predominantly composed by the intercalation of sedimentary rocks (principally sandstones) with basaltic layers, generating tabular landforms with coastal and inland steep escarpments that can reach elevations of 1500 m.

On the Brazilian coastline, a Tropical Humid climate prevails with periods of more rainfall and events of extreme rainfall concentrated in the summer. The hot and humid climate on the Brazilian coastline area favors chemical weathering of the previously mentioned rocks so that the occurrence of thick saprolite mantle is common in the hill domain. On the other hand, on steep mountain slopes and coastal massifs, there is no stability to allow the development of thick saprolites and frequent landslides are responsible for the exhumation of rocky escarpments. At the base of rock escarpments, thick slope deposits (talus and colluvia rich in blocks) may accumulate, as well as in concave up topographic hollows at headwater valleys. Even under a relatively inferior slope of around 17°, which permits human occupation and the growth of cities on the slopes, colluvial deposits are susceptible to movement, thus raising the risk of disasters during rainy periods.

Landslide conditions and problems are highly complex involving functional relationships among natural elements (climate, topography, rocks, soil, vegetation and fauna) and artificial elements (related to land use) of the landscape, which may change over space and time. Critical landslide hazards have prevailed along the highly populated coastal region, especially in the upper portion of steep mountain slopes. In this chapter we draw special attention to case studies of historical landslide disasters in the most populated area of Rio de Janeiro State, and also include some other case studies in southeastern and southern Brazil.

2. Interaction between Rainfall and Landslides

Along the SE-S coastal region occur the highest pluviometric indexes in the country, which can surpass 5000 mm of annual rainfall in some mountainous localities. More intense rainfall is associated with the action of Tropical cyclones, and recently in March 2004 the first occurrence was registered of a hurricane between the Rio Grande do Sul and Santa Catarina coasts. According to the National Oceanic and

Atmospheric Administration (NOAA), the so-called Hurricane Catarina attained level 1 on the Saffir-Simpson scale (winds between 124 and 164 km/h) and destroyed 20,000 houses along the small coastal cities. Meteorologists, however, emphasize that some predictive models indicate the possibility, however remote, of hurricane occurrence along the Brazilian coast between the states of Rio Grande do Sul (RS) and Rio de Janeiro (RJ). Some locations in this hurricane risk area have mountainous reliefs with high population density, especially concentrated in the surrounding lowlands.

The relationship between rainfall and landslides in SE Brazil has been examined by Pichler (1957); Nunes (1969); Guidicini and Iwasa (1976), and others. Guidicini and Iwasa (1976) showed that rainfall of more than 250–300 mm could trigger landslides under all conditions. In order to predict landslide risk, they proposed the so-called Cycle coefficient (*Cc*), coefficient episode (*Ce*), and coefficient final (*Cf*), as indicated by following equations:

$$Cc = \frac{\text{accumulated rainfall up to the intense rainfall event (mm)}}{\text{mean annual rainfall (mm)}} \quad \text{(eq.1)}$$

$$Ce = \frac{\text{rainfall event (mm)}}{\text{mean annual rainfall (mm)}} \quad \text{(eq.2)}$$

$$Cf = Cc + Ce \quad \text{(eq.3)}$$

Accumulated rainfall up to the intense rainfall event that caused landslides was applied for SE Brazil; accumulated rainfall was estimated after July 1. The sum of *Cc* and *Ce*, called the "coefficient final" (*Cf*), was plotted against the months of the year to indicate four landslide risk conditions (A, B, C and D). Condition A represents greater landslide risk: the amount of accumulated rainfall attains greater values than mean annual precipitation in association with intense rainfall (e.g., Fig. 12.3). The Guidicini and Iwasa coefficients have been acceptable for predicting landslide risk, but they are based on data from sites far apart and related to rainfall of distinctive nature. Therefore, their findings do not necessarily apply to one isolated rain-gauge station, as found by Pedrosa (1994).

Landslides are usually induced by less frequent, intense, and spatially nonuniform rainfall events, especially during the summer rainy periods. Landslide magnitude may vary for similar rainfall inputs due to landscape component variables and related structures at surface and subsurface levels. Two characteristic rainfall regimes that might express rainfall-landslide relationships in southern and southeastern Brazil are shown in Table 12.1. The first concerns the rainfall regime after the middle of the year marked by less intense events of longer duration, in regular years, with a gradual increase of rainfall toward the rainy season. Given these conditions, the following mass-movement types may occur: deep rotational failures along channel banks and road cuts, colluvium deposit creep along valley bottoms, and shallow translation failures in Lithosols. The second regime is characterized by extreme rainfall, especially in January and February when heavy rainstorms tend

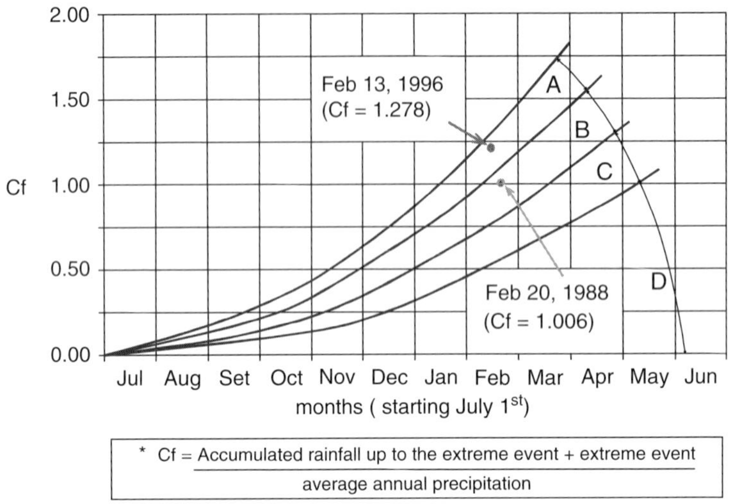

Figure 12.3 Landslide-rainfall curve given by Guidicini & Iwasa (1977); coefficient applied to two extreme events at the Tijuca Massif in Rio de Janeiro: the 1988 event fits well with type B, and the 1996 event shows higher coefficients ($Cc = 1.064$, $Ce = 0.175$ and $Cf = 1.239$) associated with type A. These two cases will be discussed in this chapter.

Table 12.1 Rainfall Regimes, Location, and Major Landslide Types

Regular Rainfall Regime		Extreme Rainfall Regime	
Location	**Landslide Type**	**Location**	**Landslide Type**
Valley bottoms and thick colluvium	Creep	Rock escarpments	Rock falls
Steep slopes and Lithosols	Translational failures	Steep slopes and river channels	Debris flow and rock debris Avalanches
Channel banks and road cuts	Deep rotational failures	Shallow soil or shallow root zone	Shallow slides or translational slides

to be less frequent but with extreme rainfall amounts. The main types of mass movements provoked include debris flow convergence at the adjacent valley bottom; shallow slope and translational failures on grassland slopes; and rock falls on vertical scarps. The spatially nonuniform distribution of landslides may also reflect the role of different local environmental conditions in controlling slope stability at a certain site, or in increasing susceptibility to landslides.

3. Mechanisms and Conditions Initiating Landslides

As mentioned earlier, major landslide types on steep slopes can be placed in the following categories: rock fall, debris flows, and shallow translational slides; rotational slides are common along road cuts or riverbanks. Table 12.2 summarizes the main landslide types, their respective initiation mechanisms, and environmental trigger conditions. Rock falls may occur near the mountain-peak zone, particularly when composed of granite or very coarse gneiss rocks associated with sheet and vertical joint settings, thus showing that underlying structural elements play an important role in generating this type of mass movement that usually occurs wherever forest cover has been degraded or already lost. Also, a water pressure increase in the void-joints tends to trigger rock falls, especially at the end of the wet season, during extreme rainfall intensities.

Debris flows are characteristic of steeper mountain slopes, as previously shown by Avelar (1996); Bovis and Dagg (1992), and Sassa (1985), and their initiation is related to two conditions. The first condition occurs during or immediately after the highest one-hour rain intensity, especially at the end of a wet season, when pore-pressure suddenly increases at the base of loose colluvium deposits, and the second occurs where there is forest degradation or deforestation since debris flow

Table 12.2 Mechanisms for Landslide Initiation and Related Environmental Conditions for Major Landslide Types along the Brazilian Coastal Zone

Landslide Type	Mechanism for Initiation	Environmental Conditions
Rock fall	Increasing water pressure in the joints	Late rainy season and intense rainfall Mountain peak zone Granite or very coarse gneiss Subvertical joint system Degraded forest cover
Debris flows	Sudden increase in pore pressure at the base of the colluvial mass	Late rainy season and extreme rains Steep and long slopes Loose colluvium
	Mass shocking against loose colluvium	Topographic hollows
		Forest degradation and root strength deterioration
Shallow translational failures	Rapid loss of suction	Wet season and extreme rainfall High and midslope Permeable and shallow soil Soil–bedrock boundary Base of dense root zone

initiation risk rises due to deteriorated root strength. Usually, these landslide scars reach the summit slopes so that the adjacent valley bottom is fed extensive rock debris avalanches along the main river channel. Another cause of debris-flow initiation is associated with the rock slab impact or failure masses in quasi-saturated colluvial deposits at the valley bottom (Avelar, 1996).

The last type of mechanism was described in Japan and Canada, respectively, by Sassa (1985) and Bovis and Dagg (1992), and it was experimentally demonstrated by Avelar and Lacerda (1997) using a geotechnical approach in undrained triaxial shock tests on colluvium samples. Avelar and Lacerda indicate shear behavior and strength parameters in the colluvium samples obtained from conventional triaxial tests (consolidated isotropic undrained, or CIU, tests), as presented in Figure 12.4 and summarized in Table 12.3. The data show the results for a friction angle equal to 29.9° and for cohesion equal to 9.2 kPa. However, according to undrained triaxial shock tests, the loads (dynamic loads) necessary for inducing soil to failure is equivalent to only 25% of the static failure load. The trend of the stress path in diagram P′-Q (Fig. 12.5) indicates that the liquefaction condition could have been reached. Hence, shock of rock blocks or a previous failure mass against the colluvium can lead to a liquefaction of the soil wherein debris flow initiation occurs and progresses downslope in a "domino" effect.

Shallow translational slides commonly occur at the mid-upper portion of mountain slopes on any slope geometry associated with shallow hydraulic discontinuities at the soil–bedrock contact zone or at the base of dense root systems (relative to forest or grass cover). Root systems favor preferential water flow paths (Jansen, 2000; Silveira et al, 2005), so that saturation or near- saturation conditions can be reached at the root base. Yet these flow paths provide rapid loss of suction and trigger slope failure during extreme rainfall events, especially at the end of the wet season. Materials move down slope into the main channels and run out as flow, showing the close relationship between slide-flow transformations as another important cause of down valley catastrophic effects. This mechanism has been reported recently by Takahashi (2000) and Iverson et al. (2000); these authors point out that shear strain in the lower parts of the soil profile can lead to liquefaction. Local and isolated translational failures do not have the large soil volume that would cause important human damages. Usually, these local slides are shallow and are generated by a loss or a decrease in suction just below the root zone of the soil profile during intense rainstorms.

Rotational slides are observed mainly along road cuts in saprolitic soils that are more than 5 m thick. In most cases the soil matrixes are block-rich and sand-clayey or clay-sandy. These failures are not frequent, but the volume of collapsed materials is usually very large. Major damages are associated with road cuts.

Several works by geomorphologists and geotechnical engineers stand out in their relevancy to landscape elements and the parameters that play important roles in controlling landslide mechanisms in south and southeast Brazil, such as Avelar and Coelho Netto (1992), Fernandes and Amaral, 1996, Oliveira et al. (1996), Vieira et al. (1997), Lacerda (1997), Cruz et al. (1998), and Coelho Netto et al. (2005). Table 12.4 summarizes major variable-controls for landslides, especially in mountainous areas.

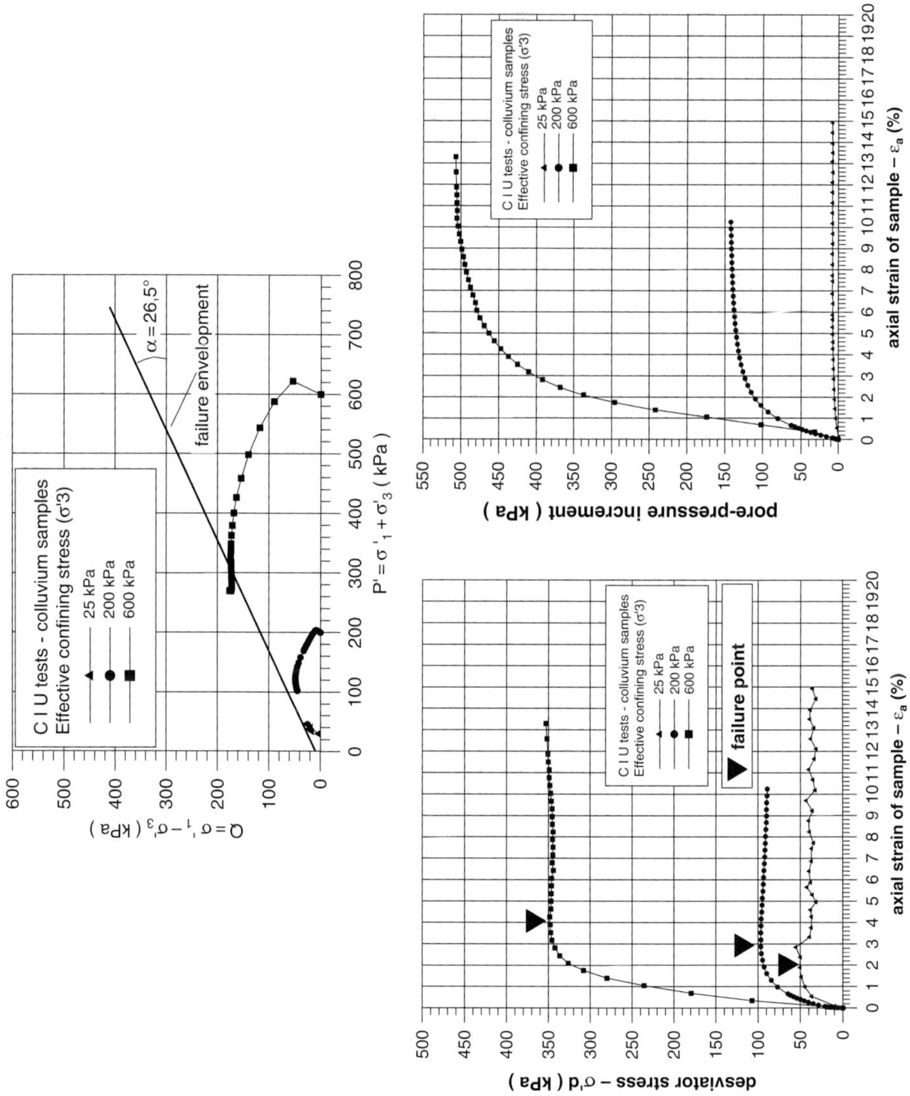

Figure 12.4 Colluvium sample shear behavior in CIU tests (Avelar, 1996).

Table 12.3 Shear Strength Parameters at the Failure in Colluvium Samples (Avelar, 1996)

σ'_3 (kPa)	σ'_d (kPa)	ε_a (%)	ϕ' (degrees)	c' (kPa)
25	50	2.0		
200	100	3.0	2.,9	9.2
600	350	4.0		

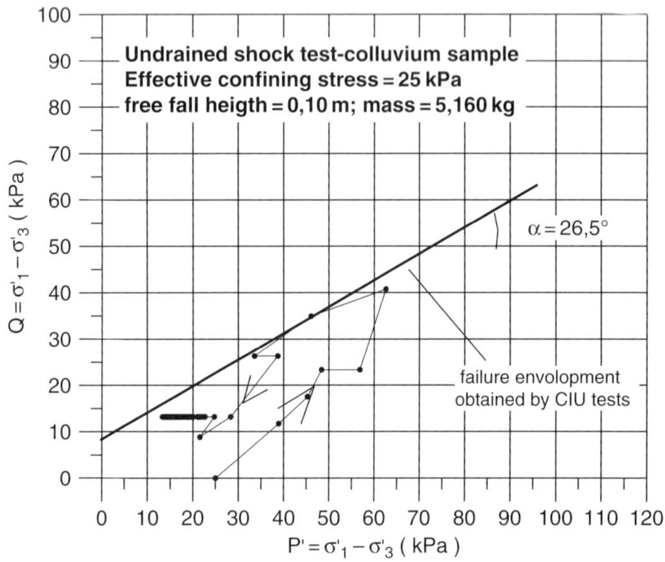

Figure 12.5 P′-Q diagram for undrained shock test in colluvium (Avelar, 1996).

4. Landslides and Disasters

The occurrence of landslides is expected in the mountainous steep slopes, but they can also be induced or potentiated by human activity, and especially in urban areas, causing diverse damage to the population, public authorities, and private institutions. It is important to emphasize that landslides can be initiated far away from the urban site and propagate down to the dense populated urban areas to cause serious damages and losses. A recent case (December 10, 2002) at Areal village, in the coastal municipality of Angra dos Reis (Rio de Janeiro), illustrates this spatial interaction (Fig. 12.6). An extensive slide occurred in a hanging valley head still covered by the rain forest and produced large amounts of sediment that drowned the adjacent creek. In a short time, water retention upstream from this sediment barrier collapsed and fed a block-rich debris flow that reached the adjacent *baixada* in a few minutes during the night. Other minor cases related to the same rainfall event occurred in this area. In total, more than 90 houses were destroyed, and serious damage affected another 936 houses; 1500 inhabitants lost their houses and 39 died.

Table 12.4 Major Variable-Controls for Landslides in Mountainous Areas of Southern and Southeastern Brazil

Landscape Element	Relevant Controls
Drainage basin	Hierarchical order Gradient (ΔZ/length of main axis) Drainage density (Σchannel length/area) Slope geometry (convex, straight, concave) Slope profile (straight or stepped)
Rock type and structure	Mineralogy Texture Joint system Bedding discontinuities
Soil (residual or transported)	Grain size curve Thickness Clay content Hydraulic conductivity Organic content Morphology and fabric
Vegetation	Type and density State of conservation Root system (depth and density)
Land use	Road cuts Urban building Artificial drains (flow paths) Slums
Hillslope hydrology	Rain-infiltration ratio Flow paths Water pore pressure Artesian flows Flow discharge at seepage faces

Even acknowledging that retention works could resolve the problem satisfactorily, it is also known that even in contained locations or in the immediate surroundings of landslide scars, new landslides continue to occur. Moreover, it is important to note the weak power of spatial and temporal prediction of landslide hazards whatever the scale may be, and even more so in cities with complex geology, geomorphology, and occupational and economic situations as is the typical case of Rio de Janeiro and similar landscapes. The ruptures of these civil engineering works could be related to a lack of adequate boring, bad samples, low quality and quantity of trials, incorrect execution of work, or another unfavorable political or economic parameter. In all, the failure of a work, or the difficulty of predicting landslides, can probably be linked to the lack of scientific understanding of the environmental elements that compose the slope, and, principally, to the

Figure 12.6 Landslide case in the coastal steep slopes of Serra do Mar at Areal village in Angra dos Reis Municipality, Rio de Janeiro, December 2003. A: block-rich debris flowing on the upper steep slopes; circle: location of the initial slide; B: landslide routing down into the *baixada* urban area. (Photo: Paulo Ney de Mello).

difficulty of knowing the processes that direct the interaction between those elements. In this way, it is only possible to raise the level of knowledge of this theme if detailed studies on initiation processes of these occurrences are undertaken.

4.1. Landslides and Disasters in the Municipality of Rio de Janeiro

Over the last 50 years the most critical landslide events were recorded in the mountainous geoecosystem of the *Tijuca* massif in the east of the municipality (see location in Fig. 12.2). The upper *Tijuca* massif is the wettest area of Rio as recorded at the *Capela Mayrink* rain-gauge station (470 m.a.s.l.) located at the basin outlet of the upper *Cachoeira* River (3.5 km^2) within the *Tijuca* National Park forest reserve.

Annual rainfall totals range from 1300 to above 3000 mm. Intense rainfalls are mostly associated with the warmest month, February (Coelho Netto, 1985; Silveira, 1998), due to the strong thermal impact of the cold Polar air mass against the warm and wet Atlantic Tropical air mass. In the upper portion of the massif, the average monthly rainfall is around 100 mm. In the high summer of regular years, mean monthly rainfall is around 250 mm. Extreme monthly totals may reach more than 900 mm, particularly due to an increasing frequency of intense rains above 100 mm/day.

This massif is composed of Precambrian gneiss and granite rocks striking NE-SW (Heilbron, 1995). Main geological structures are rock-bedding planes, sheeting joints, and subvertical joint settings, most of which strike NW-SE and NE-SW. The interconnection between subvertical joints and sheeting joints favors the *in situ* origin of rounded blocks and rock slabs. The round-shaped rocky peaks (sugar loaf or bornhardts), coupled with long and steep rock escarpments (slope angle >50°), are conspicuous morphological features that operate as recharge zones deep into the soil profile and joint system (Coelho Netto, 1985).

Local bedrock knick points produce high waterfalls and provide stepped river valleys so that relative lower topographic gradients will prevail in minor hanging valleys. Block-rich slope deposits of the Late Quaternary and recent times get thicker at the foot of the rock escarpments and along gentler inclined valley bottoms. Shallow regolith prevails on steeper valley-sides and becomes thicker toward the axis of the concave-up topographic hollows. Lithosol, Cambisol, and Latosol prevail in the mid-upper portion of the massifs; Podzolic soils occur in the lower portions.

Mass movements of great dimension causing damage to the city of Rio de Janeiro began in 1962 with the landslide event of the Morro do Querosene (Barata, 1969). Starting in this period, growing slope occupation provoked by the expansion of the city made mass movements greatly effective, more frequently in the summer, causing material damage and loss of life.

During the summers of 1966 and 1967, the city suffered various extremely grave events caused by highly intense rainfalls of long duration. Barata (1969) pointed out that these rainfalls surpassed all previous maximum levels recorded in the city up to then, still provoking the aggravation of being preceded for various days by moderate rainfall. Meis and Silva (1968) analyzed the previous 11 years of rainfall and showed that the precipitation total for January 1966 reached 617.6 mm, of which 472.0 mm were recorded over 72 consecutive hours on days 11, 12 and 13. For February 1967, these authors described a total of 432.0 mm, with 299.5 mm concentrated in 48 hours on days 19 and 20.

Extreme events also occurred in February-1988 and in February 1996. Figure 12.7 shows average monthly rainfall distribution at the *Capela Mayrink* rain-gauge station for the period 1982–1996 in comparison with the total rainfall in February 1988 and 1996 and respective previous monthly precipitation. The highest total amount (967.7 mm) was recorded in February 1988, and a higher number of rainy days was reported at that time than in February 1996 (776.3mm/month), when rainfall reached a very extreme 24-hour-intensity on February 13.

Both February 1988 and February 1996 followed a previous six-month period that was relatively wetter than the average monthly totals for the whole recorded period. In February 1988 the critical rainfall/day reached 177.6 mm/day, while in

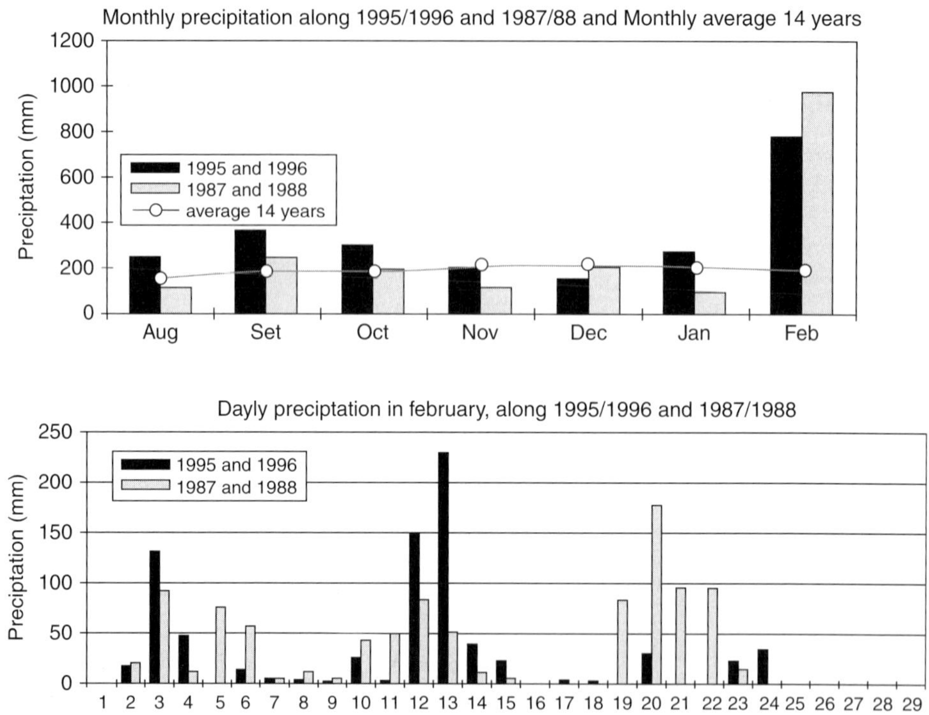

Figure 12.7 Average monthly rainfall distribution at the Capela Mayrink rain gauge station (1982–1996) in comparison with the total rainfall in February 1988 and 1996 and the respective previous monthly precipitation (Coelho Netto, 1996).

February 1996 the critical rainfall attained 380 mm in less than 24 hours. As shown previously in Figure 12.3, the 1988 event fits well with the Guidicini and Iwasa coefficient of type B. Guidicini and Iwasa's coefficient applied to the extreme event of February 1996 gives a relatively higher coefficient ($Cc = 1.064$, $Ce = 0.175$, and $Cf = 1.239$) and fits well with "type A," as shown in Figure 12.2. These events followed a long period of irregular air temperatures (above 40 °C) due to the effects of El Niño.

4.1.1. Long-term Case Study on the Soberbo Slopes: January 1966 to 1990

In Rio de Janeiro this long-term landslide case was the most detailed one studied, constituting an outright field study about landslide mechanisms and conditions. Its focus was an extensive mass movement that started on the Soberbo slope in response to January 1966 rainfall. This movement was primarily described by Meis and Silva (1968) using a geomorphological approach. Soon thereafter, Fonseca (1969) gave a detailed geotechnical view of the case, with information in common with the previous work showing the importance of a geomorphological and geotechnical interface for better knowledge and understanding of slope problems.

In 1983 a new dislocation was observed on this slope. Detailed geotechnical studies were undertaken by Lacerda and Sandroni (1985) and later by Pedrosa, Soares and Lacerda (1988). These works led to the elaboration of new field and laboratory studies, as presented in the following articles: Lacerda and Schilling (1991), Schilling et al. (1992), Silveira and Lacerda (1992), and Barros et al. (1992). More recently, Avelar (1996, 2003) and Avelar and Lacerda (1997) widened the historical investigation on this case and through experimental studies sought to demonstrate the mechanisms of initiation of this landslide case, which assumes catastrophic dimensions throughout its course and at the bottom of the adjacent valley.

According to Barros et al. (1992), the first sign of mass movement on the Soberbo slope was observed around 6 P.M. on January 14, 1966, and it was described as an avalanche of great proportions. It ran about 800 m and was generated after a great rainfall when a block of rock became dislocated by gravity from the highest part of the slope at about 430 m at the summit of the Soberbo Road. In accordance with Figure 12.8, this initial movement started in region A, with a partial rupture of the Soberbo Road. Barros et al. (1992) obtained site information from local inhabitants relating that 48 hours after the first avalanche, two new mass movements of distinct manner occurred in regions B and C. In region B, the propagated translational movement was

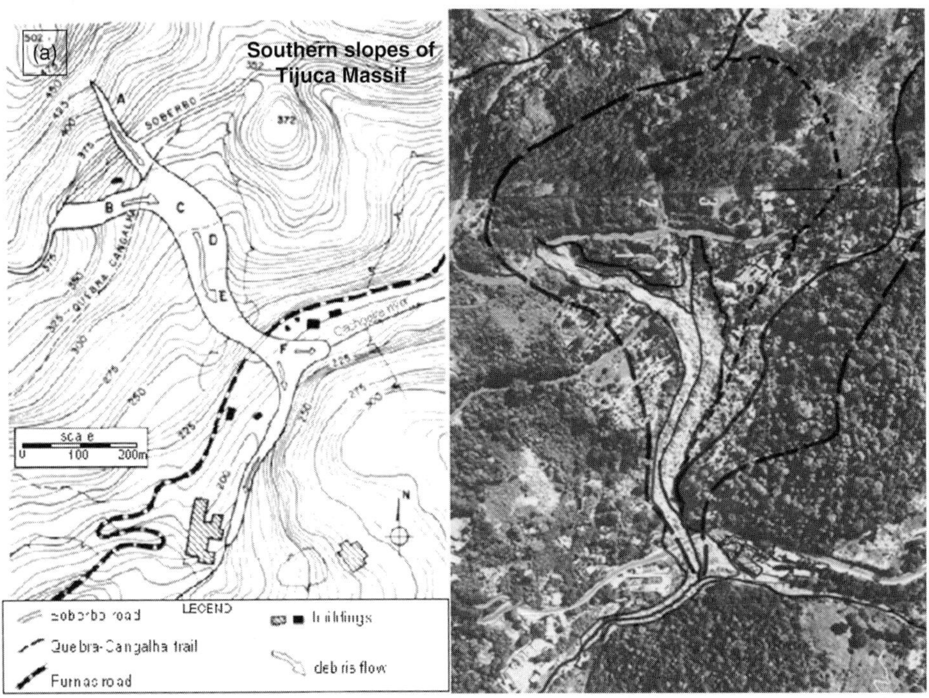

Figure 12.8 The topographic map (A) shows the location of the Soberbo mass movement in the southern slopes of Tijuca Massif and buildings and roads destroyed by this event (Pedrosa et al., 1988). The drainage area contour (dashed line) and land use around the landslide are shown in the adjacent aerial photograph.

rising up, while in region C the terrain suffered degradation on a narrow strip about 4 m wide that originated from loss of material.

Pedrosa et al. (1988) suggested that the region of mass movement is the concentration zone of superficial and subsuperficial water fluxes. Taking into account that there was a heavy rainfall of 675 mm over the three days preceding the movement, certainly colluvium mass saturation occurred, causing a sudden rise in pore pressure along the surface of the rupture.

Among the principal consequences of this great mass movement described by all previously cited authors was the destruction of the Franco-Brazilian Paper Company factory located on the margins of the Cachoeira River, which suffered a blocking raising the water level to almost 8 or 10 m, inundating, destroying, and killing people in the adjacent residences. The Furnas Highway (Fig. 12. 6) was totally obstructed by deposited landslide materials. The Quebra-Cangalha Path (or Santo André Path) that crosses the principal axis of the mass movement at the meeting point of regions A and B also suffered a dislocation downslope of about 50 m, showing a grade difference of about 5 m. Well tubing and trees also showed intense dislocations. Soberbo Road, together with region A, was damaged by mass movement. It was reconstructed in 1967 with a concrete curtain-wall anchored at a 5-m height. (Pedrosa et al., 1988).

Barros et al. (1992) indicated that after starting in 1966 the mass movement propagated slowly and climbed more than 50 m along region B reaching the Soberbo Road in 1983. During this year, the local prefecture built an anchored concrete curtain-wall almost 20 m long, but still, superficial draining gutters were destroyed due to continued movement. After completion and with the elapse of movement, sinking of the roadway caused by the loss of material under the curtain-wall foundation was observed. To reinforce the structure and stabilize the road, another panel was built 10 m long by 4 m high.

According to Schilling et al. (1992) and Schilling (1993), however, the concrete curtain-wall anchored in region B was built in 1977, having been partially destroyed in 1987 and totally so in 1988. Moreover, the authors relate the destruction of drainage gutters as well as the complete obliteration of instruments installed in 1987 (Pedrosa et al.., 1988). Despite the destruction of these instruments in 1988, Schilling (1993) installed new ones in a wider range to regions A, B, C, D, and E and continued studies between 1990 and 1992. With this, the quantification of mass movement velocity was enabled: from oscillation in subsuperficial water level, variation in piezometric levels, and precipitation intensity.

The dislocations registered by Pedrosa et al. (1988) via superficial marks attained magnitudes of 80 mm on the uphill part of region B, up to 300 mm and the downhill part where regions B and C meet. In the subsurface, inclinometers showed dislocations of about 90 mm close to a depth of 9 m in the most uphill part, 300 mm at 6 m in the central part, and 250 mm at 3 m on the downhill part where B and C meet. These movements were registered for a period from March to August 1987 (158 days).

The marks installed by Schilling (1993) registered superficial movements varying from 10 mm in the most uphill part up to 4200 mm where regions B and C meet, with readings from August 1990 and September 1991. The inclinometers showed dislocations up to 150 mm at about 7 m depth and at the meeting point of regions B and C, as well as other minor ones in regions A, C, and D. The piezometric level observed by Schilling (1993) attained artesian conditions in regions C and D not only for the piezometers installed in the soil and in rock, which is explained by the blocking effect provoked by the presence of dikes of diabase deposits transversally positioned in these two regions (see Fig. 12.8A). This same author observes that in region B a dike is oriented parallel to the movement axis, making it easier for subsuperficial water to flow, avoiding artesian pressure buildups. Still in agreement with this last author, the diverse piezometers installed on the slope showed a tendency to respond to rainfall accumulated over 25 days, a fact that Pedrosa et al. (1988) had already observed.

For the period 1986–1990, Schilling et al. (1992) and Schilling (1993) analyzed the probabilities of new ruptures on the Soberbo slope using the Guidicini and Iwasa (1977) graph. They showed that from July 1990 the tendency to rupture attained the band of 50 to 100%. To avoid such a rupture, an official alert was sent to the Geotecnical Bureau of the City of Rio de Janeiro stating the necessity of implementing deep subhorizontal drainage works. These works were actually implemented, causing the piezometric level to drop and diminishing the mass soil movement efficiently, permitting the slope to be stabilized.

4.1.2. Catastrophic Event in the Western Tijuca Massif, February 1996

On February 13, 1996, a rainfall of "200 years" recurrence period fell on the city of Rio de Janeiro. The rainiest area was located in the uppermost, western portion of the Tijuca massif. A local rain-gauge station recorded 380 mm in less than 24 hours, concentrated in two major storms: first, in the morning (reaching 150 mm); the second reaching 230 mm, and the maximum rain intensity of 50mm/h occurring around 7:50 P.M. and 8:50 P.M. (Coelho Netto, 1996).

The first storm triggered several minor landslides along the Canoas and Furnas roads (southern slopes). Catastrophic landslides spread simultaneously in response to the second storm (around 8:40 P.M.), beginning very close to the crest-zone or near the shoulder of lower interfluves (Fig. 12.9). Extensive rock debris avalanches (>4 km length) reached the adjacent flatlands of Jacarepaguá, causing serious environmental damage and property losses, human diseases, and epidemics, in addition to 70 human deaths (Fig. 12.10).

Landslide sites were spatially nonuniform within the western Tijuca massif, mirroring the role of local environmental conditions in controlling slope stability at a certain place, or providing an increasing susceptibility to landslides. Using aerial-photos (1: 10,000) coupled with field observations; we have distinguished some landscape elements that might have affected the spatial pattern of landslide scars, as described below.

In the crest zone affected by landslides, between Papagaio peak and Taquara hill, highly jointed, granite rocks and many *in situ* blocks held in place by tree roots are dominant features. The crest zone drains down toward the Cachoeira River, in the southern Tijuca massif, where only a few landslide cases have occurred probably because of two aspects: (1) the forest vegetation is well preserved (Tijuca National

Figure 12.9 Above: landslide scars at the western slopes of Tijuca Massif in February 1996 and the main routing of a block-rich debris avalanche following the Quitite River channel. Below: Detailed view of landslide scars within the rectangle shown above.

Park), and (2) the upper Cachoeira River is hanging; so its average slope gradient is around 18° and other minor hanging plateaus occur. Landslides did not propagate far apart from their initial places: coarse debris load (blocks, soil, and trees) remained stored on the slope domain.

On western mountain-slopes, largely affected by landslides, we observed that long and steeper slope profiles prevail; vegetation is highly degraded and thus, root strength has been reduced. In effect, around 42% of landslide scars ($>500\,m^2$) are surrounded by degraded forest cover (less dense; high trees become sparse) or by grass cover (43%), as shown by Oliveira et al. (1996). These authors found that 193,000 trees were lost. Cruz et al (1998) observed reactivation around older

Figure 12.10 Pictures of damages caused by the landslide events in February 1996 at the footslopes of Tijuca Massif and in the adjacent Jacarepaguá lowlands. A & B: "Rich" houses were inundated at the footslopes of the massif; C & D: Flooding led to both human and property losses, especially in the *favelas*, where the poorest people live.

landslide scars, from February 1988. The authors suggest that clearing sites due to landslide scars provides microclimatic changes right on the border (edge effects), leading to forest degradation and then increasing landslide hazards.

Landslides of the debris-flow type prevailed all over the area. Initial mechanisms should be related to two conditions: mass shock against the loose colluvium (block-rich) or sudden increase of pore pressure (as demonstrated experimentally by Avelar, 1996). Debris flows propagated downslope into the main channels to form extensive rock-debris avalanches (Fig. 12.8). Velocity reached 2.8 to 5.3 m/s as indicated by Vieira et al. (1997). These avalanches led to channel incision into the fresh bedrock and left behind wider river channels. That the landslides occurred simultaneously reveals that a certain threshold was reached, before triggering landslides. In the *baixada* (lowlands), river channels became completely drowned and flooding spread throughout the area.

5. Other Catastrophic Events Recorded in Brazil

Catastrophic mass movements recorded in Brazil are concentrated in the southeastern (São Paulo, Rio de Janeiro, Minas Gerais, and Espirito Santo states) and southern (Rio Grande do Sul, Santa Catarina, and Paraná states) regions of the country (Fig. 12.1). They are predominantly related to the occurrence of rainfall that is of great intensity and short duration, and sometimes happen after rainy periods of long duration.

The main records were registered in the twentieth century and refer to the municipality of Santos at São Paulo State (March 1928, 1947, and 1956), the southern region of Minas Gerais (December 1948), the metropolitan area of Rio de Janeiro and neighboring sierras (January 1966, February 1967, and February 1988), Araras Sierra area–Rio de Janeiro State (January 1967), Caraguatatuba Sierra–São Paulo (February 1967), Tubarão municipality–Santa Catarina State (March 1974), the municipalities of Ouro Preto and Monlevade–Minas Gerais (March 1979), the Sierra region of Rio de Janeiro–Rio de Janeiro (December 1981), Rio de Janeiro municipality (February 1996) and the municipalities of Timbé do Sul and Alto Feliz–SC/RS (December 1995 and December 2000).

The catastrophic event that occurred in the Araras Sierra involved diverse soil and rock landslides with a few debris flows, and flared up on steep slopes and road cuts around an area of 150 km^2 after rainfalls occurring between January 22 and 23, 1967. In this area banded gneisses and gneisses predominate, with well-marked foliation very rich in quartz, microcline, and biotite that, due to weathering, form shallow, sandy-clayey, permeable, and low-cohesive soils that become susceptible to rupture during extreme rainy periods. The pluviometers situated in the affected area recorded pluviometric levels between 200 and 275 mm, allowing one to infer, as Guidicini and Nieble (1984) showed, that values above 200 mm isohyets in 24 hours indicate the initiation of mass movements in the Brazilian Southeast.

Slightly after this event, between March 17 and 18, 1967, landslides occurred in about 400 locations on the Caraguatatuba Sierra escarpment, an area of 180 km^2. It is also an area with a predomination of gneisses and granites that, due to weathering, form soils that are shallow and of low cohesion. In this case, the rainfall levels over 24 hours recorded around the affected area varied between 120 and 586 mm. Contrary to the Araras Sierra event, landslides predominated over a region marked by isohyets over 400 mm. This observation reinforces the idea that the volume and intensity of rainfall necessary to initiate catastrophic events depend a lot on preceding soil humidity conditions, and, in turn, are provoked by the rainfall regime after the month of July (winter) of each year.

Some important events happened in the south of Brazil, in the region that encompasses the Serra Geral. On December 23, 1995, diverse debris flows and landslides occurred in the borders between Santa Catarina (SC) and Rio Grande do Sul (RS) States on the Sierra Geral escarpment (Pellerin et al., 1996). Some years later, on the night before Christmas 2000 and close to 150 km south of the previous occurrences, near the city of Caxias do Sul (RS), similar mass movements took place, with a predomination of debris flows, translational landslides, and rock falls (Azambuja et al., 2001). Pellerin et al. pointed out that the phenomena had no recorded precedent, according to the inhabitants, and caused vast destruction of highways and bridges, removed agricultural soils, and took innumerable human victims. Although the events were considered to be unprecedented, the presence of pebbles and boulders in valley bottom deposits proved the recurrence of these phenomena on a geological time scale, as mentioned by Duarte (1995).

The two events were the results of localized and highly intense rainfalls occurring over the domains of steep slopes, with elevation levels ranging from 40 to 1200 m over horizontal distances of just 3 km situated in the sandstone and shale sequences of the Rastro and Botucatu River formations that are overlain by basalts and rhyolites of the Sierra Geral formation. In 2000, 148 mm of precipitation fell over a 2-hour period, and the event was shown to have a 250-year recurrence interval. According to Pellerin et al. (1995), these rainfall types originate at the dislocation of cumulonimbus cloud columns in mature phase with a basal height of 600 m, moving from the ocean toward the Sierra Geral escarpment, whose maximum altitude is about 1200 m.

Acknowledgment

We thank CNPq (National Council for Developing Science and Technology) and FAPERJ (Carlos Chagas Filho Foundation for Research Development of Rio de Janeiro State) for supporting our fundamental research work on this subject.

CHAPTER 13

Floods in Urban Areas of Brazil

Jose Candido Stevaux[1], Edgardo M. Latrubesse[2,3], Maria Lucia de P. Hermann[4], *and* Samia Aquino[3]

[1]*Universidade Estadual de Maringá-GEMA, Campus, Avda. Colombo 5790, Maringá, 87020-900, Brazil*
[2]*University of Texas at Austin, Department of Geography and the Environment GRG 334; 1 University Station A3100 Austin, TX 78712 USA*
[3]*Laboratorio de Geologia e Geografía Física-Universidade Federal de Goiás-CAPES, Campus II-74001-970, Goiânia-GO, Brazil*
[4]*Departamento de Geociências, Universidade Federal de Santa Catarina, Florianópolis, Campus Universitario (88040-900), SC, Brazil*

Contents

1. Introduction: Water versus Urbanization

South America can be considered the "Continent of Water." Of the 10 largest rivers in the world, in water discharge, 6 are in South America (Latrubesse et al., 2005). With more than 70% of its area in the humid and wet-dry tropics, the continent is drained by five huge hydrological river basins: the Amazonas (6,000,000 km^2), La Plata—Paraná/Paraguay (2,600,000 km^2), Orinoco (990,000 km^2), Araguaia-Tocantins (757,000 km^2), and São Francisco (634,000 km^2). While sharing its natural water systems with almost all the countries of the continent, Brazil possesses the majority of South America's hydrological

Corresponding author: *Jose Candido Stevaux, Universidade Estadual de Maringá-GEMA, Campus, Avda. Colombo 5790, Maringá, 87020-900, Brazil, jcstevaux@uem.br*

Developments in Earth Surface Processes, Volume 13

ISSN 0928-2025, DOI 10.1016/S0928-2025(08)10013-X

potential, with 65% of the Amazon Basin, 45% of the Paraná-Paraguai Basin, and the entire São Francisco and Araguaia-Tocantins basins.

Populations and rivers have been closely interrelated throughout human history. Since the beginning of the nineteenth century, the world's population has intensively migrated to urban areas. In 1800, only 1% of people lived in cities and villages. However, during the twentieth century, urban populations came to account for 47% (2850 million) of the global population, with the trend predicted to continue, reaching more than 50% by around the year 2010 (Guglielmo, 1996). In South America, however, urbanization has been more aggressive, and today it is the leading continent in terms of urban population, with 77% of its people living in urban conglomerates (Table 13.1).

In the mid-twentieth century, only eight cities had more than 5 million inhabitants (New York, London, the Ruhr area, Tokyo, Shanghai, Paris, Buenos Aires, and Moscow), constituting about 7% of the world population. In the year 2000, about 15% of the global population lived in cities with more than 10 million inhabitants. As can be seen, the second half of the twentieth century experienced intensive growth in urban populations, especially in developing countries. In 1950, six of the eight largest cities in the world were located in Europe and the United States. However, according to Guglielmo (1996), at the end of this century, of the 35 largest cities in the world (those with more than 5 million inhabitants) 23 were located in developing countries, and of the 15 most populated cities in the world, 4 were in Latin America (São Paulo, Mexico City, Buenos Aires, and Rio de Janeiro) (Table 13.2).

Curiously, in contrast to what occurred in other parts of the world, none of the large Latin American cities mentioned above have large fluvial systems (Buenos Aires, for example, has a broad estuary that has a more "coastal" geography and

Table 13.1 Distribution of Urban Population by Continent

Continent	S. America	N. America	Oceania	Europe	C. America	Africa	Asia
% of total population	77	76	75	74	53	35	35

Source: FNUAP, home page.

Table 13.2 Populations of the Largest Cities in the World

City	Population ($\times 10^6$)	City	Population ($\times 10^6$)
Tokyo	27.8	Mexico City	16.3
Bombay	18.0	Peking	14.2
São Paulo	17.8	Jakarta	14.0
Shanghai	17.0	Lagos	13.5
New York	16.6	Los Angeles	13.0

Source: FNUAP, home page.

Table 13.3 Urbanization in South America

Country	Urban population (millions)		Total population (%)	
	1980	1997	1980	1997
Argentina	23.3	31.6	83	89
Bolivia	2.4	4.8	42	62
Brazil	80.5	130.6	66	80
Chile	9.0	12.3	81	84
Colombia	18.2	29.4	64	74
Ecuador	3.7	29.4	64	74
Paraguay	1.3	2.7	47	60
Peru	1.2	1.7	65	72
Uruguay	2.5	3.0	85	91
Venezuela	12.0	19.0	79	86

Source: Bertoni and Tucci (2003).

human–nature interaction). Large rivers are not a hazard in South America and do not produce catastrophic floods such as those produced by the Yangtze River in China and the Ganges-Brahmaputra rivers in India and Bangladesh, or, occasionally, the Mississippi River in the United States. Flood hazards in Latin America are related to small to medium-sized fluvial basins and to the additional problems created by urban drainage.

The urbanization process in the cities of developing countries, especially in South America (Table 13.3) forms large population concentrations in small areas. This growth is attributed to (a) natural population growth and (b) large-scale migration promoted by shifts from rural to urban areas (Bertoni and Tucci, 2003). These large cities are experiencing problems in the areas of transportation, water and sanitation, air and water pollution, and floods, with severe reductions in the general quality of life and health.

In the last decade, a reduction in the population growth rate has been observed in all large cities around the world, including those in South America. The manufacturing industry is no longer the main source of employment, having been overtaken by the communications, financial and economic, and service industries, and, in some cases, technological poles. In spite of this radical change, in South American countries, the sharp contrast between rich districts and the poor marginal satellite towns tends to be more evident. The majority of the suburban areas have no regulated buildings and no coordination with regard to sanitation, electrical distribution, street paving, or urban garbage collection services. A large proportion of the population live in slums settled in areas at risk of floods or landslides, and normally without basic infrastructure such as organized fresh water supplies or domestic waste treatment, with the waste draining directly to creeks and rivers. This is the case for Caracas (another large city without access to a large river) and São Paulo, in which 50 and 17% of the population, respectively, live in slums.

Urban growth in South America is characterized by the irregular expansion of suburban areas without any urban regulation. Bertoni and Tucci (2003) pointed out the consequences of this tendency:

1. Populations with low incomes and occasional economic crises.
2. A lack of planning and little public investment in urban expansion. In such areas, the price of land is normally lower than the cost of providing infrastructure.
3. Restrictive laws that do not reflect economic realities. For example, the Water Source Protection Law in Brazil forbids a landowner from building on areas next to a water source. However, the government does not buy back these areas, so the owner has to preserve the area and pay public taxes on it without any support from the government. This situation leads to poor people invading theses areas, thereby increasing the number of illegal and uncontrolled urban areas

2. Floods in Brazil

As mentioned earlier, in Brazil, urban growth increased during the last century and generated large metropolitan areas around the Brazilian state capitals. The country was first colonized in the sixteenth century, with the main population settlements located in the coastal areas. By the end of the nineteenth century, all the largest cities in the country were on the Atlantic coast, except for São Paulo on the bank of the Tamanduatei and Tietê rivers and Manaus at the confluence of the Negro and Amazonas rivers.

In this chapter, three particular cases of floods in Brazil are analyzed (Figure 13.1). The first describes the effect of floods on the population in the industrial and economic core of the country, the city of São Paulo; the second describes floods in Santa Catarina State in southwestern Brazil; and the third discusses the floods in the capital of the State of Acre in southwestern Brazilian Amazonia. Additional discussion on small-catchment dangerous floods in areas around Mar Mountains (Serra do Mar), such as Rio de Janeiro and Santa Catarina, are discussed in this book as well as by Netto and Avelar (Chapter 12).

2.1. Floods in the Metropolitan Region of São Paulo-SPRM

With the development of industrialization during the twentieth century, the need for cities drifted inland, especially in the southeastern states. Areas of high demographic concentration were formed in the interior of São Paulo State, at the western headwater of the Paraná River Basin (Fig. 13.1). Intensive investment in coffee and cotton production in these areas generated a great deal of labor along the main tributaries of the Paraná River Basin, such as the Tietê, Grande, Paranapanema, Sorocaba, Mogi Guaçu, and Piracicaba rivers. All the roads and railroads constructed during the twentieth century had a radial pattern converging on the city of São Paulo, the main megalopolis of the country. Of all the large cities of South America, São Paulo was by far the most important. With 14 million

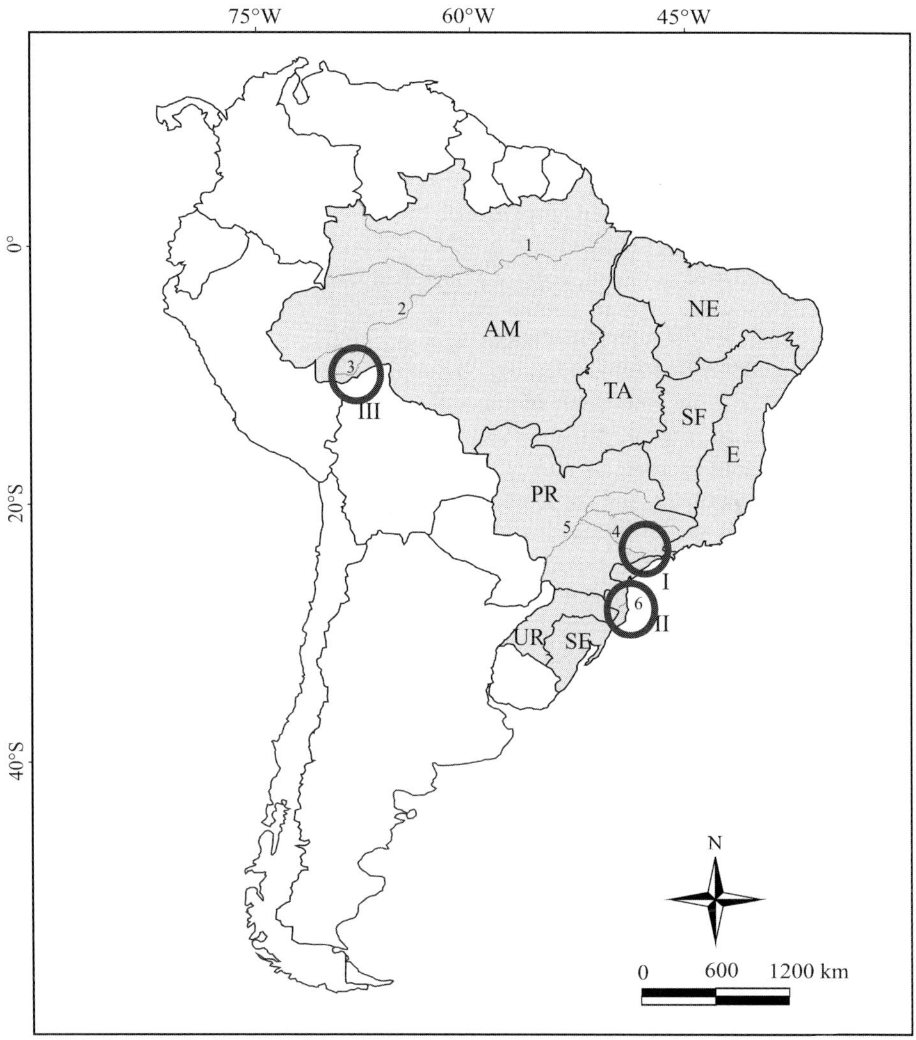

AM- Amazon Basin
TA- Tocantins-Araguaia Basin
SF- São Francisco Basin
NE- Northeastern Basin
PR- Paraná Basin
UR- Uruguai Basin
E- Eastern Basin
SE- Southeastern Basin

1- Amazon River
2- Purus River
3- Acre River
4- Tietê River
5- Paraná River
6-Itajaí River
I- São Paulo Metropolitan area and Upper Tietê River Basin
II- Santa Catarina State and Itajaí River Basin
III- Rio Branco and Acre River Basin

Figure 13.1 Location of the case study areas. 1—São Paulo Metropolitan area and Upper Tieté River Basin; 2—Santa Caratina State and Itajaí River Basin; 3—Rio Branco city and Acre River Basin.

inhabitants and 24 million people within its Sao Paulo Metropolitan Region (SPMR), São Paulo was one of the largest cities in the world and the most important industrial center in South America.

São Paulo is located in a very problematic place in terms of its geomorphological aspects. The SPMR is in the Upper Tietê River Basin, compressed between the slopes of the Mar (980 m.a.s.l.) and Cantareira (1200 m.a.s.l.) Mountains, and a large part of the city lies in the alluvial plains of the Tietê River and its tributaries, the Tamanduatei and Pinheiros rivers (Fig. 13.2). Moreover, the urban area is located in the Atlantic Forest, a dense tropical rain forest extending over the whole of the Brazilian coastal area, with an annual average rainfall of 1660 mm.

The Upper Tietê River Basin has an area of 5985 km^2 and a mean discharge of 104 m^3/s^{-1} (CBHAT, 1999), 30 m^3/s^{-1} of which is effluent. In some months, the river discharge is only one-sixth of the effluent discharge! The urban area of the Upper Tiete Basin increased from 190 km^2 in 1930 to 1900 km^2 in 2001, which equates to a tenfold increase in 70 years. Flooding in São Paulo was first recorded in the seventeenth century; the most severe and most long-lasting floods occurred in 1924, 1983, 1991, and 2005, with the last one occurring after the implementation of the latest flood-control project. The flood problems with the Tietê and its tributary, the Tamanduatei River, began at the end of the nineteenth century, when their floodplains began to be urbanized following the construction of railroads and factories. The first flood-control project was developed in 1894 with the objective of straightening some of the reaches of the Tiete River in order to maintain a peak flood discharge of 174 m^3/s^{-1}, without causing flooding. Thirty

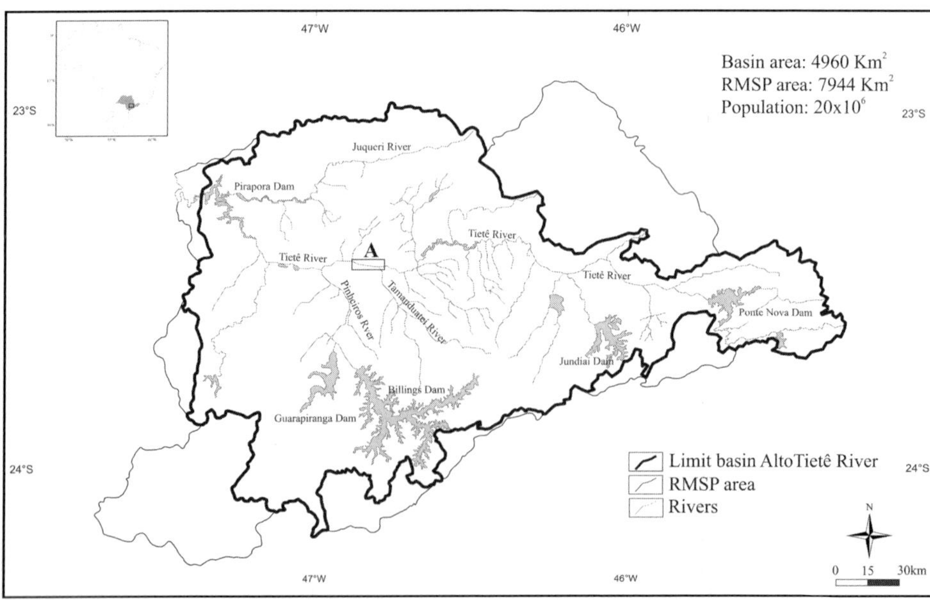

Figure 13.2 Upper Tietê River Basin and the São Paulo Metropolitan Region-SPMR (CBHAT, 1999).

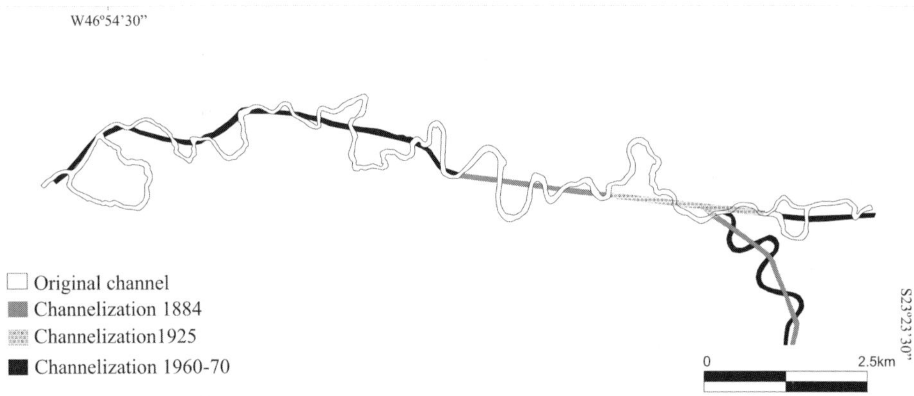

Figure 13.3 Changes in the Tietê River channel in the SPMR since 1894. (Modified from EMPLASA, 2001).

years later, in 1925, the percentage of the impervious area of the city had increased so much that a new channelization project was carried out in order to improve the peak flood discharge to $650\,m^3/s^{-1}$. The last channelization project, executed in two phases during 1998–2000 and 2002–2005, reached its economic and technical threshold and can drain a peak flood discharge of $1048\,m^3/s^{-1}$, almost six times that of the first peak discharge project, and has a recurrence interval of "100 years" (Fig. 13.3). The greatest amount of precipitation recorded before that of 2005 happened during the ENSO year of 1983. In this event, precipitation in the basin varied from 50 to 160 mm, with an average of 110 mm in 24 hours and a peak flood discharge of $860\,m^3/s^{-1}$.

2.1.1. The Flood of May 24–25, 2005

During the period from October to March, the SPMR experienced floods because of intensive summer rain. According to Andrioli and Barros (2005), there are about 500 critical points of inundation in the city, generated by different causes due to a lack of planning and management of problems in pluvial drainage projects (for example, rain-water collectors that are undersized or blocked by urban garbage). May is normally not a rainy month, having an average monthly precipitation of only 46 mm for the last 10 years (Fig. 13.4). However, between May 24 and May 25, 2005, an atypical precipitation occurred. A rainfall of 106.3 mm was recorded over a 16-hour period, homogeneously distributed over the entire basin. It was very different from the typical summer rain, characterized by a small volume of precipitation over a short period of time in small and well-defined areas. The Tamanduatei, Pinheiros, and Tietê rivers (Fig. 13.2) were the most seriously affected, overflowing at many points.

Reports on meteorological conditions in the days before the event, as collected by the National Institute of Meteorology (INMET), consisted of only a well-localized precipitation of 28.9 mm on May 24 (Table 13.4). In that week, air humidity was relatively low and temperature very high for the season. Under

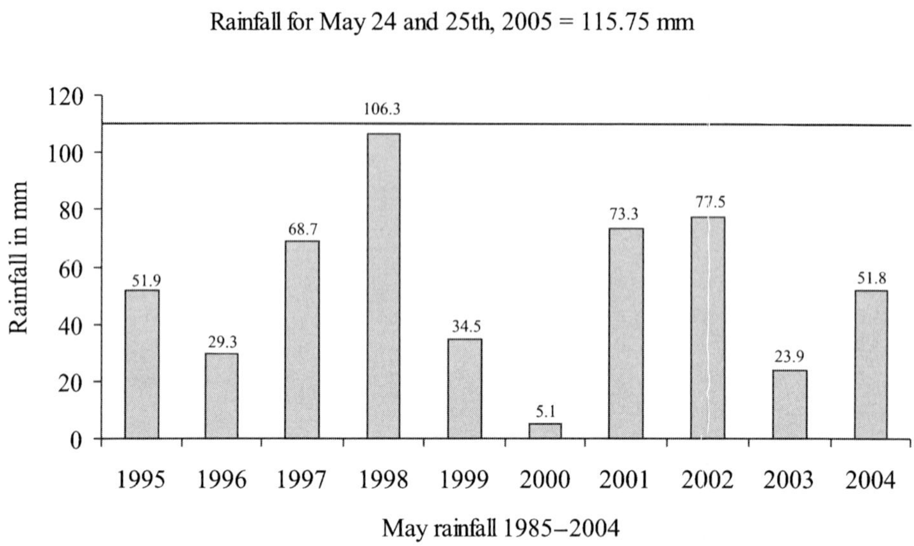

Figure 13.4 May precipitation between 1995 and 2004 (columns). The horizontal line is the 24th precipitation on May 25, 2005.

Table 13.4 Meteorological Data between May 17 and 24, 2005

May 2005	Maximum temperature (°C)	Relative humidity (%)	Accumulated precipitation (mm)
17	26.9	47	0.0
18	29.0	47	0.0
19	28.0	36	0.0
20	29.0	46	0.0
21	26.6	56	0.0
22	20.0	87	28.9
23	21.1	63	0.0
24	22.3	80	0.0

Source: INMET—National Institute of Meteorology website.

these conditions, the soil of the urban basins was dry, and the rivers produced a discharge near that of the annual minimum, as is common for May.

The rain event on May 24–25 occurred as a result of the association of a heavy, very intense cold front from the south and the hot and humid air of the area. The rain began at 3:00 p.m. on May 24, continued for 16 more hours, and ended at 7:00 a. m. on May 25. The average accumulated rainfall in the Upper Tietê Basin was 111.0 mm, with a homogeneous spatial distribution throughout the basin for the full 16 hours of rain, and with a rainfall of great intensity and volume between 4:00 p.m. on May 24 and 04:00 a.m. on May 25.

A statistical analysis of the precipitation performed by Andriloi and Barros (2005) showed a return time of 35 years for average precipitation and 100 years for the highest precipitation.

The marginal roads along the Tietê River in the city of São Paulo have some of the most intense traffic in South America, carrying about 700,000 vehicles per day. In the first hours of May 25, the river produced 18 flood points along these marginal roads. The huge traffic jam this caused reached back more than 150 km and lasted throughout the day. The Campo de Marte Airport was completely flooded, and all operations there were suspended. The largest agricultural trade center in the country, the CEAGESP, had its worst losses in its history. According to *O Estado de São Paulo* (May 26, 2005), on May 25 alone, the estimated loss was US$ 2.06 million.

2.2. Floods in Santa Catarina State

Santa Catarina State is located in southeastern Brazil. The state lies on 95,346 km^2, and by 2007 its population was estimated at 5,866,000 inhabitants (IBGE, 2007) (Fig. 13.5). With a Human Development Index (HDI) of 0.822, Santa Catarina is considered the second most developed state of the country. This index is higher than the Brazil average of 0.766 (PNUD, 2000). The area was culturally influenced by the large German and Italian immigration during the past century.

Santa Catarina has a hilly landscape dominated by the Serra do Mar Mountains and the coastal area on the Atlantic Ocean. Tropical rainfalls that trigger floods and landslides as well as storms affecting the coastal area drive the main natural disasters. In addition, this area suffered an exceptional climatic episode in Brazil: Hurricane Catarina, which impacted the southern coast of Brazil (see also Chapters 2 and 16).

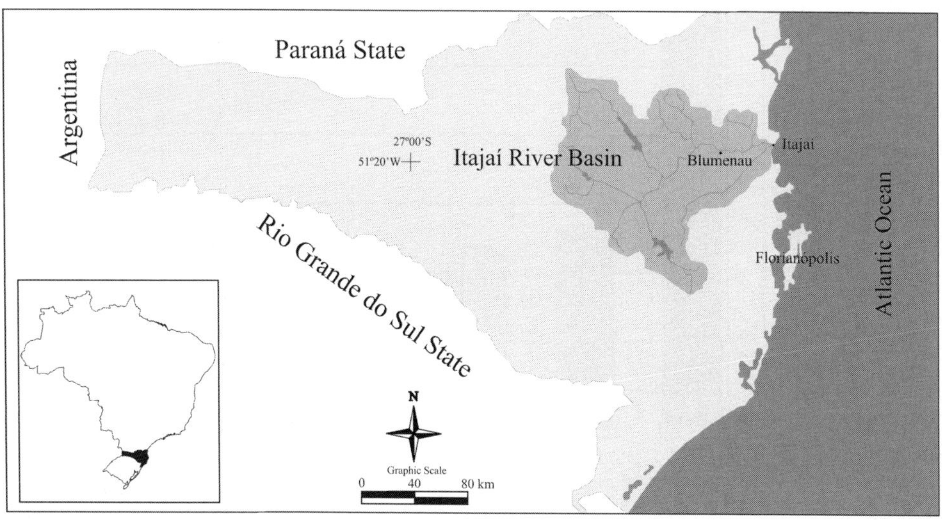

Figure 13.5 Location map of Santa Catarina State, Itajaí River Basin, and main localities.

In Santa Catarina, the years 1983 to 2003 witnessed 1293 gradual floods, 555 flash floods, 140 landslides, 495 drought episodes, 342 hailstorms, 502 tropical storms, and 43 twisters. In 1998 also were recorded 26 events of coastal storms (Fig. 13.6). In 293 municipalities of Santa Catarina State, 75% suffer frequent floods (Fig. 13.7). Floods are concentrated mainly in basins that drain from Serra do Mar to the Atlantic Ocean. In these basins, human pressure on natural resources as well as urban population increased significantly during the last decades. Blumenau Municipality, for example, ranks first in flood frequency, having experienced more than 20 episodes from 1980 to 2003. Floods can be divided into gradual floods produced by relatively long periods of seasonal rainfall affecting a good part of the region, and flash floods of more local effect produced by short-duration storms of high intensity.

Urbanization seems to be a coadjutant factor in terrain impermeabilization of small basins (Tucci et al, 2003). The frequency of urban floods is related more to the urban effect and to the spread of urban settlement in hazardous areas than to changes in climatic trends because floods can also be produced by low-intensity rainfall (Fig. 13.7).

The 2004 to 2008 period saw more than 203 episodes of flash floods in Santa Catarina. The floods of November 2008, however, had had dramatic consequences, affecting more than 50 municipalities that were declared in national emergency and public calamity. The largest volumes of rainfall accumulated in 24 hours were recorded in São Francisco do Sul (295.6 mm), Blumenau (337.4 mm and 283.1 mm), Balneário Camboriú (252.4 mm and 233.8 mm), Joinville (232.1 mm), Itapoá (213.2 mm and 195 mm), Luis Alves (192.6 mm), Itajaí (186.7 mm), and Florianópolis (160.1 mm) (Cliram/Epagri, 2008). That is, in a

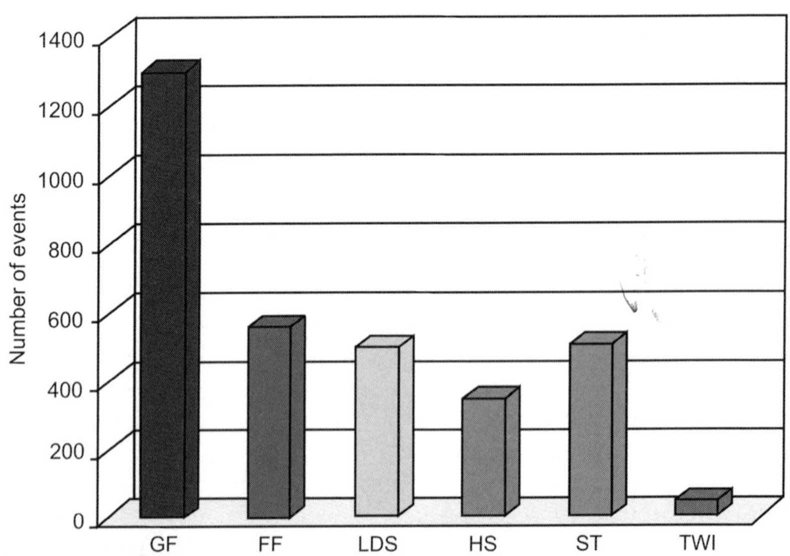

Figure 13.6 Main occurrence of disasters in Santa Catarina State from 1980 to 2003. (GF) Gradual flood; (FF) Flash Flood, (LDS) Landslide, (HST) Hail Storm, (ST) Tropical Storm, (TWI) Twister or Tornado. *Source*: Hermann, 2007.

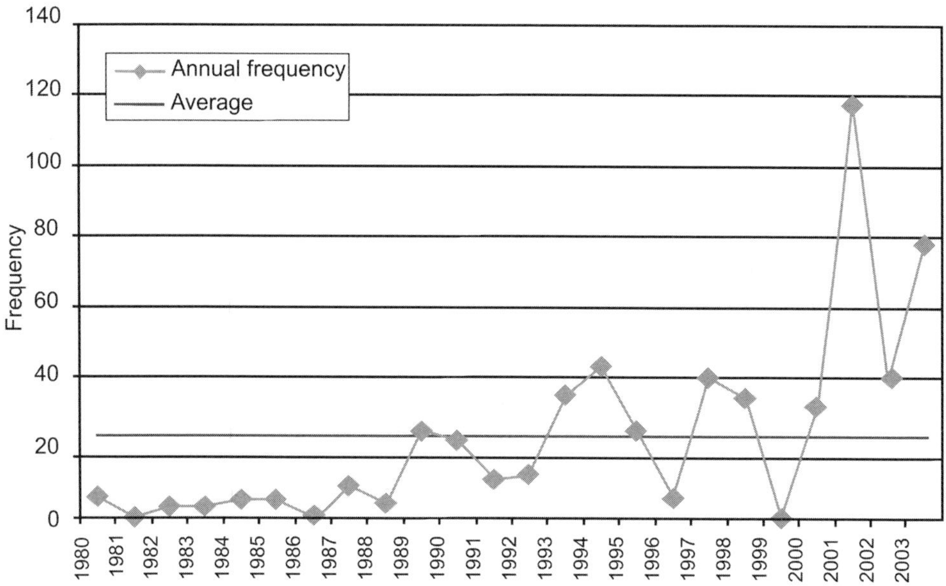

Figure 13.7 Frequency of disasters in Santa Catarina.

single day rainfall was more than the rainfall monthly average. From November 21 to 25, precipitation reached 524 mm in the eastern part of the state. The mean monthly rainfall of November 2008 reached values as high as 1143 mm in São Francisco do Sul, 940 mm in Joinville, 912.4 mm in Blumenau, 880 mm in Itapoá, and 687.3mm in Itajaí. This represents values 350 to 400% higher than the average for the northern coastal area and 270% higher than the average in Florianópolis region.

From the beginning of September to November 19, the area suffered constant rainfall of low to moderate intensity, but from that date wet air masses from the coast were entering close to the surface because of the effect of a high-pressure system on the South Atlantic that increased precipitation on the coastal area (Minuzzi and Camargo, 2008). Between November 21 and 23, a cyclonic vortex of low pressure in intermediate atmospheric levels intensified instability and cloudiness, producing strong rainfall and triggering regional flooding and hundreds of landslides. By November 27, a total of 78,707 people had been evacuated. By December 30, a total of 135 people were killed mainly by landslides, 63 municipalities were in a state of emergency, and 14 municipalities decreed a situation of public calamity. In 2008 the World Meteorological Organization (WMO) and the United Nations (UN) declared this event to be the worst catastrophe ever recorded in the history of the region and one of the worst and most serious in the world. The economic loss was not well estimated. For example, the Itajaí Harbor, responsible for 4% of the Brazilian exports, was unable to operate during the disaster, and it produced an estimated loss of more than US$ 340 million. The federal government declared an emergency plan of more than US$ 400 million.

2.2.1. The Itajaí River Basin

The Itajaí River fluvial basin was one of the more seriously affected during the floods of 2008 (Figs. 13.8 and 13.9). With a drainage area of 15,000 km^2, it is the largest basin in Santa Catarina, running through 200 km from the mountain area to the Atlantic Ocean and occupying 16.15% of the state (Frank and Pinheiro, 2003) (Figs. 13.5 and 13.8). The basin sustains 47 municipalities and nearly one million inhabitants, being 80% concentrated in urban centers along the valley (IBGE, 2000). The main industrial cities are Blumenau, Itajaí, Rio do Sul, and Brusque, which are located in the lower valley, and their economic activities are related dominantly to textile, metal-mechanic, sea products, and tourism industries.

The upper basin developed on Paleozoic and Mesozoic sedimentary rocks of the Parana Basin, and the fluvial courses generate small canyons. In the middle course, the river erodes metamorphic rocks of the Santa Catarina Granulite complex (Proterozoic/Archean). The lower basin is in general below 100 m.a.s.l., and the floodplain of the Itajaí River is more developed there, but the river is still surrounded by a hilly landscape (Fig. 13.8). Floods have affected mainly the lower course since colonial times. One of the more affected cities is Blumenau, which is located only 13 m above sea level and 34 km off the Atlantic coast. The mean annual discharge (Q_{mean}) of the Itajaí River at Blumenau (Adolfo Konder Bridge gauge station) is 140 m^3/s, and, along the urban area the channel width oscillates from 50 to 150 m. Low discharges can be as low as 15 m^3/s, but large floods can overpass 5000 m^3/s. At Adolfo Konder gauge station, civil defense considers a bankfull stage of 8.5 m when

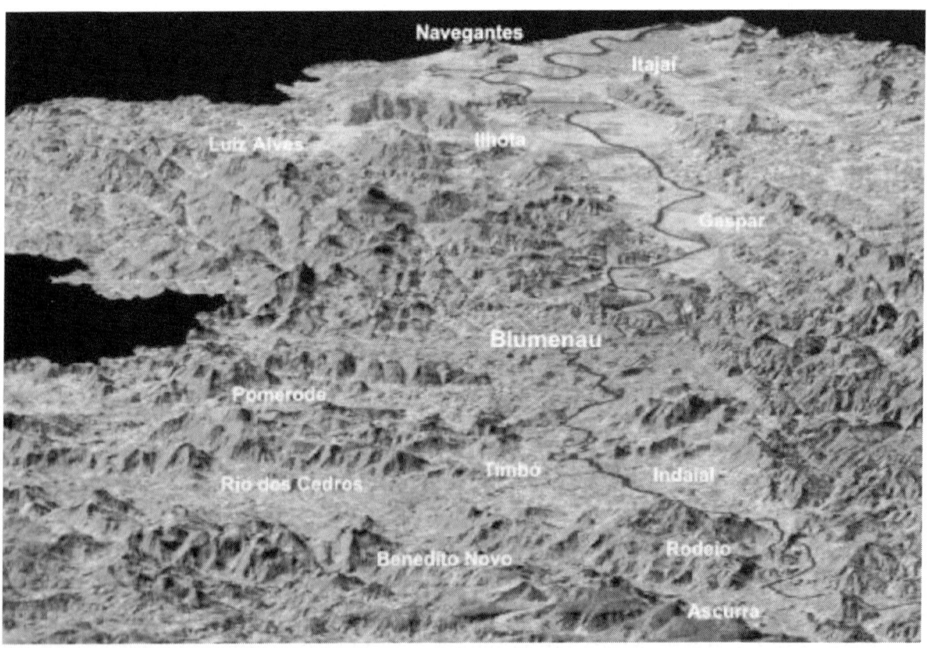

Figure 13.8 DEM of the middle and lower Itajaí fluvial basin and main cities. *Source*: From REFOSCO, 2003.

Figure 13.9 The Itajaí River during the floods of November 2008. (A) Lower reach at Itajai, (B) Blumenau.

floods start to affect the area. The river reached 15 m in 1852, 1880 (the maximum known record of 17.10 m), 1911, 1983, and 1984, and 12.8 m in 1992. During the floods of 1983, Blumenau was strongly flooded; more than 50,000 people were affected, and 8 were killed. In 1984, 70,000 people suffered the consequences of floods (~40% of the population of Blumenau) (Hermann, 2007) (Table 13.5). During the floods of 2008, the maximum level was 11.24 m on November 24, and water discharge was roughly estimated to be ~3500 m^3/s. The floods affected more than 20,000 people and killed 20 persons (Fig. 13.9B).

The water level oscillates quickly, and the natural dynamic of flood is of short duration (Table 13.6). However, urban structures delay the water flow and increase the time duration of floods as well as the effect of the disaster on the city. In addition, landslides affect the hilly slopes of the valley and increase the damage (Fig. 13.10).

Table 13.5 Consequences of Floods in Blumenau, 1983 to 2008

Date	Total of affected people	Killed	Total population of Blumenau	Percent of population affected by floods
May/1983	10,000	02	170, 490	5.86%
July/1983	50,000	08		29.3%
Dec/1983	5,000	01		2.93%
Aug/1984	70,000	-	175, 445	39.96%
Jan/1990	594	-	~210,740	0.26%
Oct/1990	1,310	20		0.59%
Nov/1991	8,528	10	212 025	4.02%
Jan/1992	21	-	~216,420	0.01%
May/1992	35,000	02		16.17%
Jan/1995	600	-	225,556	0.26%
Jan/1997	353	-	231,401	0.15%
2001	396	-	261,808	0.15%
Nov/2008	22,800	24	296,151	7.6%

Source: Herrmann, (2007).

Table 13.6 Stages (in meters) of Itajaí River at Blumenau from November 22 to 28, 2008, Recorded at 8AM and 3PM

Day	08 AM	03 PM
22/11/08	3.38	6.16
23/11/08	9.09	10.06
24/11/08	11.24	9.85
25/11/08	8.25	6.73
26/11/08	4.91	4.91
27/11/08	4.52	4.72
28/11/08	4.38	4.24
29/11/08	3.68	3.67

Source: EPAGRI (2008).

Figure 13.10 Landslide in Blumenau during the floods of November 2008 (photograph by Adriana Franciosi).

2.3. The Amazonian City of Rio Branco and the Acre River

Population growth in Amazonia during the last few decades has taken place in the main Amazon cities. The case of Rio Branco, the capital city of the State of Acre, illustrates this problem (Figs. 13.1 and 13.11). The population of Rio Branco has exploded during the last three decades, growing from about 90,000 inhabitants in the 1980s to 315,000 in 2006 (IBGE, 2006). The land was occupied in a chaotic way, without appropriate urban planning and environmental studies. As a result, thousands of families settled on the alluvial plain of the Acre River, which crosses through the city and divides it into two parts. This population is regularly affected

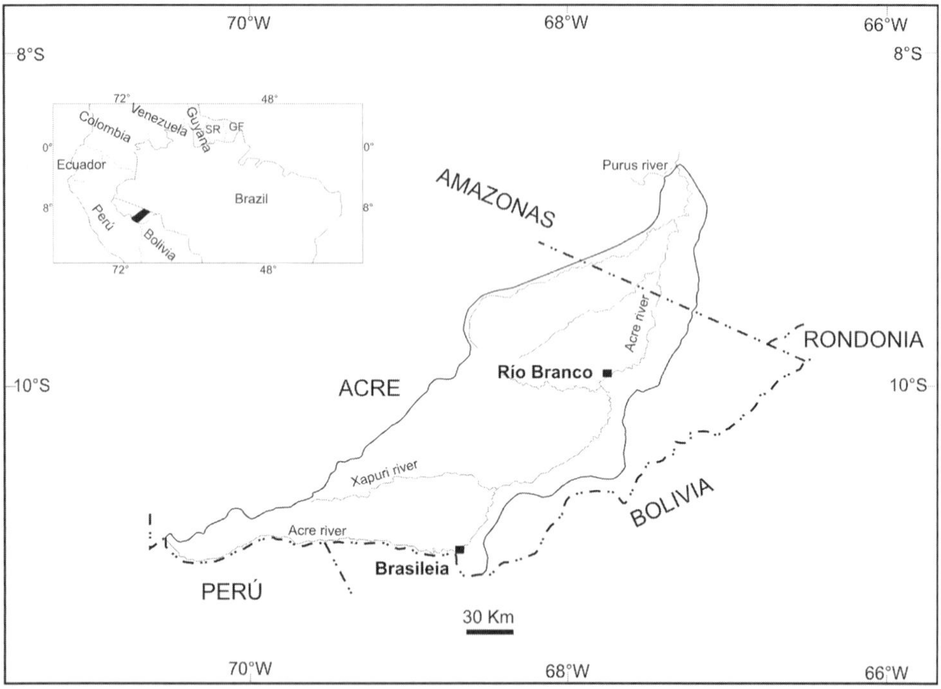

Figure 13.11 Location map of Rio Acre Basin.

by floods and mass movements, which produce large economic losses and social problems.

River and climate background: The Acre River is a tributary of the Purus River, and its basin is located on the Tertiary claystones, siltstones, and fine sandstones of the Solimões Formation (Upper Miocene-Pliocene), draining an area of 33,000 km^2. Its headwater is located at 350 m.a.s.l. in Peruvian territory (Fig. 13.11). The Acre River is a typical river of southwestern Brazilian Amazonia, with single and sinuous channels and asymmetric and complex meanders alternating with straight segments (Latrubesse et al, 2001). As is true of the majority of southwestern Amazonian lowland rivers, it carries about 98% of its solid discharge as suspended load (Gibs, 1967; Mertes, 1985).

The alluvial belt is formed by three terrace levels. The two older terraces are at 15–30 and 34–38 m above the river's lower water level, and the youngest terrace is 8–12 above. In the Rio Branco area, the lower terrace is formed by a complex paleomeander belt at three different stages of abandonment, which constitute a poorly drained area with hydromorphic soils and swamps (Fig. 13.12). The active channel of the Acre River is nearly 75 m wide, and in the urban area of Rio Branco, it forms a large meander cut into the Tertiary sediments of the Solimões Formation, which crop out at lower parts of the banks.

Daily discharge records for the Acre River (1971–2006) from gauging stations in Rio Branco (Basin area = 22,670 km^2) and Brasileia (Basin area = 3299 km^2,

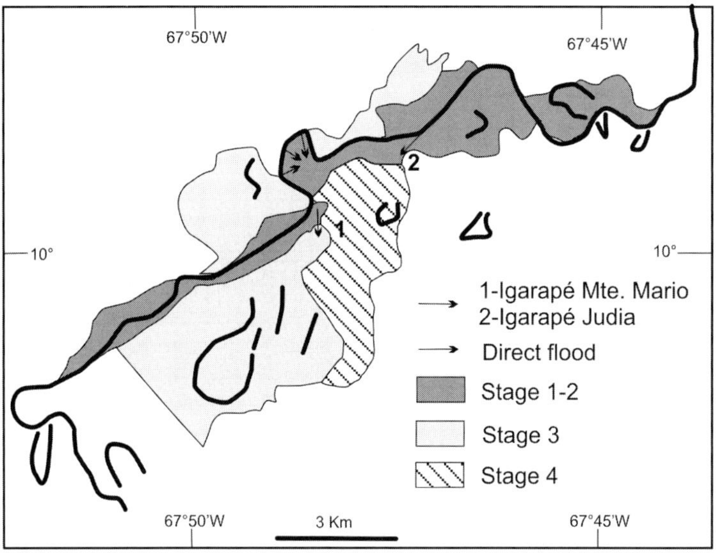

Figure 13.12 Geomorphologic units of the Rio Acre alluvial plain at Rio Branco area.

230 km upstream from Rio Branco) were analyzed. The Acre River has clear seasonal peaks and troughs of high and low discharge, and a mean annual discharge of 353 m^3/s^{-1} in Rio Branco (Fig. 13.13). High flows are concentrated between January and April, with some secondary peaks also occurring during this period. The system has pronounced annual variability between high and low discharges. The ratio of maximum and minimum daily discharge for the historical series is ~72. The maximum average discharge is 1469 m^3/s^{-1}, and the minimum average discharge is 35.5 m^3/s^{-1}. The range in stage is also large: the average difference in water surface elevation between high and low stages is approximately 8.75 m.

The recurrence interval of flooding was determined for a period of 23 years. The interval used to determine the mean flood was 2.33 years (Leopold et al., 1964). The Acre River shows low flood variability from year to year (Fig. 13.14). The flood of 2006, which reached 1940 m^3/s^{-1} (the largest value recorded), is only ~30% larger than the annual average flood discharge. A 13.5 m stage for bankfull is used by civil defense as the alert stage for flooding. This stage is close to the value of bankfull discharge, with a recurrence interval of 1.5 years in the annual series.

The flow duration curve was developed from the daily discharges from 1983 to 2006 (Fig. 13.15). Floods have a short duration compared to the long period during which low discharges are characteristic. Over 60% of the time, the river has a discharge equal to or less than the mean annual discharge of 353 m^3/s^{-1}, and bankfull discharges are present less than 5% of the time. The channel of the Acre River is deep and narrow. The width:depth ratio was determined by relating bankfull width and mean depth, and was relatively low (8.8).

In spite of its high annual precipitation of 2000 mm, southern Amazonia has a pronounced dry season between June and September. This variability is related to the

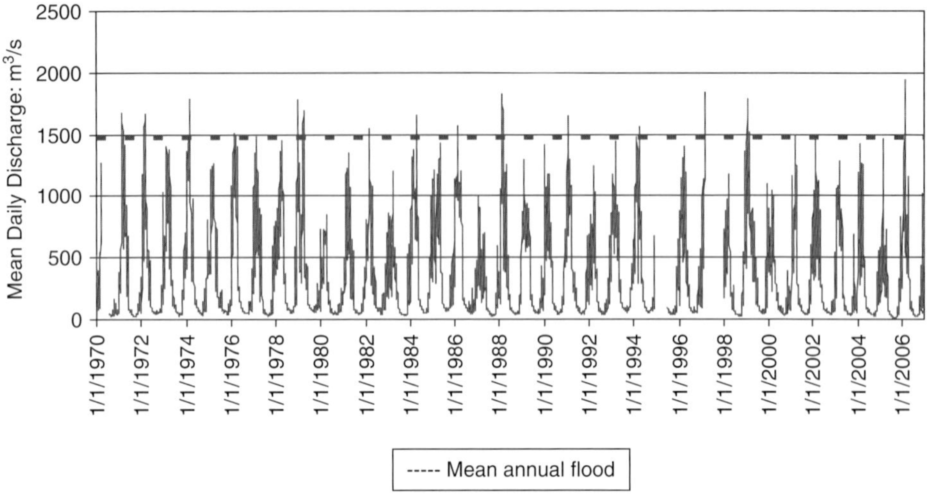

Figure 13.13 Hydrograph of mean daily discharges at Rio Branco gauge station, Rio Acre (1970–2006). Dotted line indicates the mean annual flood for the temporal series ($Q_{maf} = 1.468.99\ m^3/s$).

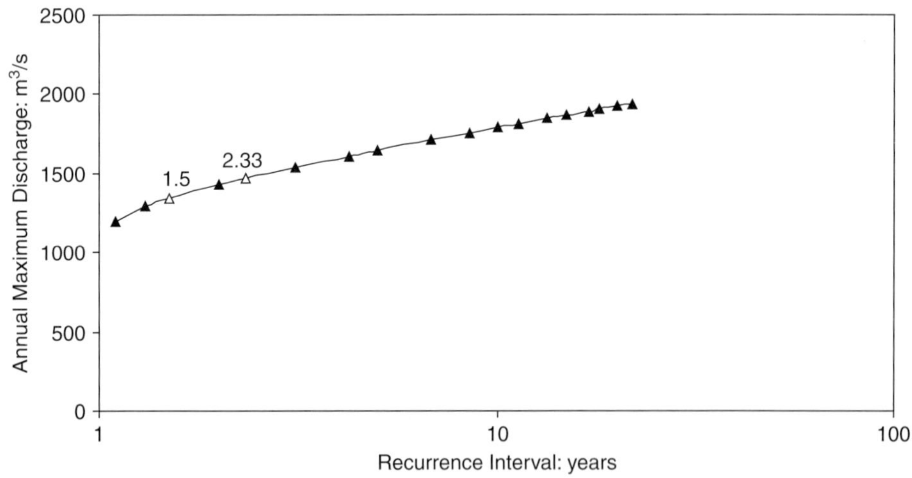

Figure 13.14 Recurrence of floods (1970–2006) at Rio Branco gauge station, Rio Branco. Bankfull discharge (1.5) and the mean annual flood (2.33) are indicated.

shift of the intertropical convergence zone (ITCZ). During the Southern Hemisphere summer, the ITCZ is located between 10° and 15° S. Southern Amazonia receives most of its rainfall during this period. The ITCZ shifts northward, reaching its extreme north position in Venezuela and Colombia from July to August. During this period, rainfall drops significantly in southwestern Amazonia, leaving it with a well-defined winter "dry season." Records from the Rio Branco station at the Federal University of Acre (from 1970 to 2006) show that precipitation for June to August is 115 mm, a very low value for a tropical humid area that supports a rain forest.

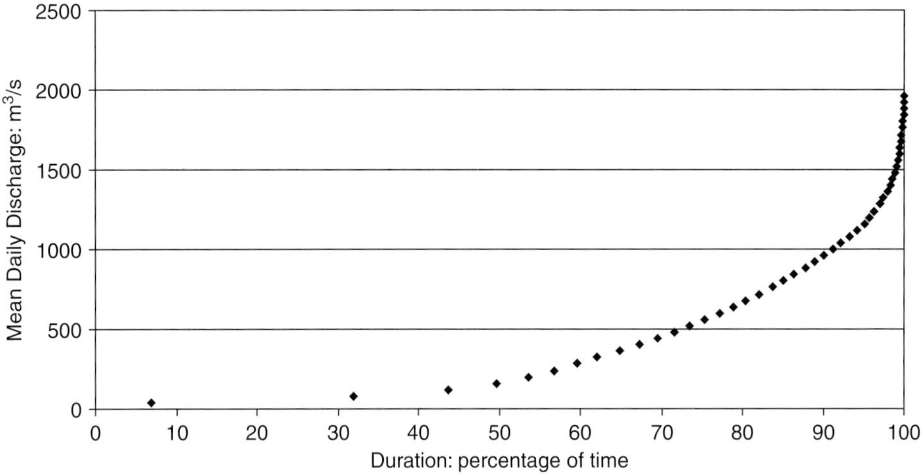

Figure 13.15 Duration flow curve of mean daily discharges at Rio Branco gauge station (1970–2006), Rio Acre.

The high and low discharge peaks roughly coincide with the periods of high and low precipitation (rainy and dry seasons). The river response to increased discharge during the rising stages occurs nearly two months after the beginning of the rainy season. This response is very different during falling stages. The abrupt decrease in precipitation during the beginning of the dry season is accompanied by a strong drop in discharge/stage values. Third, differential peaks (known in Brazil as *repiquetes*) of high discharges, which are out of phase between the Rio Branco and Brasileia stations, reflect water inputs from tributaries (the Japuri River and others). This could be indicative of differences in precipitation within short distances, which differentially affect the subbasins and produce the *repiquetes*.

2.3.1. The Floods

Floods affect the city of Rio Branco to different degrees every year. The most important properly recorded floods were those of 1988 (water stage of 17.11 m and $Q_{max} = 1{,}825\ m^3/s$), 1997 (water stage of 17.66 m and $Q_{max} = 1845\ m^3/s$) and 2006 (water stage 16.72m and $Q_{max} = 1940\ m^3/s$). The flood dynamic of the Acre River does not correspond to general overbank flow. During ordinary floods, only a small part of the lower terrace is inundated. Similar occurrences happened during the large flood of 1997 and 2006. The main damage was produced by the reflux of water into the lower terrace through the small tributary channels named Mario and Judia Igarapes ("Igarape" is the regional name for tributary creeks). In its lower course in the lower terrace, the Monte Mario Creek runs in a paeochannel of the Acre River. The occurrence of a narrow levee on the Acre River's bank impedes direct flooding over a large portion of the lower terrace. Floodwater penetrates upstream from the Acre River to the Monte Mario and Judia Creeks, flooding stage 1 and 2 and part of stage 3 of the lower terrace (Fig. 13.12). According to available records, floods have not affected the higher stage 4.

2.3.2. Floods, Land-use Changes, and the Urban Population

It has been estimated that at least 50% of annual rainfall in this region is the result of evapotranspiration from the rain forest (Salati et al., 1978). The rain forest plays another important role from a hydrological point of view: it stores water and can sustain evapotranspiration when air temperature increases and moisture is reduced. The process of deforestation for cattle farming replaces forests with pastures, which are poor at storing water. However, it is very difficult to establish the relationship between deforestation and river responses in areas with incomplete discharge and rainfall records. It is also difficult to calculate how much water is stored in the floodplain and the effect of this water on discharge. In these tropical rivers, the vegetation and floodplain play the role of interactive "sponges" that can regulate or affect fluvial dynamics. Forest loss is quite important in the Acre River Basin (Fig. 13.16). Deforestation occurred as a result of the establishment of *fazendas* (farms) and large cattle ranches located mainly along State Highway BR 317. Nearly 13% of the basin has experienced rapid deforestation during the last two decades. Although it is not possible to draw quantitative conclusions with regard to how much deforestation has affected the fluvial system, it is reasonable to consider that decreasing or reversing deforestation in the basin would reduce the impact of floods in the future.

The expansion of cattle ranches, the abundance of *seringais* (rubber plantations), and the arrival of immigrants from midwestern and southern Brazil has led to rapid population growth, especially since the 1970s. In 1970, the city of Rio Branco had about 20 neighborhoods and 35,000 inhabitants. By 2000, it had increased to more than 150 neighborhoods and 315,000 inhabitants, with a significant expansion of its

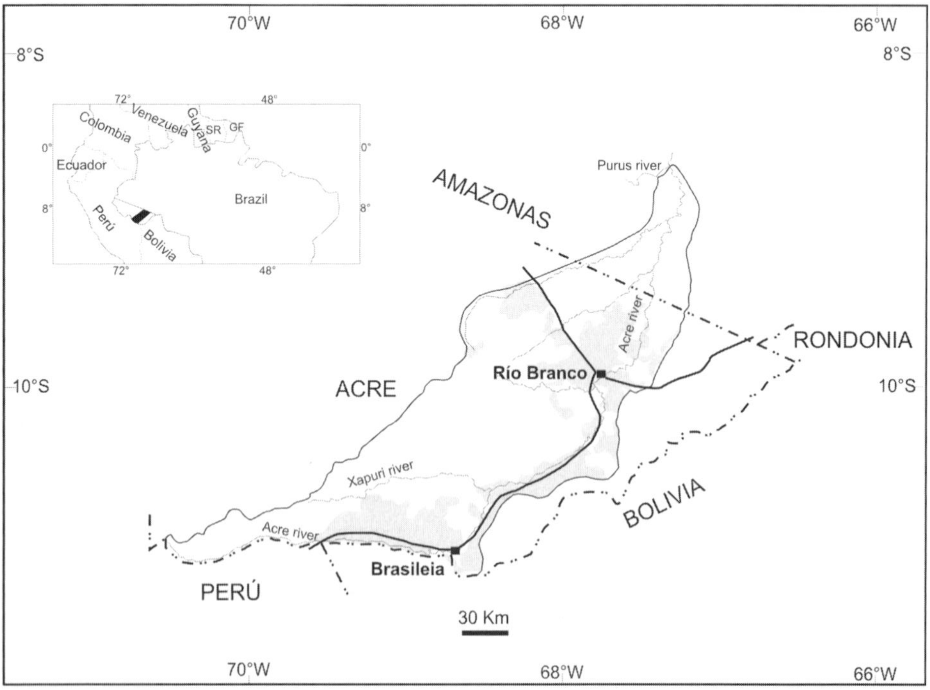

Figure 13.16 Deforested area of the Rio Acre Basin. Shadow areas have been deforested.

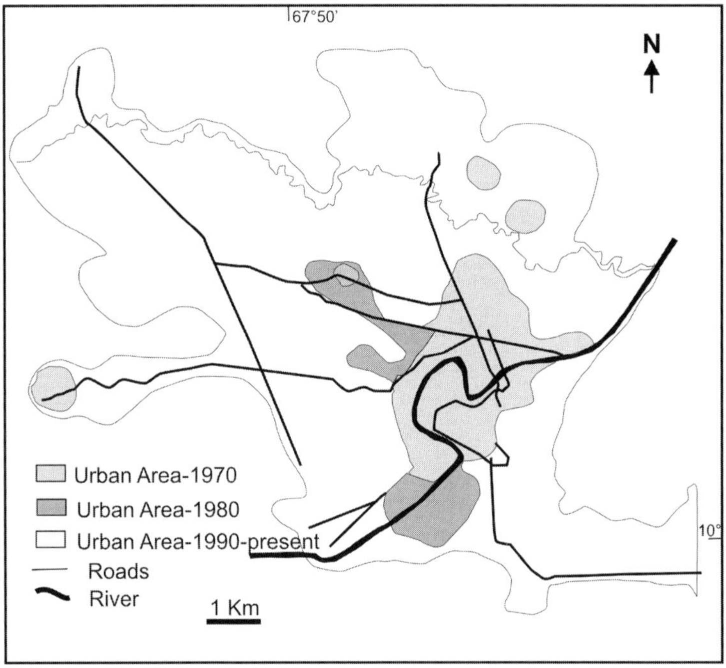

Figure 13.17 Urban expansion of Rio Branco city since 1970.

urban area (Fig. 13.17). This acceleration in urbanization, combined with a lack of urban planning, resulted in the settling of areas of low slope gradients, the headwaters of drainage systems (*igarapés*), and along the floodplain and lower terrace of the Acre River—areas that are periodically or occasionally flooded. This occurred despite the recommendations by the City Planning Directory of Rio Branco County in 1986, that these areas should not be occupied.

During the flood of 1988, 23 neighborhoods along the fluvial belt of the Acre River, as well as Judia, Monte Mário, and São Francisco creeks, were affected. It is estimated that approximately 20,000 people were impacted, to a greater or lesser extent. During the flood of 1997, approximately 70,000 people and 40 neighborhoods were affected. However, in 1997 the discharge was only about 5% greater and the water stage 0.49 m higher than in 1988, with the area of the lower terrace affected by flooding being only slightly larger than during the 1988 flood. The flood of 2006 was a little more minor than the 1988 flood with regard to the water stage, but 33,500 inhabitants from 27 neighborhoods were affected, as well as more than 8300 houses and other buildings (Fig. 13.18). This illustrates the types of circumstances that need to be considered (and thereby avoided) in the process of urban planning. The 1988 flood event provided a good example of the impact that could be expected from the fluvial dynamics of the basin. Failure to consider this possible impact when planning the expansion of the city resulted in much greater damages in 1997 and in 2006, which could have been easily avoided or reduced by applying simple geomorphologic knowledge and tools during the planning of the new settlements.

Figure 13.18 Area of Rio Branco city affected by floods on February 19, 2006.

3. Final Remarks

In the three cases studied in this chapter, it is evident that although Brazil possesses many of the largest rivers in the world, the problems it faces concerning fluvial floods are induced more by lack of planning in urban areas on the banks of medium to small-sized rivers than by flood events of large rivers. Railroads, roads, and waterways were the sites of urban concentration, and the majority of large Brazilian cities developed in such medium to small-sized rivers. At São Paulo, the extensive urbanization around and within floodplains introduces problems concerning rain-water drainage. The flat topography, associated with channel rectification and pluvial drainage systems, increases surface runoff and, consequently, water concentration in low parts of the city. On the other hand, obstructed and undersized rain-water collectors are responsible for many of the small, but common, localized floods.

In Santa Catarina's mountain basins with landslides, flash floods are the response to intense tropical storms, and indiscriminate and poorly planned floodplain occupation along the valleys are the main factor triggering disasters. Even in the case of Rio Branco in southeastern Amazonia, floods are more a problem of planning than of nature. Floods affect the population settled on the densely occupied floodplain. In the three cases, the number of deaths connected directly to the floods is small when compared with countries like Bangladesh, India, or China. However, the number of people affected is relatively high, and the urban problems brought by the floods are significant.

CHAPTER 14

Seismic and Volcanic Hazards in Argentina

Laura P. Perucca[a,*] *and* Stella M. Moreiras[b]

[a]*Consejo Nacional de Investigaciones Científicas y Técnicas (CONICET) - Gabinete de Neotectónica. INGEO. Facultad de Ciencias Exactas, Físicas y Naturales. Universidad Nacional de San Juan*
[b]*Consejo Nacional de Investigaciones Científicas y Técnicas (CONICET). Instituto Argentino de Nivología, Glaciología y Ciencias Ambientales (IANIGLA) - CRICYT*
e-mail:moreiras@lab.cricyt.edu.ar

Contents

1. Introduction

The analyses of seismic and volcanic hazard made during the last few decades have progressively been improved with the use of geologic data that permit the reconstruction of the seismic history of active faults and the volcanic activity in the past. The historic and instrumental records cover too short a time span to establish precise seismicity values for a long term and for a correct evaluation of earthquake hazard.

* **Corresponding author**: *Laura P. Perucca Consejo Nacional de Investigaciones Científicas y Técnicas (CONICET) - Gabinete de Neotectónica. INGEO. Facultad de Ciencias Exactas, Físicas y Naturales. Universidad Nacional de San Juan. Tel: (54-0264) 4265103. Fax: 54-0264-4265103. e-mail: lperucca@unsj-cuim.edu.ar*

Developments in Earth Surface Processes, Volume 13
ISSN 0928-2025, DOI 10.1016/S0928-2025(08)10014-1

The western part of South America has a complex morphological pattern, with an active western margin, as shown by its relief and seismicity produced by the drift of the Nazca, Antarctic, and South American plates. The convergence of the Nazca and South American plates began about 200 million years ago, with the subduction of the oceanic plate beneath the continental one. The subduction surface has an eastern slope, and the ocean-continent boundary shows a displacement to the west at an absolute rate of 2.2 cm/year (Uyeda and Kanamori, 1979). This displacement rate should have been different in magnitude and direction during the geologic past, thus giving rise to a complex mosaic of movements, which now appear to be static because of the observation time span and brief in relation to the total time during which these movements took place. Therefore, Quaternary faulting and volcanism in Argentina is the result of ongoing subduction of the Nazca Plate beneath the South American Plate and a consequence of differing geometries of the Nazca Plate (Barazanghi and Isacks, 1976, Jordan et al., 1983). For that reason the distribution of neotectonic strain and volcanism are not homogeneous along the Andean belt. The northern region between 18° and 28° south latitude is characterized by a high seismicity and an active record of earthquakes. The Nazca Plate is being subducted at a 30° dip angle. A recent and active volcanism is present, and several Quaternary faults are described affecting alluvial fans, *bajadas* (bajada (or bahada) is a series of alluvial fans, typically laid down by ephemeral flows, which has coalesced along the foot of the piedmonts to form a gently slope plain of fine and unconsolidated sediments), and salt lakes. The most important historic earthquake recorded in this region is located in the Province of Salta (Talavera del Esteco, 1692) with a reconstructed magnitude of 7.3 (INPRES, 2006).

Between 28° and 32°S, a series of geologic and tectonic factors occurs, which seem to be related to surface seismic activity, since at these latitudes the Nazca Plate is subducting subhorizontally. A substantial part of the Quaternary deformations in Argentina is concentrated in this region. Here the main seismogenic sources can be located and defined with certitude. Such sources show different degrees of activity. Generally, they are subparallel faults, predominantly N-S trending.

It is in this intraplate setting that the most important destructive earthquakes took place. They were related to surface ruptures, such as the earthquakes of 1894 (Ms 7.5), 1944 (Ms 7.4), and 1977 (Ms 7.4). At 32° S the main Quaternary deformations are faults and folds, related to high seismicity but without clear evidences of rupture surfaces during the earthquakes that affected the region. The only exception could be La Cal Fault in northern Mendoza, which was probably ruptured during the earthquake of 1861 (Ms 7.2) (Mingorance, 2000a,b). At the western flank of Precordillera of San Juan, neotectonic also is present. In the lglesia valley, the El Tigre Fault is a more than 120-km-long right-lateral strike-slip fault along which geomorphic features provide evidence for its Quaternary activity. (Withney and Bastias, 1984; Bastias et al., 1990; Siame et al. 1997).

Eastward of the Central Andes, the Sierras Pampeanas are basement blocks that have been uplifted and tilted during the Neogene, bounded by west-verging reverse faults. Small to moderate earthquakes affected the area during the twentieth century, the 1934 Ms 6.0 Sampacho earthquake, for example, but no rupture surfaces have been recognized.

South of 33°S, the subduction angle of the Nazca Plate varies, and from here southward the seismicity diminishes notably: only low-magnitude earthquakes are recorded in this area. In the Isla Grande de Tierra del Fuego, at least three earthquakes during the middle past century have been recorded, because of the interaction of the South American, Scotia, and Antarctic plates.

The volcanic activity in the western border is also directly related to the subduction of the Nazca Plate beneath South America. This activity takes place in a discontinuous volcanic arc, from 16° to 28° latitudes, which includes the southern portion of the Central Volcanic Zone (to which we will refer in later pages). From 33° to 46° S, a set of active volcanoes, which extend to the south of the Tupungatito volcano and which is known as the southern volcanic zone, is associated with a subduction angle of 30° of the Nazca Plate. Finally, from 49° to 56° S extends the Austral Volcanic Zone, with isolated eruptive centers associated with the subduction of the Antarctic Plate. The volcanic activity is interrupted from 28° to 33° 15′S, coincidentally with the increase in the superficial seismicity, which is ascribed to the progressive subhorizontal subduction of the Nazca Plate since the upper Miocene (Barazanghi and Isacks, 1976).

After having analyzed individually the different tectonic aspects, the regional geomorphologic features, and the seismic and volcanic activity in the past and at present, it is important to consider their interrelations. The occurrence of these phenomena could indicate the regions in the Earth crust that are now being submitted to higher deformations and then, releasing stresses; this understanding could be important in carrying out studies of risk.

2. Seismotectonic Setting

The seismic area in Argentina extends along the Andes, from the Province of Jujuy in the north to the Isla de Tierra del Fuego in the south (Fig. 14.1). Nevertheless, there is but scarce documentation of the earthquakes in this area. The first historic mention of destructive earthquakes is about the event that destroyed the village of Talavera del Esteco, in the Province of Salta (northwest of Argentina) on September 13, 1692; which reached intensity IX of the Modified Mercalli scale (MMI).

There exist references about an earthquake in the locality of San Nicolás, Province of Buenos Aires, on October 31, 1527, according to accounts from the Sebastian Cabot expedition (Volponi, 1976), although this cannot be confirmed. Other destructive earthquakes occurred in the Province of Salta (1782, 1844, 1871, 1948, 1959, 1973); Sampacho, in the Province of Córdoba, with a reconstructed intensity of VII (MMI) during 1934; Province of San Luis, with intensity VIII (1936); Tierra del Fuego, with intensity VII (1949); the provinces of San Juan-La Rioja, with intensity IX (1894); and the Province of Catamarca in 2004. Earthquakes have also taken place in regions of the

Figure 14.1 Simplified map indicating the seismicity of Argentina (M > 5). (Modified from USGS/NEIC, 2002).

country considered to have very low seismicity, such as the one that occurred in 1888 in the Río de La Plata (Volponi, 1976) between Buenos Aires and Montevideo (Uruguay) (Fig. 14.2).

In northwestern Argentina, the angle of the subduction surface of the Nazca Plate beneath South America is 25°, between 21° 30′ and 23° 30′ south latitudes, and 19° between 23° 30′ and 27° 30′ latitudes. In this northern region the present-day surface seismicity is low, the volcanism is important, and no evidences of active faults are recorded. On the contrary, between 24°

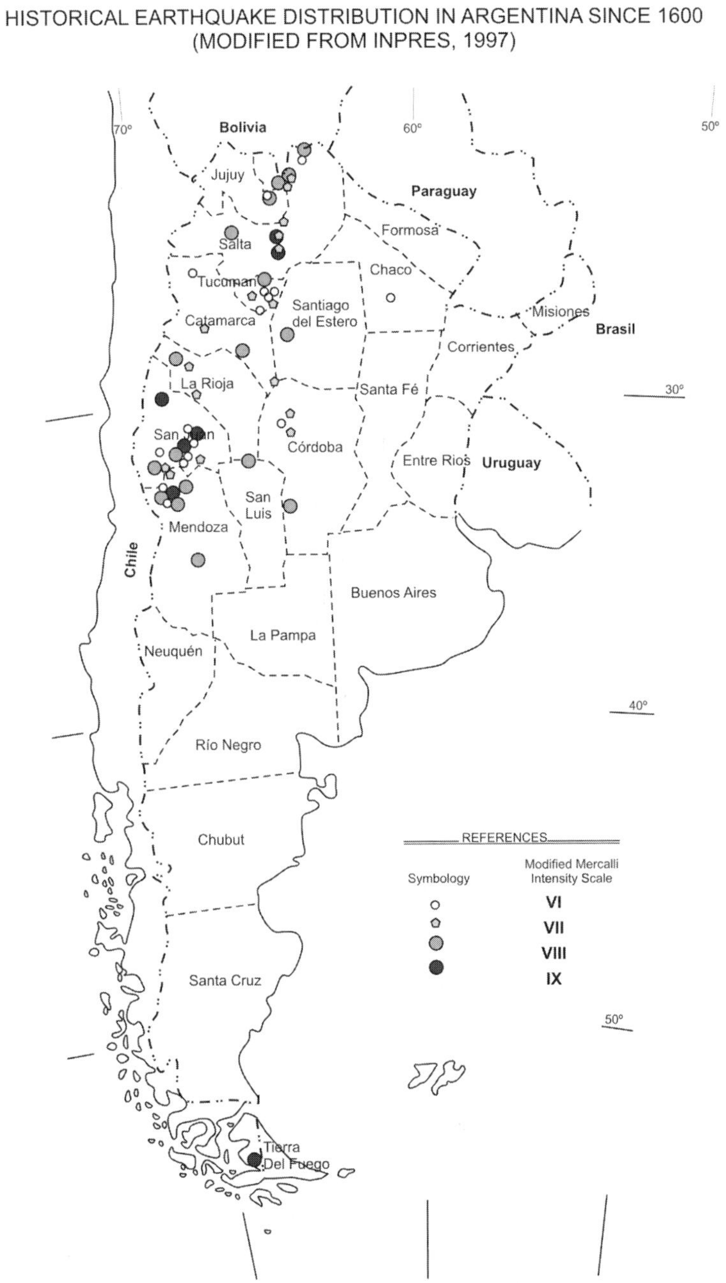

Figure 14.2 Map with the historical distribution of earthquakes.

and 28° south latitudes the surface seismic activity increases, the volcanic activity continues, and some extensional tectonic evidences have been found (Araujo et al., 1999) (Fig. 14.3).

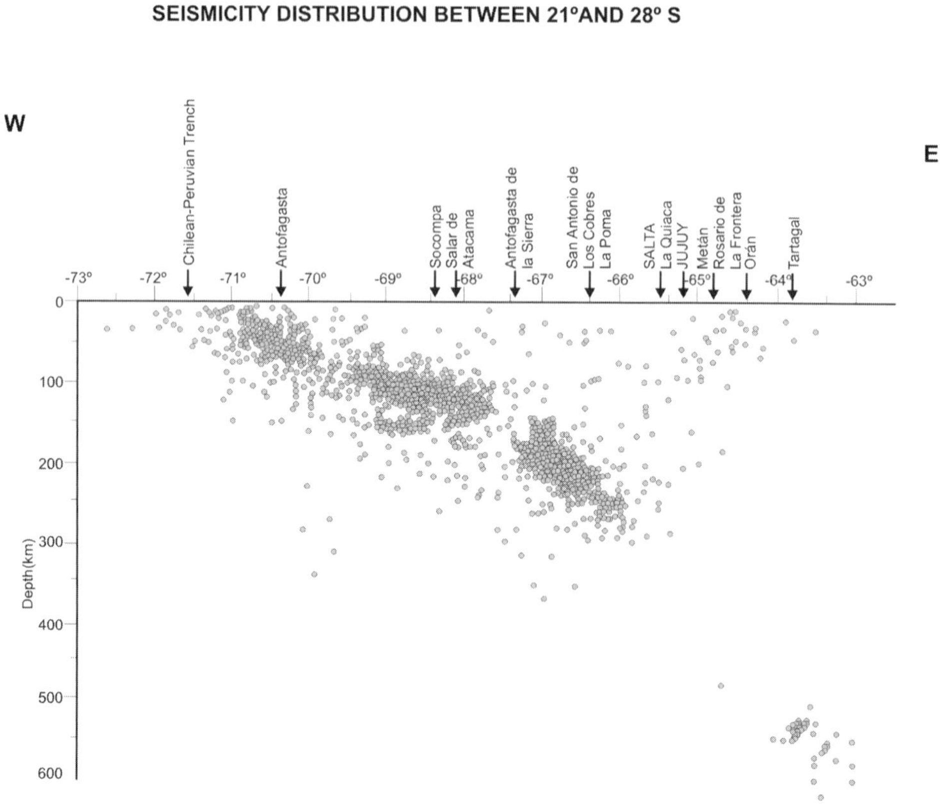

Figure 14.3 West–East seismic cross section between 21°and 28° SL.

But the region with higher seismicity is in the provinces of San Juan and Mendoza, in central-west Argentina, between 28° and 34° latitudes. Here the subduction angle of the Nazca Plate under the South American Plate tends to be subhorizontal and is at about 100 km depth (Ramos, 1988) and extends near horizontal several hundred kilometers eastward before descending into the mantle. The flattening of the Nazca Plate began at 8–10 Ma (Jordan and Gardeweg, 1987; Kay et al., 1991). This situation is linked to the shutoff of the volcanic arc and the uplift of the Sierras Pampeanas basement. A large number of earthquakes with depths up to 90 km occurs, but the most destructive earthquakes are the moderate to superficial ones that happen between 5 and 50 km depth. Seismic activity is lower from 90 to 150 km depth (INPRES, 1993) (Fig. 14.4).

The provinces of San Juan and Mendoza have a very short historical record of earthquakes (Bastias et al., 1993). The oldest accounts are from the seventeenth century, after the Spanish conquest in 1561. However, during the last 150 years, this region experienced no less than six earthquakes with magnitudes higher than 7.0, such as those in 1861, 1894, 1927, 1944, 1952, and 1977,

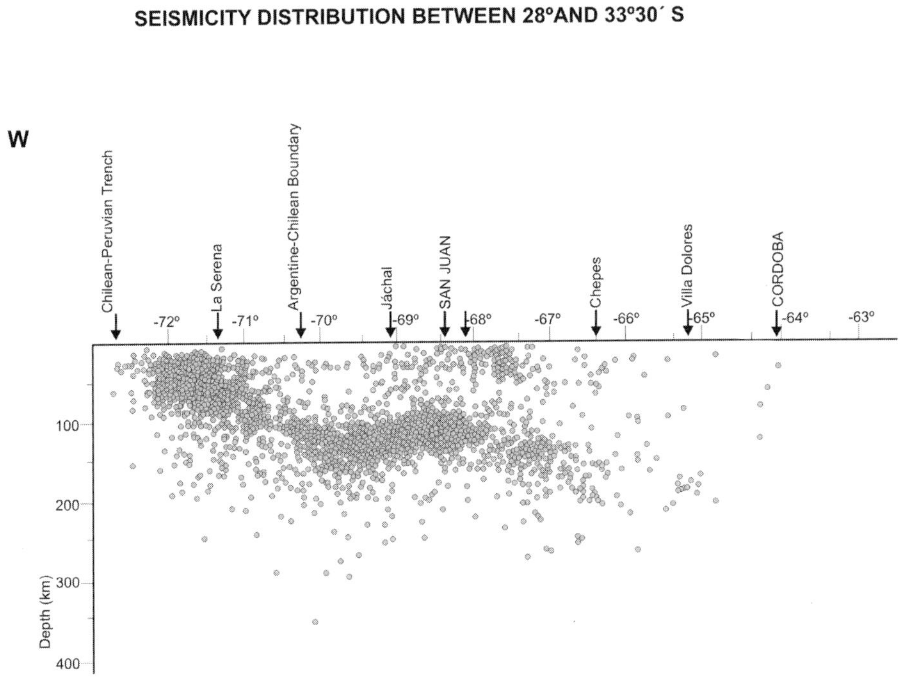

Figure 14.4. West–East seismic cross section between 28° and 33° 30′ SL.

which caused many casualties, destruction of towns, and heavy economic losses. Other important earthquakes took place in 1929 (Ms 6.5) and 1941 (Ms 6.7), among the effects of which numerous liquefaction phenomena and landslides processes with mass movements were recorded (Moreiras, 2004; Perucca and Moreiras, 2006).

Although historical earthquakes are recorded, the decrease in seismic activity south of 37° S (Fig. 14.5) of surficial earthquakes, seems to be related to the Liquiñe-Ofqui fault system. The activity increases in the area of the triple junction point of Chile, at the Taitao Peninsula area (~46° 30′S). From 37° to 41°S the shallow seismic activity increases offshore with powerful earthquakes and tsunami generation, such as those of 1562 (M = 8.0) and 1985 (M = 7.8). From this point to the south, the oceanic part of the Antarctic Plate is transported under the South American Plate to a velocity of 2 cm/y (Fig. 14.5). In spite of this situation, only three destructive earthquakes have been recorded; the first in 1879 and two on December 17, 1949 (Ms 7.8) (Schwartz et al., 2002). It should be pointed out that the seismic records in this region are very incomplete because of the scarce population centers located in it, but in recent years, the population has increased as a result of European immigration. The current seismic activity is low.

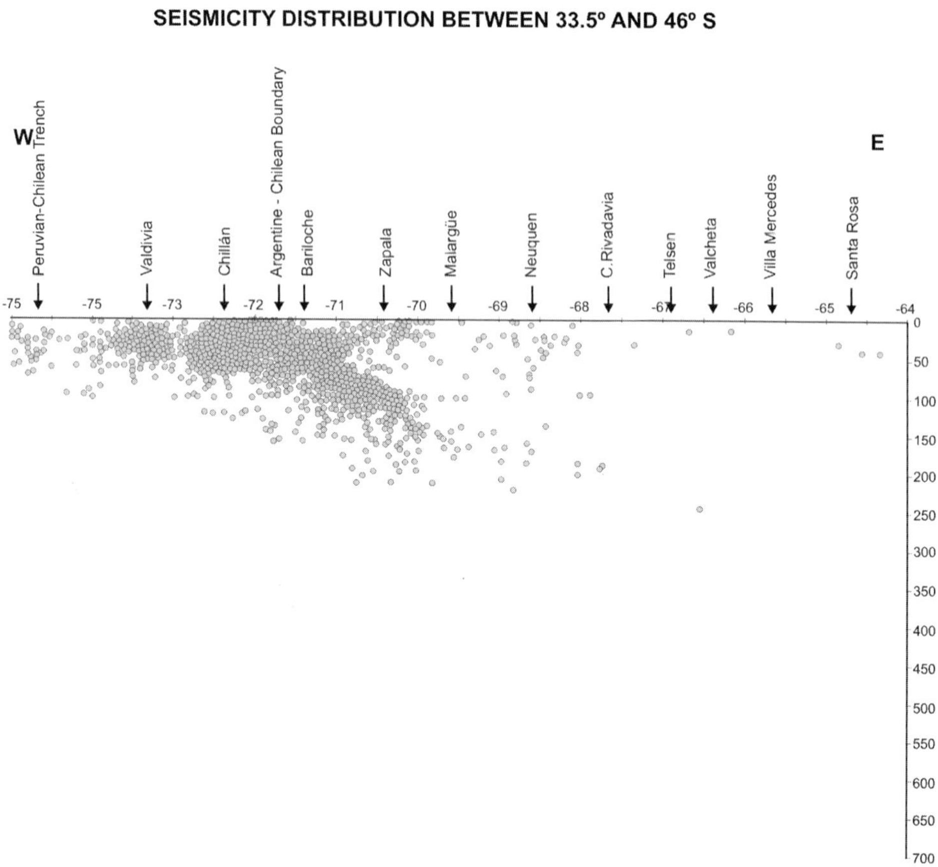

Figure 14.5 West–East seismic cross section between 33° 30'and 46° SL.

According to the regulations established by INPRES-CIRSOC (National Institute for Seismic Prevention- Research Center of the National Safety Regulations for Civil Works) 103; five areas with increasing degrees of seismic hazard (0 to 4) have been defined (Fig. 14.6). For instance, the Cuyo region (La Rioja, San Juan, and Mendoza provinces) is the most hazardous (high to very high), and the provinces of Salta and Jujuy are slightly less hazardous (high to moderate). The Isla de Tierra del Fuego is in a region with high seismic hazard, whereas northwestern Patagonia (Province of Neuquén and the western part of the Province of Chubut) has a moderate seismic hazard. In Patagonia, however, the seismicity has only been partially investigated and the main reasons thereof are few seismological observation points and geophysical researches.

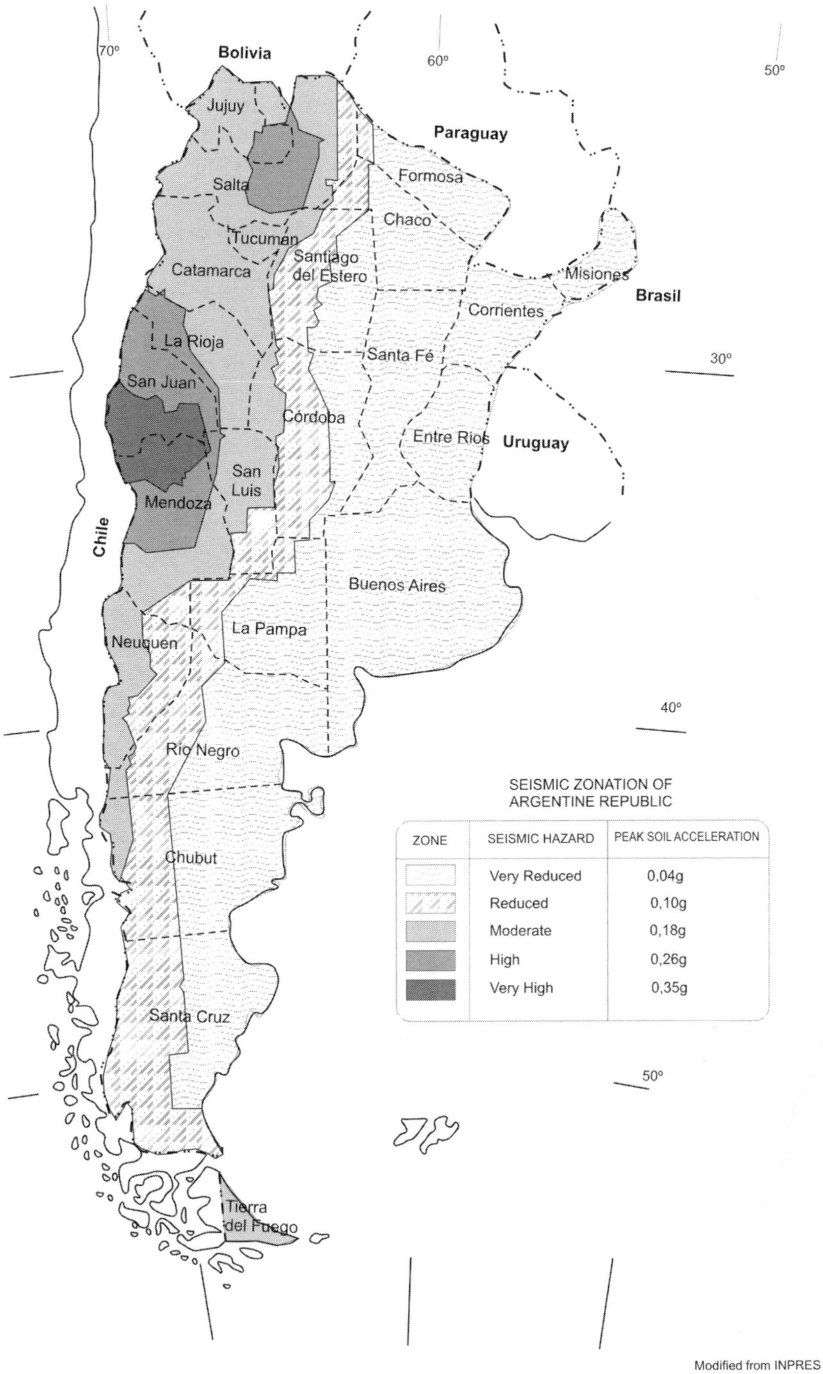

Figure 14.6 Simplified map of the main Seismic Zonation of Argentina. (Modified from INPRES, 2006).

3. Neotectonics and Historical Seismicity

The word "neotectonic" is used in Argentina for the study of the crustal deformations that took place after the end of the Pliocene, in coincidence with the last orogenic events in the western part of the country. This term is sometimes not clear, since it lacks precision and indicates only recently formed structures in a general sense. For this reason, it is preferable to characterize the deformations by referring them to the last movement determined, that is, to refer to Pleistocene or Holocene structures.

One of the most frequently used definitions of an active fault is the one given by Slemmons (1977, 1981), which states that an active fault is the one that registers historical or seismologic antecedents of activity and/or has an expectation of occurrence in a time interval similar to human life expectation. It is thought that a fault that has undergone displacement during the Pleistocene and/or Holocene can move again in the future. For this reason, the paleoseismic study of such structures, which are considered potential seismogenic sources, is a fundamental tool for evaluating the seismic hazard from a geologic viewpoint.

The term *active fault* is used to designate faults, fault segments, or fault systems along which relative displacements or rupture surfaces took place, together with moderate or intense earthquakes, during the last 10,000 years (Holocene). These faults are potential sources of destructive earthquakes and rupture surfaces because they experienced displacement during the present-time seismotectonic conditions, have evidences of activity shown by stratigraphic evidences, and the regional tectonic setting suggest the probability of future displacement.

The parameters that can be determined by the investigation of active faults by using empirical relations are:

- Recurrence
- Magnitude of the earthquake to be produced in the structure
- Yearly average movement

From the viewpoint of the seismicity, the Argentine territory almost entirely lacks pre-Hispanic earthquake information. There is also little information about earthquakes during the 300 years of Spanish colonization for many regions in the country. This lack of information affects the seismologic record interval. For this reason geologic studies related to neotectonics are essential in Argentina for extending the paleoseismic record in the country.

The Argentine territory can be divided into areas where the main Quaternary faults occur, even though most deformations are in the Precordilleran sector of the provinces of La Rioja, San Juan, and Mendoza (Fig. 14.7). In northwestern Argentina, evidences of Quaternary tectonic activity are numerous. For instance, in the Argentine Puna, the tectonic activity is so recent that the faults affect Holocene evaporite crusts or Pleistocene lacustrine sediments situated in tectonic terraces. Fault scarps

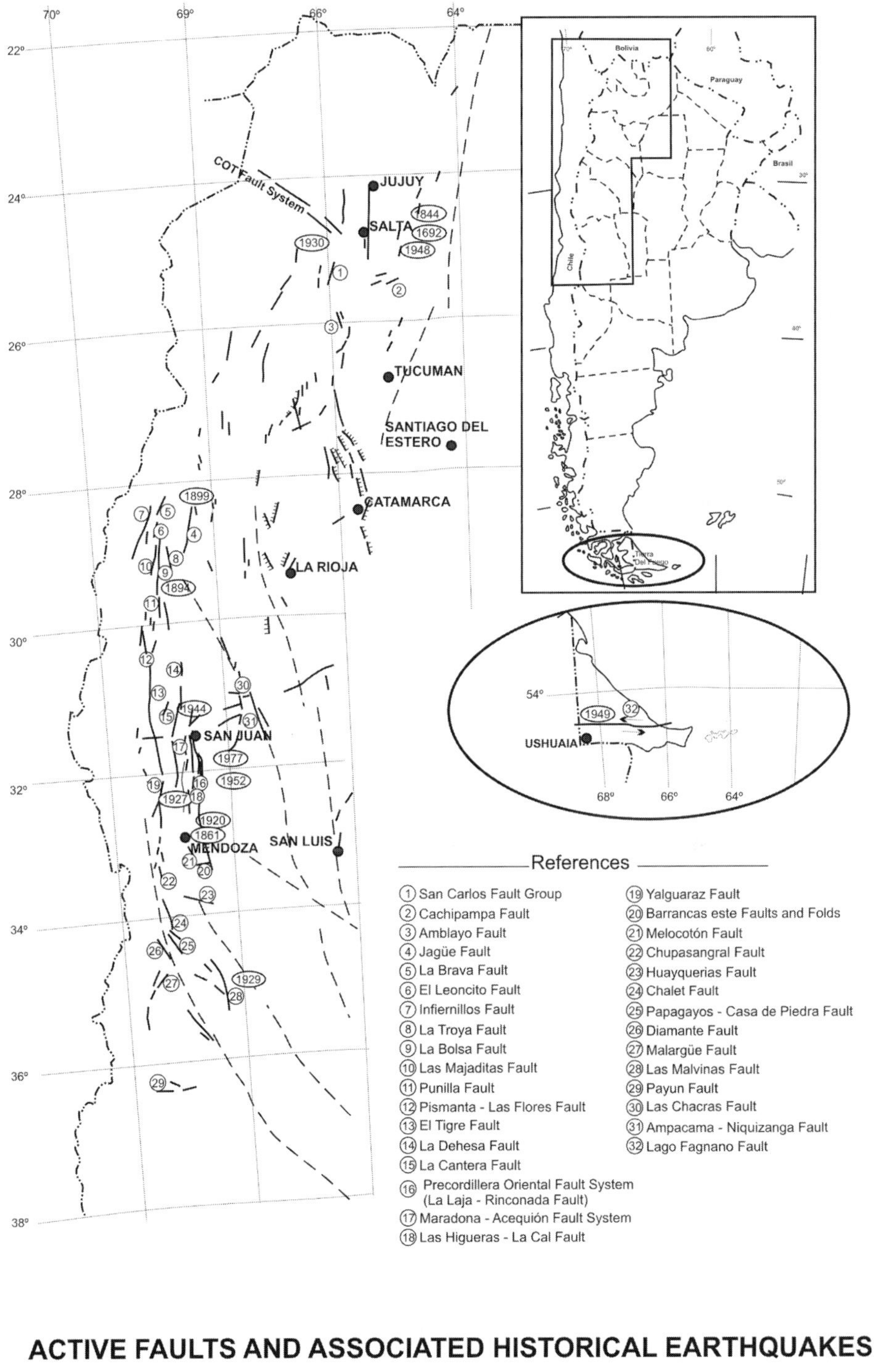

Figure 14.7 Simplified map showing active faults and associated historical earthquakes.

occur in alluvial fans where regional fractures cross the foot of mountainous blocks. Here triangular or trapezoidal facets indicate Quaternary faulting. Water runoff directions toward many *salars* (saline playas) indicate tilting of the blocks where these saline depressions are located (Igarzábal, 1999). Even though many faults with Quaternary tectonic activity occur in this area, in which many earthquakes took place, as previously mentioned, no evidences of ruptured surfaces associated with this activity have yet been found. Cortés et al. (1987) and Allmendinger et al. (1989), however, described some deformations of Quaternary age that affected alluvial units north of 28°S in the Puna,in the Eastern Cordillera thick skin setting, and in the Subandean Ranges folded structures (Ramos et al., 2003). The most important earthquakes (Ms > 5.4) that took place in northwestern Argentina are listed in Table 14.1.

North of 28°S, in the Puna-Eastern Cordillera thick-skin setting and in the Subandean Ranges folded structures, neotectonic affected Quaternary alluvial units (Cortés et al., 1987, Allmendinger et al. (1989, Ramos et al., 2003). Some of the most important faults are Cachipampa, Amblayo, and the San Carlos Fault Group (Bastias et al., 1995) (Fig. 14.7). They are located in the Eastern Cordillera, which is a Paleozoic orogen uplifted during the Neogene. The Cachipampa Fault is about 20 km long and trends N30°E (Wayne, 1994). The age of the last movement of this fault is Quaternary, but the sense of movement of this fault is not known. The Amblayo Fault is about 22 km long and trends nearly N-S. It shows east-facing scarps in the alluvium (Wayne, 1994). The San Carlos fault Group is 33 km long, and its average trend is N48°E (Wayne, 1994). The most important earthquakes (Ms > 5.4) that took place in northwestern Argentina are shown in Table 14.1.

In the Province of La Rioja (the central-western region of Argentina), numerous faults with Quaternary tectonic activity have been identified (Costa et al., 2000). They are potential seismogenic sources, for which reason they should be studied, even though no evidences of ruptured surfaces have been detected. This is the case of the Jagüé Fault, whose length is about 58 km and trend is N20°E (Fig. 14.7). Evidences for Quaternary activity found in this fault are springs, sag ponds, and vegetation aligned along the fault trace. The trace, in spite of being formed against the slope, which makes it easily erodable by the action of streams, is well preserved. It faces to the west and affects the alluvial fans coming down the mountainous zone (Perucca and Navarro, 2005).

The best evidences of Quaternary faulting are found in the provinces of San Juan and Mendoza (Costa et al. 2000). The El Tigre fault system extends from the Province of La Rioja in the north to the Province of Mendoza in the south. It runs parallel to the western boundary of the Precordillera, along an estimated distance of 600 km (Bastias et al., 1984, 1990). This fault has typical features that indicate dextral sense of movement, with normal component, and it is possible that the big earthquake that occurred in 1894 (Ms 7.5) was associated with it (Bastias et al., 1990, Perucca et al., 1999); (Fig. 14.7), even though no historical rupture surfaces have been recognized.

Table 14.1 Historical Earthquakes of the Northern Region of Argentina. SL (South Latitude), WL (West Longitude) D (depth in km), Ms (Magnitude), I (MM intensity)

Date	Locality	SL	WL	D	Ms	I	Characteristics
13-09-1692	Talavera del Esteco (Salta)	25° 24′	64° 48′	30	7.0	IX	Liquefaction, landslides, and several victims.
04-07-1817	Santiago del Estero	28°	64° 30′	30	7.0	VIII	Liquefaction and damages in Tucumán, Jujuy, and Salta
19-01-1826	Las Trancas (Tucumán)	26° 12′	65° 15′	30	6.4	VIII	Damages in Tucumán and Santiago. del Estero
18-10-1844	Salta	24° 48′	64° 42′	30	6.5	VII	Liquefaction and damages in Salta, Tucumán, Catamarca, La Rioja, etc.
14-01-1863	Jujuy	26° 36′	65° 00′	30	6.4	VIII	Houses and public building collapses
09-10-1871	Orán (Salta)	23° 06′	64° 18′	30	6.4	VIII	The city of Orán completely destroyed
17-11-1906	Tafí del Valle (Tucumán)	26° 45′	65° 42	30	6.0	VII	Some damages in buildings
11-08-1907	Monteros (Tucumán)	27° 12′	65° 30′	30	5.5	VI	Material damages
05-02-1908	Metán (Salta)	25° 12′	64° 42′	30	6.0	VII	Material damages
06-11-1913	San Miguel (Tucumán)	26° 48′	65° 06′	30	5.5	VI	Material damages
24-12-1930	La Poma (Salta)	24° 42′	66° 18′	30	6.0	VIII	Liquefaction, landslides, material damages, and fatal victims
03-04-1931	El Naranjo, Tucumán	27° 00′	65° 00′	180	6.3	VII	Material damages
12-02-1933	Raco (Tucumán)	26° 36′	65° 21′	30	5.5	VI	Material damages
25-08-1948	Anta (Salta)	24° 54′	64° 48′	50	7.0	IX	Liquefaction, victims, material damages
12-05-1959	San Andrés (Salta)	23° 10′	64° 39′	100	6.8	VIII	Landslides, material damages
19-11-1973	Santa Clara (Jujuy)	24° 34′	64° 35′	12	5.4	VII	Material damages
07-09-2004	Los Angeles (Catamarca)	28°34′	65°50′	22	6.5	VI	Material damages

Other important faults with Quaternary activity are: La Cantera (Bastias et al., 1984; Bastias, 1986; Mingorance, 1998); the Maradona-Acequión fault system (Perucca, 1990; Bastias et al., 1984, 1990), and the Precordillera Oriental fault system, with which the 1944 (Ms 7.4) and 1952 (Ms 7.0) earthquakes are associated (Fig. 14.7). The Precordillera Oriental fault system is located in the eastern piedmont of Precordillera Oriental range. It is about 120 km long (Bastias et al. 1990; Tello and Perucca 1993; Martos 1995, 1999a, 1999b). Their maximum magnitudes calculated by Martos (1995) and by Perucca and Paredes (2003, 2004) vary from 6.4 to 6.9. La Laja Fault, with a length of 19 km and sense of movement reverse, had historical surface rupture during 1944, with a maximum slip at the surface of 60 cm (Castellanos 1945; Bastias 1986).

For the Maradona-Acequión fault system, Perucca and Paredes (2004) estimated a maximum magnitude between 7.0 and 7.2. Table 14.2 summarizes the most important earthquakes that took place in this central-western region of Argentina.

The Quaternary structures of the piedmont zone of the cerro La Cal area (Province of Mendoza), are the La Cal and Melocotón faults (Fig. 14.7). The La Cal Fault (also known as the Salagasta Fault) is reverse, dipping to the east, and it has a sinuous trace that can be recognized in the field along a 32-km length. Here exist evidences of recent displacements along 8 km, immediately to the north of the cerro La Cal, probably related to the 1861 earthquake. The maximum probable magnitude for this fault was estimated between 7.2 and 7.6 (Bastias et al., 1993). Melocotón fault is reverse, with an East-facing fault scarp that affects Pleistocene alluvial fans. It has a low angle and dips to the west.

Further south, the Las Vacas-Tupungato fault system has a sinistral sense of displacement and extends several hundreds of kilometers, trending NNW-SSE, and its southern prolongation, the Diamante and Papagayos faults affect Quaternary deposits (Fig. 14.7).

South of Mendoza Province, the Malvinas Fault (Fig. 14.7) trends nearly N-S and shows a left lateral and normal sense of displacement, affecting Quaternary deposits (Bastias et al., 1993). This fault has been related to the earthquake that occurred on May 13, 1929 (Loos, 1929; Lunkenheimer, 1930).

Other parallel faults are the Piedemonte Andino fault system, rimming the eastern piedmont of Cordillera Frontal, where are located the Chupasangral, Chalet, and Malargüe faults (Fig. 14.7). These faults trend nearly N-S, parallel to the mountain ranges front (Bastias et al., 1993; Cortés, 2000). There are no historical rupture surfaces related to these faults. The Payún Matru Fault (Fig. 14.7), which is in the south of the Province of Mendoza, near the boundary with the provinces of Neuquén and La Pampa, is located in the extensive volcanic field known as Payunia (Bastias et al., 1993).

The Sierras Pampeanas basement, which is situated to the east of the Central Andes, is considered to be made up by uplifted blocks of the broken Andean foreland, thus being another feature of the Pampean flat subduction segment (Jordan et al., 1983; Jordan and Allmendinger, 1986). From the neotectonic viewpoint, the evolution of these mountain ranges is related to the Neogene Andean orogenic processes, in which the response to crustal shortening of the Andean foreland was the inception of a thick-skinned

Table 14.2 Historical Earthquakes of the Central-Western Region of Argentina. SL (South Latitude), WL (West Longitude), D (depth in km), Ms (Magnitude), and I (MM intensity)

Date	Locality	SL	WL	D	Ms	I	Characteristics
22-05-1782	Santa Rita (Mendoza)	32° 42′	69° 12′	30	6.5/7	VIII	Damages in buildings but no victims were registered
20-03-1861	Mendoza	32° 54′	68° 54′	30	7.2	IX	Liquefaction and landslides. About 12,000 victims. The city of Mendoza was completely destroyed.
27-10-1894	North of San Juan	29° 45′	69° 00′	30	7.5	IX	It was the largest historical earthquake in Argentina, causing approximately 100 victims and severe damage in San Juan, La Rioja, Córdoba, and Mendoza. Liquefaction and landslides.
12-04-1899	Jagüé (La Rioja)	28°39′	68° 25′	30	6.4	VIII	Destroyed Jagüé and caused important damages in several La Rioja towns.
12-08-1903	Uspallata (Mendoza)	32° 06′	69° 06′	70	6.3	VIII	The earthquake intensity produced water flow ejection. Fissures and cracks appeared in several places.

(*Continued*)

Table 14.2 (*Continued*)

Date	Locality	SL	WL	D	Ms	I	Characteristics
26-07-1917	Panquehua (Mendoza)	32° 20′	68° 54′	50	6.5	VII	Several damaged buildings and liquefaction phenomena were recorded in Mendoza.
17-12-1920	Costa de Araujo (Mendoza)	32° 42′	68° 24′	40	6.3/6.8	VIII	Liquefaction and landslides, several victims and building damages
14-04-1927	Uspallata (Mendoza)	32° 24′	69° 18′	>60	7.1	VIII	Liquefaction and landslides. The government buildings and several schools were completed destroyed
30-05-1929	Las Malvinas (Mendoza)	34° 54′	68°	40	6.5	VII	Building damages and several victims. Liquefaction
15-01-1944	La Laja (San Juan)	31° 24′	68° 24′	30	7.4	IX	The city of San Juan was completed destroyed, and there were 10,000 fatalities. Liquefaction and landslides occurred.
11-06-1952	Rinconada (San Juan)	31° 36′	68° 35′	30	7.0	VIII	Liquefaction, landslides, and building damages
24-10-1957	Villa Castelli (La Rioja	28° 54′	68	37	6.0	VII	Building damages
26-01-1985	Lunlunta (Mendoza)	33° 06′	68° 30′	12	5.9	VIII	Several damages in buildings

tectonic style, with less deformation rates than those occurring in the Andes (Costa, 1999). In general, the blocks show an asymmetrical east-west profile, usually with a steeper western slope affected by 30° to 70° east-dipping faults. In these western slopes, Costa (1996, 1999) and Costa et al. (2000) described faults with Quaternary tectonic activity. The fault scarps generally are not exposed, except in the river ravines, where the basement-piedmont contact and the faults themselves are exposed. Most faults situated in the main range fronts have had little activity during the Quaternary. This activity is migrating toward the piedmont (Costa, 1999). For these reasons the Sierras Pampeanas have been considered an area with moderate seismic activity, with little superficial deformation.

Nevertheless, the 1977 earthquake that took place in Caucete, Province of San Juan (Ms 7.4), gave origin to secondary ruptured surfaces related to normal faults in the eastern piedmont of the Sierra de Pie de Palo, the Ampacama-Niquizanga fault system (Volponi et al., 1977; Bastias, 1986; Bastias et al., 1990) (Fig. 14.7). It is also probable that superficial deformation existed during the Sampacho earthquake (Ms 6.0) in the Province of Córdoba, which occurred in 1934 (Mingorance, 1991; Sagripanti et al., 1999; Costa et al., 2001). The Ampacama-Niquizanga fault system extends along approximately 70 km, segmented in many stretches and not surpassing 8 km in length, along a N20°E trend (Fig. 14.7).

Evidences for Quaternary fracturing are found in the neighborhood of the eastern mountainous front, where their effects on alluvial fans and terraces can be observed (Bastias, 1986).The scarps face westward, and, during the 1977 earthquake, a new displacement was produced, as revealed by geodetic leveling. This caused permanent deformation of the ground amounting to 1.20, while the displacement of the scarps was not higher than 0.30 m (Bastias, 1986). Another important earthquake took place in 1941 (Ms 6.3) in the same region, probably related to the same fault system, but no ruptured surface was found.

The most important earthquakes that occurred in this region are listed in Table 14.3. The calculations based on empirical relationships made by Bastias (1986) to determine the maximum magnitude of a probable earthquake gave values from Ms 6.8 to Ms 7.1, with a recurrence interval of about 15,000 years.

The seismic activity in the Patagonia region is related to a deformation and fracture belt that, at the latitude of Lago Fagnano, extends E-W for over 600 km, from one ocean to the other, and is known as the Magallanes-Fagnano Fault System (Fig. 14.7). All along this active fault, the South American continent moves very slowly to the west with respect to the Tierra del Fuego Andean region. The border between the South American and Scotia plates extends for more than 3000 km, from the western section of North Scotia ridge, in the Chilean Southern trench, at 50° S, through the Isla de Tierra del Fuego. The Magallanes-Fagnano fault system, with a sinistral E-W displacement, evolved as a component of relative plate movements between the southern end of South America and the Antarctic Peninsula. This regional alignment can be seen mainly in the eastern branch of the

Table 14.3 Historical Earthquakes of the Sierras Pampeanas. SL (South Latitude), WL (West Longitude), D (depth in km), Ms (Magnitude) and I (MM intensity)

Date	Place	SL	WL	D	Ms	I (MM)	Characteristics
11-06-1934	Sampacho (Córdoba)	33° 30′	64° 30′	30	6.0	VIII	This city was partially destroyed, with several victims.
03-07-1941	Caucete(San Juan)	31° 54′	67° 54′	30	6.7	VI	Liquefaction. Damages were mainly characterized by cornices collapses and walls fissures.
23-11-1977	Caucete (San Juan)	31° 02′	67° 45′	17	7.4	IX	Liquefaction, 70 victims, about 200 injured, and substantial damage to buildings
28-05-2002	La Rioja	28° 56′	66° 48′	22	6.0	VIII	Several damages in buildings

Magellan Straits, along the northern shore of Lago Fagnano and along the Atlantic coast (Winslow, 1982; Winslow and Prieto, 1991). It has a length of 165 km and trends N89°W. The present deformation, measured at stations located on both sides of the main faults of the South American and Scotia plates, shows a sliding rate of around 0.5 cm/yr (Del Cogliano et al., 2000). Furthermore, the seismicity along the entire fault system is low (Ms < 3.5), mainly superficial, and the focal mechanisms indicate a tensional component and a strike-slip feature (Pelayo and Wiens, 1989). The Isla de Tierra del Fuego was affected in 1949 by two earthquakes of Ms 7.8 that caused widespread strike slipping on the shores of Lago Fagnano, and a local tsunami in the western branch of the Magellan Straits (Jaschek et al., 1982). Surface ruptures and displaced fences were reported (Costa et al., 2000).

From 1969 to the present date, more than 400 superficial earthquakes of magnitude over 4 have been registered in the Argentinean Patagonia. Even through there are poor historical seismic data in the Santa Cruz Province, and because of its proximity to regions with earthquakes of magnitude higher than 7 (Isla Grande de Tierra del Fuego), González Bonorino (2002) assumed that the eastern region of the province was moderately seismic.

In short, although the magnitude of earthquakes in Patagonia is generally moderate in strength, most of these events were superficial and related to active faults, which indicate potential future seismogenic sources. As shown in the previous paragraphs, the relatively large number of records corresponding to the second half of the past century, in comparison with records from previous years, is an indicator of the scarcity of seismological stations in the region and not the absence of seismic activity in the past. The historical record of Isla Grande de Tierra del Fuego is also very brief, owing to the relatively recent European settlement and its low population density. The seismic activity record in Tierra del Fuego goes back to 1879 and continues up to the present date, recording 1600 events, although most of them are of a very low magnitude. An ancient earthquake took place before the European colonization, according to a Yaghan (indigenous Fuegians) legend mentioned by Lucas Bridges (2000). The first recorded earthquake occurred on February 1, 1879, at 5 a.m. (local time). It was described by Thomas Bridges (1879), an Anglican missionary, the first European settler in Tierra del Fuego. Its intensity was of grade VI in the MM (Modified Mercalli) scale, and its epicenter was located at 54° S and 65° W (Fig. 14.2).

4. Seismic Hazard and Risk

Even though many earthquakes in Argentina were recorded during historical times, no clear relation exists between their epicenters and the main Quaternary structures. In most cases, no historical rupture surfaces have been observed associated to these epicenters, with the exception of those observed in La Laja Fault during the 1944 earthquake, the Fagnano Fault in the 1949 event, and the Niquizanga Fault in the 1977 earthquake. It is also probable that many prehistorical earthquakes of high magnitude have been characterized by deformations in folds and secondary fractures rather than by surface ruptures.

Costa et al. (1999) indicated that these types of paleo-events, even the destructive ones, may not have left evidences in the Quaternary morphostratigraphic record, thus complicating the estimation about parameters, such as recurrence and movement rates. In other cases, the evidences of tectonic activity during the Quaternary suggest the occurrence of destructive prehistoric earthquakes recurrent during the Holocene and related to Quaternary structures.

Earthquakes with magnitudes higher than Ms 6, with subsurface hypocenters shallower than 30 km generally produce deformations in the surface, thus constituting a major risk for the residents. Among the primary effects produced by an earthquake are surface ruptures and deformations that occur at the epicenter or near it. It is here where greater damages are produced in constructions. On both sides next to the fault, no constructions must be built because this is the zone with surface displacements hazard.

The secondary effects associated with liquefaction processes and slope instability can cause greater damage than the earthquake shake itself, even in areas distant from the rupture zone when local conditions favor this hazard. A second effect concomitant with the majority of the earthquakes in western Argentina is soil liquefaction. The fissures, lateral spreading, sand volcanoes and spills, have been very abundant in areas up to 260 km far from the epicenter and affected areas up to 4000 km^2, such as during the 1977 earthquake (INPRES, 1977). The damages produced by liquefaction include soil collapse under buildings, roads, railroads, irrigation channels, water reservoirs and houses, due to the lack of bearing strength. Alluvial plain, paleochannels, and playa deposits of Holocene age, due to their predominant lithology (sand) and superficial water table, among other conditions, were strongly affected by liquefaction (Perucca and Bastias, 2005, 2006; Perucca and Moreiras, 2006). Moreover, numerous landslides occurred during most historical earthquakes in the provinces of Mendoza and San Juan (Moreiras, 2004).

5. Volcanism and Volcanic Hazard

The Andean Cordillera is segmented into three portions where the present-day volcanic activity is concentrated (Fig. 14.8): (a) the northern segment, the southern portion of which includes the Central Volcanic Zone (CVZ) that extends from the Province of Jujuy to the Province of Catamarca, (b) the central-south segment located in the Province of Mendoza and the

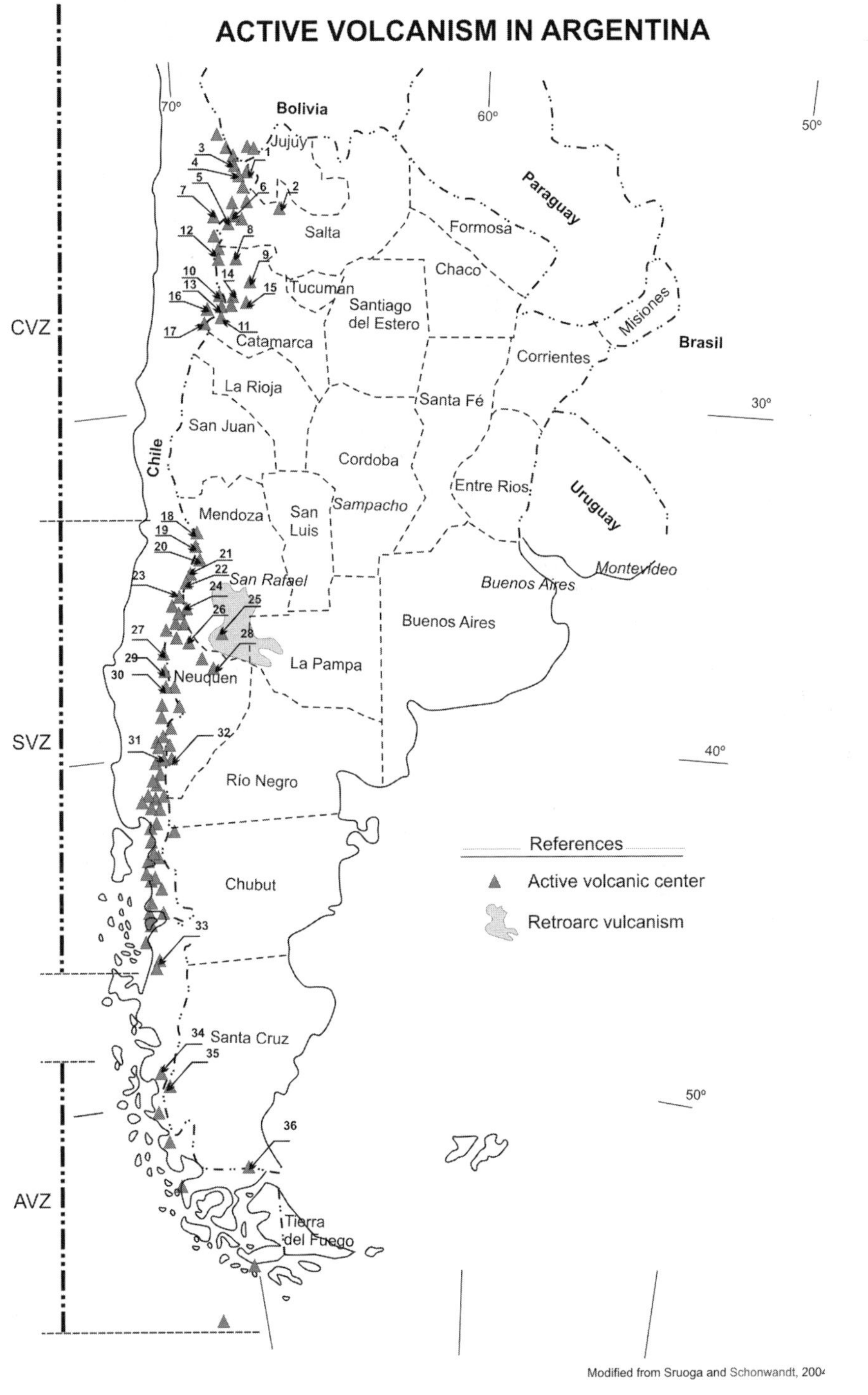

Figure 14.8 Simplified map of Argentina with the active volcanism. (Modified from Sruoga and Schonwandt, 2004).

northern part of the Province of Chubut and that has been named South Volcanic Zone (SVZ), and (c) the austral segment in the Province of Santa Cruz, which makes up the Austral Volcanic Zone (AVZ).

The eruptive centers in these segments are huge stratovolcanoes, andesite-dacite calderas, and basalt plateaus (Sruoga et al., 1993), younger Pliocene, Pleistocene, and Holocene in age (less than 2 Ma) related to the last great uplifting and faulting events in the Cordillera de Los Andes. The lavas and pyroclastic deposits associated with these volcanic events extend over wide regions and attain great thicknesses in all western Argentina. These materials are predominant in the northwestern portion (Puna) and from the middle part of the Province of Mendoza to the southern portion of the Province of Santa Cruz. They are not restricted to the Andes, but also extend over the extra-Andean part of the provinces of Mendoza, Neuquén, La Pampa, and Santa Cruz.

The historical record, in spite of being scarce, points to eruptions in some volcanoes in the bordering zone with Chile, as the eruption of the Hudson volcano in 1991 and of the Descabezado-Quizapu volcano in south Mendoza in 1932. The active cordilleran magmatic arc has migrated in recent geologic times, in most cases to the west, nearer the subduction zone. For this reason, the greater part of the active and latent volcanoes is along the international boundary or few kilometers to the west, in Chilean territory. During the twentieth century, several eruptions occurred with different consequences in some regions of the Argentina. They were explosive in most cases and were of plinian, pelean, phreatomagmatic, and vulcanian types; they took place preceding earthquakes and related to this earthquakes (Sruoga and Schonwandt, 2004). Among the volcanic eruptions that occurred during the twentieth century, the most remarkable ones are those of the Descabezado-Quizapu (1932), Tupungatito (1952, 1980, and 1986), Peteroa (1937, 1991), Hudson (1991), Copahue (1992), and Lascar (1989, 1993); of which all of them were characterized by ash rain.

Because the active volcanic centers are situated in zones with scarce or even no population at all, the direct impact of lavas and pyroclastic flows was of relatively low significance. Nevertheless, the indirect effects were potentially much higher, especially on the farming and animal husbandry activities due to the accumulation of ashes. For instance, the eruption of the Hudson volcano in 1991 caused the death of a great part of the sheep flocks in the Province of Santa Cruz, due to the considerable destruction of the vegetal cover (Clarín, 1991). This phreatoplinian eruption affected a region greater than 300,000 km^2 and accumulated a layer of ash surpassing 1m in thickness in the areas near the volcano. Bitschene (1995) estimated that this eruption implied, in the provinces of Chubut and Santa Cruz, direct losses estimated at US$ 10,000,000 in animals, wool, houses, cleaning of water supply systems, cleaning of airports and roads, communication and electric equipment, and so on. The Lascar volcano, situated in northern Chile, next to the boundary with the provinces of Salta and Jujuy, had eruptions of the plinian type near the end of the twentieth century, with emission of ashes which affected the north of Argentina, especially air routes.

One of the few regions where eruptions can have direct effects on the population is the zone of the Copahue volcano, in the neighborhood of which are situated the villages of Copahue and Caviahue. Also, the Lanín volcano, near the tourist towns of San Martín de Los Andes and Junín de Los Andes, shows indications of Holocene activity and is closely related to active volcanoes in Chile. Other impacts can be mentioned, such as strong earthquakes like those that took place in the city of Bariloche, with casualties and material damages, caused by the eruption of the Calbuco volcano in 1961 (Klohn, 1963).

5.1. Central Volcanic Zone (CVZ, 10°–28° S)

This segment, the Central Volcanic Zone, has some of the higher active volcanoes of the world: Ojos del Salado and Llullaillaco, both surpassing 6000 m.a.s.l at the boundary of the provinces of Catamarca and Salta with Chile (Fig. 14.8).

The southern portion of the Central Volcanic Zone, between 22° and 27° S, is an uplifted tableland known as Puna with an average altitude of 3700 m, and N-S trending mountain chains encircling closed depressions. The volcanic activity in this region began in the Oligocene-Early Miocene up to the present (Coira et al., 1993; Allmendinger et al., 1997). The location of the volcanism in this area is conditioned by the regional tectonic structures (Riller et al., 2001), as indicated by the existence of a north-south trending magmatic arc and volcanic chain oriented obliquely to the Andean trends. This chain coincides with regional alignments running from NW to SE (Viramonte, 1984; Salfity, 1985).

The Calama-Olacapato-Toro (COT) lineament is a seismically active fault zone (Schurr et al., 1999) delineated by a 170-km-long volcanic belt between the Incahuasi volcano and the Río de Las Burras volcanic complex (Hongn et al., 2002). In this fault zone there are more than 20 active volcanic centers, which include domes, volcanic fields, stratovolcanoes, collapsed calderas, and small volcanic centers, all of which indicate magmatic activity from the Miocene to the Pleistocene and, in some cases, to historical times.

The volcanic activity in this region is also confined to two WNW-ESE trending faults (Marret et al., 1994; Petrinovic, 1999), connected by normal faults and with en echelon horizontal displacement. Therefore, the fault activity must have had an important role in the origin of these volcanoes.

To the north and to the south, respectively, of the COT lineament, the Puna seems to have had different volcanic evolutions. North of this lineament the volcanic activity notably diminishes from the upper Miocene-Early Pleistocene, so that ignimbrite mantles are affected locally by faults (Seggiaro, 1994). To the south, volcanism and deformation continued during the Pliocene and Quaternary (Mon et al., 1988; Marret et al., 1994).

The potentially active volcanoes occur mainly in the N-S trending magmatic arc, between 22° and 27° S, and their composition is principally andesitic to basaltic. They have produced restricted eruptions, with very short lava flows (1 to 8 km) and associated ash deposits. Table 14.4 outlines some of the most important active volcanoes of the CVZ.

Nearly all of these volcanoes are far from populated zones or infrastructure works, and so their risk is low. Nevertheless, the risk associated especially with ash rains must not be underestimated in the provinces of northwestern Argentina, principally in Salta and Catamarca.

5.2. South Volcanic Zone (SVZ, 33° to 43°S)

To the south of 33° 15′S, a Plio-Quaternary volcanism developed associated with two tectonic settings related to the subduction of the Nazca Plate beneath the South American Plate. The present-day volcanic arc is situated along the Cordillera de Los Andes and the backarc, to the east, in the extra Andean region.

One example of high risk in the zone (in Chilean territory) is the Descabezado Grande-Quizapu volcanic field, which was fairly continuously risky in historical times. Even though these volcanoes are located 30 km west of the Argentina-Chile boundary, owing to the direction of the winds, from west to east, they have affected Argentine territory mainly with a fall of tephra and ash. The volume of ejected tephra has been estimated at 20 km^3. The ash fall covered more than 2 million km^2 of South America, reaching Santiago, Montevideo, Brazil, and even the coast of South Africa. In the city of Buenos Aires, located 1200 km east of the volcano, the rain ash reached 150 ton/km^2 (González Ferran et al., 2003).

The Tupungatito volcano has erupted 18 times during the last 160 years. For this reason, it is highly dangerous and a risky center for the north zone of the Province of Mendoza, where the majority of the population lives. Moreover, Loss (1929) mentioned an eruption of the Maipo volcano in 1882, which affected the city of San Rafael (in the south of Mendoza).

The Peteroa volcano has also had eruptions in historical times. In 1937 it covered the city of Mendoza with a significant accumulation of ash. The 1991 eruption produced difficulties for air navigation. The potential occurrence of lahars and fall of ejecta in the case of a new eruption of this volcano is high (Haller et al., 1991).

In summary, in view of the high rate of eruptive activity during historical times, the area between 33° and 36°S is the most hazardous region in the Azufre-Planchón-Peteroa and Descabezado Grande-Quizapu volcanic centers.

Besides their predominantly explosive character, the action of ash rains could produce great damage, especially in the San Rafael oasis. For instance, during the eruption of the Quizapu volcano in 1932, the most important changes were reduced temperatures because the ash rain produced an immediate total darkness, high turbidity in the rivers, livestock dispersal, damming of water sources, and the death of animals and plants within a 40-km radius around the volcano. The population was highly alarmed by quakes and subterranean noises (González

Table 14.4 Active volcanoes of the Central Volcanic Zone. SL (South Latitude), WL (West Longitude)

Volcano	SL	WL	Summit elevation(m)	Characteristics
Lascar (1)	23° 22’	67° 34’	5550	Stratovolcano that consists of an elongate series of six overlapping craters, trending roughly northeast, with the active, fuming crater located near the centre of this cluster. It has registered permanent episodic and explosive fumarolic activity. During 1986 an eruption occurred and the ash arrived to the city of Salta.
C. Tuzgle (2)	24° 03′	66° 29′	5500	It is the easternmost young stratovolcano of the central Andes Many youthful-looking Holocene lava flows were erupted from the well-preserved summit crater. Inactive.
Aracar (3)	24° 15′	67° 46′	6082	Well-preserved lava flows are found at the base of the volcano. Prior to a report of possible ash columns from the summit in 1993, the volcano was not known to be active
Socompa (4)	24° 24′	68° 15′	6051	Stratovolcano. During the Holocene produced a 600 sq km debris-avalanche deposit that is one of the largest and best exposed of the world. No historical eruptions are known. Current fumarolic activity.
Llullaillaco (5)	24° 43′	68° 32′	6739	Stratovolcano with well preserved cone. Two explosive eruptions and another that may have included lava effusion were reported from Llullaillaco in the 19th century. Latent activity
Co. Escorial (6)	25° 05′	68° 22′	5447	Stratovolcano that has young-looking lava flows and a well-preserved crater. It is considered inactive.
Co. Bayo (7)	25° 25′	68° 35′	5401	It is a complex volcano of partial Holocene age with lava flows well preserved. Currently inactive
Antofalla (8)	25° 32′	68°	6100	There are mentions of smoking in 1901 and 1911. De Silva and Francis (1991) did not see evidence for Holocene activity but noted that the scoria cones and lava flows are very well preserved, indicating ages of a few thousand to ten thousand years. Latent activity.

(*Continued*)

Table 14.4 *(Continued)*

Volcano	SL	WL	Summit elevation(m)	Characteristics
Antofagasta de la Sierra Volcanic field (9)	26° 05′	67° 30′	4000	Contains the youngest volcanic vents of the Argentinian Puna region (Hormann et al., 1973). The area includes several "extremely youthful" scoria cones, few thousand years old (de Silva and Francis, 1991)
Co. El Cóndor (10)	26° 37′	68° 21′	6532	Strato-volcano that contains several ash cones and has been a source of lava flows of Holocene age (de Silva and Francis, 1991). Latent activity.
Falso Azufre (11)	26° 48′	68° 22′	5890	Consists of overlapping craters that contain small composite cones and lava domes of recent activity of the complex (de Silva and Francis, 1991).
Cordón del Azufre (12)	25° 20′	68° 31′	5463	Small volcanic complex with a cluster of lava flows and a chain of vents. An older edifice was covered by younger Holocene lava flows. Latent activity.
Lastarria (13)	25° 10′	68° 30′	5697	Although no historical eruptions have been recorded, the youthful morphology of deposits suggests activity during historical time. Fumarolic activity occurs.
Co. Peinado (14)	26° 37′	68° 09′	5740	This stratovolcano is the source of well-preserved lava flows of Holocene age (de Silva and Francis, 1991) and is one of the youngest volcanoes in the region. Fumarolic activity occurs
Nevado de Incahuasi (15)	27° 02′	68° 17′	6621	Stratovolcano, the youngest one is capped by a crater, and lava flows radiate down the volcano's flanks. The fresh-looking morphology of the youngest products suggests a Holocene age (González-Ferrán, 1995). Fumarolic activity.
Tipas (16)	27° 12′	68° 33′	6660	This volcanic complex of craters, cones, lava domes and lava flows covers an area of 25 sq km. It displays a youthful morphology, and its latest eruptions were considered by de Silva and Francis (1991) to be of Holocene age. Fumarolic activity
Nevados Ojos del Salado (17)	27° 07′	68° 33'	6887	The most recent eruptive activity involved formation of lava flows and cones, lava domes, and explosion craters. No confirmed historical eruptions have been recorded, but the volcano has fumarolic activity.

et al., 2002; Hildreth and Drake, 1992). Also felt in this region were the seismic movements associated with the Azul volcano eruption in Chile in July 1924.

Between 33° and 34° S, the eruptions are effusive in character. Less evidence of volcanic activity is found here than in the former areas, but still this activity could produce lahars and mass movements due to the glaciers covering the volcanic centers. Lahars were documented in the Cerro Overo volcano, near the Atuel River, in January 1942. In view of its proximity to population centers, this presents a substantial hazard not only in Chile but also in Argentina.

Between 34° and 35° S, hazards are related to the potential activity of the Maipo volcano (Sruoga et al., 1993, 2000), which could generate an eruption like the one that resulted in the Diamante caldera. If a similar phenomenon were to occur, cities such as Santiago and localities in the south of the Province of Mendoza could be affected. The ash rains would affect not only the Province of Mendoza, but also a great part of the Argentine territory (Sruoga et al., 1993).

In terms of the backarc volcanism, such as the Payun Matru complex, the volcanic hazard is low because of the low population density and the few cultivated areas and natural woods that could be affected by its basaltic magmas with low explosive character. But the roads could be affected, and rivers deviated or dammed, as has happened in recent years (Delpino, 1993).

Currently, the southern segment (37° to 45° S) has the highest volcanic activity and presents the greatest concern because of the large population in this area as well as many tourists. Generally, the volcanoes in this region are not characterized by explosions of great magnitude, with the exception of the Hudson volcano, which is in Chilean territory (Sruoga and Schonwandt, 2004). Table 14.5 describes some of the most important volcanoes in this region. But even phreatomagmatic explosions with small volume, such as the one that took place in 2000 in the Copahue volcano, could cause substantial economic losses, because of the increasing urban development in the area. The studies made on this volcano have shown the occurrence hazard of pyroclastic flows, tephra falls, and lahars (Delpino and Bermúdez, 2002; Naranjo and Polanco, 2004). The presence of an acid lake in the active crater represents one of the principal risks that could affect the tourist locality of Caviahue, populated by 800 to 1500 inhabitants depending on the season of the year.

5.3. Austral Volcanic Zone

Volcanoes located to the southern extreme of Argentina and Chile are characterized by calc-alkaline magmatism, or adakitic composition, and are grouped in the Andean Austral Volcanic Zone (AVZ; Stern, 2004). The magmagenesis of adakitic volcanoes is attributed to dehydration melting of oceanic crust on the surface of subducting young and hot slab, the phenomenon called slab melting (Defant and Durmmond, 1990).

In the south of Chile, near the Lautaro volcano, the tephra deposited on the glacier could be entirely swept out in about 60 years. If so, based on simple calculation, the tephra of the 1959–1960 eruption should be totally gone by 1990, and the recent tephra layers near the ice surface could originate from

Table 14.5 Active Volcanoes of the South Volcanic Zone (SVZ, 33° to 43°S). SL (South Latitude), WL (West Longitude)

Volcano	SL	WL	Summit elevation (m)	Characteristics
Tupungatito (18)	32° 24′	69° 48′	6000	Stratovolcano with lava flows from the northernmost vent that traveled down the NW flank breach. Tupungatito produced frequent mild explosive eruptions during the nineteenth and twentieth centuries
San José (19)	33° 47′	69° 54	5856	Stratovolcano active during the Holocene. Mild phreatomagmatic eruptions were recorded from San José in the nineteenth and twentieth centuries.
Maipo (20)	34° 09′	69° 50′	5264	It has a youthful appearance, and ash fall deposits overlie glacial ice. Lava flows from one of the cones blocked drainages in 1826 inside the caldera, forming Lake Diamante on the eastern caldera. Also erupted in 1882. Latest eruption occurred during 1908.
Caldera del Atuel (21)	34° 39′	70° 03′	5189	Caldera with a group of cones overlies the SW rim. On the NE flank of the caldera there is a group of young cinder cones of possible historical age
Planchón Peteroa (22)	35° 14′	70° 34′	4107	Stratovolcanoes. Peteroa has been active into historical time and contains a small steaming crater lake. Historical eruptions have been dominantly explosive, although lava flows were erupted in 1837 and 1937. Last known eruptions in 1991 and 1998.

Descabezado Grande-Quizapu (23)	35° 35′	70° 45′	3953	Stratovolcano. A lateral crater that formed in 1932, shortly after the end of the major 1932 eruption from nearby Quizapu volcano on the north flank of Cerro Azul, was the site of the only historical eruption of Descabezado Grande.
Caldera del Maule (24)	36° 01′	70° 35′	3092	The latest activity produced an explosion crater on the east side of the lake and a series of Holocene lava domes and blocky lava flows that surround Laguna del Maule
Payun Matru (25)	36° 25′	68° 12′	3680	It is a massive Hawaiian-style shield volcano. According to oral traditions, native tribes were present at the time of the latest eruption.
Domuyo (26)	36° 35′	70° 25′	4709	Stratovolcano of Holocene age. At least 14 dacitic lava domes and other eruptive centres were constructed within the caldera, and others lie outside.
Cochiquito Group (27)	36° 46′	69° 49′	1435	Stratovolcanoes of estimated Pleistocene-Holocene age
Tromen (28)	37° 08′	70° 02′	3,978	Holocene vents are located in the lower NE flank. There are some reports of an 1822 eruption (and supposed eruptions in 1820, 1823, 1827, and 1828) as being from Tromen, but these are not confirmed.
Trocon (29)	37° 45′	69° 53′	Unknown	This dome complex has two summit craters and a pyroclastic cone that fed lava flows that traveled to the east. González-Ferrán (1995) estimated a Pleistocene-Holocene age.

(*Continued*)

Table 14.5 (*Continued*)

Volcano	SL	WL	Summit elevation (m)	Characteristics
Copahue (30)	37° 51′	71° 10′	2997	Acidic hot springs occur below the eastern outlet of the crater lake. Infrequent mild-to-moderate explosive eruptions have been recorded since the eighteenth century. Twentieth-century eruptions from the crater lake have ejected pyroclasts rocks and chilled liquid sulfur fragments.
Lanin (31)	39° 38′	71° 30′	3747	A small lava dome at the summit of Lanin fed blocky lava flows about 2200 years ago. It was reported active after an earthquake in 1906, but this is not sure. L\The last known eruption was dated in 560CE ± 150.
Huanquihué Group (32)	39° 52′	71° 33′		There are compound cinder cones. The Escorial lava is a youthful flow of about 200 years ago that diverted local drainages and formed new lakes. Oral histories described this eruption.
Hudson (33)	45° 54′	72° 58′	1905	The ice-filled caldera was recognized during the twentieth century eruption in 1971. An eruption about 6700 years ago was one of the largest known in the southern Andes during the Holocene; another eruption about 3600 years ago produced also more than 10 km^3 of tephra. An eruption in 1991 was Chile's second largest of the twentieth century.

unknown eruptions after 1970. Lliboutry (1999) cited two possible eruptions, in 1972 and 1978, based on oral communications of local farmers. Thus, it is possible to infer multiple eruptions of the Lautaro volcano after 1970.

An eruption in 1988 confirmed the presence of a postulated subglacial vent in the Patagonian Icefield NW of Viedma Lake. The Viedma volcano eruptive center is a subglacial dacitic volcano beneath the Patagonian Icecap, with only part of the older edifice rising above the surface of the icecap (Fig. 14.8). However, these volcanoes are far from towns and infrastructure, and so they present only low risk. Occasionally, ash rains could affect areas far from these volcanic centers, as happened during the eruption of the Hudson volcano in 1991. The risk of pyroclastic flows and lahars is less probable, in spite of the existence of a cap of ice over the volcanoes, because there is little or no population in the area. Tables 14.6 summarizes some important volcanoes in this zone (Chilean and Argentine territories).

6. Final Remarks

Study of the seismic, tectonic, and morphological characteristics in Argentina leads to the following conclusions. There is a marked lack of uniformity in the distribution of earthquakes and anomalous areas, due to the excess of or absence of notable seismic events in relation to what may be considered a normal value. For example, in Patagonia, many earthquakes have been localized in Chilean territory, whereas to the east seismic activity is either low or nonexistent. There seems to be a close relation between the great structures and the site of seismic events, especially in regard to the separation of environments with different levels of seismic activity. These large structures would delimit the Neogene volcanic environments as well. Works done on neotectonic activity in Argentina indicate the need to develop an extensive research field, not only as regards the potential hazards of earthquake, but also the architectural point of view of Neogene tectonism.

We have only scarce pre-Hispanic historical information on the provinces of northwestern Argentina, as well as the majority of those located in seismic regions of the country, and very little data for the three centuries of Spanish rule as well. This lack of historical records was likely due to the area's low population density, to political events that led to the foundation of urban centers east of the Cordillera, and to colonists' ignorance of the culture and habits of the indigenous populations. This deficiency must be corrected with geologic research on the active faults in order to provide viable strategies for accurate territorial planning in accordance with the potential seismic hazard that could affect numerous regions in Argentina.

Most valleys are located the principal towns of western Argentina and are liable to undergo liquefaction, because the sediments filling those valleys are mainly alluvial and unconsolidated. Therefore, even when constructionis carried out according to the rules against seismic hazard, lack of soils studies at the town sites could cause great economic losses involving not only buildings, but also communications, water wells, roads, gas or oil lines, electric lines, and the like. In many of

Table 14.6 Volcanoes of the Austral Volcanic Zone. SL (South Latitude), WL (West Longitude)

Volcano	SL	WL	Summit elevation (m)	Characteristics
Lautaro (33)	49° 01′	73° 33′	3607	Stratovolcano. Ash deposits from eruptions of this dominantly dacitic volcano are visible in aerial photos. At least three layers of fallout deposits have been recognised in parts of the surface of the glacier, which may correspond to the latest known eruptions.
Viedma (34)	49° 21′	72° 10′	1500	It is a subglacial volcano. The 1988 eruption deposited ash and pumice on the Patagonian Icecap and produced a lahar that reached Viedma Lake.
Pali-Aike (35) volcanic field	52°	70°	282	The most recent volcanic event produced scoria and spatter cones and fresh lava flows not covered by soil. Ejecta covers prehistorical artifacts.

the earthquakes recorded in historical time in Argentina, soil liquefaction was produced because the physical characteristics of the soil and shallow water table are factors of local damage amplification and reassert the need for carrying out detailed studies in those regions of Argentina where cities have been founded in areas prone to liquefaction.

The finding of at least one historical displacement in the Fagnano Fault, the last one having occurred in 1949, offers the possibility of new working labs, such as those in San Juan and Mendoza provinces, where destructive earthquakes have become valuable tools for neotectonic studies in that part of Argentina. The studies carried out in the Cuyo region illustrate the methodologies to be used for detailed analysis of each area, such as trench making and calculation of seismic parameters associated with each structure in order to estimate seismic risk values in any part of the Patagonia region. The occurrence of earthquakes with Ms $>$ 7.5 in the Isla de Tierra del Fuego and western border of Patagonia in historical times constitutes clear evidence that an earthquake with similar characteristics will likely take place in the future. The recurrence of magnitude 6.5 earthquakes, taking as an hypothesis the distribution of the movement in the time, is calculated as around 500 years and more than 10,000 years for magnitude $7\geq$ earthquakes.

One of the most difficult aspects of these evaluations involves establishing the window of time possible to carry out detailed studies and obtain absolute ages of the deposits affected by faults. The seismic potential of the western portion of South America and the Isla de Tierra del Fuego varies between moderate to high; this is an important consideration since correct planning represents the best strategy for reducing the impact of a destructive earthquake. Governments and planners should take into account ideas about reducing the seismic risk by means of detailed studies, thereby demonstrating that the occurrence of an earthquake in the area is a dynamic and probable process. This will allow the existence of big infrastructure building sites and expansion of cities in the southern continent.

In comparison with other regions, the volcanoes do not pose an important threat to human activities, but there is substantial geological evidence pointing to the significance of volcanism in recent times. In the northern region, the absence of population in the neighborhood of the volcanoes permits the assumption of a low risk for the area. This risk probably is restricted to the damage arising from ash rains that originated in the west falling on the cities of Salta and Jujuy; of special concern is the problem created for air traffic.

In the central region, the rivers with catchment basins in the high Cordilleran ranges flow through the piedmont and wash the flat zone where the main towns or the region are located. These water courses can be important channels for lava, pyroclastic flows, and lahars.

The presence of strong winds over all of western Argentina will also facilitate the spread of fine pyroclastic material expelled by the volcanoes over wide, distant areas in a short time, as happened during the Hudson and Peteroa eruptions. The velocities and directions of the winds are very variable in the region; therefore, an individual study must be done for each case, in order to determine the dynamics and evolution of the ash plume.

In the south of the country, the villages of Baños Copahue and Caviahue also demonstrated their extreme vulnerability to hazards related to the Copahue volcano. Any larger magnitude event would seriously impact them, partly due to the inhabitants' low awareness of the potential danger posed by the volcano. It would be very difficult to persuade both the local people and authorities to make contingency plans for mitigating the consequences of eruptions. Thus, an educational program is urgently needed.

The main hazards related to volcanic activity seem to be limited to those areas where eruptions of acid to mesosilicic magmas are predominant, since these magmas are viscous and have a high gas content. It is precisely this type of volcanism that is dominant in the Andes Cordillera at the latitudes of Argentina. On the other hand, the volcanic risk increases in those areas where large population lives near the volcanic centers. The existence of an abrupt relief or water bodies (lakes, rivers, ice caps) could promote lava and pyroclastic flows and lahars occurrence. Particles ejected from volcanic eruptions and transported by winds arising from west, play an important role too, concerning their effects on human health and environment placed at the east of volcanic centers. Finally, it is important to note that the absence of civil protection programs in Argentina increases the potential impact of volcanic hazards.

CHAPTER 15

Landslide Processes in Argentina

Stella M. Moreiras[1,*] *and* Andrea Coronato[2]

[1]*Instituto Argentino de Nivología, Glaciología y Ciencias Ambientales (IANIGLA-CONICET).*
[2]*Centro Austral de Investigaciones Científicas (CADIC-CONICET). B. Houssay 200 (9410) Ushuaia, Argentina.*

Contents

1. Introduction

During the early twentieth century, landslides in Argentina were vaguely mentioned in technical reports on the main earthquakes as "*secondary effects*" (Forbes, 1861; Bodenbender, 1894; Loos, 1926, 1928; Lünkenheimer, 1929; Harrington, 1944) or by railway workers when railway lines were affected

* **Corresponding author**: *Instituto Argentino de Nivología, Glaciología y Ciencias Ambientales (IANIGLA-CONICET). Av. Dr Ruiz Leal s/n. Parque (5500). Mendoza. ARGENTINA. Casilla Correo 330. Tel: +54-261-5244256. Fax: +54-261-5244201. moreiras@mendoza-conicet.gov.ar*

Developments in Earth Surface Processes, Volume 13
ISSN 0928-2025, DOI 10.1016/S0928-2025(08)10015-3

(Coleman, 1949, in González Díaz et al., 2001). Kilt (1939) was a pioneer in this area of study, reporting landslides affecting mountain roads independently of their triggering mechanisms. Later, Polanski (1966) studied these processes in arid regions and proposed a classification for debris flows and floods based on their deposit characteristics. The number of geomorphological studies describing prehistoric landslides increased in Argentina 10 years ago, but these studies misjudged current events. At present, landslide research aims to understand the probable causes and chronology of these paleo-landslides, but actual landslide hazard and risk assessment studies are starting to be included in the government's land-use planning efforts.

This chapter reviews data on landslides in Argentina in order to explain their activities and their implications for landscape evolution and thereby analyze both the vulnerability of mountain communities and the potential risk landslides pose for regional society and economies. Earlier works on this subject are Schuster and Highland (2001) and the Special Issue of Argentinean Association of Applied Geology and Engineering (*ASAGAI, 2004*).

2. Geographical Location of Landslides

Although landslides are widespread in Argentina, the most catastrophic ones have affected mountain regions, which comprise one-third of the country, extending from northern Jujuy to southern Tierra del Fuego, and involving mountains, ranges, tablelands, and sea-cliffs where precipitation varies widely. For this reason, slope instability processes are mainly analyzed on the basis of Argentine geological provinces according to Ramos (1999) (Fig. 15. 1).

Atypical rainstorms and seismic events affecting geological structures and weathered rocks have been identified as the main forces triggering landslides in Argentina. Heavy summer rainfalls occur in the north and northwestern arid mountains, where the abundant deep gorges and gullies moved debris flows to the lowlands and valley floors. The Patagonian Andes, where precipitation is produced by the winter westerlies, also affect the fall and flux mechanisms of landsliding. Excessive snows as well as fast-melting snows during exceptionally warm spring days cause avalanches and debris flows. In addition, volcanic activity has promoted the formation of lahars.

Landsliding triggered by seisms can occur all along the Andes, especially in its central and northern geological units, where the seismic hazard is high due to the activity of Nazca and South American tectonic plates. Landslides occur over the volcanic-sedimentary of the Patagonian tablelands, and are most commonly triggered by fluvial undercutting after heavy summer storms, wave erosion at the footslopes, or groundwater-level variations. Anthropic activities, as well as exotic fauna that modify rivers along the mountain slopes, should also be considered to be factors triggering landslides. Table 15.1 summarizes the main landslide processes recognized all over Argentina, and reports on events that occurred in each region also appear in this chapter. Only mudflow/debris-flow processes seem to be climatically conditioned in specific regions, whereas the other processes have a wide distribution.

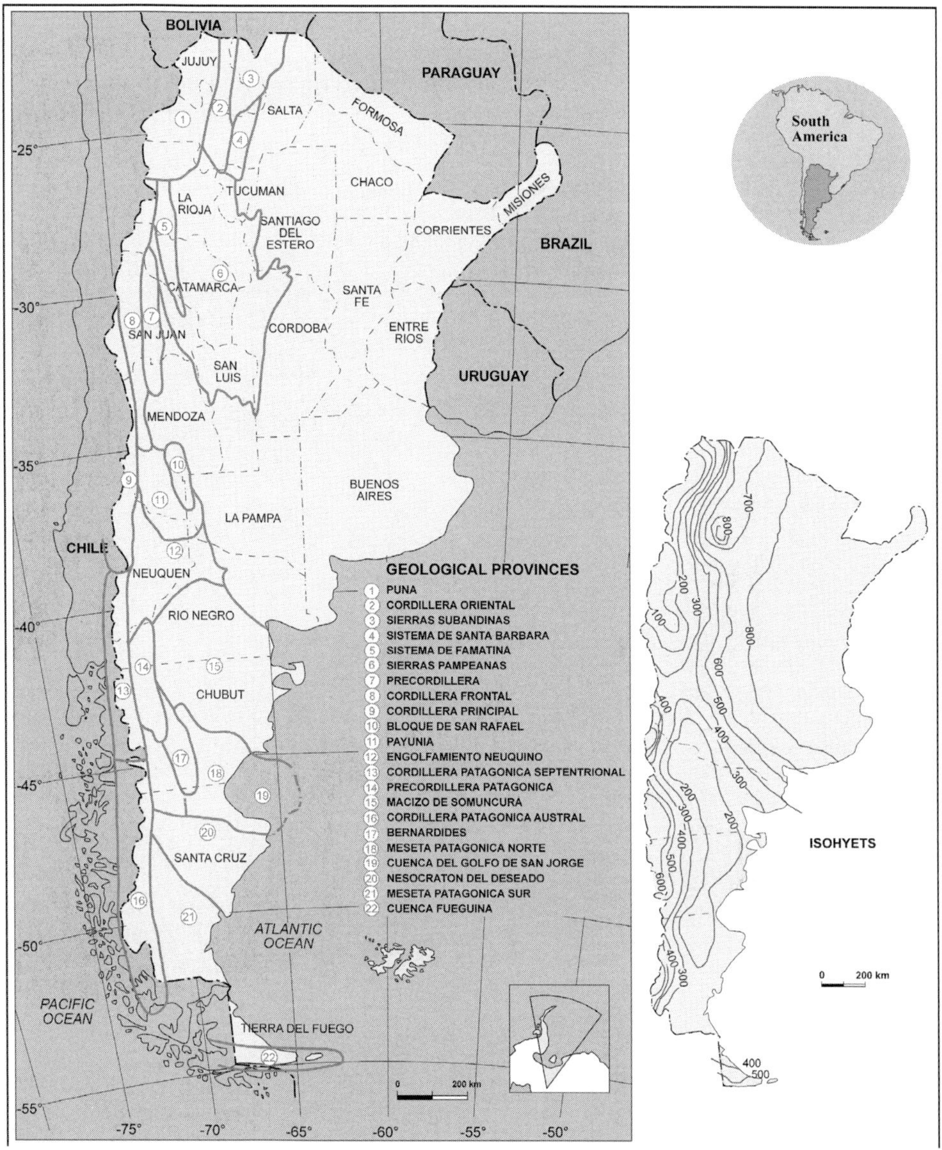

Figure 15.1 Location of Argentinean provinces and geological provinces (represented by numbers 1, 2, 3, etc.) where landslides are common geomorphologic processes. Mean annual precipitation in Argentina is also shown. The steep west-east rainfall gradient is produced by the topographic effect of the Andes barrier.

Landsliding does not affect the Argentine population as much as other geomorphologic processes do because only about 30% of the country's total population inhabits provinces with mountain ranges (Table 15.2). Moreover, very few cities are located in mountainous areas, the most populated among them being Salta,

Table 15.1 Summary of Landslide Processes Identified in the Main Geological Provinces

		Debris flow	Mudflow	Landslide	Soilslide	Rockslide	Rockfall
Northern Argentina	Puna	x	x				
	Cordillera Oriental	x	x	x		x	
	Sierras Subandinas	x	x	x	x	x	x
Northwestern Argentina	Sierras Pampeanas	x	x	x		x	x
Central-western Argentina	Cordillera Principal	x	x	x			x
	Cordillera Frontal	x	x				x
	Precordillera	x					x
Southern Argentina	Cordillera Patagónica Septentrional and Austral, Cordillera Fueguina	x		x			x
	Meseta Patagónica Norte and Sur	x		x		x	x

Table 15.2 Population values of the Argentine provinces grouped into geological regions where landslides processes occur, in relation to the total population of the country

Geological regions	Political provinces (from N to S)	% of population related to the whole country	Population density (inhabitants/km^2)
Northern Argentina	Jujuy	1.69	11.5
	Salta	2.98	6.9
Northwestern and Central Argentina	Tucumán *	3.69	59.4
	Catamarca	0.92	3.3
	La Rioja	0.80	3.2
	San Juan	1,71	6.9
	San Luis	1.01	4.8
	Córdoba *	8.46	18.6
	Mendoza *	4.36	10.6
Southern Argentina	Neuquén *	1.31	5
	Río Negro *	1.52	2.7
	Chubut	1.14	1.8
	Santa Cruz	0.54	0.8
	Tierra del Fuego	0.28	4.7
	Total	30.41	- - - - -

Data are provided from the last national population census carried out in November 2001 (Instituto Nacional de Estadística y Censos, Buenos Aires: www.indec.gov.ar). Note that the values include rural and urban populations. Provinces with * have high rates of urban population concentration. See geological regions and provinces location in Figure 15.1. See also main regional landslide occurrences in Table 15.1.

Jujuy, San Carlos de Bariloche, Comodoro Rivadavia, and Ushuaia. Western cities such as Mendoza and San Juan are located in piedmont areas, so they are not directly affected. Because only small towns or villages, mainly rural, are established in mountain regions, neither the general population nor the government gives great consideration to landslides.

2.1. Northern Argentina: Puna—Cordillera Oriental—Sierras Subandinas

Despite its arid climate, debris flows or mudflows in the high basinal plateau of Puna have been full described; however, landslide activity is generally more important in Cordillera Oriental (Fig. 15.2). Debris-mudflows show a great spatial distribution along the Río Grande Basin (Jujuy Province) (Chayle and Agüero,

Figure 15.2 Northwestern Argentinean region comprising the Puna, the Cordillera Oriental, and the Sierras Subandinas geological provinces.

1987; Solis and Rivero, 1994; Solis and Orosco, 1996; Azarevich et al., 1999). Earlier, Harrington (1945) described the "Volcán" debris flows that occurred in January and February 1943, which ran 10 km until reaching the Humahuaca valley. They moved 3×10^9 m^3 of material with an estimated velocity of 10-15 km/h, resulting from water saturation of soils due to vigorous rainfall on the Peñorco peak and the later mobilization of blocks from the Arroyo del Medio headwater. Annual occurrence of these events, which usually damage the railways, is well known; the most destructive events occurred in 1923 and 1930. Similar debris flows channeled into the Río Grande destroyed houses, disrupted roads, and interrupted railroads in Humahuaca village in January 1986 and March 1990; whereas Purmamarca village was affected in January 1984 (Wayne and Alonso, 1991). Furthermore, extreme events have dammed the River Grande, crossing to the opposite valley hillslope, such as occurred in 1945 when Volcán village was flooded.

Atypical summer rainstorms have also generated debris flows in the Huasamayo River (tributary of the Grande river) affecting Tilcara village. In 1984, a flow damaging the river's defenses destroyed 30 homes, and village inhabitants had to be evacuated. Later, a debris flow on January 6, 1998 deposited 5-m-thick sediments on Road 9. On January 20 and February 2, 1999, cultivated fields were damaged on the river's alluvial plain. In addition, the Grande River was dammed on March 22, 1999 and again in January 2000, inundating neighboring sectors (Azarevich et al., 1999; Solís et al., 2004). Planar slides related to high-dip Cambrian rocks have also been reported, and rock avalanches have been observed along the Arroyo Santa Rita (Palma Sola). In 2001 a mudflow was recorded in Tunalito, next to the Maimará locality. These processes are related to important erosion rates in the Tumbaya Grande valley locality (Solís et al., 2004).

Kittl (1939) reported the occurrence of an active Volcán landslide from Cumbres del Obispo (Salta Province) that was originated by the fall of moraine boulders reaching a volume of 3×10^7 m^3 (observe that this event differs from the above mentioned events). Kittl noted that the involved mass had been moving several millimeters per year. In addition, rockfall debris flows and mudflows were described by Igarzabal (1971) along the Quebrada del Toro stream; Igarzabal pointed out that the rockfall debris flows were the most common, as flows resulting from soil saturation were mainly generated during the wetter seasons. Furthermore, Sánchez et al. (2005) identified paleo-landslides along the Pascha-Incamayo River draining into the Toro River; however, they observed that debris flows and rockfalls are common at present. A debris flow disrupted the Belgrano Railway connecting Salta city with Chile on January 10, 2005, covering it with 7-m-thick sediments. The most catastrophic event in this region was a debris flow triggered by a rainstorm in January 1976, which caused overflooding of the Escoipe River. As a result, the town of San Fernando de Escoipe was completely destroyed, being covered by 3 m of rocks and mud (Igarzabal, 1978; Wayne, 1987, Alonso and Wayne, 1992). Subsequently, damaging debris flows have regularly occurred in this region (Marcuzzi et al. 1994, Chayle and Wayne, 1995).

Catastrophic dense flows have been reported for the western margin of the Valle Calchaquí (Salta Province), between San Carlos and Angastaco localities. These flows occurred in 1964 and 1967 in the Quebrada San Lucas valley area. In 1967 they caused fatalities in Santa Rosa and San Martin, as well as in Arcadia, Las Barrancas, Quebrada del Tonco, Quebrada de la Calderilla, and Quebrada de la Bajada (Solís et al., 2004).

Tartagal village (Sierras Subandinas) was severely damaged in March–April 2006 due to a 1311.9-mm rainfall (mean annual 692.9 mm). More than 5000 people had to be evacuated, and half a village stood isolated as a bridge was destroyed by a raised river stream that also generated many landslides along river margins, collapsing many homes. Road 34 and several bridges were damaged. Moreover, during this powerful event, a creek 6 m deep was eroded until reaching 20 m in depth. Also, the vicinity of the Yacui locality was completely evacuated when a natural dam collapsed.

2.2. Northwestern Argentina: Sierras Pampeanas—Sierras de Santa Barbara—Sistema de Famatina

Initially, Lafone Quevedo (*La Nación,* 1880) reported a flood of the Andalgalá River (Catamarca Province) caused by several debris flows in the Sierra Aconquija (Sierras Pampeanas), which most impacted the Choya valley and damaged the Hoyada and Negro valleys (Fig. 15.3). Kühn (1915) portrayed the occurrence of a debris flow that was channeled into a creek, impacting Andalgalá village on November 16, 1913. Four debris flows were recorded in December 1915 due to a 166-mm precipitation (Kantor, 1916). They raised the Andalgalá River to 1400 m^3/sec (normal 0.7 m^3/sec), flooding the whole village, and also damaging houses, vineyards, and cultivated fields. Twelve fatalities and twenty-four missing persons were reported.

Mudflows have also impacted Valles Calchaquies (Tucumán Province), such as those recorded in Colalao del Valle locality (1968–1970), the Amaicha locality (1966), and the Tafí del Valle (1987) (Suayter, 1997). On March 13, 1964 a rapid debris flow triggered by an intense rainfall (117 mm followed by 22 mm in the next two days) was channeled into the Chumbicha Arroyo (Catamarca Province) (Gonzalez Diaz,1972). Rainfall had begun during the previous days on the upper mountainous basin, and debris avalanches, debris slides, and soil slides were generated upstream. When this violent flow moving toward the east left the creek, it formed an alluvial cone and split into two tracks. One of them affected Chumbicha village, destroying houses, streets, and cultivated fields. A railway infrastructure stopped one of the currents, saving several buildings in town. Nonetheless, wagons were covered by 1.5 m of mud. The other flow destroyed bridges and affected uncultivated fields.

Along Sierra Chica (Córdoba Province), rockfall mainly concentrated along a regional fault, and block slides and soil slides occurred. Moreover, a rotational slide near the La Falda locality affected the route connecting La Falda with Río Ceballos localities (Beltramone, 2005). Moreover, in the Nogolí locality (San Luis Province), a slide was generated over alluvial fans with a steep slope after a violent rainfall in 2001, damaging provincial Road 9, the construction of which could promote instability of these materials (Giardini et al., 2003). One slide, with rotational and translational components, was documented as well for the Intihuasi peak, near the

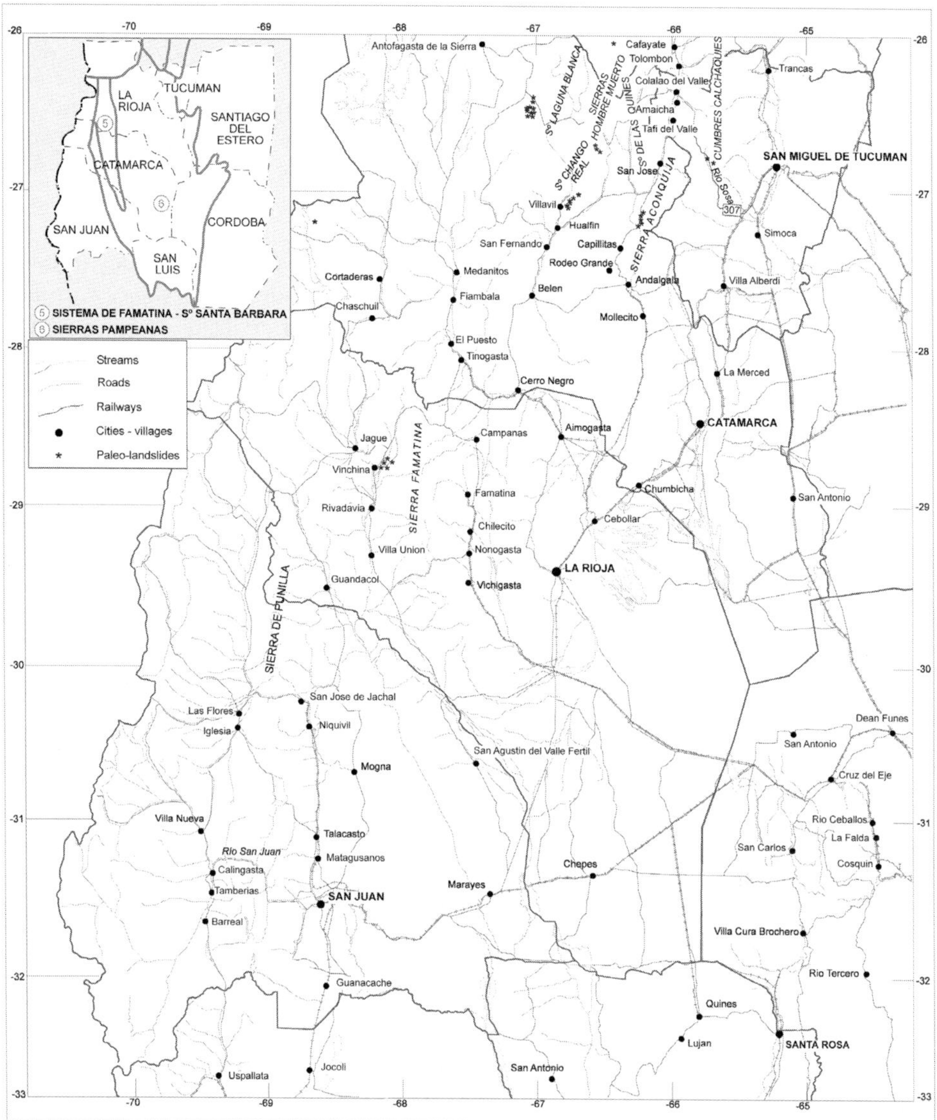

Figure 15.3 Northwestern Argentinean region comprising Sierras Pampeanas, Sierras de Santa Barbara, and Sistema de Famatina geological provinces.

La Carolina locality, in 1978. This process started in a traction crack system on weathered volcaniclastic rocks 120 m long associated with a subvertical fault plane with 6- and 20-m displacements. This slide could have affected the cultural site “Inti Huasi cave,” which was located on the opposite slope. Moreover, unstable conditions exist in this rock outcropping and slide blocks (Sales et al., 2002; Fauqué and González, 2004).

2.3. Central Western Argentina: Cordillera Principal—Cordillera Frontal—Precordillera—Bloque San Rafael

At 32°S, since the beginning of nineteenth century several travelers crossing from Mendoza city to Santiago de Chile have documented about hazards relating to sudden and violent rockfalls (Miers, 1826; Brandt, 1828; Forbes, 1861; Lemos, 1884, Proctor, 1920) along the valleys of the Mendoza and Cuevas rivers, where the international road to Chile was installed later. In 1819, while on one of his travels, Miers (1826) found a fallen block in Cortaderas that caused the death of the muleteer, Santiago Molina, in 1790. Then, Brand (1828) described the occurrence of a rockfall on February 27, 1828 at the same place. Miers (1826) also reported several rockfalls in the La Jaula and Las Cuevas localities and referred to the occurrence of dangerous debris flows in the Precordillera causing severe damages. Afterward, Verdaguer (1929) mentioned damage caused by debris flows triggered by rainstorms in the Precordillera during the summer of 1824. In 1834, Osculati (in RJEHM, 1987) mentioned some fatalities caused by falls, extreme cold, and snow avalanches along this trail. In 1890 the Los Andes newspaper reported serious damages caused to the Transandino Railway, also connecting Mendoza with Santiago de Chile, during its inauguration (Fig. 15.4).

Salomón (1969) identified several active mudflows in his geomorphological study along the valleys of Las Cuevas and Mendoza rivers, comprising Cordillera Principal and Cordillera Frontal geological provinces. This author also noted a huge ancient rockslide at the Las Cuevas locality sourced on the Tolosa peak, producing deposits 1.5 km long and 50 m thick, reaching 1×10^7 m^3 in volume and with boulders >7 m in diameter (Pereyra, 1995). Oral tradition holds that further events took place here during the eighteenth and nineteenth centuries following earthquakes (Miers, 1826). Moreover, rockslides periodically damaged the Transandino Railway in Cuevas locality, until 1993 when it was finally closed.

Furthermore, similar events, whose ages are unknown, were identified in Quebrada Tolosa Oriental valley and Cristo Redentor locality; whereas minor deposits were observed in the Quebrada Matienzo and Paramillos de Cuevas valley areas and near Puente del Inca. The last named involved a 6-km^2 area and has recently been interpreted as a deep-seated gravitational slide associated with subsequent debris flows (Fauqué et al., 2005a). Moreover, dangerous rockfall areas were identified in Cristo Redentor, Punta de Vacas, Cerro Juan Pobre, La Jaula, and Cortaderas (Espizúa et al., 1993).

The landslide inventory map elaborated by Moreiras (2002, 2003) along the Mendoza River valley shows more than 250 events, 82.5% of them corresponding to debris flows and 5.7% to rockfall. Moreover, the magnitude of the debris flows, expressed as deposit area, can be as large as 3.15×10^6 m^2 and as small as 800 m^2 ($X = 0.27$ km^2; SD $= 0.29$ km^2), despite the fact that slides and complex events involve greater areas, 10 km^2 and 17.5 km^2, respectively. In addition, 48% of landslides show evidence of activity either in the scar or at the toe (Moreiras, 2004a, 2004b, 2005a). In addition, substantial historical data were gathered on rockfall and debris flows from historical sources and interviews with local people. Based on these data, Moreiras (2006a) established the recurrence interval as 4 and 15 years in

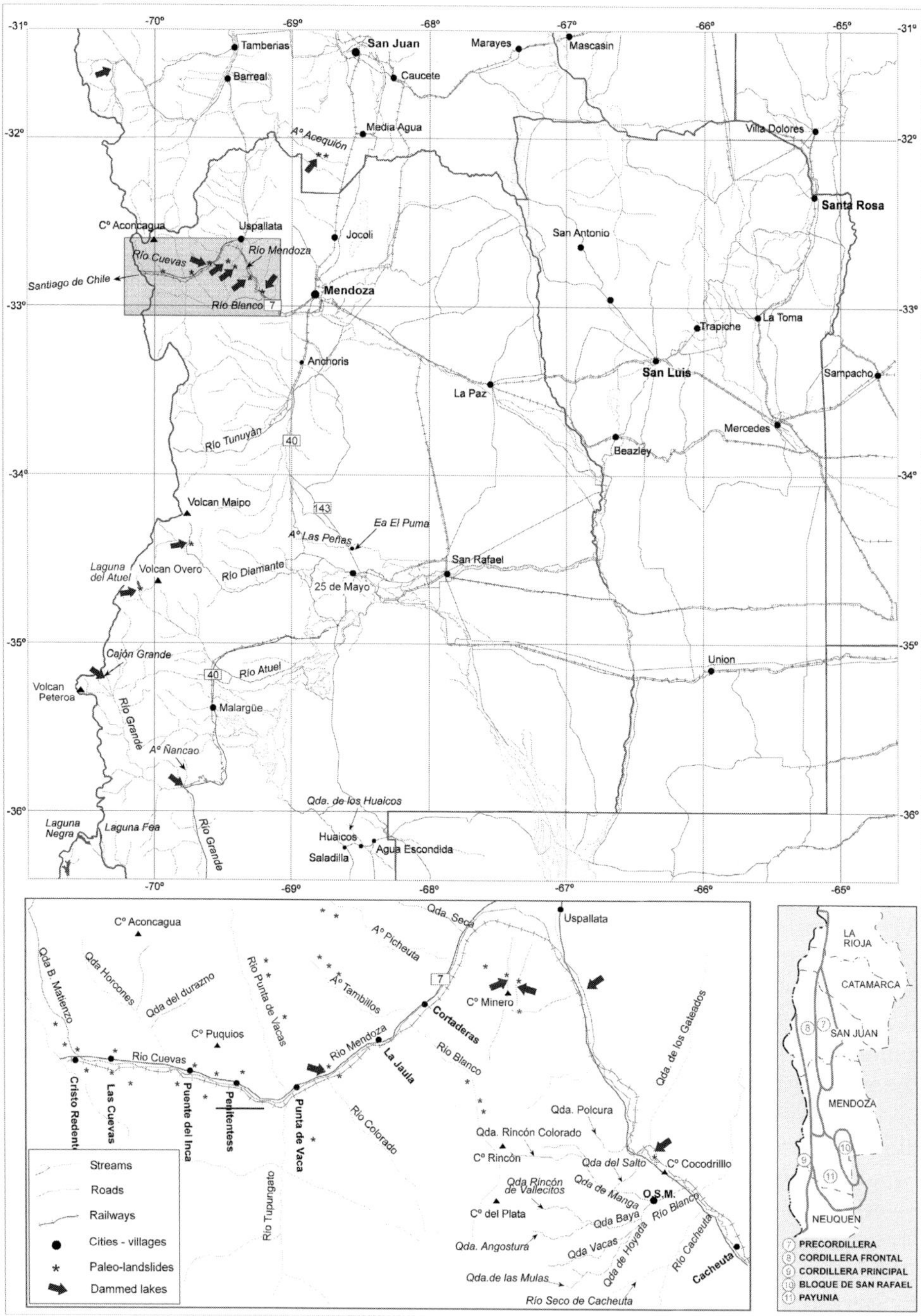

Figure 15.4 Central Western Argentinean region involving the Cordillera Principal, the Cordillera Frontal, the Precordillera, and the Bloque San Rafael geological provinces. A detailed map of the Mendoza River Basin is represented.

Figure 15.5 Debris flow triggered by a rainfall in November 2005 from weathered granite outcroppings along the Río Mendoza valley. It blocked the international road to Chile in the Guido locality.

Cordillera Frontal and Precordillera, respectively. Debris flows near the Picheuta River caused interruptions in international traffic in 1975, 1998, 2003, and 2006; dangerous flows from the Quebrada Seca and Quebrada Camino stream-valley localities are also well known. A car was dragged into the water by one of these flows in 1968, and similar events happened in 1976, 1980, 1999, and 2000. Others coming from the Quebrada de la Polcura or Quebrada del 60 and Guido valleys also inundated this road in 1999, 2001, and 2005 (Fig. 15.5). In addition, rockfalls are very common, producing blockages on tunnels for roads or railways, as during the Ms 3.9 earthquake of 1987. A train fell into the river in 1976 as a consequence of a blockage, and a block fell over the railway in the Cacheuta locality in 1956. Similar events were recorded in 1976, 1982, 1984, 1989, 1990, 1992, 1997, and 2000. Recently, several rockfalls produced by a persistent winter rainfall were observed along the whole international road in July 2006; the most important one blocked the road near Uspallata. Recurrent topples, blockfalls, and blockslides have been observed in steeply dipping Triassic sediment of the Bayo and Cocodrilo (Potrerillos locality) peaks where weaker levels are eroded, exposing massive and compact conglomerate or sand-layer levels prone to instability.

Along the River Blanco Basin, intense rainfall has triggered many damaging debris flows in the arroyos of El Salto, Quebrada La Angostura, Las Mulas, and Quebrada Colorada, adversely impacting the local population mainly established at the Potrerillos, El Salto, and Vallecitos localities. These debris flows, which began at the Cordón del Plata Range, are commonly channeled into the Blanco River, increasing its stream flow and sometimes reaching the Mendoza River valley. They generated overflows on the Blanco River in 1942, 1945, and 1957. Damaging torrents reported in January 1946 destroyed a bridge in Chacritas, and in January 1947, damaged a school at Potrerillos village. Intense rainstorms in December 1954 and 1955 produced debris fall along the QuebradaColorada stream, closing off access to the Vallecitos ski tourist complex. Along this glacial valley, boulders commonly fall where morainic deposits are water-saturated. Moreover, a debris flow again cut access to the complex on February 7, 1967.

In the Tunuyán Basin, Polanski (1966) noted historical debris flows during summer seasons caused by a rainstorm in Anchoris (Mendoza Province) at 33°S, remobilizing great amount of fine material, reaching 1–2 m/sec in velocity. Polanski also mentioned a debris flow in February 1962 that was sourced from weathered volcanic rocks in Huaicos and Salinillas de la Escondida (Mendoza Province). These flows were channeled into the Huaicos valley, reaching the locality of Jagüel and the Amarillo Peak area after traveling 12 km, and deposited a 1- to 1.5-m-thick layer of unsorted sands. In addition, Polanski detailed a sudden debris flow that occurred on December 15, 1951 in Las Peñas creek (El Puma farm) that reached a velocity of 5 m/sec and rose 1.6 m high. This event, the largest recorded in 10 years, dragged a 1-ton pickup truck along for 500 m. Smaller debris flows caused by rainstorms were reported for 1951 and 1952.

At more southern latitudes, Espizúa and Bengochea (2002) produced a landslide inventory map along the Grande River, concluding that slides and flows are the phenomena with the largest dimensions and that flows are the most common. However, they give only two historical data: a mudflow resulting from a torrential rainstorm on March 25, 1990 in Ñancao creek damaging a humble home; and a reactivated slide in Cajón Bayo creek, which occurred between March 1985 and February 1986, as was inferred by satellite images. González Díaz et al. (2000) reported recent flows and rock avalanches along the Barrancas River. Moreover, rotational slides and earthflows caused by river lateral erosion over moraine deposits have been observed along the Varvaco River and Manchana Covunco River (Neuquén River Basin) (González Díaz et al., 2003). These processes are frequently related to ephemeral impounding lagoons. Twenty-five years ago, a debris flow channeling into La Tregua creek (Neuquén River Basin) dammed the Buta Mallin Arroyo and generated a tiny lagoon (González Díaz et al., 2005) (Fig. 15.6).

2.4. Southern Argentina: Cordillera Patagónica SeptentRivernal—Cordillera Patagónica Austral-Engolfamiento Neuquino—Bernardides-Meseta Patagónica Norte—Meseta Patagónica Sur—Cordillera Fueguina

Further south, along the Patagonian Andes (Fig. 15.7), landslides are very common in the slope-modeling process. Rapid snow melting during warm spring days, sometimes linked to heavy precipitation events, produces channeled debris flows along the medium and stepper slopes. Debris is composed not only of channel deposits but also of forest material that frequently erodes routes, blocks sewage systems, or damages bridges, which usually are not properly designed for the volume of material moved by these processes.

In the city of Bariloche (Río Negro Province), a violent debris flow that occurred in June 1976 affected the Melipal neighborhood and settled on the

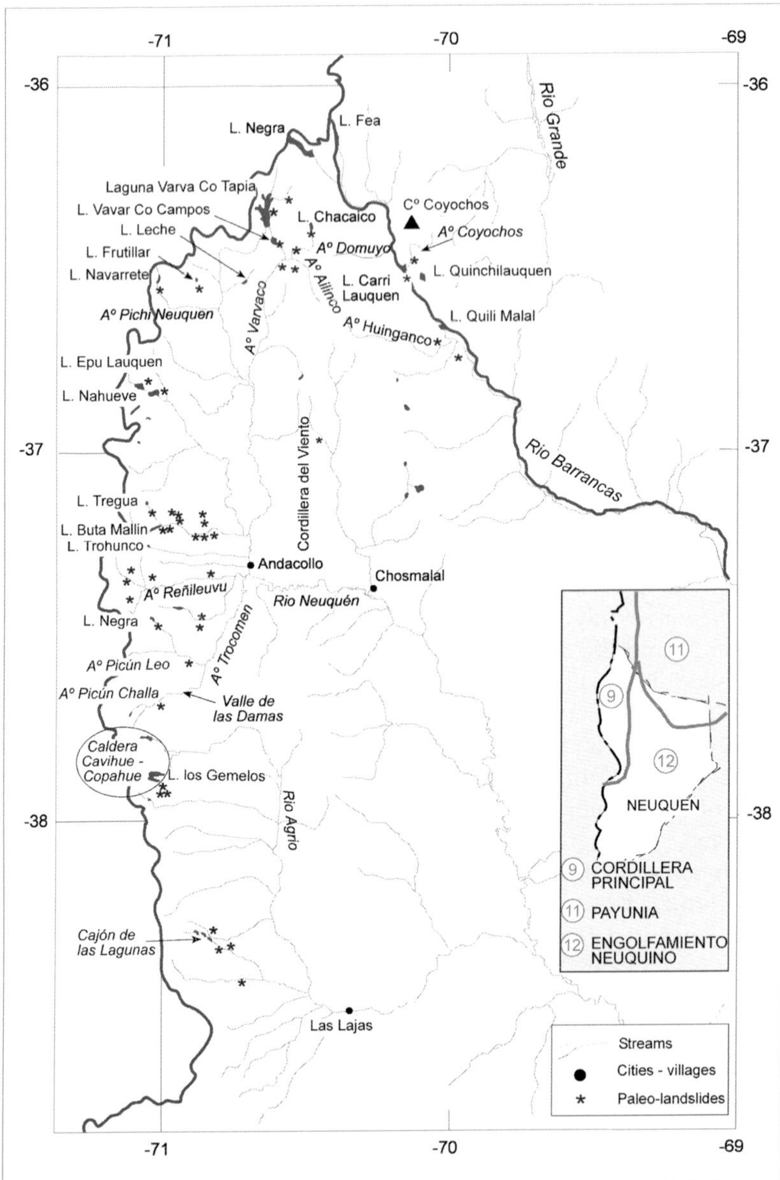

Figure 15.6 Cordillera Principal southern limit in Neuquén Province, where several seismic-triggered landslides have been recognized.

northern slope of the Otto peak. This process destroyed houses, bridges, roads, water reservoirs, and gas tube systems. It was generated by an abrupt increase in discharge, with a theoretical frequency of 10 to 25 years, as a consequence of high precipitation values (higher than the mean of the last 24 years) and high temperatures producing accumulated snow melting. This discharge moved huge glacial

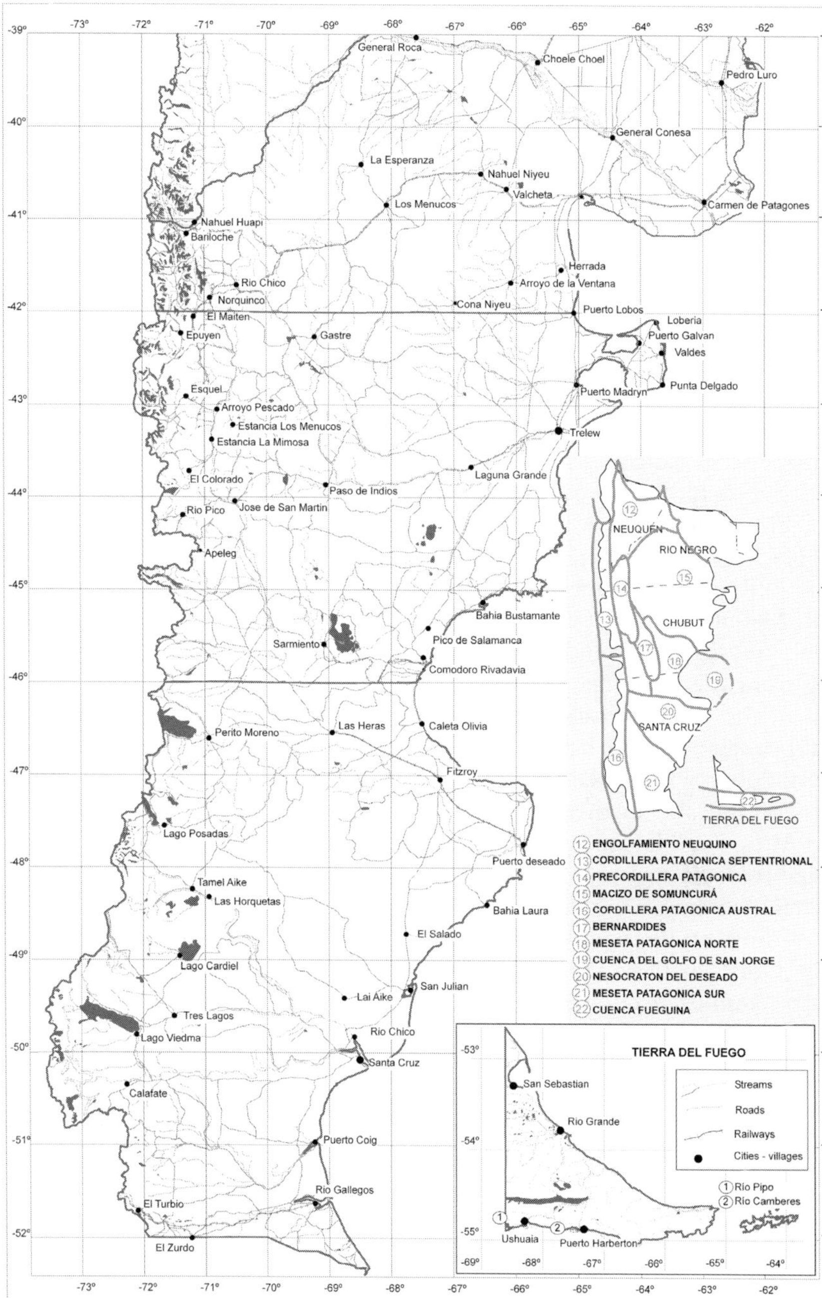

Figure 15.7 Southern Argentinean region involving the Cordillera Patagónica Septentrional, Cordillera Patagónica Austral, the Engolfamiento Neuquino, the Meseta Patagónica Norte-Meseta Patagónica Sur, and the Cordillera Fueguina geological provinces.

boulders and fallen trunks off the channel bed, promoting bank erosion (Dominguez et al., 1981). Several recommendations were proposed to the local government in hopes of mitigating future risks in this increasingly urbanized area; one of the recommendations was to restrict urbanization on the generated alluvial cone. Dominguez and Rabassa (1981), who analyzed slope instability along the Río Ñireco, recommended not using this area for urban purposes. However, the neighborhood is now already fully urbanized. Besides landslides, the Ñireco River often floods its alluvial plain, which is amply occupied by houses. During June 1977, precipitation values were also higher than the mean monthly value; therefore, landslides on oversaturated morainic deposits occurred in the downtown area, producing large economic losses (Dominguez et al., 1981).

In Cordillera Fueguina (Tierra del Fuego Province), landslide processes modified the slopes of glacial valleys after the general ice recession, which would have occurred after 12 ky BP (Coronato, 1995). Morphological evidence of earthslides, avalanches, and rockfall has been found on the slopes, but no detailed data or proper geomorphologic knowledge of these processes are available. Undoubtedly, former glacial and periglacial processes provided fine and low permeable sediments, whereas the present cryogenic fragmentation of schists and volcanic outcrops offers debris material for landsliding. During a heavy precipitation in March 2000, the Cambaceres River, near its mouth at the Beagle Channel, was blocked, and its iron bridge was totally curved by tree trunks coming from the upper basin (Fig. 15.8).

East of the Andes, inside the edges of sedimentary and volcanic Patagonian tablelands, a stepped relief has developed due to ancient rotational slumps associated with earthflows. Several examples could be seen along the Limay, Negro, and Santa Cruz rivers and also in the tablelands where the the Posadas, Stroebel, Cardiel, Buenos Aires, San Martin, Viedma, and Argentino lakes (Pereyra et al., 2002) were excavated by former glacial or tectonic processes.

Lateral spreading is observed in Comodoro Rivadavia city (Chubut Province). Frequent slides on the highly urbanized Chenque peak have also been reported in this city (González Díaz, 2004; Tejedo et al, 1999). In the city's littoral neighborhoods, six slumps occurred during 1990–2000 (Hirtz et al., 2000). The most important of them was in February 1995, severely damaging the urban infrastructure and cutting the traffic on National Road 3 (Codignotto, 2004), cutting the city off for several days. Along the tableland scarp mudflows, debris-creep, and debris fall are produced in claystones, whereas block falls are present in sandstones. Urbanization has advanced on the surrounded sedimentary hills, generating landslide reactivation (Hirtz et al., 2000).

In the extra-Andean tablelands, rotational slides sourced from basaltic plains are common; however, these events have ended as debris flows in the distal parts. González Díaz (2004) mentions that several tablelands such as Meseta de Somuncurá in the Negro River and *mesetas* (plateaus) in Castillo, Salamanca, and Montemayor in Chubut show many successive landslide scars. Moreover, retrogressive rotational slides have been identified in Cañadón del Pilar, disturbing the Sierra de la Victoria landscape.

Figure 15.8 *Nothofagus sp.* Tree-trunks of the Southern Hemisphere beech moved downstream along the Río Cambaceres (Tierra del Fuego) after a heavy autumn rain in March 2000 in the upper basin. Trunks reached the main river because of landslides that occurred in the slopes, 20 km upstream. They occupied the one-third of the channel width producing riverside erosion and flow obstructions and breaking the iron-made bridge.

3. Paleo-Landslides

Even though there is no evidence of historical landslides, a broad record of paleo-landslides in Argentina is available. Their occurrence reveals their relevance as modeling processes and emphasizes the actual landslide hazard in Argentina.

In northern Argentina, 10 rock avalanches related to high-magnitude earthquakes have been reported for Sierra Laguna Blanca (Puna) as neither regional nor global warm periods nor interstadial periods were found. Numerical dating ranging from 262 ± 36 to 172 ± 13 Ky was determined for six of these superposed deposits showing a high scarp height to travel distance (H/L) ratio (Hermann et al., 2000). The Brealito landslide associated with a paleo-dammed lake was identified in Las Lagunas peaks (25° 17′ S) (Hermann and Strecker, 1999a). Prehistorical rock avalanches were also noted at the abrupt fault scarp that defines the eastern limit of Sierra La Palca Range (Catamarca Province) (González Díaz, 1972b).

Along the valley of Quebrada del Toro (at 24° 40′ S), six rock avalanches sourced at the Choro Range and Pascha Range have been described (Hermann and Strecker, 1999b; Hermann et al., 2001). Two of these events dammed the Toro River near the Golgota locality, which is evidenced by thick lake sequences. ^{14}C dating of rare fresh water snails for both paleo-lakes provided ages of 30,050 ± 190yr BP and 26,080 ± 130yr BP. For that reason, landslide occurrence was linked to the Minchin wet period (25–40 Ky) (Trauth and Strecker, 1999; Trauth et al., 2000).

In the Santa Maria-Las Conchas Basin (25° 58′ S), huge landslide deposits (10^8 m^3) related to paleo-dammed lakes were described near the Yesera River (Torres, 1985; Gallardo, 1988). These deposits were then called El Paso I, El Paso II, Casa de Loros I, and Casa de Loros II by Hermann and Strecker (1999b), who established, based on stratigraphic relations, that El Paso I is the oldest event and the youngest is Casa de Los Loros II, which, moving at 294 km/h, dammed the de las Conchas River again. A minimum age of 28,990 ± 150 yr BP was then obtained for the El Paso I rock avalanche by dating bivalve shells sampled from the lake sediments. Hence its occurrence was related to the Minchin wet period as well (Trauth and Strecker, 1999; Trauth et al., 2000). Nearby, five rock landslides were identified in the Tonco River valley (25° 30′ S), whereas at 26° 06′ S further extraordinary landslide deposits outcrop in the areas of Paranilla peak, Sierra Carahuasi, and Sierra de Vazquez (Hermann and Strecker, 1999b). Remnants of tephra overlaying the Paranilla peak and Tonco River deposits were dated 723 ± 89 Ky and >33 Ky, respectively (Hermann et al., 2000).

The Alemania rockfall, exceeding 6.5×10^7 m^3 in volume, derived by erosion during intense rainstorms from the Quitilipi peak (Salta Province). Radiocarbon dating of shells from sediments related to an impounding lake on Las Conchas River showed an age of 5180 ± 50 yr BP (Wayne, 1999). Similar ages (4910 ± 40 yr BP, 5,140 ± 40 yr BP, and 4840 ± 40 yr BP) were obtained from further ^{14}C datings on lake sediments (Trauth et al., 2000). In addittion, González Díaz and Mon (1996) interpreted the Lagunas de Yala (Jujuy) to be a water-filled depression of a massive rock avalanche deposit.

In the Sierras Pampeanas, seven giant landslide deposits (>10^6 m^3) collapsed from the southern mountain front of the Sierras Calchaquies during the Late Pleistocene (Fernández, 2005). Fernández noted that the town of Tafi del Valle is advancing on these deposits as several settlements and cultivation fields are established at present.

Southward (27° 10′ S), eight rock avalanches with volumes ranging from 5×10^6 to 6.5×10^7 m^3 were recognized in a 10-km-wide strip in front of Sierra Aconquija at Loma de la Aspereza, Zarzo, and Loma Redonda localities (Catamarca Province). Because of the carbonate stage (III–IV) developed on these deposits, a Middle Pleistocene age (0.6–1.2 My) was suggested (Fauqué and Strecker, 1998). Moreover, six rock avalanches (10^6 m^3) sourced from Sierra Chango Real at 26° 37′ S moved toward the El Rincón valley (Hermann and Strecker, 1999a). And 10 rock avalanches with volumes of 10^7 - 10^8 m^3 were identified in the Hualfin Range next to the locality of Villavil (Catamarca Province- 25° 30′ S). Three of them obstructed the Bolsón River; thus organic material preserved in one of these lakes gave a ^{14}C age 1432 $\pm$ 132 yr (Fauqué and Tchilinguirian, 2002). Seismic shaking related to regional faults is suggested in both areas, and the potential hazard concerning Villavil rockslides has been highlighted because 350 inhabitants live in this locality.

Eight rock avalanches along the Segovia stream on the southwestern slope of Sierra de Famatina (La Rioja province) were identified by Fauqué (1987). In addition, avalanche deposits were noted in Sierra de Umango and Sierras de Sañogasta-Vilgo (Fauqué and Gónzalez, 2004). Moreover, two landslides identified in Sierra de la Punilla seem to be related to the neotectonic activity of the Las Majaditas and Bolsa faults (Perucca, 1995).

In addition, two Pleistocene rock avalanches were identified in Sierras de San Luis: Potrero de Leyes and Las Cañas, with a volume of 6×10^8 m^3 and 2.5×10^7 to 3.7×10^7 m^3, respectively (González Díaz et al., 1997; 1998). Dating by cosmogenic isotopes on blocks showed ages of 32.7 $\pm$ 6.7 Ky and 40.8 $\pm$ 8.6 Ky (Potrero de Leyes), and 52.9 $\pm$ 10.8 Ky and 66.1 $\pm$ 14.1 Ky (Las Cañas) (González Diaz et al., 1999). González Diaz et al. suspect $M > 6.2$ earthquakes as the triggering mechanisms.

In the Precordillera, two Holocene rock avalanches were studied by Perucca and Moreiras (2003). Both events dammed the Acequión River, which is evidenced by a 20-m-thick sequence of fine lake sediments. Organic matter preserved in the younger paleo-lake gave cal. 7,497 $\pm$ 157 yr ^{14}C BP. A regional fault system associated with the historical rupture of the 1952 earthquake (Ms 7) affects the region, and lake deposits show liquefaction features evidencing at least three paleo-earthquakes ($M > 5$).Therefore, a seismic shaking ($M > 6$) has been proposed as the triggering mechanism (Perucca and Moreiras, 2006).

In the Mendoza River valley, Salomón (1969) identified two paleo-landslides, one sourced at the southeastern hillslope of the Juan Pobre peak and the other at the south margin of the river. They were called Amarillo and Negro landslides, respectively. The Amarillo landslide covered a 0.8-km^2 area, and the Netro comprised 1.25 km^2, with an estimated volume of 3.44×10^8 m^3. According to stratigraphic studies, the Amarillo landslide preceded the other, occurring during an interglacial stage from 125 ky to 75 Ky. Even so, a seismic source was suggested by Espizúa (2005).

At Cordón del Plata Range (32° S), at least six closed rock avalanches were identified: one in the Soltera valley called Tigre Dormido (TD), two in Placetas Amarillas (PA-1 and PA-2), and three in Piedras Blancas (PB-1, PB-2, and PB-3)

(Fauqué et al., 2000). However, Moreiras (2004a, 2006b) reclassified the PA-2 deposit as a rockslide due to discrete material involved and to moderate displacement. Moreiras then established that PB-1, PB-2 and PB-3 deposits resulted from the same event renamed as the PB rock avalanche (Moreiras, 2006c). According to Moreiras (2004a, 2005e), the TD rock avalanche is at least Middle Pleistocene in age as its deposit is eroded by a Pleistocene moraine younger than an ash level earlier dated as 360 $\pm$ 70 Ky. This event reached a volume of $\sim 1.7 \times 10^9$ m^3 and traveled more than 10 km (H/D = 0.22). Moreover, the PB-1 rock avalanche, with an estimated volume of 1.6×10^9 m^3 (H/D = 0.11), is younger than an ash layer dated 350 $\pm$ 80 Ky by the Ar^{39}/Ar^{40} method (Moreiras, 2006b). Based on geochemical analysis and field observations, a similar age was established for PB rock avalanche ($\sim 8.9 \times 10^8$ m^3, H/D = 0.2) since an ash layer covering its deposits was correlated by geochemistry with one dated previously (Moreiras, 2006c).

The occurrence of an ancient channeled mudflow in Quebradade las Vacas creek, which is a tributary of the Blanco River (Mendoza Province), was noted (Mikkan, 1992). Several agricultural villages such as Las Vegas, Valle del Sol, and Tierras Blancas have been established over these deposits because of its good quality as water reservoir.

Between 36° and 38° S, more than 50 paleo-landslides have been identified in Cordillera Principal. In many cases, however, neither their chronology nor their stratigraphic position has been determined. Initially, Groeber (1916) reported a huge ancient landslide in the Barrancas River area, damming the current Carri-Lauquen Lake. This event was later classified as a rock avalanche by González Díaz et al. (2001), and its age was established by cosmogenic dating at 2 Ky (González Díaz et al., 2005). González Díaz also described two prehistoric landslides in the nearby Coyochos and Huingaco *arroyos,* with uncertain age. Southward, in the upper basin of the Neuquén River, two prehistoric rock avalanches were identified along the Vavarco River (González Díaz et al., 2000). As they dammed present Varva Co Tapia and Varva Co Campos lakes, they were identified with the same names as the lakes. The Varva Co Tapia rock avalanche, covering 15.35 km^2 and traveling 5.5 km, reached $\sim 1.07 \times 10^8$ m^3 in volume. The Varva Co Campos rock avalanche moved 7.5 km with a velocity of 310 km/h, and its volume has been estimated as 3.51×10^9 m^3. According to Gonzalez Diaz et al. (2005) ,the isotopic cosmogenic ages of Varva Co Tapia and Varva Co Campos rock avalanches are 60 Ky and 30 Ky, respectively. A younger successive slide named Malvarco was also identified in this area, and landslides damming secondary valleys were also reported for Cajón Chacaico, Cajón de la Crianza, and Laguna de la Leche (González Díaz et al., 2000).

In the Ailinco River surroundings (Río Neuquen Basin), González Díaz et al. (2003) recognized two rock avalanches, Ailinco and the Papas peaks, which Groever (1947) initially mapped as till deposits. The first moved 90 km toward the opposite valley hillslope, reaching $\sim 2.84 \times 10^7$ m^3 in volume; whereas the Papas peak rock avalanche, $\sim 8.24 \times 10^7$ m^3 in volume, moved 9 km. These authors also recognize the Olletas slide, 1.77×10^6 m^3 in volume, in this region. At 37° 10′ S, the neighboring Los Cardos rock avalanche, covering 24 km^2 and another six smaller rock avalanches have been identified in the region between the Nahueve and Reñileuvu rivers (tributaries of the River Neuquén) (González Díaz et al., 2005). Earlier, Escosteguy et al. (1999) identified the Moncol rock avalanche covering 8.3 km^2 that dammed the Reñileuvú River and generated a paleo-lake, evidenced by 1-m-thick lake sediments. Nearby, further events were identified: the Piche Moncol rock avalanche along the Reñileuvú Arroyo, the Guañaco peak rock avalanche, the Coronal peak rock avalanche, and a rotational slide in the Ñireco Arroyo that involved 2.4 km^2 and dammed the current Lauquén Mallin Lake (González Díaz and Folguera, 2005). Finally, four landslide deposits were also distinguished along the Pilun Challa Arroyo, one of them corresponding to the Pilun Challa rock avalanche, with an estimated volume of $1.46 \times 10^7\, m^3$, which dammed Las Damas valley (González Díaz et al., 2006). The Pilun Challa rock avalanche is younger than a glaciation dated before 30 Ky. Another landslide in the Pilun Challa Arroyo covering 2.8 km^2 impounds Los Maderos Lake, and another one impounds the current Lagunas Negras Lake. Though numerical dating is still lacking, the occurrence of all these events during postglacial times is suggested. Anyway, a seismic triggering mechanism is presumed based on the seismic-tectonic conditions of the area (González Díaz et al., 2000, 2001, 2003).

Later, Hermann et al. (2006) carried out ^{3}He cosmogenic dating determining the ages of 3,700 $\pm$ 350 yr and 2700 $\pm$ 700 yr for the rotational slide of the Ñireco Arroyo and the slide identified in front, respectively. They concluded that the younger slide should have fallen into the Lauquén Mallin Lake, causing a lake tsunami downslope. In addition, these authors established the age of the Moncol rock avalanche as 3,800 $\pm$ 650 yr, supporting its age as contemporary with the Ñireco Arroyo rotational slide and suggesting a common seismic origin.

At 38–45° S, in the surroundings of the Caviahue-Copahue volcanic edifice, González Díaz (2003, 2005) recognized the Compul and the Cajón Chico rock avalanches in the Hualcupén valley. These were initially described as terminal moraines by Groeber (1925) and recently defined like rock avalanche deposits similar to those described above (Pillún-Challa, Picún-Leo, and Trocoman or Las Damas valleys). The Compul rock avalanche dammed the valley, forming the Compul Lake (also known as Hualcupén Lake). The Holocene age is attributed to both rock avalanches, considering that they should have occurred after the Holocene glacial retreat, perhaps favoring slope instability. The authors also suggest that seismic movements related to any eruption might have triggered them. In addition, an ancient landslide with uncertain age is recognized south of Cavihue Lake. At 38° 30′ S, Escosteguy et al. (2005) described another six landslides; two of them correspond to rotational slides, and four are rock avalanches. The rock avalanche identified in Romero valley reached $1.13 \times 10^8\, m^3$; whereas those occurring along the Codihue valley are $1.5 \times 10^7\, m^3$, $1.03 \times 10^6\, m^3$, and $8.5 \times 10^6\, m^3$ in volume, respectively. The last named is related to an impounded paleo-lake. Escosteguy et al. concluded that these events would have happened after the Last Pleistocene glaciation (30–640 Ky), assuming that during the deglaciation, steeper slopes were more susceptible to instability due to loss of ice support and increase in precipitation. A seismic cause is discounted because the seismicity decreases south of 38° S (see Perucca and Moreiras in this volume).

Paleo-landslide processes have been mapped as rock avalanches or earthslides in different glacial valleys of Tierra del Fuego (Coronato, 1994; Coronato and Roig, 1999). Many of the huge earthslides are interpreted as the consequence of the loss of pressure after the ice recession. In the Roca glacial lakes, postglacial landslides shape its northern shore. Avalanche chutes are very well developed from 700 m.a.s.l. to the bottom of the valleys, almost 120–200 m.a.s.l. above the treeline. Thus, the forest has been completely removed during violent processes, and many of them remain without vegetation. Few scars of earthslides are also visible in many steep slopes, but no data about their origin and date are available.

4. Outburst Floods Caused by Dammed Lake Failure

Landslides generally distort hydrological systems and generate dammed lakes. Nonetheless, these lakes commonly have a short life on a geological timescale; their sudden collapses resulting in overflows/overburst floods are potential catastrophes for humans. González (1979) warned of the probable failure of the Laguna del Atuel (the headwater of the Río Atuel) impounded by a prehistoric landslide, which is over soluble karst gypsum. Moreover, other causes for a probable collapse are moraine erosion in the western sector, another landslide sourced at Overo volcano falling into the lake, a sudden advance of the surge glacier draining into the lake, and a seismic shaking.

A natural dam located at the Río Santa Cruz (Blanco River's headwater, San Juan Province) collapsed on November 12[t,] 2005 generating an outburst flood on the San Juan River that reached 1000 m^3 (Fig. 15.9). The resulting violent flow

Figure 15.9 a. The 57-m-high natural dam located at the Río Santa Cruz (Blancós River headwater, San Juan Province). b. Outburst flood/debris flow resulting from the dam failure on November 12, 2005 (Photograph by J. P. Ceballos).

caused severe damages downslope destroying bridges and roads, and isolating many people in the mountain areas. The just-built Caracoles dam was severely damaged, as was the Ullum dam reservoir, located 180 km from the failure area. Three minutes later, the flood reached the Caracoles dam increasing water turbidity and affecting the potable water availability of San Juan city.

Perhaps the most drastic outburst was recorded in 1914 when the Carri Lauquen Lake collapsed, generating an extraordinary outburst flood on the Barrancas River. This lake was formed by a Pleistocene landslide that had dammed the river. Groeber (1916) reported that overnight 2 billion m^3 of water were drained; as a consequence, the original 21-km-long lake was reduced to 5.6 km, and its surface was lowered about 95 m. The resulting debris flow/flood dammed the Quili-Malalal River, forming a tiny lagoon 200 m long and 25 m wide, and flowed more than 300 km devastating downstream valleys. Several animal farms completely disappeared, and fields of wheat, corn, and alfalfa were buried by debris. In addition, two small towns in this valley were devastated. This debris flow/flood also wiped out farms, railway stations, railway lines, and roads located along the valley of the Colorado River, which is formed by the union of the Barrancas and Grande rivers. The overflow also triggered several landslides. In all, 175 fatalities were reported, and more than 100 persons disappeared. Groeber (1933) noted that 20 years later the farmland in both valleys still had not recovered.

Impounded lakes are not always related to catastrophic failures, however; some of them maintain their equilibrium between inflow and outflow, and so they can be preserved for long periods. Such are the cases of existing Varva Co Tapia and Varva Co Campos lakes (Neuquén) related to the Varva Co Tapia and Varva Co Campos rock avalanches, respectively (González Díaz et al, 2000). Nearby, the Coyochos slump earthflow and the Huinganco landslide, respectively, damming Coyochos and Huiganco creeks, have experienced a gradual drain. Even so, they represent a potential hazard during periods with increased precipitation or spring thawing. Both kinds of water sources may increase the lake volume, causing dam failure; for this reason a periodic monitoring of the water level is suggested. The Carri-Lauquen Lake, which collapsed drastically in December 1914 during the late spring season, used to drain slowly to the Barrancas River. The local people present during the event reported that the cause of the failure was the increased inflow resulting from extreme thawing of high snow accumulation during the 1914 winter season. Native inhabitants remember common increases in the Barrancas River flow during spring time but at lower levels (González Díaz et al., 2001).

Temporary impounded lakes have also been common. On February 15, 1961 the Diamante River was dammed by a landslide at 35 km downstream of the Laguna Diamante Lake. Rainfall increased the volume of this impounded lake (200 m high and 2000 m long), reaching 259,300 m^3 in five days. Water flowed by seepage, but the lake lasted until the 1970s, amazing populations downstream. A dammed landslide-lake was observed in the Cajón Bayo valley (Grande River Basin) on June 6, 1986, but its failure is uncertain. A new landslide dammed the valley again in this sector in 1992. González Díaz et al. (2001) also mentioned an ephemeral impounding lake caused by a landslide in Menucos Arroyo, a tributary of the Barrancas River, in 1999.

Dammed paleo-lakes have occurred in the past, lasting for long periods as evidenced by thick lake sequences. Several paleo-lakes have been reported along the Las Cuevas and Mendoza rivers at Las Cuevas, near Puente del Inca and near Colorado River localities (Salomon, 1969). The last one, resulting from the Amarillo landslide, was 7 km long, and its volume was estimated at $4 \times 10^8\,m^3$ (Fauqué et al., 2005). At Los Arboles place, another rock avalanche dammed the Mendoza River, evidenced by an 18-m-thick sequence of fine material. This paleo-lake could have been $\sim 1.72 \times 10^6\,m^3$ in volume. Ancient dammed lakes have also been reported along the Quebrada Horcones valley (Fauqué et al., 2005). Those Pleistocene rock avalanches identified in the Cordón del Plata Range have also generated paleo-lakes (Fauqué et al., 2000). The TD rock avalanche dammed the Mendoza River; the resulting barrier-lake could have an estimated volume of $1.6 \times 10^9\,m^3$ (Fauqué et al., 2005). Whereas the paleo-lake generated by the PA-1 rock avalanche reached $0.6\,km^2$, the lake sequence is 38 m thick (Moreiras, 2006a). The paleo-lake caused by PB rock avalanche left a 30-m-thick relict sequence reaching an extent of $\sim 0.7\,km^2$ (Moreiras, 2006d).

5. Main Landslide Trigerring Mechanisms in Argentina

Seismic movements are a major factor triggering landslides, and for that reason historical landslides have been reported during high-magnitude earthquakes affecting seismic areas of Argentina, as mentioned earlier. Similarly, more than one hundred events were documented for northern Mendoza Province, which has had historical earthquakes $Ms > 7$. According to these data, rockfalls began to be recorded with an earthquake magnitude ≥ 3.9 (Moreiras, 2004a). The maximum distance between events and epicenter does not exceed 250 km; still the relation between earthquake magnitude and epicenter distance is uncertain (Moreiras, 2004a, 2006a).

Torrential summer rainstorms are also widely related to landslide occurrence, mainly in arid climate regions. Debris flows in northern Argentina occur from January to March when storms drop large amounts of rain in a few hours on saturated slopes. Moreover, more than 150 events were caused by rainfall along the Mendoza River valley from 1952 to 2002. They were mainly related to crack filling of water in rock outcroppings or water saturation of sedimentary material (Moreiras, 2004b, 2006a). Kittl (1939) observed that hillslope instability in the Volcán landslide (Salta Province) depends on material cohesion that varies with humidity condition forced mainly by the occurrence of rainfall or wetter periods. However, the precipitation threshold value for triggering landslides has rarely been determined. Moreiras (2005c) statistically established that at least 6 to 12 mm precipitation ($X = 17.5$ mm) per day is needed to generate rockfalls and debris flows along the Mendoza River where mean annual precipitation does not exceed 250 mm. Nonetheless, several limitations in this value exist due to the scarcity of meteorological stations in the region, the meso-scale rainstorm distribution, the lack of precipitation duration, and uncertainties about previous land conditions. Antecedent precipitation also plays a considerable role in slope instability (Moreiras, 2005c).

In addition, the temporal distribution of landslides triggered by precipitation is forced by extraordinary cyclic climatic events. The Intihuasi slide (San Luis Province) in 1978 occurred during a period of elevated precipitation. Moreover, the slide that occurred at Nogalí (San Luis Province) may be related to an increase in precipitation during January 2000–April 2001 (Giardini et al., 2003). In addition, the debris flow that destroyed San Fernando de Escoipe in 1976 occurred in the years affected by the El Niño Southern Oscillation (ENSO) phenomenon (Alonso and Wayne, 1992). In addition, Moreiras (2005c) maintains that landslide activity in the Cordillera Frontal increases during warm ENSO episodes (*Niños*), concluding that this geological province is mainly influenced by the Pacific anticyclone. In contrast, in the Precordillera region no significant differences in landslide activity have been identified for both warm and cold phases of the ENSO phenomenon. However, a greater number of landslides occurs during wet periods influenced by the Atlantic anticyclone, whereas fewer events are observed during the dry periods. Thus, the occurrence of landslides triggered by precipitation in the Precordillera seems to be mainly forced by the Atlantic anticyclone.

Water saturation resulting from the rapid thawing of snow cover is also well known as a landslide-triggering mechanism. González Díaz (2004) noted that water resulting from snow ablation may swap the action of rainfall in the Patagonian extra-Andean tablelands. For this reason, those years associated with greater snow accumulation in mountain areas show greater landslide activity, as can be demonstrated for the Mendoza River when an El Niño ENSO phase exists.

Further events are caused by lateral erosion of rivers, in both mountainous areas and plains (González Díaz, 2005; Moreiras, 2005a, 2005c). In rivers fed by snow thawing, landslides caused by this mechanism are common during the spring (October, November, and December). Similarly, slope instability may be greater after an ENSO warm phase as a consequence of greater snow accumulation (Moreiras, 2005b). Volcanic processes are also related to mass movements. Dangerous lahars were generated on the Copahue volcano by snow thawing caused by fallen hot pyroclastic material in the modern eruption of 1992 (Gonzalez Diaz, 2005). Furthermore, Sruoga (2004) reported falls in the oriental sector of Planchón volcano caused by strong explosions in March 1994. Wind action has caused debris falls along the Mendoza River valley; clasts and rubble generated in the weathered granite are easily removed, damaging the windshields of vehicles moving on International Road 7 (Moreiras, 2004b, 2005c). This process also occurs in the Cordillera Fueguina during sunny spring days, when volcanic and metamorphic rock fragments, cryogenically weathered, suddenly fall on the National Road 3 injuring vehicles and people alike.

Far away from mountain areas, along the Patagonian coastal zones, several landslides including slumps and rockfall have been generated by intense rainfall, water-table increases, pore-water saturation in sedimentary rocks, fluvial undercutting, and wave erosion (González Díaz, 2004). Traffic on the coast avenue, running along the Las Grutas beaches (Rio Negro Province), was closed owing to fissures that might be related to potential landslides. Moreover, on December 28, 2003, a creek 5 m deep was eroded along 1 km of the beach, damaging much of the infrastructure in Villa Gessel city, one the most important beaches of the Buenos Aires Province (Codignotto, 2004).

In the Cordillera Fueguina, the invasive beaver species *Castor canadensis*, with semiaquatic habits, builds dams with tree-sticks, impounding creeks or streams (Coronato et al., 2003). These dammed lakes fail when they lose their input-output equilibrium because of heavy autumn rainfalls or late spring snow and rains in a short time. When these small lakes are suddenly empty, the flow starts to transport not only fine sediment from the bottom of the lake, but also sticks and branches, which jump over and out of the channel bed, causing erosion on the margins and damaging the forest. When creek flows reach the modified slopes due to route construction, the flow velocity decreases and the debris material is deposited over the road. A debris-flow event occurred in Monte Susana, close to Ushuaia city, during a late snow storm in November 2001 (Fig. 15.10). A beaver pond located at 500 m.a.s.l. was overflooded and the dam broke, generating a violent flow composed of mixed fine sediment, glacial boulders, rock boulders, beaver sticks, and trees. This debris flow, carrying a heavy load, knocked the ground and the

Figure 15.10 In November 2001, the failure of a beaver dam built in one of the steep slope creeks of Monte Susana, Tierra del Fuego, generated a debris flow that destroyed a road and dammed the Pipo River for several hours. This biotic-triggering mechanism is only present in the Tierra del Fuego Mountains where beavers were introduced in the 1940s. (Photographs by A. Coronato and M. Chiaradia). (A): Rocky near vertical slope of Mount Susana eroded extensively by the debris flow; (B and C): beaver dam tree debris and sediment spread over the road.

vegetation, destroying the surrounding forest. Both flood and jumping marks could be seen at 0.80 m and 2 m, respectively, over the ground, damaging forest trees with deep scars. Downslope, the mixed load crossed the road, damming the main valley river for many hours. Upstream, a golf course, a camping place, and a tourism railway station could have been affected if the damming had lasted several more days. The road was closed, isolating many people on the other side of the valley for many hours.

The influence of human activity on landslide activity cannot be ascertained in Argentina due to lack of historical data and/or a monitoring system. In any case, landslide activity seems to be greater along those roads where vertical slope cuts have been made. After the wider reconstruction of the Paso de Jama (Jujuy province) several rockfall have affected this route. (Solís et al., 2004). Moreiras (2005b) noted that construction of the international road to Chile and the Trans-andino Railway favored slope instability by slope cutting and detonations, resulting in weak rocks. Small earthslides and rock landslides have been generated along National Road 3 along the Cordillera Fueguina, after two years of the talus modification due to road work. Unfortunately, planned pavement works did not take into consideration either the talus stabilization or rockwall net protection; thus landslides and rockfall were activated, generating dangerous processes for people and vehicles. In addition, the present pit activity in the higher gravel and sand beaches along the Santa Cruz and Chubut coasts is causing talus instability on National Road 3 (Lapido and Pereyra, 2002), which in many sections is threatened by rockfall and landslides from the lateral tablelands. In Tucumán Province, changes in slope geometry that occurred during the construction of Road 307, as well as fire on Selva Montana, enhanced unstable conditions (Ferreira Centeno and Giambasti, 1992). Catastrophic events (rockfall, slides, and debris flows as well as floods and river obstructions) occurring at La Falda (Córdoba Province) in 1995, which were generated by an intense rainstorm, were especially damaging because of previous land conditions resulting from forest fires, route construction, and improper soil uses (Beltramone, 2005). Moreover, the flood of 1992 affecting San Carlos Minas destroyed the river defenses and advanced on this village. It was especially damaging because the overflow of the Río Noguinet was partially dammed by a very low bridge (Ferreira Centeno and Giambasti, 1992).

Along the western margin of the Río Neuquén, one of the largest rivers in northern Patagonia, located 11 km north of the city of Centenario(Neuquén Province), the edge of the sedimentary rock tableland shows that recent landsliding processes have been triggered by drop irrigation (Fig. 15.11). Fruit orchards have been developed on top of the tableland, which require intensive planned irrigation in semiarid climates. Infiltration of a larger water volume oversaturates lower clay levels, generating slumps on hillslopes. Downslope, slow mudflows are generated with the groundwater source intruding on the higher fluvial terraces where there are fields and homes. This landslide mechanism induced by anthropogenic activities is common in areas where no land-use planning has been established.

The violent debris flows that damaged the town of Tartagal in 2006, as well previous events, were facilitated by fires and intense deforestation on Yungas and the tropical forests of Tucumán-Salta provinces. Deforestation began in the early

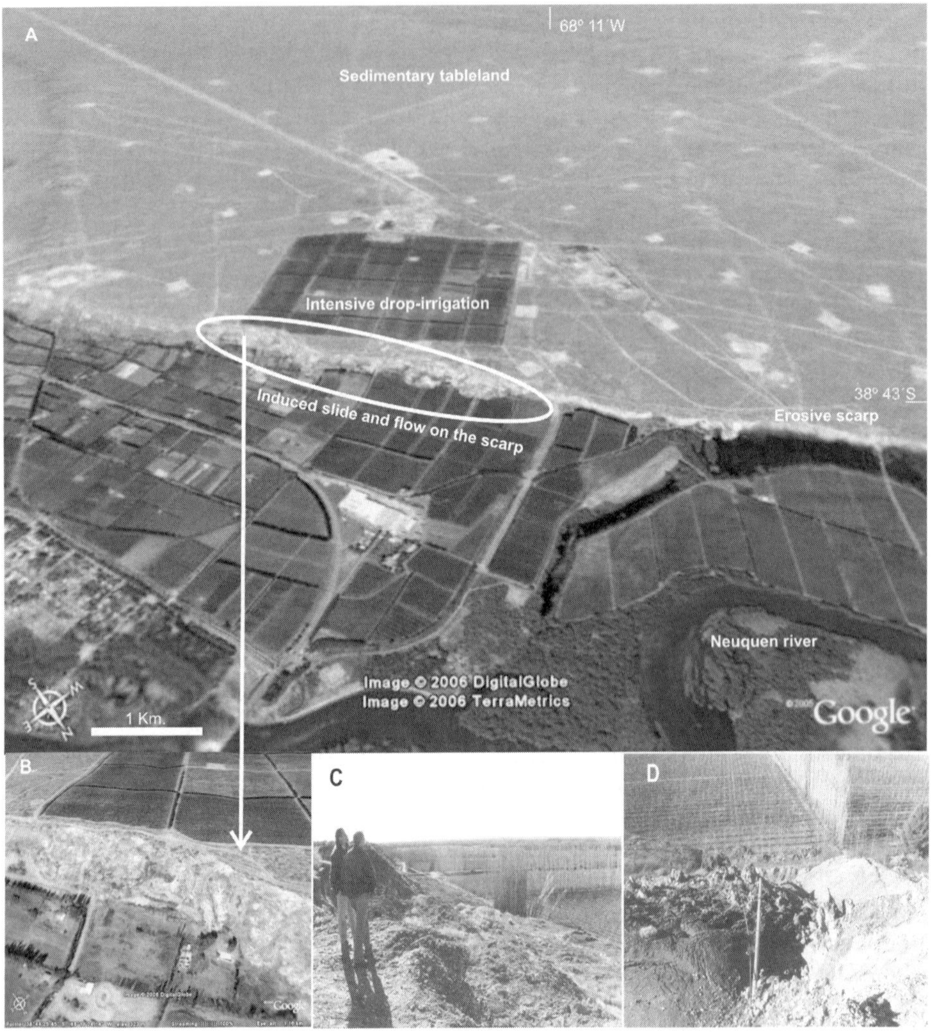

Figure 15.11 Satellite image of the lower Río Neuquén eroding its valley in a tableland landscape (A). Artificial drop-irrigation in fruit plantations developed over the arid tableland promotes the groundwater saturation in claystones, moving down through the upper permeable rock layers to generate rotational slumps and flows. The earthflows are invading the fields in the upper fluvial terrace (B), generating economic losses and damaging the high electrical power system (D). Tensional faults are developed along the scarp (C). Satellite image (2006) taken from Google-Earth free software and photographs by A. Coronato (2001).

1960s, when oil companies made seismic lines for exploration, and it increased in the 1990s. Noninfiltrated rainfall water is channeled into these paths, reaching higher velocities. Moreover, during the last eight years, hundreds of sawmills have been established in these forests. According to the Forest Direction, 94,087 hectares (10.84 %) of Yungas and 56,664 hectares in Chaco forest were deforested during 1984–2001.

6. Losses, Economical Impacts, and Physical Land-Use Planning

Economic analysis of losses or the impacts of landslides still remains to be conducted in Argentina. Landslides have been assessed only from the perspective of their effect on human activities or infrastructure.

It is mostly communication systems that have been affected by these natural processes, as discussed in Kilt's early work (1939). The international road to Chile (No. 7) and the Transandino Railway along the Mendoza River have been pervasively affected by landslides (Moreiras, 2005a; 2006a). A debris flow triggered by rapid snow thawing affected Road 220 along the Atuel River. In addition, in 1982 a debris flow dragged a truck and a car into the Avispas Arroyo, causing four fatalities, and similar events occurred in 1997. Earlier, in 1978, a rockfall cost two people their lives at tunnel 7. Harrington (1945) mentioned damages caused to the railroad by recurrent debris flows reported for Humahuaca (Jujuy Province) in 1923, 1930, and 1945. Debris flows with a recurrence, that rarely exceeds 10 years, have affected Barreal village (San Juan province) implying a potential hazard (Ceballos, 2005). This kind of events usually damage the road along the Río San Juan that connects Barreal with San Juan city.

Landslides affecting towns and villages have damaged homes, governmental buildings, streets, and churches, and during extreme events fatalities have been reported. Debris flows reported in Andalgalá (Catamarca Province) have sometimes been devastating, as happened in 1913 when the village of Andalgalá was flooded, killing at least 40 people. Severe damage to irrigation systems was also reported, and streets were eroded 1.2 m in depth (Kantor, 1916). The debris flow portrayed by Kühn (1915) demolished a railway bridge in this locality after overwhelming its defenses.

Further examples of these violent flows can be found in the River Blanco Basin. On February 11, 1945, a channeled debris flow in the El Salto Arroyo increased the stream flow of the Blanco River, destroying a bridge over this river, carrying away a bus with tourists, and razing the Civic Center of Potrerillos as well as hotels in this locality. A new catastrophic debris flow, triggered by a summer rainstorm on January 12, 1960, in 40 minutes overwhelmed homes, roads, and three cars. The debris flow drained into the Blanco River, carrying away a bridge and covering houses and hotels on the riverbank with mud and boulders. On this occasion, the Mendoza River stream flow reached 250 m^3/sec at the Cacheuta station.

An intense summer rainfall on December 25, 1967 caused a debris flow in the Las Mulas valley, 12 km upstream of Potrerillos village, which damaged temporary settlements, overwhelmed a bus with 20 tourists, a car, and a motorbike, and completely destroying a road. A new debris flow from this creek was triggered by a rainfall on February 13, 1970, overwhelming six cars and a jeep with four persons inside, and destroying several homes and a police building. Potable water pools of Obras Sanitarias were completely destroyed.

Then, on December 27, 1982, flows coming from the El Salto, de las Mulas, and Burrito Arroyos drained into the Blanco River also damaging Obras Sanitarias facilities, and upon reaching the Mendoza River damaged the Transandino Railway and the international road to Chile. When the debris flows reached the Mendoza River, they damaged facilities installed in this valley. The electric power plant at Cacheuta was severely damaged in 1913 and again in 1957. Also, two fatalities were reported, and the thermal baths of the Cacheuta Hotel were affected when the Mendoza River stream flow reached 346 m^3/sec in 1920.

In some cases, cultivated fields and livestock were lost. The violent flow in 1880 that affected the village of Andalgalá also impacted desolate areas of the Choya, Hoyada, and Negro valleys. This event carried away a local man who drove a mule herd (*La Nación,* 1880). During the 1990s, rotational slides at Volcán (Jujuy Province) were caused by lateral river erosion; one block of this town was destroyed completely, and many cultivated fields were lost (Solís et al., 2004).

Finally, economic losses stemming from lost revenues from tourism and the country's historical heritage must be taken into account. Mountain areas are important tourist attractions, as are the valleys surrounded by these ranges, where pre-Hispanic settlements are found. Much of Argentina's historic heritage is associated with the high landslide hazard zone in Uspallata valley (Mendoza River Basin) where Inca monuments and Inca routes have been found. Diaguita hieroglyphics 1300 years-old and Huarpe remains were also discovered in this valley (Moreiras, 2004d, 2005d, 2006a). Other places such as the village of Tilcara, which is pervasively affected by debris flows, is an international tourist spot, being one of the main pre-Hispanic settlements of the Omaguaca group in northern Argentina. The localities of Humahuaca, Purmamarca, and Maimará have similar historic importance. Debris flows have also damaged many tourism infrastructures. For example, the thermal baths of Fiambala, located at Sierra Fiambala (La Rioja Province), were severely damaged by a debris flow emanating from the Quebrada de los Baños valley on February 2, 2003 (González, 2003).

In any case, during the last decade some applied studies on land-use planning have included analyses of natural hazards. Along several sections of the provincial roads E-55 and A-73, Córdoba province, a landslide hazard map has been created (Bejerman, 1993; Bejerman and Giraud, 1993) and the Landslide Possibility Index (LPI) has been applied in order to evaluate risks from natural hazards. In the E-55 road, 73% of rocky talus has high and middle LPI values; whereas the highest LPI value was assigned to 76% for the A-73 road. Landslide occurrences have been identified in several road sections built in mountainous areas of San Juan Province (Salinas et al., 1995). The incidence and susceptibility of landslides have been analyzed for the Mendoza River (Moreiras, 2004c, 2005a), and a probabilistic hazard assessment and risk map was elaborated (Moreiras, 2004a, 2004d). In the Lasifashaj valley, Tierra del Fuego, a geomorphological hazard map has been prepared as a tool for ecological landscape analysis and land-use planning (Coronato and Roig, 1999). In the most hazardous zone, earthslides, rockfalls, debris falls, and avalanches occupy more than 50% of the area.

7. Final Remarks

During the last two decades, an increasing number of geomorphological studies on landslides have appeared. Although currently there is a boon of paleo-landslide dating for evaluating landscape evolution, calculating erosion rates, and extending paleoseismicity, the occurrence of present-day landslides has not been studied at all. Moreover, assessments of landslide hazard and risk are lacking in Argentina. Such assessments are essential for proposing proper territorial planning and future preventive measures.

Such studies would help to reduce future economic losses: population growth and expanding urbanization are increasing the potential devastating impact of landslides in most of Argentina's mountain areas. The Patagonian Andean cities, which are generally built over glacial deposits and over steep slopes, are expanding owing to increases in tourism. Urbanization and road construction have increased significantly over the last three decades, transforming the landscape. Consequently, improving the criteria used to identify hazards associated with natural events is even more critical. However, neither threshold precipitation nor minimum earthquake magnitudes producing landslide occurrences have been determined for different Argentine regions.

Both local people and the authorities should be educated about landslides. Intensive educational plans should especially be designed for the communities that could be most affected. People should learn to recognize which places are dangerous to settle in and to develop economic activities. Governments should develop careful monitoring systems for seismic, volcanic ,and climatic processes, including the monitoring of beaver dams, to avoid any surprises connected with landslides. Alarm systems have not yet been developed in the affected areas, although they could save lives and properties. In addition, the lack of evaluation studies for the slope instability associated with volcanic eruptions should also be remedied, for lahars sourced from sudden ice or snow melting often result in catastrophic events.

Hazards related to impounded lakes also have to be taken into account. Hermann et al, (2003) established that at least 90 % of Pliocene-Holocene landslides identified along the Andes between 36° and 38° S have dammed lakes, and about 23 of them remain until the present day. Moreover, instability in the boundaries of dammed lakes should be evaluated, as rocks show a different behavior when they are saturated with water Even though impounded lakes drain slowly, they represent a potential hazard because periods of increased precipitation or spring thawing may increase the lake level, causing dam failure.

Although landslides do not affect most of the Argentine population, they are one of the most important processes in modeling Argentina's mountainous areas, which is clearly evidenced by all recognized paleo-landslides. Adding to their potential significance is the fact that several deposits earlier identified as of glacial origin have been recently been reinterpreted as landslides. Moreover, the frequency of landslides may increase in the future in response to climate change and extreme climate phenomena (Paolini et al, 2005; Moreiras, 2006a,

2006c). Hence, landslides need to be better known in Argentina, not only for academic purposes, but especially for the economic development of mountain regions.

Acknowledgments

We want to express our gratitude to Edgardo Latrubesse for inviting us to participate in this book and to Mauro Soldati for his review and important suggestions and comments. We also thank Jack Shroder for comments and suggestions that improved the manuscript. We thank Rafael Bottero (IANIGLA) for figure design and Nelly Horak (IADIZA) for checking the English version. We are also indebted to Juan Pablo Ceballos for photos of dam failure on the River Santa Cruz (San Juan provinces) and to Marcelo Chiaradia for photos of the Monte Susana debris flow over the National Route 3 in Tierra del Fuego.

CHAPTER 16

Floods in Argentina

Edgardo M. Latrubesse[a] *and* Daniel Brea[b]

[a]*University of Texas at Austin, Department of Geography and the Environment, GRG 334; 1 University Station A 3100, Austin TX 78712 USA*
[b]*Instituto Nacional del Agua, Programa Hidráulica Fluvial, Universidad Nacional de La Plata, Área Departamental Hidráulica, La Plata, Argentina*

Contents

1. Introduction

Because of its geographical position in the southern part of South America, its latitudinal distribution of approximately 3700 km from north to south, and its complex relief formed by a variety of landscapes and complex geology, Argentina possesses a unique variety of fluvial systems. As in many other countries, some of the large Argentinean cities, as well as power generation facilities and agriculturally productive areas, are connected to rivers. Floods affect some of the population and infrastructure, and also have a strong impact on agricultural productivity, one of the main economic activities of the country.

Argentina's water resources have an unequal distribution. The humid regions cover 24% of the total surface area of the country but support 68% of the population. On the other hand, the dry areas extend over 61% of the country but support

Corresponding author: *Edgardo Latrubesse, Centro de Investigaciones Geológicas-CIG and LATYR-CONICET, Universidad Nacional de La Plata, Calle 1 N^{0} 644, (1900) La Plata, Argentina, latrubesse@yahoo.com.br*

Developments in Earth Surface Processes, Volume 13
ISSN 0928-2025, DOI 10.1016/S0928-2025(08)10016-5

only 6% of the population. Despite these regional differences, water availability per inhabitant is very high, reaching 22,000 m^3 per inhabitant per year. This value is 22 times higher than the "hydric stress level" adopted by the United Nations (Lopardo et al., 2003).

Practically all the rivers in Argentina drain into the Atlantic Ocean because of the location of the Andes to the west, which act as a natural frontier with Chile, and because of other geotectonic units that extend north-south, such as the Pampean and the Pre-Cordillera Mountain Ranges. However, large areas of the country behave as endoreic/arreic systems (Fig. 16.1).

The fluvial systems in Argentina can be classified into three main groups: large tropical rivers, torrential rivers with headwaters in mountain areas, and flat-plain rivers with insufficient drainage efficiency. The torrential rivers group can be subdivided into two categories: (a) Rivers fed by torrential rains and (b) rivers fed by rain and snow.

2. Torrential Rivers with Headwaters in Mountain Areas

2.1. Rivers Fed by Torrential Rains

Almost two-thirds of Argentina is made up of semiarid to arid environments, and many of its fluvial systems are connected to mountain environments, such as the Pampean and the Main and Frontal Cordillera Ranges, as well as the Sub-Andean and Eastern Cordillera Ranges, which receive strong and concentrated rainfall during the summertime. Generally, flash floods on these rivers cause damage to urban centers and transport infrastructure (roads, railways, bridges), and, in some cases, human loss.

The main rivers draining the Pampas Ranges of Cordoba and San Luis are named Cuarto and Quinto (that means: fourth and fifth respectively), which are characterized by low discharge in winter and flashy floods in summer. The Rio Quinto is discussed below together with other Pampean rivers. Other important rivers come from the Andes crossing the Chaco region such as the Juramento/Salado River (discussed below), Bermejo, and Pilcomayo rivers.

Many of these rivers have unstable channels in piedmont areas generating alluvial fans and fluvial fans. The Pilcomayo River (drainage area = 240,000 km^2), a tributary of the Paraná Basin, with a mean annual discharge of nearly ~210m^3/s, carries more than 140 million tons/year of fine sediment, but a large quantity of the sediment is stored in the Chaco plain before their arrival at the Paraguay River, a main tributary of the Parana River. The Pilcomayo River, due to its extreme avulsive activity, has generated several problems in the border delimitation between Paraguay and Argentina for decades.

In addition, remote geomorphologic and ecologic systems, such as the Paraná delta, depend on the sediment transport of the Andean rivers. The Andean tributaries of the Paraná River carry high amounts of suspended sediment. With a mean annual discharge of 145 m^3/s, the Bermejo River contributes ~50% (48 million tons/year) of the total suspended sediment transported by the Paraná River (Q_{mean} = 18,000 m^3/s) (Amsler and Prendes, 2000). The sediment introduced to

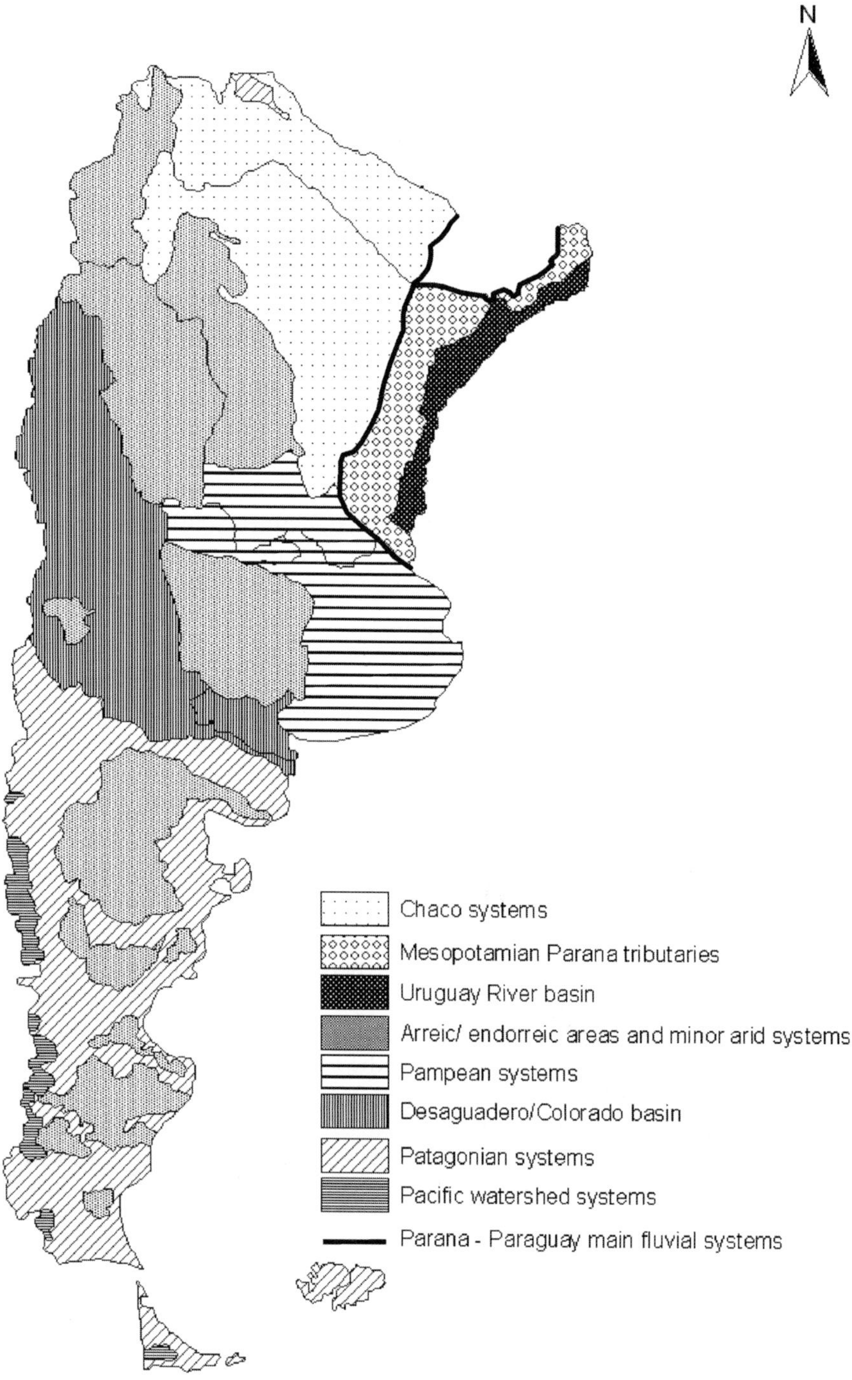

Figure 16.1 Main Argentinean hydrographic areas.

the Paraná Basin through the lower Paraguay River by the Bermejo River, for example, is a main source that feeds nutrients to the aquatic ecosystems in the Paraná alluvial plain and delta.

Channel shifting in Chaco rivers also has created several problems damaging infrastructure, and they demand immediate strategic and specific plans for river management because of their dynamic behavior. Furthermore, sediment production has been increasing in response to land-use changes in the Sub-Andean zone of Bolivia and Argentina, and the environmental consequences are not clear. The effect of the geomorphologic processes on the maintenance and dynamic of the hydrophysical and biotic environment of Chaco is practically unknown.

The smallest tributaries of the Andean zone provide huge amounts of sediment load to the main Chaco systems and affect towns and structures with floods and other damages. Many of these rivers have unstable channels in piedmont areas where alluvial and fluvial fans develop. The floods in the city of Tartagal, located in Salta Province, in 2006 serve as an example of a destructive flood in a city located at the apex of an entrenched piedmont alluvial fan. Tartagal has 47,000 inhabitants, and the Tartagal River, a tributary of the Bermejo River (the Paraná fluvial basin) crosses through it. The city is located at 500 m.a.s.l., but the upper basin upstream is just 75 km^2 and is spread along a very steep relief of the Sub-Andean Chain, which is approximately 1300 m a.s.l. In addition, the mean annual precipitation there oscillates between ~800 and 1000 mm/y, and the area is covered by submontane to montane forest (Yungas). The fan is entrenched, and the river has a meandering pattern from the apex to its downstream reaches. This gives a false perception of a low level of activity and of channel stability. For this reason, the city of Tartagal has spread over the fan lobe.

Between December 2005 and March 2006, however, rainfall exceeded the mean monthly precipitation by more than 400 mm, producing landslides that dislodged the chemically weathered regolith formed under tropical conditions. The torrential flooding of the river, fed by mass movements of fine sediment, reached approximately 1000 m^3/s and devastated parts of Tartagal city. However, the main problem was not overbank flooding, but the lateral erosion of the river channel in the soft sediment of the former alluvial fan. The river today is entrenched in up to 20 m of alluvial deposits, and because of the high energy of the flow, the channel shifted laterally, destroying 200 houses and causing the evacuation of 2000 inhabitants. Several pieces of infrastructure built in 1982 and 1984 were destroyed-for example, a pedestrian bridge and various canalizations-were or put at risk, such as a bridge along the road (Fig. 16.2).

On the morning of February 9, 2009, a water flow overloaded by detritus, mud, and logs coming from the mountains through the Tartagal River once again affected the city. The event was triggered by a small convective/high-intensity local cell located in the subandean zone that produced more than 95 mm in near 45 minutes. The flood lasted just two hours, but 10,000 people were affected, 1100 of whom were evacuated, 3 were killed, and 8 went missing, and a bridge, houses and vehicles were destroyed.

2.2. Rivers Fed by Snow and Rain

Some of the Andean tributaries from Mendoza Province to the south are rivers that are fed by snow and rain. The most characteristic of these are the

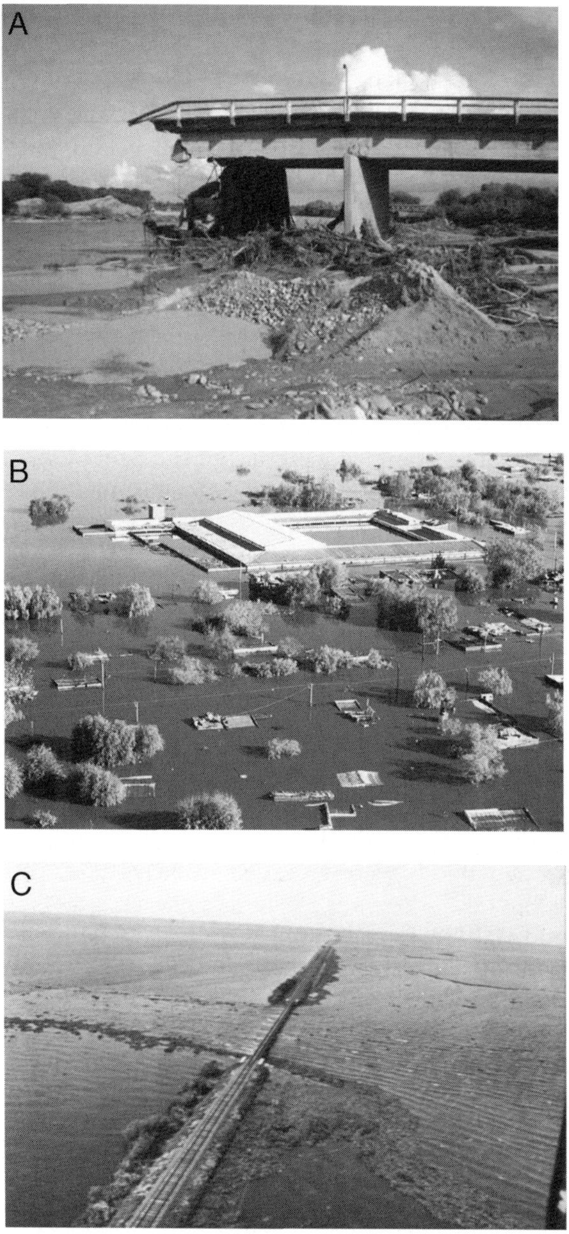

Figure 16.2 Structures destroyed or affected by floods: (A) bridge in Tartagal River, Salta; (B) Santa Fe during the floods of the Salado River, 2003; (C) the Paraná River floodplain near Santa Fe during the big floods of 1983.

Patagonian rivers, predominantly the headwaters in the South Andes, practically all of which drain into the Atlantic Ocean, with the exception of a small area drained by the upper courses of the Manso and Futaleufú rivers. Eight major river systems drain the Patagonian tableland toward the Atlantic; they are, from north (36° S) to south (50° S), the Colorado, Negro, Chubut/Chico-Senguer, Deseado, Chico, Santa Cruz, Coyle (or Coig), and Gallegos rivers. The combined drainage area of these rivers is approximately 325,000 km^2. Closed basins, some of them with beds below sea level, occupy approximately 247,000 km^2.

The rivers have different channel patterns, but rivers such as the Colorado, Negro, and Chubut have predominantly meandering patterns. The Chubut River ($Qmean = 42\,m^3/s$) is a meandering river with a floodplain up to 8 km in width in the lower course, which is completely occupied by agriculture. The Chico Basin also meanders. The Rio Negro ($Qmean = 858\,m^3/s$) is a meandering river with a floodplain occupied by agriculture, which reaches 16 km in width at some points, while the Colorado ($Qmean = 131\,m^3/s$) also develops meanders, but only in its narrowest fluvial belt. Several of the southernmost rivers, such as the Deseado, Coyle, and Gallegos, show a braided pattern, but the Santa Cruz River has a low, sinuous, single-channel pattern with a tendency to meander.

The population density of Patagonia is low because of the arid climate, and human activities and urban centers have concentrated on the floodplains of the major rivers. Floods affect some of the cities, towns, and cultivated areas along the floodplains. Trelew is one of the cities most affected by floods (Fig. 16.3). Located on the lower course of the Chubut River, the city has been undergoing floods since its foundation. The oldest flood records go back to 1901, 1902, and 1904. The floods covered the valley and destroyed houses, irrigation channels, and other structures. A severe flood was also recorded in 1945, with an estimated discharge of 253.8 m3/s at Gaiman (Maza and Ruiz, 2006). The Ameghino Dam was built in the middle course of the river in 1964. The dam offered a false sense of security with regard to flooding, and floodplain use increased downstream from then on. However, more recent floods were recorded in May 1992, October 1995, and April and May 1998.

During the flood of 1998, rainfall reached 251 mm in 54 hours, resulting in the evacuation of 15,000 inhabitants from the cities of Trelew, Rawson, and Puerto Madryn. The flood affected 80% of the city of Trelew and caused economic losses of nearly US$ 5 million (Maza and Ruiz, 2006). The flood was caused by sporadic storms that occurred in the lower reach, downstream of the Ameghino Dam, because of the persistence of a low-pressure center close to the Atlantic coast.

3. Flat-plain Rivers with Insufficient Drainage Efficiency

Huge plains, with an historical record of drought and floods, dominate 35% of the land area along the Chaco and Pampean areas, spreading over approximately 950,000 km^2. These areas are characterized by spectacularly flat landscape, where local relief can be an impediment to water drainage, being even more important than the regional slope. Typical slopes can be as low as 10^{-3} to 10^{-4} (Sala et al.,

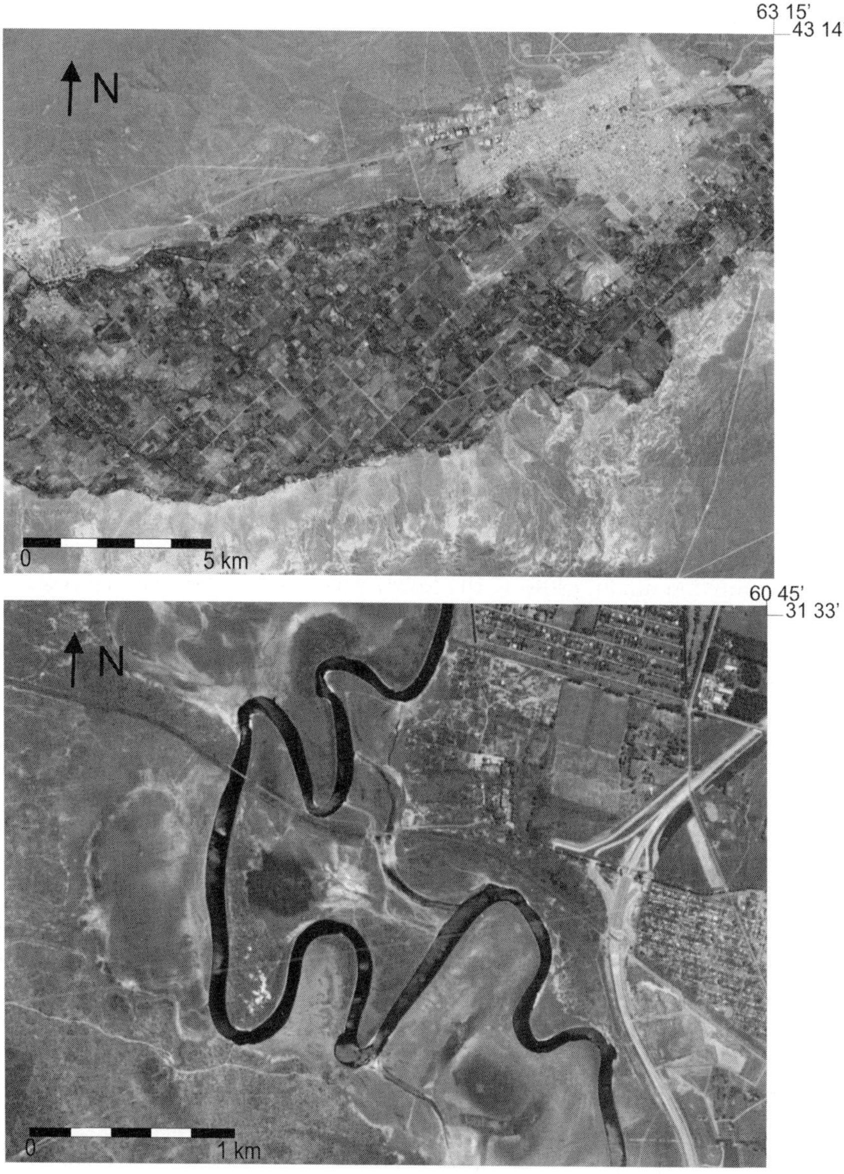

Figure 16.3 (A) The Chubut River at Trelew; (B) the Salado River near Santa Fe.

1983). In these extreme conditions, drainage systems have severe difficulties in organizing well-structured drainage networks and frequently lack defined basin limits. Vertical variables (infiltration, evaporation) dominate over horizontal flow (runoff), producing water storage in surface areas and low lateral hydraulic gradients. In addition, large areas of the Pampean plain consist of inactive aeolian landscapes belonging to the so-called Pampean Sand Sea, which has an area of around 300,000 km^2. It originated during the Late Pleistocene period and suffered several aeolian reactivations through the Holocene period. This particular geomorphology contributes to the deficient drainage and the generation of some areic areas.The main flat areas that suffer flood processes are the Sub-Meridional lowlands, the Central Pampa and the Salado River Basin.

3.1. The Sub-Meridional Lowlands and the Salado River (Santa Fe)

These depressed areas extend over approximately 80,000 km^2 in the provinces of Santa Fe, Chaco, and Santiago del Estero. The floods here are mainly caused by local rainfall and water table saturation. Nevertheless, unplanned artificial drainage, changes in land use, and a lack of urban and road planning have all contributed to the occurrence of worse flooding in local areas. The floods have a recurrence interval of approximately two to four years and a duration of 30 to 180 days.

The Salado River has a drainage area of 125,659 km^2 and is a tributary of the Paraná River. The headwaters are located in the Eastern Cordillera (Nevado de Ancay), and the upper reaches flow downstream through the provinces of Salta, Catamarca, and Tucumán, crossing the Sub-Andean zone. The middle reach takes a NNW-SSE direction and alternates between sections with well-defined channels and sections with more anarchic channel patterns associated with swamps, shallow lakes, and flooded areas such as the "Bañado del Copo," which act as sink areas for sediment and water. The broad area of sedimentation is, in reality, part of a complex Quaternary alluvial megafan, 650 km long and 150 km wide, which characterizes the Middle Salado.

Downstream, the system recovers a single-channel pattern because of the water sources provided by small tributaries and by the swampy area described above. The lower reach crosses the Sub-Meridional lowlands and enters the Paraná River alluvial belt. The lower section of the river receives small discharges from the middle reach. The main tributary is the Calchaqui River, which drains part of the Sub-Meridional lowlands.

The lower reach is characterized by a poorly developed drainage network in a very flat landscape. Downstream, at the confluence with the Calchaqui River, the Salado flows on a well-developed floodplain. The river shows an irregular (non-harmonic) meandering pattern alternating with short, low-sinuous reaches and is bordered by narrow levees and backswamps. The meanders shift slowly at rates of 0.7m/y, and the river is bordered by well-developed natural levees along its entire course, 2.5 m higher than the general level of floodplain. Backswamp areas occupy 70% of the floodplain.

In this area (Ruta Province N 70 gauge station), the mean annual discharge is 145.6 m^3/s. This is in agreement with the tendency of increasing precipitation, recorded since the 1970s, in the region (see "Climatic Trends and Floods" below and García et al, chapter 1, in this volume). An analysis of monthly discharges indicates that the mean of the monthly maximum discharges can be 6 to 10 times the value of the mean annual discharge. The worst scenario for floods is produced when the input from floods on the Calchaqui River tributary is combined with intense local rainfall in the lowermost reach of the Salado, which completely saturates the flat landscape.

With 489,500 inhabitants, the city of Santa Fe is the main urban center in the Salado Basin. The city has expanded close to the river mouth on the left bank, where the Salado River meets the Paraná River. Historically, the city has suffered floods, though of low magnitude.

The most catastrophic flood in the Salado Basin started in April 2003. Heavy rain fell on the lower basin, especially on April 23 and 24. During those days, a frontal system of semistationary warm air was located over the region, causing strong convective rainfall. At this time, the terrain was already saturated because of the summer rainfall and the existence of a high water table in the region as a consequence of the wet period the area had been experiencing from since the 1970s. The rainfall during February, March, and April was similar to the mean annual precipitation. The water discharge grew strongly over seven days in April, increasing from 700 m^3/s to approximately 3800 m^3/s (INA, 2003). Within a few hours, one-third of the city was flooded, affecting nearly 140,000 inhabitants (IARH/CAI, 2004) (Fig. 16.2B). For the purpose of comparison, before the 2003 flood, the maximum discharges recorded in the area were 2596 m^3/s in 1972-1973 and 2672 m^3/s in 1997-1998). The flood caused economic losses of approximately US$ 1,000 million (CEPAL, 2003).

But one can ask why such a disaster happened in a relatively easily predictable and low-energy fluvial system. To protect the city from floods, the state government build several embankments topped with roads on the west side of the city in 1943 and again, more recently, between 1994 and 1998. The artificial levees constrained 70% of the area of the floodplain, encouraging the human occupation of risky areas (FICH-INA, 1998).

However, the northern part of the embankment was never completed. The floodwaters entered along the northern area and covered parts of the city with a water sheet of up to 4 m in depth. A badly designed bridge also contributed to the catastrophe. Located directly upstream of the mouth of the river, on the Santa Fe-Paraná Highway, the bridge only had a small water bypass (8% of the width of the alluvial valley), which created a backwater effect and increased the water stage upstream. Paradoxically, since the beginning of the 90s, the State of Santa Fe had been carrying out a detailed study of flood hazards in its territory, where hazards and vulnerable areas (the same as those affected by the 2003 flood) were correctly assessed and mapped (INCYTH, 1991).

3.2. Central Pampa and the Salado River Basin (Buenos Aires)

The Central Pampa region is located in the southern part of the provinces of San Luis, Cordoba, and Santa Fe, the northern part of La Pampa Province, and the northwest sector of Buenos Aires Province, comprising an area of 60,000 km^2. The area includes several creeks in south Cordoba, the lower course of the Quinto River, which has its headwaters in the San Luis Mountain Range, and several lake basins (lagunas). The most important lakes are La Picasa, La Salada, del Siete, and the Hinojo–Las Tunas Complex, which is connected to the headwaters of the Salado River Basin.

During floods, the Rio Quinto spreads its excess water over the southern part of Cordoba in a depressed area known as the La Amarga swamps (Bañados de La Amarga). However, human intervention in La Pampa Province has conducted the excess flows of the Rio Quinto to the northwestern area of Buenos Aires Province. This region is also known as the Sandy Pampa, because the landscape is dominated by fossilized aeolian landforms. Huge lineal dunes, Late Pleistocene in age and extending more than 100 km in length and 5 to 7 m high in relation to the general relief of the plain, are a formidable obstacle to the development of a well-organized drainage network. The interdune areas are occupied by permanent and nonpermanent bodies of lentic water, and the whole area is predominantly arreic (Fig. 16.4).

The surficial water sheet is shallow and flows slowly along the interdune depressions, moving toward ancient large aeolian pans now occupied by shallow lakes such as the Hinojo-Las Tunas Lake Complex, Salada Lake, and some paleochannels obliterated by eolian sediments that still act as depressed water-collection areas.

In this kind of extremely flat landscape dominated by a cover of soft Quaternary sediments, the drainage network is highly dependent on the interrelations between rainfall and the water table, and the impediments generated by local relief. Figure 16.5 shows the particular situation produced by an extremely wet event in March 1999 where, in some areas, rainfall reached 450 mm (Kruse et al., 2005). In several places, the water table rose more than 1.2 m, coming very close (1 to 2 m in depth) to the terrain surface, flowing into depressions in the landscape and increasing flooded areas.

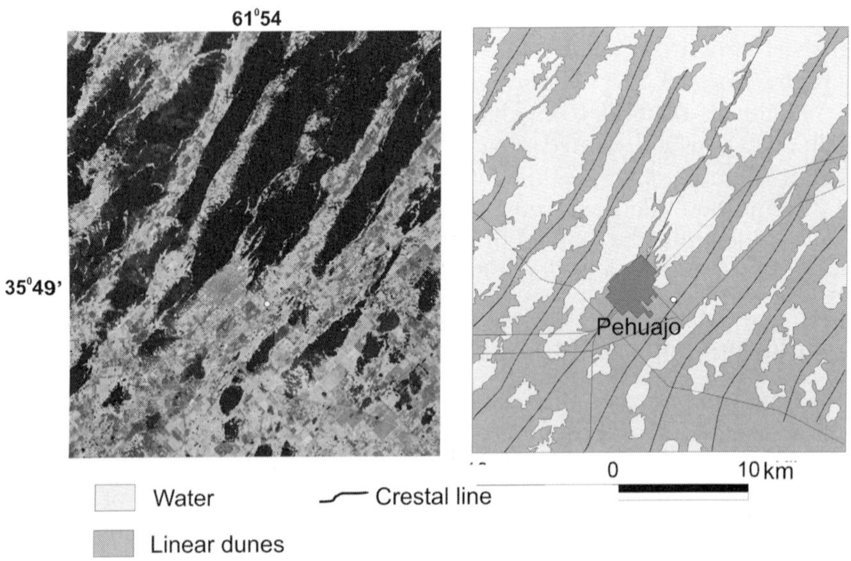

Figure 16.4 The poorly drained area occupied by large Pleistocene linear dunes at the northwest of Buenos Aires Province.

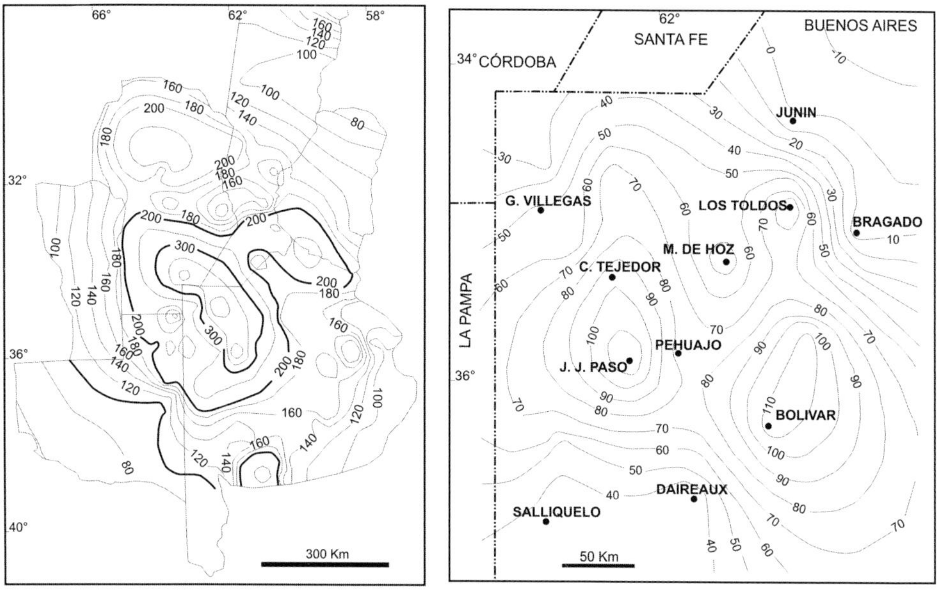

Figure 16.5 (A) Rainfall in the Pampean region during March 1999; (B) elevation of the water table as a response to high rain precipitation in 1999. (From Kruse and Laurencena, 2005).

The Salado River Basin is the most important fluvial system in Buenos Aires Province and at the core of the country's agricultural and farm production. Originally, the drainage area was 90,000 km^2, but because of human intervention (canalizations and water diversion), the drainage area expanded to approximately 170,000 km^2. The main human interventions that expanded the Salado Basin drainage area include the channel in the northwestern area of Buenos Aires Province that drains surplus waters to the Salado system during floods, channels draining waters from the areic area of the Sandy Pampa, and channels draining a complex of lakes located to the south of the province, named Encadenadas. In addition, farmers built unplanned secondary channels to drain their own ranches and farms. However this tremendous expansion in drainage area was made without increasing the channel capacity of the Salado River channel. For that reason the channel is overpassed in its flow capacity during flows. The poorly developed floodplain and the channel formed dominantly by cohesive sediments restrict natural hydraulic geometry adjustments of channel during peak discharges favoring floods.

Today, this system drains half of the territory of Buenos Aires Province, which is occupied by 1.3 million inhabitants and is, from a socioeconomic point of view, the most important area in the country, being responsible for 25 to 30% of national cereal and meat production.

The river discharges into the Atlantic Ocean at Samborombom Bay. The mean annual discharge oscillates through time from 340 to 430 m^3/s at the Guerrero gauge station (Table 16.1). With the exception of some low mountain ranges to the

Table 16.1 Mean Annual Discharge at Guerrero Gauge Station for Different Periods

Year	Discharge (m^3/s)
1980	1798
1993	1741
1985	1549
2001	1304
1959	1035
1967	947

south (the Tandil Range, approximately 500 m.a.s.l., and the Ventana Range, approximately 600 to 700 m.a.s.l. with peaks up to around 1200 m.a.s.l.), the basin extends over a very flat area with elevations of less than 100 m.a.s.l., and a large portion of its flow is below the hypsometric line of 20 m.a.s.l. From the coast to a distance of 100 km inland, the regional slope is extremely low: on the order of 1:10000.The network is characterized by a very low drainage density. The more particular and unique pattern is recorded in the middle reach where the river is characterized by a sinuous channel that connects large aeolian pans occupied today by lakes (Fig. 16.6).

In general terms, the river does not have the required capacity to drain the water of its large and flat basin. Aeolian deposits and pans are obstacles that the river tries to pass over with a particular sequence of autigenic adjustments.

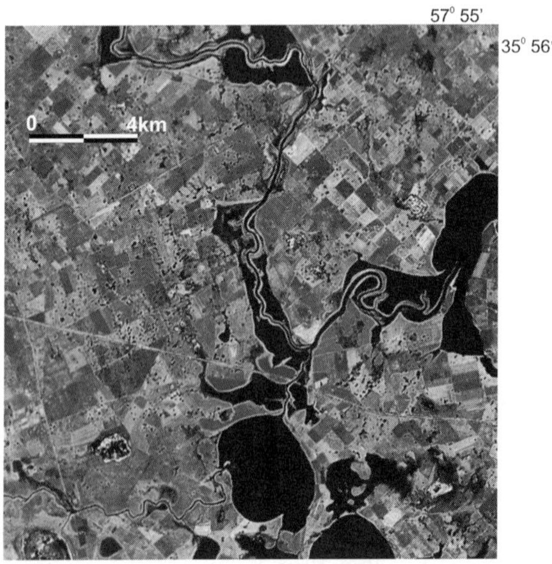

Figure 16.6 The Salado River (Buenos Aires) upstream the Guerrero gauging station.

Table 16.2 Main Floods (Daily Discharges) of Salado River Recorded at Guerrero Gauge Station

Guerrero gauge station (periods)	Mean annual discharge (m^3/s)
1932–1995	340.2
1959–1995	386.4
1967–1995	429.7

Table 16.2 shows the main floods (peak daily discharge) recorded from the Guerrero gauge station in the basin. The main floods occurred in 1980, 1985, 1986, 1987, 1993, and 2001. The floods have increased since the end of the 1960s/beginning of the 1970s, following a tendency of increasing rainfall in the area (see "Climatic Trends and Floods" below). This trend is also reflected also in the record of mean annual discharges (Table 16.1).

In general, the floods do not affect the Salado Basin in the same way. For example, during the 1980 flood, discharge increased in the Lower Salado because of flows coming from the Vallimanca and Las Flores creeks (*arroyos*), which are located in the southern part of the basin, in the Tandilia Mountain Range. In 1993, with a similar discharge to that recorded in 1983 (at the Guerrero gauge station), the floods were caused by rains in the middle and upper basin. In 1985, the floods were generated by a more uniform distribution of rainfall along the basin. The duration of the floods was four to five months for the 1980 and 1985 floods, and three or four months in 1993, with this last flood mainly being concentrated in the lower Salado. The floods in 1993 began in October, but the others started in March/April, and continued until autumn/winter.

The floods cause economic losses in agriculture and cattle farming, affect the infrastructure of private ranches and farms, and cause damage to state rural infrastructure. Table 16.3 shows the potential relationship between the increase in floods and the main regional crop productions (PNUD/ARG 02/006; Halcrow and Partnerss, 1999). Crop production has been characterized by wheat, corn,

Table 16.3 Estimated Cropped Area Affected by Floods and Economic Losses in Relation to Flood Recurrence Intervals (data from Halcrow and Partners, 1999)

Recurrence (years)	Flooded area (km^2)	Cropped area affected by floods (hectares)	Economic losses (in millions USD)
2	10,561	76,406	~20M
5	22,499	232,093	~60M
10	34,515	316,802	~83M

sunflower, and soy; however, during the last years, soy has become a dominant crop. Lost production on cattle ranches in the northwestern region of Buenos Aires Province was estimated to vary between US$ 33 and 83 million, depending on the magnitude of the flood event, and the damage to rural infrastructure was estimated to vary from US$ 203 million for an average flood to US$ 570 million for a severe flood (CENTRO, 2004).

4. Large Tropical Rivers

With a huge drainage area of 2,400,000 km^2 and a mean annual water discharge in the middle reach of around 17,000 m^3/s, the Paraná River drains a variety of landscapes and climatic regions in Brazil, Bolivia, Paraguay, and Argentina. The Paraná River is responsible for approximately 75% of the total water discharged into the Atlantic Ocean along the coast of Argentina.

Almost all of the water discharged comes from two subbasins: the Upper Paraná and the Paraguay River basins. With a drainage area of 1,150,000 km^2, mainly in Brazilian territory, and an absolute maximum rainfall of more than 2250 mm, the Upper Paraná Basin contributes 70% of the total water discharged, due to the heavy rainfall produced in south Brazil. With a drainage area of 980,000 km^2, the Paraguay River Basin contributes nearly 30% of the water discharged, coming mainly from the Upper Paraguay River, where the Pantanal of Mato Grosso is located.

Along its middle reach, the Paraná River flows on a wide and complex floodplain, 13 to 40 km in width, and is inundated completely during extraordinary floods (Fig. 16.2C and Fig. 16.7). The river shows a complex anabranching pattern with multiple channels of different orders, with the main channel and branches flowing between islands formed by more stable alluvium. Secondary channels can show a braided behavior with shifting channel bars (Orfeo and Stevaux, 2002, Ramonell and Amsler, 2000), but main flows concentrate on a meandering thalweg (Amsler et al., 2005).

Several important Argentinean cities are located along the banks of the Paraná River, such as: Corrientes, Resistencia, Goya, Santa Fe, Paraná, Rosario, Zárate, and Campana. Fluvial stages and discharges have been recorded on the Paraná River since the nineteenth century (1891 to present). In 116 years, the river has reached the critical stage at which the evacuation of endangered populations has been necessary 22 times.

The most dramatic flood occurred during the ENSO-El Niño event of 1982-1983, which also affected the Uruguay River Basin. The discharge of the Paraná River reached 60,200 m^3/s in Corrientes, at the confluence with the Paraguay River, and flooded an area of more than 30,000 km^2 of the floodplain. These exceptional floods persisted for more than one year (Fig. 16.7B).

The impact on the social and economic activities of the region was huge (Pochat, 2002). More than 234,000 people were evacuated, and the damage caused losses of more than US$ 2,600 million. The floods of the 1991-1992 El Niño-ENSO event were also catastrophic, causing an estimated loss of US$ 513 million and affecting more than 122,000 inhabitants (Pochat, 2002).

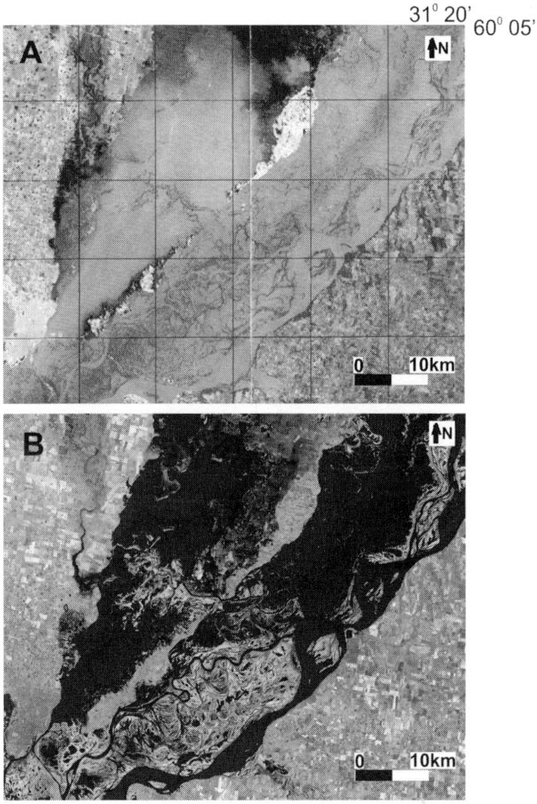

Figure 16.7 The Paraná River and floodplain at Santa Fe: (A) during the large flood of 1983; (B) during the flood of 1998.

5. Climatic Trends and Floods

Increases in mean annual precipitation have been detected from 1956 to 1991 over wide areas of northeastern Argentina, varying from more than 10% to more than 30% in some places (Castañeda and Barros, 1994). The annual precipitation trends weren very positive, without exception, from 1970 to 1971 (García, 2000).

Since the beginning of the 1970s, the stock of pluvial and fluvial water in the region has increased, and the regional water table level has risen. This positive trend has occurred throughout the whole of the northeast region of Argentina, as well as in the south of Paraguay. The increase in mean annual precipitation across the whole region suggests a displacement of the isohyets toward the west, favoring the expansion of the agricultural frontier in this direction. However, the negative consequence of these changes is repeated or continuous floods, which are also favored by the greatest recorded frequency of daily precipitation of more than 100 mm in almost all the rain gauges of the region (Canziani 2003; García et al., 2007).

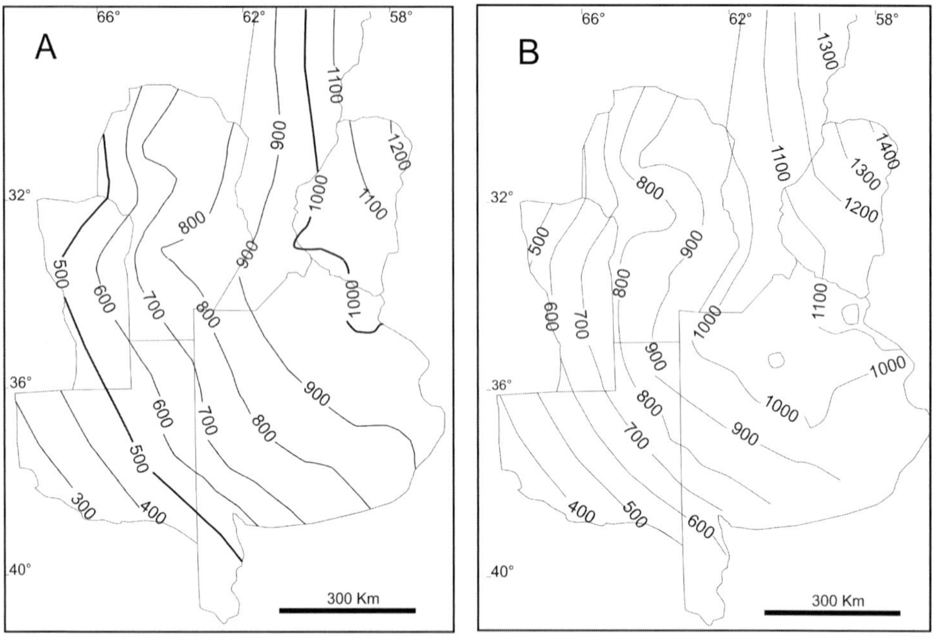

Figure 16.8 Changes in rainfall and shifting of the isohyets in the Pampean region: (A) mean annual rainfall from 1941 to 1970; (B) mean annual rainfall from 1971 to 2002. (From PNUD ARG 02/006, 2006).

Changes in the frequency of droughts and floods have also been recorded in the Pampean region. The frequency of droughts seems to have been decreasing throughout the whole humid Pampa region since 1970, with an average of one drought every three years until 1969, and one every five years from then onward (Venencio and García, 2005). Figure 16.8 shows the changes in precipitation since 1941. In some places, the increases in rainfall exceed 150 mm/year. The regional elevation of the water table has produced a decrease in the rainwater storage capacity of the terrain and lakes, increasing the region's sensitivity to floods. In addition, canalization is conducting more water, overloading the drainage network capacity and favoring floods along river tracks such as the Salado River; other human interventions, such as defenses and embankments, stem the water flow, increasing the duration of floods in some places.

6. Final Remarks

Despite the fact that a large part of Argentina has arid, semiarid, and temperate environments, the geomorphologic characteristics of large parts of its territory make the country highly vulnerable to floods. Agriculture and cattle farming have been the main activities of the national economy since the origin of the nation, with these activities concentrating in the Chaco-Pampean region, an extremely flat area with very productive land but poor drainage capacity. The most efficient fluvial corridor in the country, the Paraná River, which has many important urban centers along its banks, also acts as the main waterway for the agricultural production of the Chaco-Pampean zone.

In addition, two other factors contributed to the generation of increasingly worse flood vulnerability through the twentieth century: a climatic trend of increasing rainfall since approximately 1970, and the political instability of the country throughout practically the entire second half of the twentieth century to the present day.

The scarcity of continuous and planned public policies and programs in national and state agencies and institutions, because the changing policies applied arbitrarily for each government that assumed temporarily the govern, has eliminated any possibility of implementing plans for basin management where hazards and disasters are a serious and delicate issue. In the middle of this chaotic situation, there has been no effective plan to integrate the environmental and socioeconomic issues of the country's water resources and their development. National and state agencies, as well as universities and research centers, have been unable to maintain active research teams and monitoring systems over the years, and the blight of political crisis has triggered a continuous impoverishment of human resources in scientific and technical establishments. These actions, in addition to the nonexistence of national and state planning to harmonize land use, environmental, and socioeconomic issues, and water monitoring and management in highly productive areas such as the Pampean region, far from making the country a better place in the future, are likely to make it worse for two reasons: the additional problems caused by the high deforestation and environmental pressures some areas are experiencing, such as the Sub-Andean zone and Chaco, due to timber exploitation; and the expansion of the agriculture frontier.

Acknowledgments

We especially thank C. G. Ramonell for the critical review of this chapter and for providing us relevant suggestions.

CHAPTER 17

Desertification in Patagonia

Elizabeth Mazzonia *and* Mirian Vazquez

[a]*Universidad Nacional de la Patagonia Austral, Unidad Académica Río Gallegos. Lisandro de la Torre 1070, Río Gallegos, Santa Cruz, Argentina. E-mail: gis@uarg.unpa.edu.ar*

Contents

1. Introduction

Desertification is a deterioration process of the physical and biological environment through which economically productive soils of arid, semiarid, and subhumid ecosystems lose their renewal or regeneration capacity. In extreme cases, it results in an environment that is unable to support the communities that once depended on it. This process is associated with the overall loss of productivity of the ecosystems affected, which has a negative impact on human activities by restricting the sustaining capacity, reducing the sources of income, and deteriorating the life quality of the population (Mérega, 2000).

Corresponding author: *Universidad Nacional de la Patagonia Austral, Unidad Académica Río Gallegos. Lisandro de la Torre 1070, Río Gallegos, Santa Cruz, Argentina. gis@uarg.unpa.edu.ar*

Developments in Earth Surface Processes, Volume 13

ISSN 0928-2025, DOI 10.1016/S0928-2025(08)10017-7

The concept of desertification has varied in time. Aubreville (1949) introduced this concept in the scientific literature when he described human-induced degradation in the tropical forests of Africa. In 1976, Dregne defined desertification as the impoverishment of terrestrial ecosystems caused by human impact, whereas the Convention to Combat Desertification, implemented in the Conference on Environment and Development (UNCED, 1992), included climatic changes as one of the possible causes of desertification, which was defined as "land degradation in arid, semiarid and subhumid areas originated by several factors, such as climatic changes and human activities."

In a revision of the concept of desertification, Glantz and Orlovsky (1983) pointed out that some researchers consider desertification as a process of ecosystem change, in which different states can be observed (slight, moderate, severe, very severe), depending on the degree of alteration of biophysical conditions (Dregne, 1986, FAO and UNEP, 1984). Other authors define desertification as the result of a change process: the generation or global spread of arid conditions (Meckelein, 1980; WMO, 1980).

Despite the different ways this issue can be approached, desertification is presented as an adverse environmental process. The negative descriptors used mainly consider plant, soil, and water resource degradation. Plant degradation is evidenced in a decreasing percentage of plant cover, reduced biodiversity, and changes in physiognomic-floristic ecosystem composition: a decrease of palatable species and shrub encroachment. Soil degradation includes soil depletion (loss of organic matter, reduction of nutrients, acidification, salinization) and intensification of erosion processes. Water table deepening is one of the most noticeable signs of water-resource degradation. All these factors contribute to impoverished ecosystems and their reduced productivity (Dregne, 1976; Hare, 1977; UNCOD, 1977; Kovda, 1980; López Bermúdez, 1996, 1997). When the desertification process is incipient, these changes may be difficult to perceive, especially if no reference studies describing the initial status of the ecosystems are available (Paruello et al., 1993).

The United Nations Conference on Desertification (UNCOD, 1977) identified this process as the most important environmental issue in the world. It has been estimated that about 6100 million hectares, approximately 40% of continental areas, are affected by desertification processes, involving some 1000 people, mainly in Asia and Africa. More than 600 million hectares undergo different levels of desertification in Latin America and the Caribbean, whereas in Argentina, this process affects more than 60 million hectares (SA and DS, 2000).

2. Geographical Characterization of Patagonia

Patagonia covers the southernmost portion of the American continent, from 37° S approximately to Cape Horn, at 56° S. With an area of 790,000 km^2, it is the only emergent landmass in the seas of medium-high latitudes of the

Southern Hemisphere. Its main geographical feature is the Andes Range, which stands as a water divide and an international boundary between Argentina and Chile in vast portions. This orographical feature defines a marked climatic contrast between the lands located to the west (Pacific side) and to the east (Atlantic side), and is coupled with one of the most marked vegetation gradients in the world (Endlichter and Santana, 1988; Warren and Sugden, 1993).

Regarding latitude, Patagonia lies between the subtropical high-pressure belt and the subpolar low-pressure areas; therefore, it is fully included in the area of prevailing westerly winds in the Southern Hemisphere. Hence, the Andes stand as a topographic barrier intercepting humid winds from the Pacific Ocean and defining a narrow western band (windward), with hyperhumid to humid climates, and a wider eastern area (leeward), where subhumid, semiarid, and arid climates prevail. This band extending to the east of the Andes is known as "extra-Andean Patagonia," where desertification processes are more intense.

The extra-Andean Patagonia covers an area of approximately 550,000 km^2; it is part of the Argentine Patagonia and comprises the provinces of Neuquén, Río Negro, Chubut, Santa Cruz, and Tierra del Fuego (Fig. 17.1). The relief is characterized by mountain ranges and plateaus of gradually west-east descending slope, from approximately 900 m in the cordillera piedmont to 150 m.a.s.l. on the Atlantic coast. Geologically, it is composed of different provinces (Fig. 17.2): the Neuquén Basin (Digregorio and Ulliana, 1980), the Patagónides (Frenguelli, 1946), the Massif of Somún Curá (Stipanicic and Methol, 1972), the Massif of Deseado (Leanza, 1958), and the Patagonia Plateau (Nágera, 1939; Ramos, 1999). The Patagonia Plateau unit represents the typical physiognomy of the extra-Andean Patagonia (Fig. 17.3, photograph); it is characterized by elevated plains composed of friable sedimentary rocks, partially covered by Cenozoic basalt flows. Many plateaus have a gravel cover of Plio-pleistocene age, traditionally called Patagonian gravel (Fidalgo and Riggi, 1965). This tabular topography is dissected by wide fluvioglacial valleys of prevailing west-east direction.

Precipitation decreases abruptly from west to east (Fig. 17.4). Annual rainfall in almost the whole extra-Andean Patagonian territory amounts to less than 250 mm, with absolute values below 100 mm in the central region. Rains are irregularly distributed throughout the year; winter precipitation prevails, especially in the northern and central areas up to 46° S, where vegetation is scarce. Arid environmental conditions also increase, with permanent strong winds from the west and southwest blowing across the region, which are more intense in spring and summer and thus favor evaporation.

Because of the huge latitudinal extension, temperature variations are very noticeable in Patagonia, from temperate to cold climate conditions from north to south, with annual means ranging between 14° and 4° C, respectively. Temperature and especially precipitation characteristics determine a dominant steppe vegetation in extra-Andean Patagonia, which has adaptive features to arid conditions. The most representative vegetation types are the grassy-

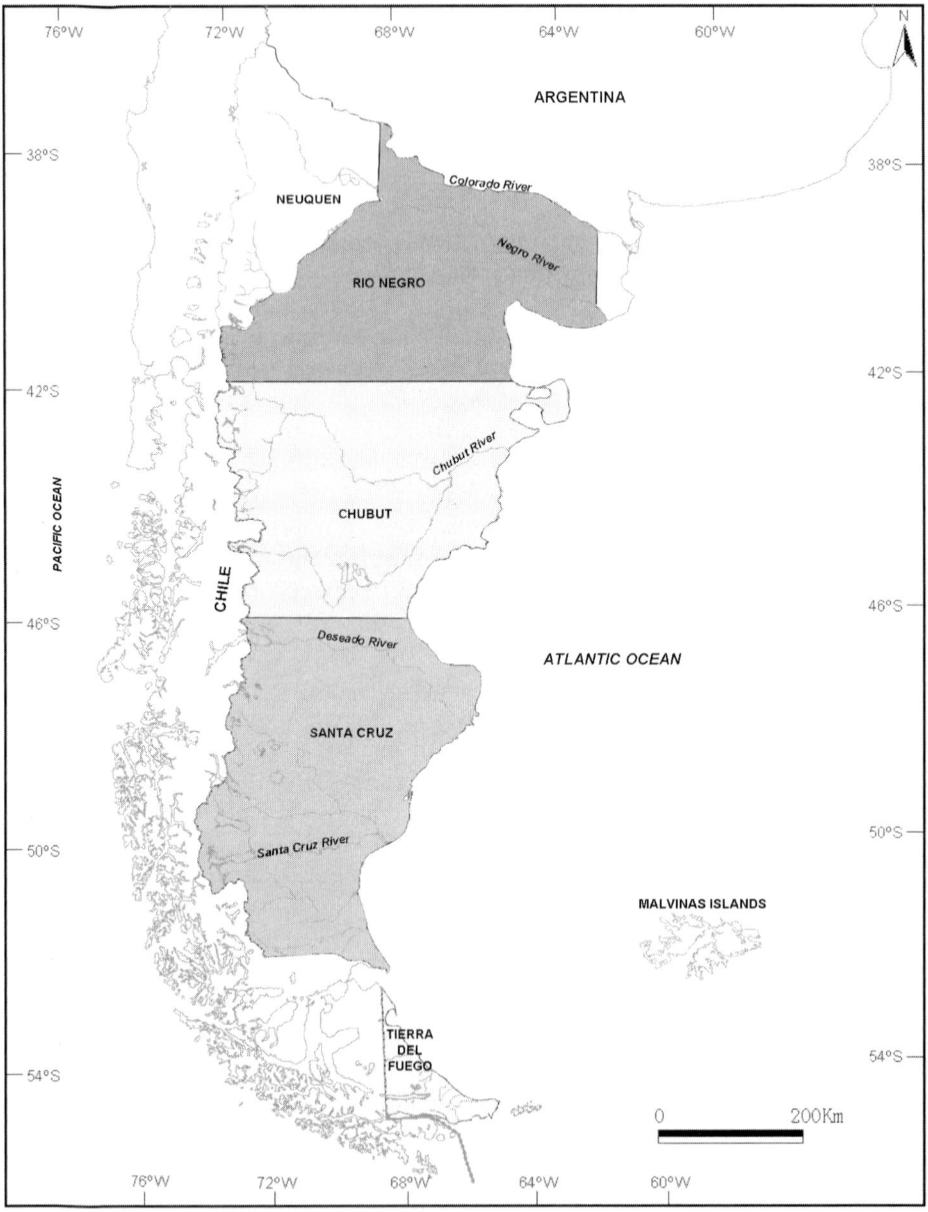

Figure 17.1 Argentine Patagonia, political division.

shrubby steppe of medium height (20–80 cm) and density (1 shrub every 6 m^2), and dwarf cushion shrubs (5 to 20 cm) with scarce grasses and very low total cover (León et. al., 1998). Vegetation in these ecosystems is distributed in a heterogeneous horizontal pattern of plant patches alternating with bare soil

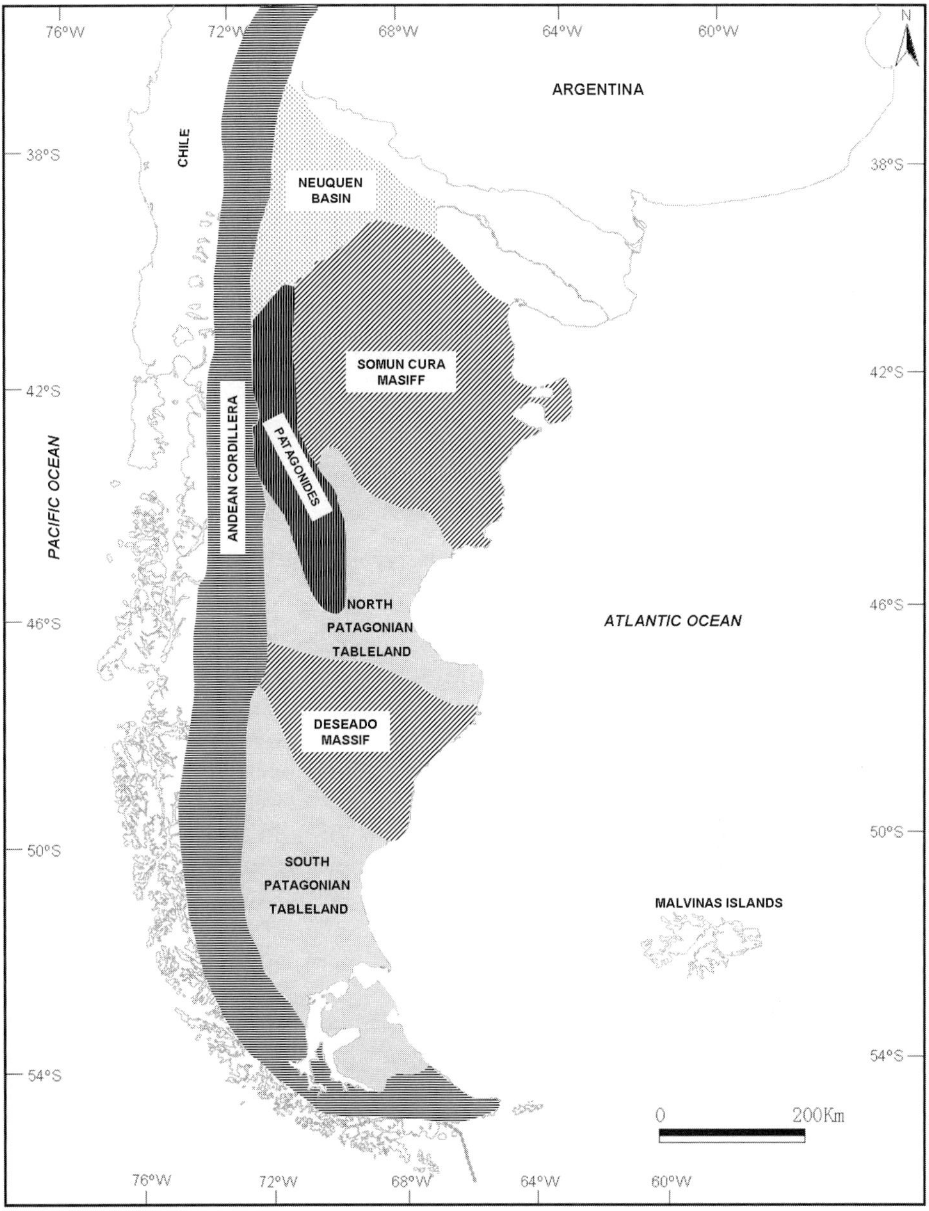

Figure 17.2 Extra-Andean Patagonian Geological Units. *Source:* Geología Argentina. Anales 29. SEGEMAR. Buenos Aires.

areas (Noy Meir, 1973). Wet grasslands (locally known as *mallines* or *vegas*) develop, associated with valleys and water outcrops, and are composed of cyperaceous and graminoids of high plant productivity (Movia, 1984a). As

Figure 17.3 Overview of the Patagonian tablelands. (Photograph by Mazzoni)

will be shown later in this chapter, these areas are heavily grazed and are therefore highly vulnerable to water and wind erosion processes.

Climatic characteristics largely influence the ecosystem structure and functioning in Patagonia, mainly through their effect on water dynamics (León and Facelli, 1981; Bertiller et al., 1995). Relative abundance of grasses and shrubs varies with the amount and annual distribution of precipitation: shrubs increase as precipitation decreases and the proportion of rain in winter increases, whereas grasses increase with increasing precipitation (Paruelo and Lauenroth, 1996). Another important functional group, herbs, does not respond to regional climatic variables; its distribution is influenced by local-scale factors, such as landscape structure, edaphic characteristics, or land-use history (Jobbagy et al., 1996).

Environmental conditions notably influence land use in Patagonia. In the extra-Andean region, the human population is concentrated mainly in a few localities situated on the Atlantic coast. Population density throughout Patagonia is 1.9 inh/km^2; however, density in rural areas is even lower, slightly exceeding 0.1 inh/km^2 (INDEC, 2001).

The main economic activities are associated with hydrocarbon exploitation, tourism, agriculture, and livestock production. Agricultural activities are concentrated in the irrigated valleys (the Negro River valley, the Chubut River valley, and other smaller valleys, like that of Los Antiguos River, in the northwest of the Province of Santa Cruz). Livestock production, especially sheep, is extensively conducted throughout the extra-Andean environment.

Rural land subdivision differs noticeably between sectors. Environmental and cultural differences influence both the extension of ranches and the land tenure system. In the southern portion, medium to large-sized ranches prevail

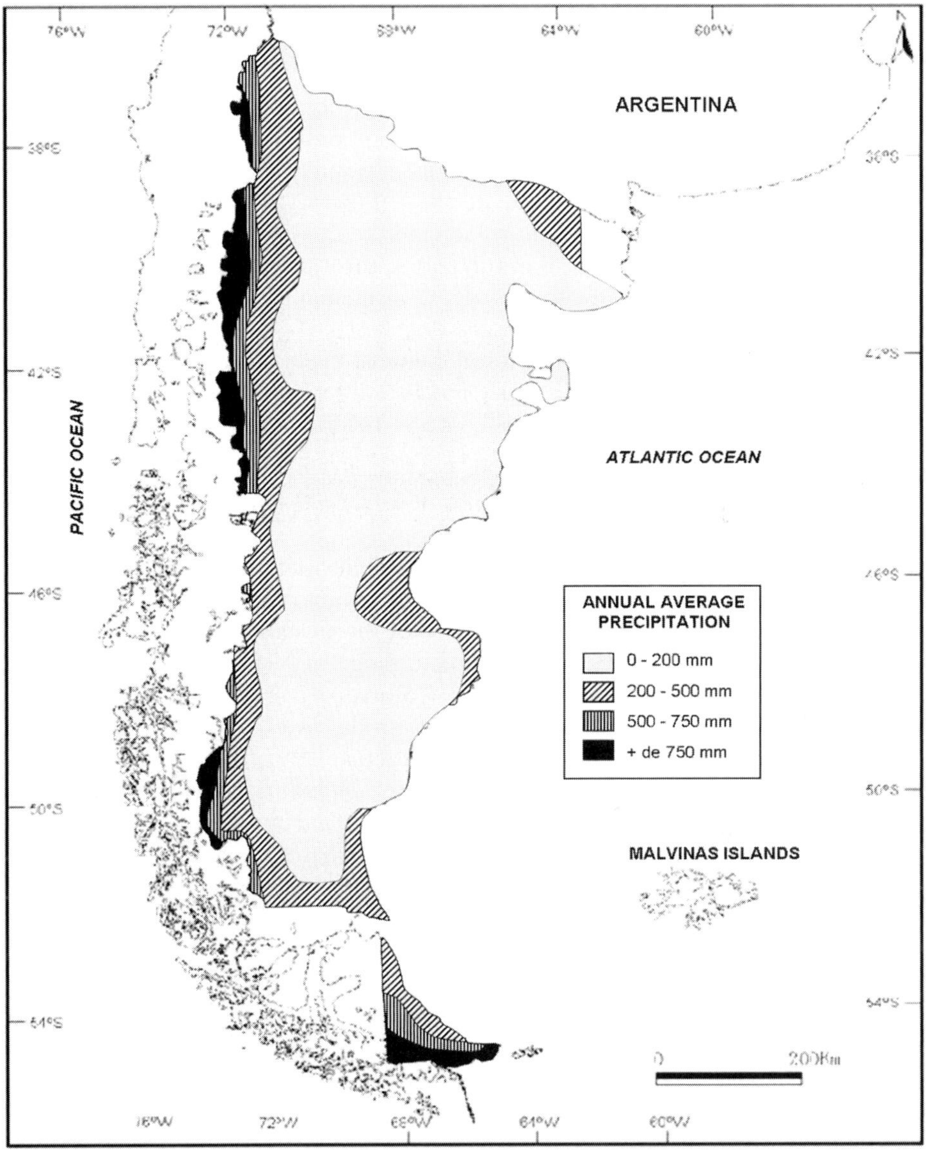

Figure 17.4 Mean annual precipitation in Patagonia. Modified from *Atlas total de la República Argentina*. Centro Editor de la República Argentina. Buenos Aires. 1982.

(15,000–20,000 ha), whereas in the northern area agricultural units are characterized by their small size—smallholdings—with aboriginal rural populations that conduct transhuman practices (Bendini et al., 1990).

3. Desertification in Patagonia

Occupation of the Patagonia rural area started by the end of the nineteenth century with the settlement of wool growers in different areas of the plateau. Between 1880 and 1950, sheep stock increased rapidly from 1,790,000 head in 1895 to more than 25,000,000 head in 1952 (Huerta, 1991). The international demand for wool at very convenient prices and the easy production and conservation processes consolidated a wool monoculture throughout the region, supported by a high profitability of sheep breeding, estimated at 30–40% (Huerta and Sarmiento, 1989).

Grazing pressure caused plant and soil degradation in a few decades (Barbería, 1995). Not only environmental fragility but also different political, economic, and administrative factors contributed to such degradation:

- Despite technicians' recommendations, land subdivision in an orthogonal pattern barely took into account the distribution of water and summering and wintering fields (high and low areas) (Barbería, 1995).
- The process of conversion of private lands. The most productive lands were given to a few influential individuals or partnerships, who settled in the area as absentee landlords. By 1900, the property of the fields located in the Magellan steppe (south of Santa Cruz), the mountain range of Neuquén, Río Negro and Chubut, the valleys of the Negro and Colorado rivers (Fig. 17.1), extensive coastal areas near harbors, and the most productive "oasis" (meadows) had already been given (Bandieri, 1990; Barbería, 1987; Dumrauf, 1992). In contrast, most rural workers and inhabitants, including aboriginal and *criollos* (native) producers, occupied the less productive and more isolated areas, with plots of insufficient size or ecological conditions to sustain their productive systems at a suitable functioning level. This generated a structural pressure that resulted in a trend toward maintaining the maximum livestock stock possible (Peralta, 1995).
- The low level of technology applied in livestock management that permitted uncontrolled grazing throughout the fields (Barbería, 1995).
- The fact that land receptiveness to ranch management was not taken into account. Current studies estimate that the livestock stock should not have exceeded 0.2 sheep/ha. However, during the most profitable times, this parameter was 0.45 sheep/ha in average (Oliva et al., 1995).

The evidence of the land degradation process induced by grazing in extra-Andean Patagonia was detected early. The first changes in the ecosystem were evident in the floristic composition and plant cover. In 1904, Clemente Onelli stated: "[I]f an excessive number of animals is allowed to graze, the results are negative; then, the fields are destroyed and sometimes need five to six years of rest to be used" (cited by Barbería, 1995:117). Likewise, in 1914 Bailey Willys described overgrazing in Patagonian fields with notable accuracy, warning communities about the situation and the need to apply rotational grazing techniques: "{I]n summer and winter, year after year, sheep are taken to graze in the same prairie; edible plants barely have the opportunity to develop or propagate, whereas weeds multiply everywhere" (Willys, 1914).

By the middle of the twentieth century, several researchers recognized the signs of land degradation and associated them mainly with overgrazing by sheep: Auer and Cappannini (1957), Soriano (1956), and Amigo (1965), among others, reported the effects of degradation on vegetation. These authors' works and many other studies on different regions of extra-Andean Patagonia led us to conclude that vegetation degradation is evidenced mainly in the gradual substitution of palatable grasses for unpalatable, woody species (Paruelo et al., 1993), combined with an increasing percentage of bare soil.

Besides degradation produced by overgrazing by sheep, shrub removal contributed to the clearing of extensive areas (Andrada, 2002). Clearing was at first done for fuel production for family use and also as an alternative source of income, such as fuel supply for cookers, heaters, and boilers for mining, oil, and railroad camps, as well as military bases or towns.

Since 1960, the development of mining and hydrocarbon activities in the region has contributed to an increased level of environmental degradation, which was also increased by the explosive growth of petroleum activities in the 1990s. Exploration, exploitation, and storage activities generated water pollution in the aquifers, especially the groundwater table, and damage to livestock production activities through broken fences, losses of animals, and depreciation of wool because of direct contact with hydrocarbons (SA and DS, 2000). Similarly, the clearing of lands with the aim of generating spaces for the establishment of petroleum locations and the wide network of roads that communicate to them have resulted in important areas of bare soil that are exposed to wind erosion, rain, and surface runoff. In these areas, "deflation tongues" are generated, which constitute one of the most active aeolian forms in Patagonia (Movia, 1972).

At present, the whole extra-Andean Patagonia territory exhibits a diverse degree of desertification. This degradation is evidenced explicitly in biophysical components of the ecosystem, but it also has social and economic consequences. Since the middle of the last century, together with these degradation processes, the livestock number has gradually decreased, jobs have been lost, and rural populations have increasingly been migrating to urban centers. Together with this regional scenario, internationally the price of wool is trending downward while the price of inputs is increasing. This has led to a strong decrease in the profitability of livestock production, the consequent closing and abandonment of ranches, and the emigration of the rural population (Andrada, 2002).

4. Evaluation of the Status of Desertification in Patagonia

Evaluation of the status of desertification in Patagonia started in the 1990s, with the aid of information provided by satellite images of different spatial resolution and physical and biological indicators observed in the terrain (Mensching,

1996; Abraham et al., 2003). The area affected by desertification processes throughout the region was mapped based on the estimation of vegetation indices on NOAA–AVHRR LAC satellite images, with a spatial resolution of 1.000 m × 1.000 m and detailed field surveys (Del Valle et al., 1998). This allowed us to obtain a final map at a scale 1:1.500.000, whose simplification is presented in Figure 17.5. Five desertification categories were defined (slight, moderate, moderate to severe, severe, and very severe) based on different indicators, adapted to methods described in FAO/UNEP (1984) (Table 17.1).

Of the whole area studied, 93.6% (73.5 million ha) shows some signs of desertification, moderate to severe and severe being the most abundant desertification categories, which cover almost 60% of the territory. The severe and very severe categories correspond to highly degraded lands, where environmental damage is irreversible for the development of most of the economic activities (Fig. 17.6). The provinces included in the two latter categories are: Santa Cruz, with 38% of its territory, Neuquén, 37%, Chubut, 31%, and Río Negro, with 26%.

A more detailed evaluation of the state of desertification was conducted in several sectors of Patagonia (Ayesa et al., 1995; Del Valle et al., 1995; Elissalde et al., 1995, Oliva et al., 1995), which describe the environmental conditions and the main degradation processes affecting different landscape units. The overall analysis of these works allowed us to make some general considerations on the spatial distribution and magnitude of the land degradation processes affecting Patagonia. Water and wind erosion are the geomorphological processes that are most strongly evidenced. Water erosion is specifically evident in the mountain environments, especially those located in the western portion of extra-Andean Patagonia (Patagónides), where annual precipitation is higher and the moderate slopes favor surface runoff (Del Valle et al., 1995). In the plateau environments, the most significant water erosion effect is evidenced in the contact zone between different terraced levels, where gullies usually of several meters in depth appear. In the valleys, which are the most densely populated areas in the region, the intensified river erosion processes and the damage generated by them are related to the management of water resources: irrigation projects, dams, construction in unsuitable places, alteration of original drainage system, and so on. Soil salinization becomes important in these geomorphological units (Mazzoni and Vazquez, 2004).

On the other hand, wind action is widely evidenced throughout the region. Landforms resulting both from aeolian erosion and accumulation are present in the different landscape units of the region under study. Their development is favored by several factors, including climatic, geological-topographic, edaphic, and land-use considerations.

Climatic factors are related to the location of Patagonia in the belt of permanent westerly winds and the shadow effect of the Andes, which capture moisture coming from the Pacific Ocean. Thus, the extra-Andean environment is subjected to the action of strong dry winds that blow across the region mainly in a west-east direction, removing the substrate particles and transporting them in that prevailing direction. The arid climate, in turn, has been an obstacle to the development of deep and structured soils.

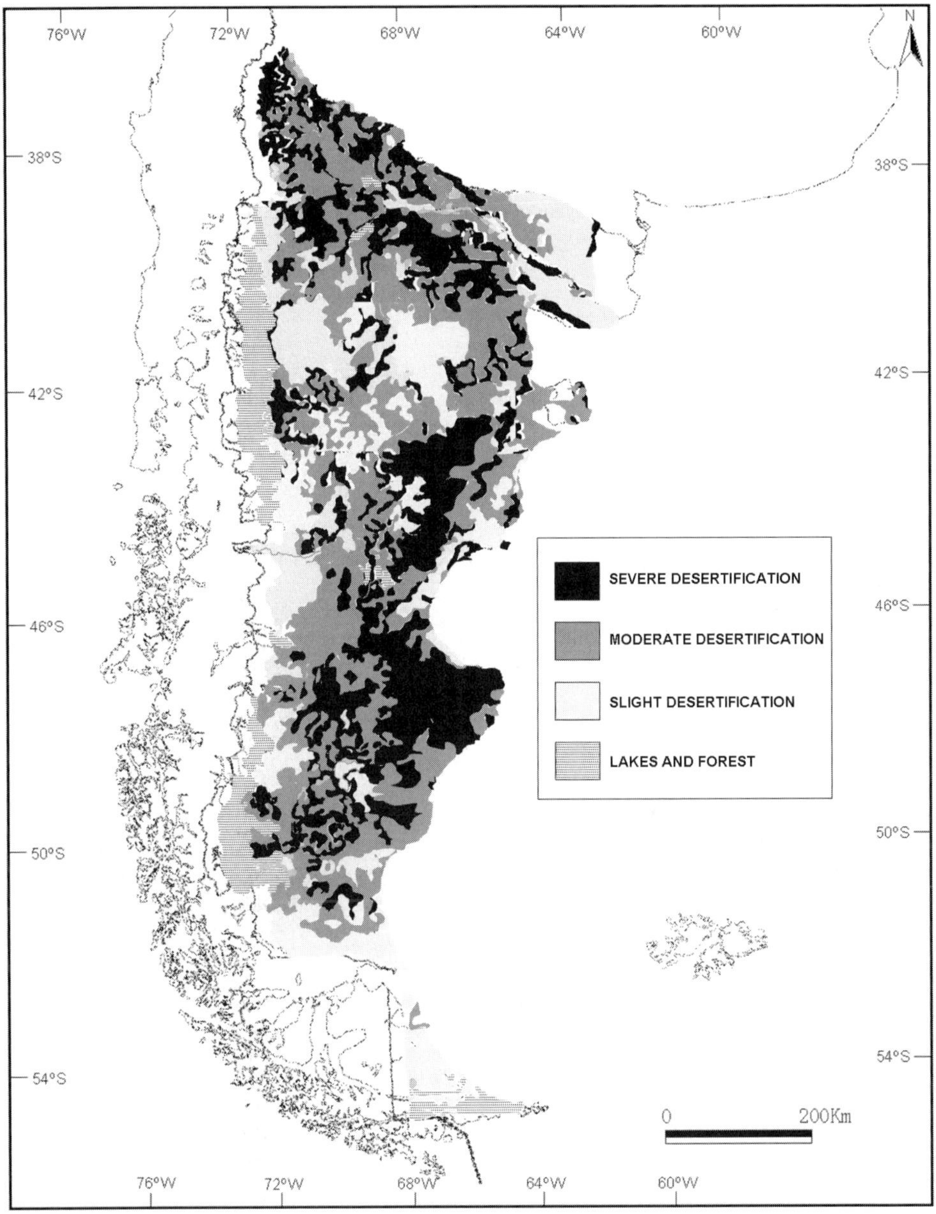

Figure 17.5 Desertification states in extra-Andean Patagonia. Modified from Del Valle, 1998; Del Valle et al., 1998

The subhorizontal relief, with a lack of significant topographic barriers, favors wind circulation. Soils of thick texture with poor plant cover provide material for transport. Human activities in this environment, mainly extractive ones, also favor the supply of materials vulnerable to wind transport. With regard to the effect of

Table 17.1 Desertification in Patagonia

Desertification States		Area	
		Ha	%
Slight	More than 50% of perennial plant cover. More than 30% of forage species. Less than 10% of bare soil. Less than 10% of area affected by slight to moderate water or wind erosion. Less than 5% of the area negatively affected by salts.	7,318,600	9.3
Moderate	Between 20 and 50% of perennial plant cover. 10–30% of forage species. 10–25% of bare soil. 10–25% of the area affected by moderate to severe water and wind erosion. 5–20% of the area negatively affected by salts.	13,503,800	17.1
Moderate to Severe	Less than 30% of the area in slight conditions and more than 40% in moderate to severe conditions. Less than 30% of the area in very severe state.	27,781,600	35.4
Severe	5–20% of perennial plant cover. 5–10% of forage species. 20–50% of bare soil. 25–40% of areas with severe water and wind erosion. 20–50% of the area negatively affected by salts.	18,235,800	23.3
Very Severe	Less than 5% of perennial plant cover. Less than 5% of cover with forage species. More than 50% of bare soil. 50% or more with very severe water or wind erosion. More than 50% of the area negatively affected by salts.	6,704,500	8.5
Total of desertified areas		**73,544,300**	93.6
Bodies of water streams, ice, and forest		**5,005,200**	6.4

Source: Del Valle, 1998; Del Valle et al., 1998; http://www.fao.org/ag/agll/lada/arg/Archivos/Degradación/Degradación.htm#Patagonia

Figure 17.6 Very severe desertification at East center of Santa Cruz Province. (Photograph by Mazzoni)

grazing, some authors estimate that this activity produces an acceleration of the natural deflation process of 100 to 200% through a decrease of plant cover and degradation of surface soil structure by trampling (Del Valle et al., 1995).

As a result, the landforms originated by wind action, both by aggradation and erosion, are present throughout the extra-Andean Patagonia. Among the former, the most spectacular forms are the "sand tongues" (Movia, 1972). The presence of accumulations in the form of mounds, nebkas, and thickenings are also frequent (Laya et al., 1984). The dune fields appear only in an isolated way; they may be made up of sand or salty silt-clay sediments that group in fields of small "lunettes" (Bowler 1973). The latter are usually associated with deflation basins. Sand dunes have been described for the southern portion of the Valdés Peninsula (Chubut), where the advance of numerous dune fronts is recorded (Del Valle et al., 2000). They cover approximately 590 km^2, and their average speed, estimated by the comparative analysis of information provided by remote sensors, was about 25 m/year.

The sand tongues, also called aeolian plumes (Mazzoni, 2001), are elongated accumulations oriented in the direction of prevailing winds. They are the aeolian landforms of greatest dynamics, which may advance up to 2 km/year. The typical shape usually consists of an active front or moving dune followed by a sand sheet, alternating with several proportions of pavements. The length varies considerably, reaching more than 50 km. When the active distal portion is very far from the origin, part of the pavement may recover its plant cover, but with species different from the original ones. In general, a loss of gramineous species and forage shrubs and an increase of unpalatable subshrubs with a percentage of cover below the original occur (Salomone and Schenkel, 2002). Sand tongues originate in several areas devoid of vegetation: usually depressions, lake and lagoon edges, degraded meadows, cattle tracks, roads, and so on. A dynamic analysis has demonstrated that they may vary in their shape and composition in time, alternating between the typical deflation and accumulation shapes, depending on whether the supply of sand varies seasonally, especially if the origin is connected with a closed basin whose water level has important fluctuations (Mazzoni et al., 2002).

Besides the typical sand tongues, other types that are recognized include the following (Movia, 1984 b):

(a) *Of dominant accumulation*: in general, they have a well-developed moving dune with a more or less deep sand sheet from the origin. They are relatively short (up to 5 km). If their size increases, they turn into a "typical" type. The front may be divided by obstacles or vegetation conditions.
(b) *Of dominant deflation*: they are composed of a gravel pavement with much reduced or almost absent dune front. They are generally wide and relatively short in shape. They usually originate in shallow depressions and coincide with highly pebbly areas with low vegetation cover.
(c) *Mixed sand tongues*: deflation forms developed on clay-silt materials with a low proportion of gravel. Dune fronts are usually absent. They are complex, since sheet erosion acts in combination with deflation. The process is accelerated by trampling. They are generically of very narrow, straight, and long shapes.

Analysis of the internal structure of deposits usually shows a sequence between sedimentary and organic material that would show alternating periods of accumulation and others of relative stability that allow vegetation growth (Mazzoni, 2001). Thickenings (Laya et al., 1984) are accumulations that present in the form of a sheet covering relatively extensive areas, generally of more than 50 cm in thickness. They may act as soil-originating material. Mounds, however, are smaller and disperse, of relative heights, generally of a few tens of centimeters. They often develop at the base of plants. Nebkas are similar to mounds in shape, but of somewhat larger dimensions. Rostagno and Del Valle (1988) explained that both aeolian and alluvial processes intervene in their formation, surface sheet wash being the main process involved. Thus, nebkas or mounds associated with shrubs would appear as a residual form of sheet erosion that occurs around tussock grasses. The presence of thick fragments (similar to those making up the intermound area) as part of the internal structure of mounds would seem to confirm such an hypothesis.

Among the forms of aeolian erosion, desert pavements appear as a result of the loss of fine soil materials transported by deflation. They constitute residual concentrations of gravel located in different landscape units, mainly on sedimentary plateaus, piedmonts, and fluvial terraces. In areas with severe desertification, like the central areas of the Province of Santa Cruz, they are several hundreds of km^2 in size (Oliva et al., 1995). At the microscale, pavements also appear in the inter-nebkas areas, with different degrees of

Figure 17.7 Extensive desert pavements in the central area of Santa Cruz Province. (Photograph by Mazzoni)

development. On these slightly concave surfaces, clasts may cover between 20 and 90% of the soil surface (Súnico et al., 1996). Among the dominant forms of erosion are deflation basins and desert pavements developed in different geological and morphological units. Clasts frequently appear as faceted and abraded (Fig. 17.7).

Loss of the topsoil by deflation or by the combined effect of aeolian processes, erosion by raindrop, and nonchanneled surface runoff (sheet wash) may reach significant dimensions. The common presence of "uprooted plants" with the root system exposed is a consequence of this phenomenon (Fig. 17.8). Súnico et al. (1996) estimated that this process is responsible for the loss of topsoil in soils classified as Typical Natrargids and Typical Haplargids in the northeast of the Province of Chubut, where the clay B horizon (argillic) appears exposed on the surface or covered by a thin sand layer. This exhumed upper limit of argillic B horizon usually constitutes the substrate of erosion pavements.

Deflation basins, usually called "bajos sin salida" (closed basins) (Fidalgo, 1972), have a wide distribution in extra-Andean Patagonia. They are depressions of variable shape and size, whose genesis is attributed to multiple processes, among which aeolian action is significant (Feruglio, 1949; Groeber, 1953; Auer, 1956; Methol, 1967; González Bonorino and Rabassa, 1973; Clapperton, 1993, Laity, 1994, among others). Their morphometry is variable, depending on their lithology. Closed basins can be several kilometers in diameter and a few tens of meters in depth. They are very important in the region because they are water-concentrating points that retain rain-snow

Figure 17.8 "Uprooted plants" and mounts, formed by combined effect of hydric and aeolic (wind) erosion processes. (Photograph by EEA Santa Cruz INTA)

precipitations. However, most of the ponds present inside them have a temporary regime. As the body of water restricts its dimensions, especially through evaporation, the base of the basin becomes a source of materials for aeolian transport, a process that contributes to the deepening of the depression. From the base of these closed basins, a great part of the aeolian plumes present in Patagonia is generated, making up a greatly dynamic "closed basin-sand tongue" system.

5. Spatial Distribution of *bajos sin salida* (Closed Basins) and the Dynamics of "Aeolian Plumes" in the South of Santa Cruz

The southern portion of Patagonia and the Andes environment have a slight level of desertification. Land degradation is evident in alterations of plant cover (Oliva et al., 1998) and in the development of deflation tongues, whose formation is associated with closed basins. These landscape traits are not as important as in other sectors of Patagonia (tens of km). However, most of them extend for several meters and make up active landforms.

Topographic depressions (*bajos sin salida* or closed basins) are a common feature in the regional landscape. Mazzoni (2001) counted 220 closed basins above 5 ha in size in an irregular area of 5600 km^2, located between 51° 14' and 52° S and between the Atlantic coast and 72 ° W. These depressions of different morphometric characteristics are present in morphological units formed by plateaus of sedimentary and volcanic origin and sediment of glacial origin (landscape of moraines) where the density of depressions is 0.05, 0.03, and 0.17 closed basins per km^2, respectively. However, the size of depressions varies significantly between units: the sedimentary plateau has the basins of greatest dimensions, whose maximum size exceeds 10,000 ha. Medium values obtained for each unit are 636, 246, and 52 ha for sedimentary plateau, volcanic plateau, and glacial relief, respectively. The modal interval in the three cases is below 25 ha, indicating the predominance of small-sized basins.

Aeolian plumes are present in 54% of the closed basins located in the plateau landscapes, whereas in glacial moraines this amount drops to 5%. The highest frequency of plumes is found to the east of 71° W, where environmental aridity decreases (Mazzoni, 2001). During 1986–2000, plume lengths measured on Landsat TM satellite images increased in 39% of the closed basins, with an average increase of 1.52 km and a maximum of 3.6 km. The length of 46% of the plumes decreased, with an average of 0.98 km, whereas the maximum reduction value was 2,4 km. The remaining 15% was stable (Mazzoni et al., 2002).

Based on this variable behavior of aeolian landforms in a climatically homogeneous environment, we can infer that a great number of aspects are involved in their dynamics. In some cases, a close relationship between fluctuations of the body of water and changes in the shape, size, and structure of the plume has been confirmed, whereas in other cases this association is not so evident. In contrast, local variables such as availability of material for transport (associated with the depth and topography of the depression) and land use may have a strong influence on the behavior of these landscape forms (Mazzoni et al., 2002).

We have observed the plume behavior in a closed basin of an area of 2600 ha located in fields of Bella Vista ranch (51° 55' S–70° 31' W) since 1999. The depression is located in an environment of volcanic plateau, formed by a vast Tertiary basalt flow dissected by meltwater from glaciations that affected the region during the Pleistocene (Mercer, 1976, Meglioli, 1982). The basalt flow covers the depression under study almost completely; its base has a height of 110 m.a.s.l., some 80 m below the plateau edge. An aeolian plume extends from the eastern border of the lake occupying the basal portion of this basin. Its length has shown a gradual increase from 1968 (oldest photographic record) to the present, ranging from 0.6 km (1968) to 4.1 km at present. Its distal portion, located leeward the basaltic border, is made up of 2-m thick sandy deposits (Fig. 17.9).

Based on the incipient plume development detected in 1968, we can infer that its formation does not date back longer than that date and is correlated with a decade characterized by low-rainfall records (Cibils, 2001). Unfortunately, there are no previous photographic records available to confirm such hypothesis.

The plume area is 250 ha. In the period studied, there have been changes in its shape and in the sediment distribution and thickness, showing the extraordinary dynamics of these landforms. It has been represented in maps showing the situation recorded in different temporal sections. In general terms, it was observed that when there is water in the lagoon or when the substrate is humid, the sediments forming the plume migrate to the east, increasing the size of the deflation area because of the limited sand supply. In contrast, when

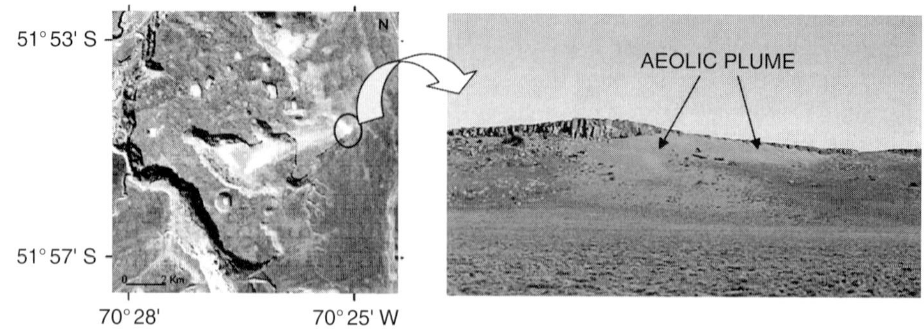

Figure 17.9 Landsat image of "Bella Vista" closed basin (left) and detail of distal portion of its aeolic plume (right).

the closed basin is empty, a notable increase of the accumulation area occurs, forming a continuous sand sheet, with surface microdunes about 10 to 20 cm in height (Fig. 17.10).

Through these observations, it was confirmed that the geomorphological behavior of this plume is closely connected with fluctuations of the body of water. Figure 17.11 shows the situation recorded between May 1999 and February 2002, when important hydrological and geomorphological changes occurred. Abundant precipitation (416 mm) occurred in the winter and summer of 2002, which almost doubled the regional mean, and stopped the accumulation process. Since that time, the loose substrate material was removed from the depression and the plume was revegetated with annual herbs and tussock grasses that have been partially covered by the sandy deposit.

During a period of strong winds recorded in spring of 1999, changes in the morphology of the plume distal portion on the eastern slope of the basaltic plateau (which has an area of 5.7 ha) were quantified. Detailed topographic surveys were conducted on October 7 and November 5, 1999, and maps and 3D modeling of the state of the deposit were generated.

Figure 17.10 Panoramic view showing the accumulation area of the accumulation of the eolic plume of the "Bella Vista" closed basin. (Photograph by Mazzoni)

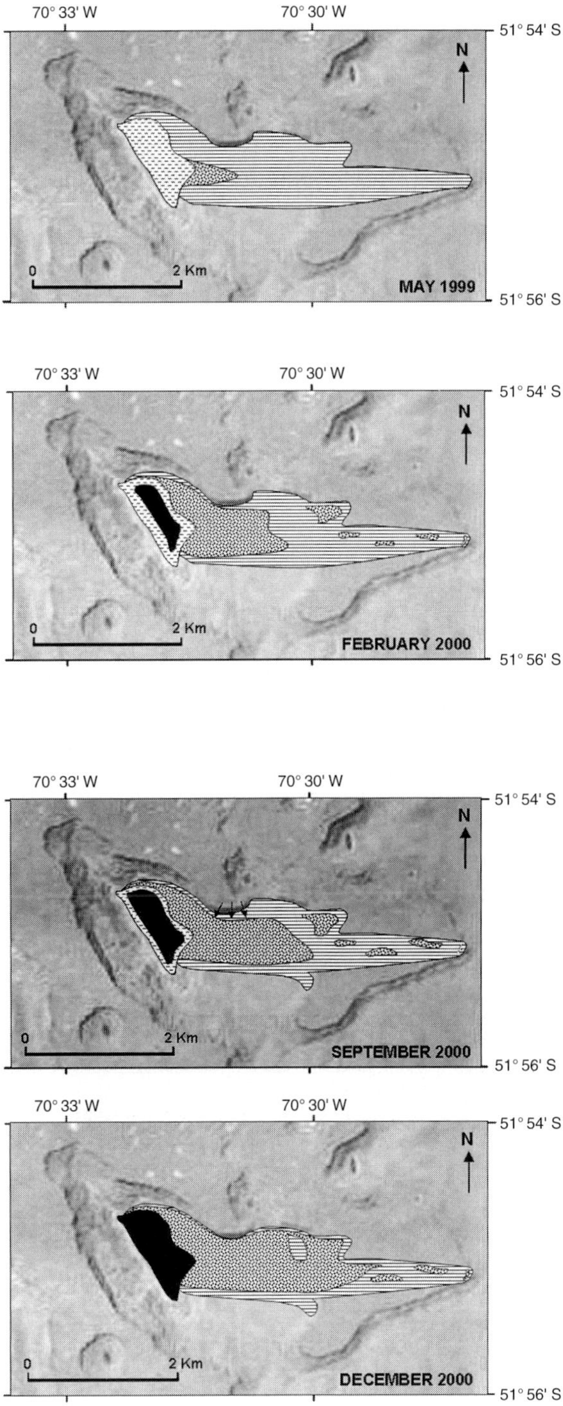

May 1999
The pond was dry. The plume was mostly composed of a sand sheet of 10–20 cm in thickness, with microdunes; it covered 4,100 m, surpassing the borders of the depression and extending across the basaltic plateau. In its distal portion, the sand deposit was 3 m thick in average and covered the basaltic scarp almost completely. The plume showed a central deflation area of 800 m in length inside the depression.

February 2000
The strong winds recorded in spring-summer 1999-2000 brought about notable changes in the landform studied:
The deflation area inside the closed basin is extended and the deposit thickness decreases. Inside the depression, the northern border is the one showing the greatest loss of materials (up to 6 cm). Thickness and continuity of deposit is reduced on the plateau.
An important loss of sand occurred in the distal portion, reaching an average of 1.5 m in thickness.

September 2000
The pond had abundant water. The deflation area of the plume continued to increase. The sand deposit is redistributed to the south. The distal portion remains stable. A notable increase in rainfalls generated rills on the slopes devoid of vegetation on the northern border of the depression.

December 2000
Continuous rainfall and permanent body of water limited sand supply, generating a strong increase of the deflation area and a drastic reduction of the sand volume inside the closed basin. Only a small accumulation area remains in the new southern portion of the plume.

Figure 17.11 Evolution of the plume of the Bella Vista closed basin.

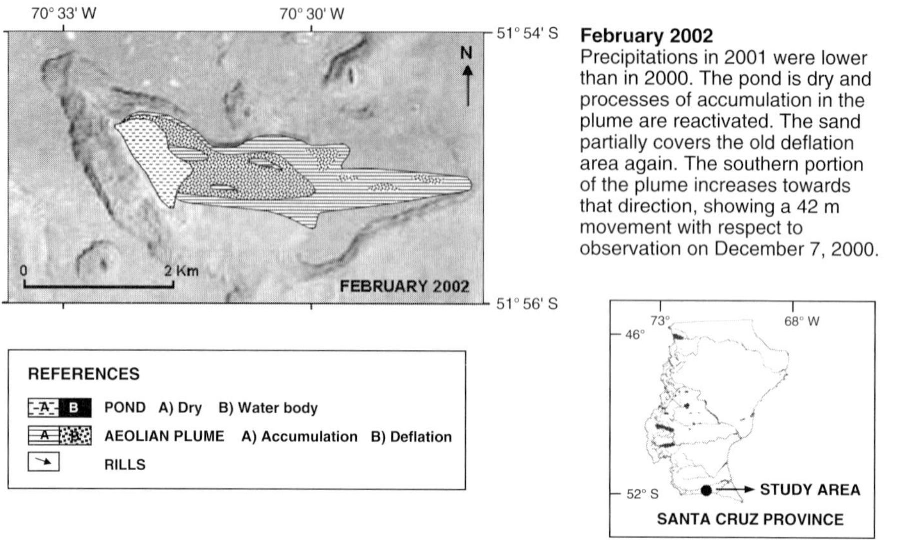

Figure 17.11 (Continued)

The automatic Meteorological Station located near the area (51° 57' S–70° 25' O) recorded prevailing winds from the west-southwest of over 35 km/h during that period, with a maximum mean speed of 70 km/h. Between October 13 and 21, wind was constant, with a mean maximum speed of 96 km/h and an absolute maximum of 122 km/h on October 15.

The results obtained are shown in Figure 17.12. As can be observed, deflation processes prevailed during those days. The total sand loss was 49,770 m^3, approximately 1.15 m in thickness in an area of 4.13 ha. Likewise, material redepositation occurred (6720 m^3) in sectors near the basaltic scarp, with a mean thickness of 50 cm.

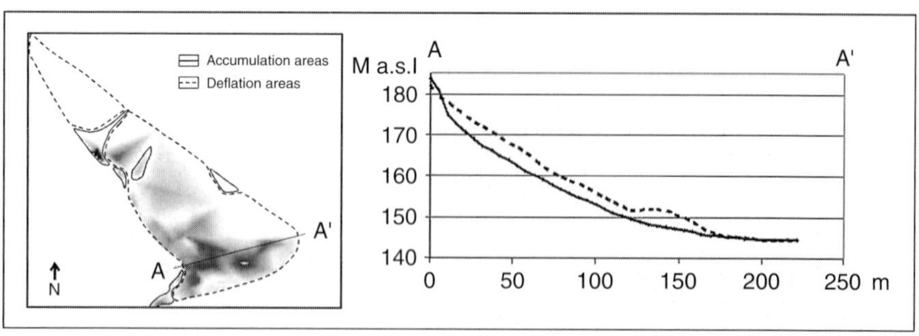

Figure 17.12 To the left, model of the distal part of the plume at the Bella Vista closed basin. Areas where deflation took place after the windstorm are shown with cut line, and those with accumulation are shown with continuous line. To the right, the cut line shown the previous topographic profile of the plume, whereas the continuous line shown the new profile after the windstorm.

6. Degradation of Meadows

The term *mallin* (meaning meadow) stands for humid grasslands of high density and floristic richness, whose genesis is associated with the presence of water near the soil surface. Meadows are landscape units with hydromorphic characteristics, generally of small size (a few hectares). Their spatial distribution is sparse, in response to particular topographic, geomorphological, and hydrological factors (Mazzoni and Vazquez, 2004).

As they are subjected to permanent or temporary flooding, these ecosystems have some ecological similarities. However, there is considerable variability in the types of meadows, depending on factors such as climate, hydrology, salinity, and plant succession (Moore, 1990). Accordingly, they have been classified in different ways. One of the most widely used typology types is based on the content of edaphic humidity: "humid meadows," generally located in Andean areas, and "subhumid meadows," situated in ranges and plateaus areas, with saline or alkaline characteristics. The latter are more vulnerable to grazing pressure (Speck et al., 1982). Water availability inside a meadow and its seasonal fluctuations may have the most influence on plant distribution and general ecosystem dynamics (Horne, 1998; Mazzoni, 2006).

Excessive grazing without suitable livestock management produces plant cover deterioration and soil erosion, occasionally leading to total ecosystem degradation. Not only grazing but also trampling pressure exerted on the soil is involved in this process, especially when the soil is saturated. Livestock weight produces soil compaction, disruption of plant cover, and rupture of the meadow's topographic profile, which generates topographic benches and scarps (Mazzoni, 2006). The bare soil is then exposed to material removal by surface runoff and/or wind action (Fig. 17.13).

Figure 17.13 Scarps developed in subhumid wet-meadows (Photograph by Mazzoni)

Likewise, the degradation process modifies water retention capacity, increasing runoff and favoring gully formation. This process has two negative consequences: on the one hand, it modifies irrigation conditions in the meadow, going from a slow sheet flow to a faster stream flow, with greater erosive potential; on the other hand, the groundwater table drops, with a resulting desiccation of the wetland surface area. This decrease in moisture, coupled with an increase in salinity, produces irreversible changes in vegetation. In semihumid meadows, this process accelerates the transition of *Festuca pallescens* communities and rushes to communities that are more tolerant to salinity, dominated by *Distichlis* sp. (Speck et al., 1982).

The complete destruction of these ecosystems as a result of degradation has been inferred in volcanic environments in the central region of Santa Cruz Province (Mazzoni, 2006). The presence of meadows on slopes of basaltic plateaus has been confirmed in different areas of Patagonia, which provide the water necessary for meadow development. However, basins with shallow ponds and aeolian plumes are observed at those sites that would have been occupied by meadows in this portion of the study area. At those sites where water outcrops occur, at the base of the basaltic edge, only relicts of the original vegetation exist at present. Water is still available, but erosion in the soil, which has lost the fertile layer, hinders the development of plant cover. It should be noted that this is one of the areas most strongly affected by desertification process in Patagonia (Fig. 17.14).

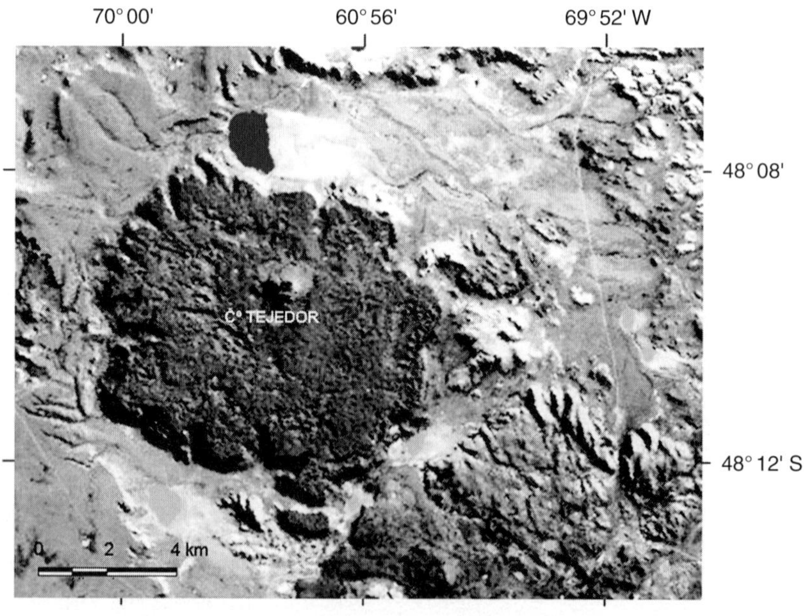

Figure 17.14 Landsat image of a volcanic area in the center of Santa Cruz Province. At the border of the volcanic plateau (Cerro Tejedor), lagoons occupying the areas left by old, now degraded meadows can be observed (Mazzoni, 2006).

7. Degradation Induced by Hydrocarbon Activities

In the areas of intensive hydrocarbon exploitation (identified as Neuquén Basin, San Jorge Gulf Basin, and Austral Basin, Fig. 17.15), degradation processes are a result of clear cutting and soil compaction produced at each

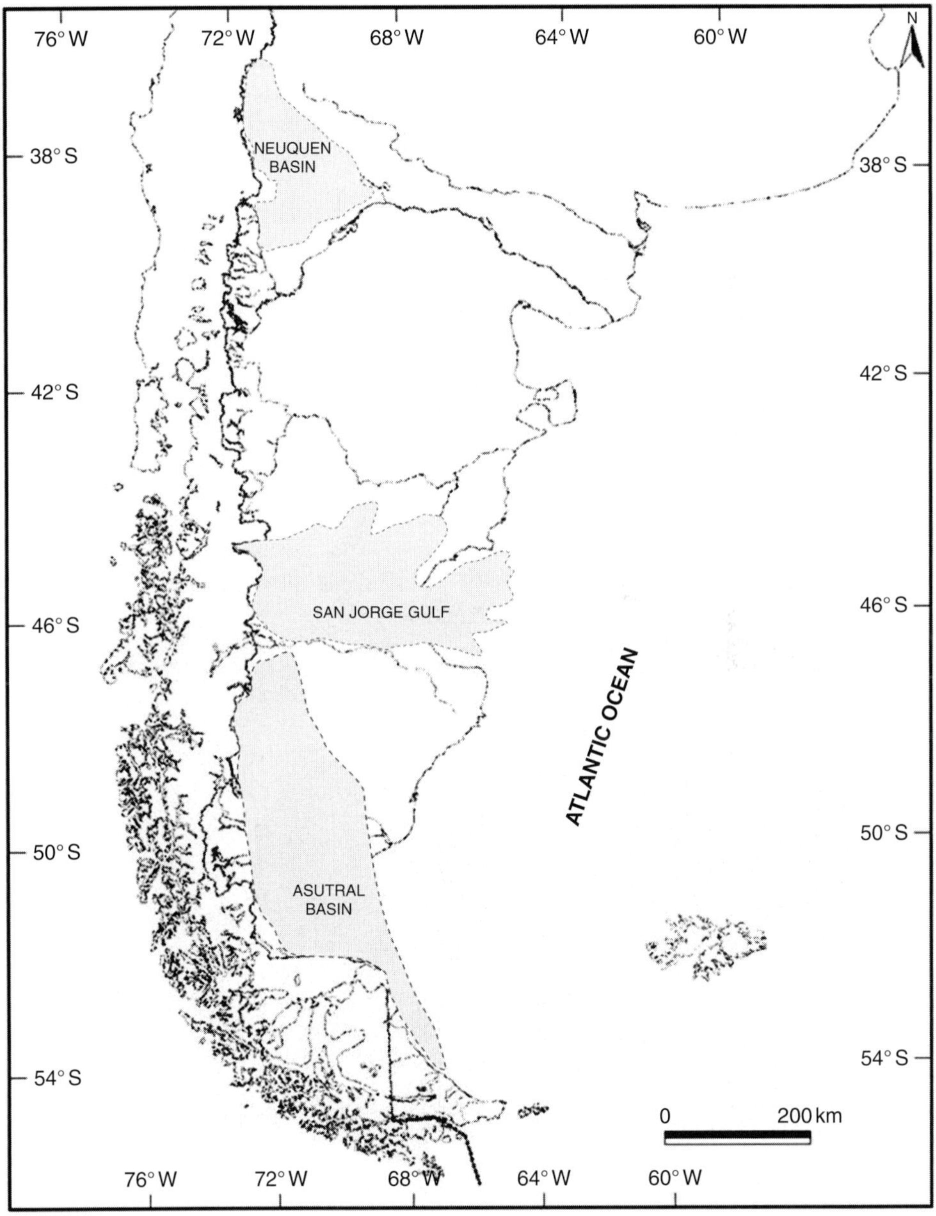

Figure 17.15 Hydrocarbon basins in Patagonia.

petroleum location and in the roads and pipes connecting them (Gambino and Vargas, 2002). Each location has a maximum area of 1 ha. The road sizes range from 10 to 20 m. Each area devoid of vegetation is a potential erosion focus, since the soil is exposed to the action of wind and surface runoff. Rostagno et al. (1999) and other authors have confirmed that in the absence of vegetation, soils tend to form crusts, which increase surface runoff by limiting infiltration, favoring the development of rills and gullies; these have already developed in the less used roads (Fig. 17.16).

Figure 17.17 shows the density of existing petroleum locations in a 250 km^2-portion located to the north of the locality of Pico Truncado, in the basin of San

Figure 17.16 Gully affecting roads. (Photograph by Mazzoni)

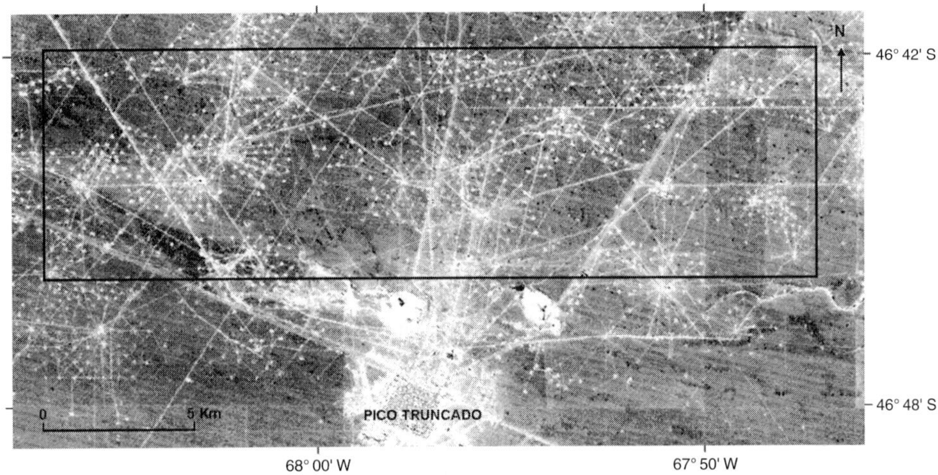

Figure 17.17 Landsat TM Image, band 5, of an area affected by oil extraction activities. The white dots indicate each petroleum location and the roads and pipes connecting them. The box indicates the quantified area.

Jorge Gulf (south of Chubut and north of Santa Cruz). The image was obtained on May 2, 2004. The white dots indicate each location and the roads and pipes connecting them. A total of 1310 locations have been identified in the area, with a density slightly above 52 locations/km^2. This is one of the most heavily exploited sectors in the region. The cleared area was quantified using the digital classification of this portion of the image, totaling 2483 ha, that is, 10% of the area. Given the short history of this activity, the image does not show the presence of aeolian plumes yet; however, they are present and active in nearby areas, associated with old pipelines (Fig. 17.18).

Figure 17.18 General view (left) and detail (right) of an aeolic plume associated with a pipeline in the Ramón Santo area in the limit between Chubut and Santa Cruz provinces. The plume extends over 4.5 km in length.

Although current degradation processes associated with this economic activity as well as with other extractive activities like gold and silver exploitation (under development at present) are still not important, they are undoubtedly a potential problem. The current legislation is considering the rehabilitation of affected areas, but its enforcement is not fully effective.

8. Conclusions

Desertification in the extra-Andean region of Patagonia is a most significant environmental problem, both due to its severity and to the area it covers. The areas classified as irreversible for the development of agricultural and livestock production activities cover 58% of the whole area. As a consequence, rural populations have been partially reduced, and ranches have been abandoned (Rial et al., 1999).

Several authors agree that the main cause of desertification in the region has been overgrazing by sheep (León and Aguiar, 1985; Bisigato and Bertiller, 1997; and others). Grazing was conducted uncontrollably for decades in a fragile environment composed of patches of vegetation surrounded by bare soil. These patches are made up mostly of shrubs and tussock grasses, or exceptionally by meadows where local water is available.

Erosion occurs naturally in the areas devoid of vegetation, but is accelerated by degradation of the plant cover generated by improper grazing practices. Accordingly, it should be noted that overgrazing may occur both by an excessive animal stock and in terms of "opportunity." Paz and Buffoni (1986) state that grazing conducted during the grass-growing season may limit grass development, generating a progressive degradation process. Besides the traditional grazing practices, hydrocarbon and mining activities have been incorporated into the system, with the consequent clearing of extensive areas. Plant removal generated by current extractive economic activities, which have partially replaced sheep production, suggests that the degradation of Patagonia lands will be reactivated in the near future.

Studies on desertification in the region have focused on the diagnosis of the situation, based on mapping the different degrees of desertification at small and medium-sized scales. They are based on the interpretation of satellite images and the evaluation of different indicators in the field. These studies provide static information on the situation, which was evaluated in the 1990s.

Most studies on modifications induced by desertification have focused on alterations in vegetation, with evaluations of changes in time. Models describing the sequence of deterioration have been developed using the "states and transitions" method (Westoby et al., 1989) for the different physiognomic-floristic units of Patagonia. Analysis of these models leads to the conclusion that, in general, the process of plant degradation is associated in most cases with a gradual replacement of palatable grasses by unpalatable woody species and a decrease in biodiversity (Bertiller, 1993; Okada, 1995).

The increase in bare soil percentage contributes to intensified erosion processes, the aeolian and hydrological processes being the most significant ones. As a whole, these processes are evident in the loss of the surface soil horizon by the combined action of sheet wash and deflation. The action of stream flow is restricted to local situations, whereas the landforms of aeolian origin are present throughout the extra Andean Patagonia region, deflation tongues being the most dynamic ones.

CHAPTER 18

Geology and Geomorphology of Natural Hazards and Human-Induced Disasters in Chile

Adriano Cecioni[a] *and* Verónica Pineda[a]

[a]*University of Concepción, Concepción, Chile, Departamento de Ciencias de la Tierra, Casilla (P.O. Box) 160–Concepción – Chile*

Contents

1. Introduction

Chile lies in a N-S direction along 4300 km, from 18°S to 56°S. Its width, between the Pacific Ocean and the Andean Range, does not normally exceed 200 km, reaching a maximum of 400 km. The total area of continental Chile is 746,767 km^2, but the Chilean territory also comprises 1,250,000 km^2 in Antarctica, Juan Fernández Island, and Easter Island (Isla de Pascua). The maximum altitude is 6892 m at the Ojos del Salado, the highest volcano in the world, the second highest elevation located at the Andean Cordillera, and the highest point in Chile.

Corresponding author: *Adriano Cecioni Universidad de Concepción Departamento Ciencias de la Tierra Casilla (P.O. Box) 160–Concepción – Chile, acecioni@udec.cl*

Developments in Earth Surface Processes, Volume 13
ISSN 0928-2025, DOI 10.1016/S0928-2025(08)10018-9

Population totals over 16 million, 86% of which is urban-dwelling; nearly half of those (approximately 6.5 million people) are concentrated in the metropolitan area of Santiago, Chile's capital.

The main morphostructural units from west to east are Coastal Cordillera, Central Depression or Central valleys, and Andean Cordillera. One-third of Chile is covered by the towering ranges of the Andes. As should be expected from the length of the country, morphoclimatic environments vary widely from north to south, from the northern Atacama Desert, the driest place on Earth, to the southern zone with glacial valleys and fiords shaped by Pleistocenie glaciations. The north is characterized, on the whole, by a dry arid climate. The central zone has a Mediterranean climate with warm and dry summers and cold winters with frequent rainfalls. This moderate climate favors agriculture in this fertile region. In southern Chile the climate is colder with frequent rainfall, especially in winter.

Due to its geologic, geotectonic, geomorphic, and climatic characteristics, Chile is exposed to a wide variety of hazardous processes such as earthquakes, tsunamis, volcanism, floods, and mass movements. As the main cities and tourist areas are located along the coast, on the floodplains of rivers, the margins of the hills and near volcanoes, the country presents a high level of risk due to geo-hazards.

2. General Geologic, Geomorphic, and Climate Characteristics of Chile

The geologic constitution of Chile determines its morphostructural features and relief as well as the processes related to lithosphere activity. Interaction between relief and atmospheric circulation controls hydrogeomorphic processes and hazards. A brief description of those features follows.

2.1. Geology, Geotectonics, and Geomorphology

The geological evolution of Chile is the result of the subduction of the Nazca and Antarctic plates, separated by the Chile Ridge, beneath the South American Plate at the Chile-Perú Trench. The dynamic interaction between the active Chile Ridge and the continental margin determines the Taitao Triple-Junction (46°–47°S). North of the triple junction, the Nazca Plate is subducted beneath the South American Plate in an ENE direction at a rate of 8.4 cm/year, while to the south, the Antarctic Plate subducts along an easterly direction at only 2.4 cm/year (Bangs and Cande, 1997) (Fig. 18.1). The most apparent effects of the subduction of young crust from the Chile Ridge are a shallow trench that is nearly devoid of sediment and the rapid narrowing of the forearc region close to the Taitao Triple Junction (Cande et al., 1987). Evidence of subduction processes along the western margin of South America since at least the Triassic has been thoroughly described (e.g., Mpodozis and Ramos, 1989; Cembrano et al., 2007).

The earthquake locations and the determination of focal mechanisms have allowed study of the geometry of the Wadati-Benioff zone beneath the western

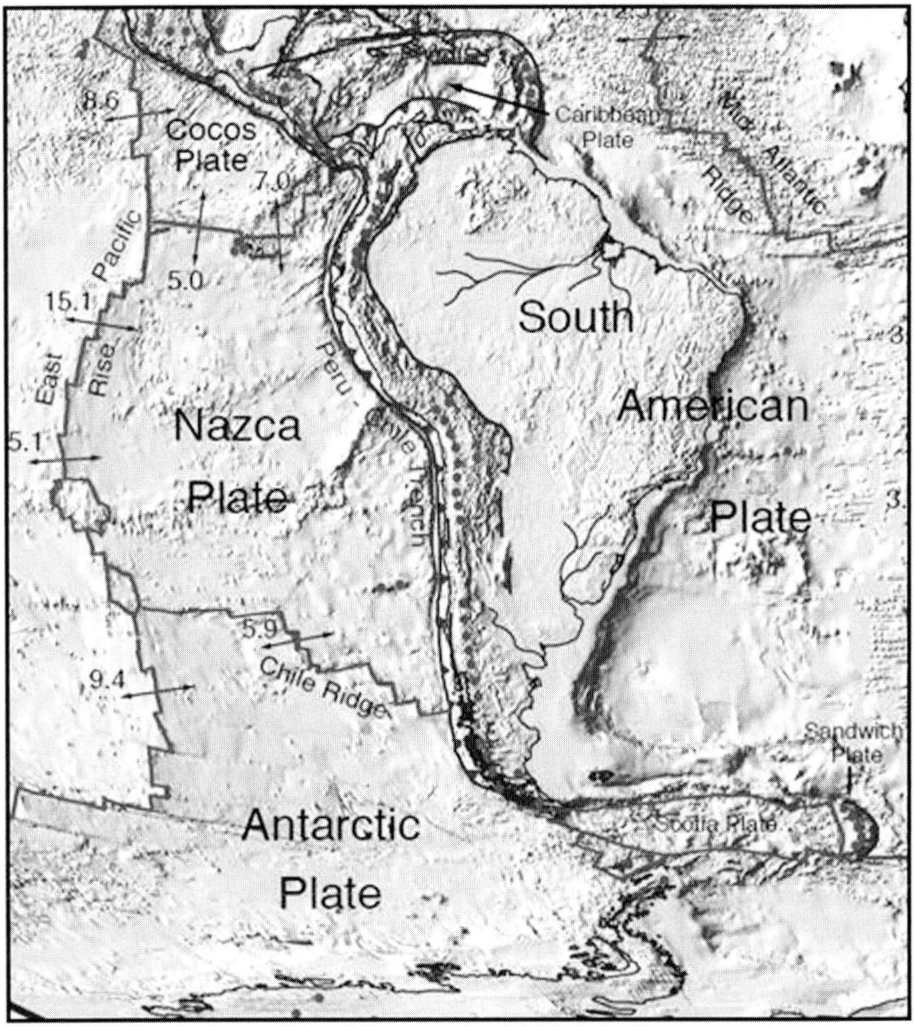

Figure 18.1 Digital tectonic setting of Chile (adapted from NASA/GODDARD Space Flight Center, 2002).

South America margin. The geometry of the subducted slab varies from flat to steep along its extent beneath the Central and Southern Andes. The region where the passive Juan Fernández Ridge is subducting beneath the continental margin (between 27°S and 33°S) corresponds to a flat-slab subduction zone. North and south of this region, the Wadati-Benioff zone is steeper (Cahill and Isacks, 1992).

In the southernmost part of the Andes, the Wadati-Benioff zone shows a lower dip-angle. The Andean Range is the result of this long subduction process, which has taken place at least since the early Mesozoic. In its Chilean part the range has a general NNE–SSW orientation that extends between 18°S and 48°S and has the following main subdivisions (Charrier et al., 2007) (Fig. 18.2).

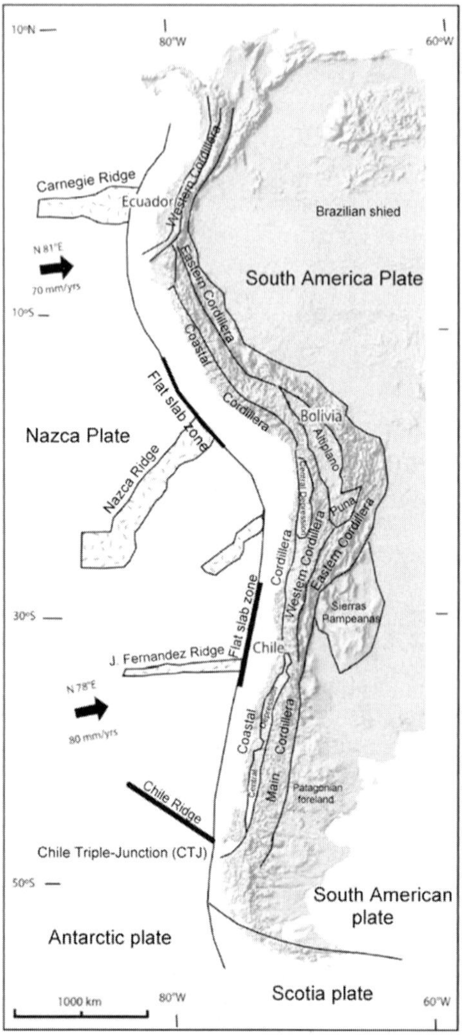

Figure 18.2 Morphostructural features of Chile (adapted from Lavenue, 2006).

North of the Taitao Triple Junction (CTJ), the main morphostructural feature of the Chilean Andes caused by the flat-slab subduction zone is the absence of the Central Depression (27°S–33°S), a morphological unit characterized by transverse river valleys that separates the Coastal Cordillera from the Andean Cordillera. North (18°S–27°S) and south (33°S-46°S) of this zone the Central Depression is well developed (Charrier et al., 2007). On the Chilean side of the Andes, the absence of the Central Depression in the flat-slab segment is associated with the lack of recent volcanic activity. Between 18°S and 27°S, from west to east, there are four main geomorphological units: the Coastal Cordillera, the Central Depression, the Precordillera, and the Western (Main) Cordillera with the Altiplano and Puna.

South of 33°S, the units that can be differentiated are the Coastal Cordillera, the Central Depression, and the Main Andean Cordillera (which south 39°S is called Patagonian Cordillera). South of the Taitao Triple Junction, from west to east, the morphostructural units are the Archipelago, Patagonian Cordillera, and the Precordillera (Charrier et al., 2007).

North of 27°S the Coastal Range is constituted by calc-alkaline intrusive, metamorphic complexes and basaltic to andesitic volcanic rocks. It presents an abrupt rise from sea level to 2500 m. The Central Depression, located between the Coastal and Andes Ranges, is structurally a forearc alluvial basin, filled with thick continental clastic, saline, and volcanoclastic rocks. The volcanic arc is characterized by rhyolithic to andesitic ignimbrites (pre-Pliocene) that are overlain by Plio-Quaternary volcanoes. These volcanoes are constituted by andesitic rocks and tephra, and reach more than 6000 m. Between 27°S and 33°S, Plio-Quaternary volcanism and the alluvial forearc basins are absent. The Coastal Cordillera is constituted by Mesozoic volcanic and intrusive rocks and the Andean Cordillera by older volcanoclastic and intrusive rocks. Altitude progressively increases from the coast up to the Andean mountain axis, where it reaches 6892 m at the Ojos del Salado volcano. Between 33°S and 41°S the altitude of the Andes gradually decreases from 5000 to 2000 m. South of 33°S, geologic and geomorphic features change abruptly due to the dip change of the subduction zone. Plio-Quaternary volcanism and forearc alluvial basins reappear. The Coastal Range reaches 1500 m and is constituted by low-grade metamorphic rocks and Paleozoic intrusive rocks, which also are common southward to about 38°S. The Central Depression is filled with up to 4000-m-thick sequences of alluvial sediments, but south of 41°S has been invaded by the sea. The basement of the Andes consists mainly of Mesozoic plutons south of 41°S, in contrast to the pre-Pliocene andesitic to rhyolithic volcanic rocks and sediments that crop out between 33°S and 41°S (Lamy et al., 1998).

2.2. Climate

The Chilean climate is extremely varied due to the abrupt relief and geographic location of the country. The geographic location determines atmospheric circulation, controlled by the influence of the Pacific Ocean and Antarctic region. Movements of Antarctic and Sub-Antarctic water currents (the Peru-Chile Current) and masses of polar air influence the whole country.

The topography of the country strongly controls temperature and rainfall patterns. The Coastal Cordillera and the Andean Cordillera constitute geographic barriers that block the maritime influence of the Pacific Ocean on their eastern slopes and in the Central Depression (Muñoz et al., 2007). Atmospheric circulation mainly involves the impact of the South Pacific anticyclone, located in front of the Chilean coast, normally between 20°S and 40°S. The occurrence of long droughts or large floods is strongly controlled by the location and persistence of that anticyclone and also by El Niño current (Muñoz et al., 2007).

The northern part of Chile (17°S–27°S) is mostly a desert, with extremely low rainfall (< 50 mm/year). The Atacama Desert, for instance, has a variability rainfall from 0 mm/year at c. 2400 m to 200 mm/year at 4000 m. Over the Altiplano Plateau there is some rainfall (200–300 mm/year) from December to March, called Bolivian Winter. Further south (27°S–32°S) the climate is semiarid, with scanty winter rainfall. Between 32°S and 38°S, the Mediterranean climate is characterized by rainfall during the winter season (50–1000 mm/year) and a dry summer season. Even further south (38°S–42°S) the climate becomes temperate and with increasing rainfall. Between 42°S and 46°S it is very cold and humid with snow and rainfall over 3000 mm/year (Muñoz et al., 2007).

3. Geo-hazards and Risks

The main geo-hazards in Chile are mass movements, volcanic eruptions, earthquakes, and tsunamis. This chapter presents an overview of the wide range of hazards that can be found in Chile, focusing on some remarkable events.

3.1. Mass-Movement Hazards

Chile is a mountainous country, and mass-movement hazards are inherent to this kind of terrain. Many of Chile's important cities are located near the coast, at the foot of mountains, volcanoes, or on valleys, and are exposed to mudflows, debris flows, rockfalls, slides, and lahars (Fig. 18.3). The most frequent causes of damage are mudflows and debris, flows triggered by intense rains, which descend along river valleys and affect cities. Slides and rockfalls due to natural (rainfall, earthquakes) and human (excavation at the talus of slopes) triggers also occur. Mass movements of different kinds have produced important damages in the past. The cities of Antofagasta, Santiago, and Concepción, among others, were affected. Also, the North-South highway (Route 5), which constitutes the main communication infrastructure of the country, has been affected on a number of occasions. This has caused considerable socioeconomic disruption, as in many cases there is no alternative. Estimates for the volume, velocity, and intensity of landslides are shown in Tables 18.1 and 18.2. Table 18.3 is an inventory of the main mass movements recorded in Chile. Some of these examples are described below.

3.1.2. Description of Some Important Landslides

3.1.2.1. Antofagasta, June 18, 1991

With 286,000 inhabitants, Antofagasta is a main city of northern Chile (Fig. 18.3). The city is located in a desert, narrow belt of land between the coastline and the coastal escarpment that reaches up to 500 m.

To the east, the coastal cordillera rises to more than 1000 m. Several watersheds spread from the Andes to the coast flowing through the urban area of Antofagasta. The rainfall is extremely low, with an average of less than 4 mm/year. The El Niño phenomenon, however, produces anomalous rainfalls in the north of Chile, including the Atacama Desert (Vargas et al., 2000).

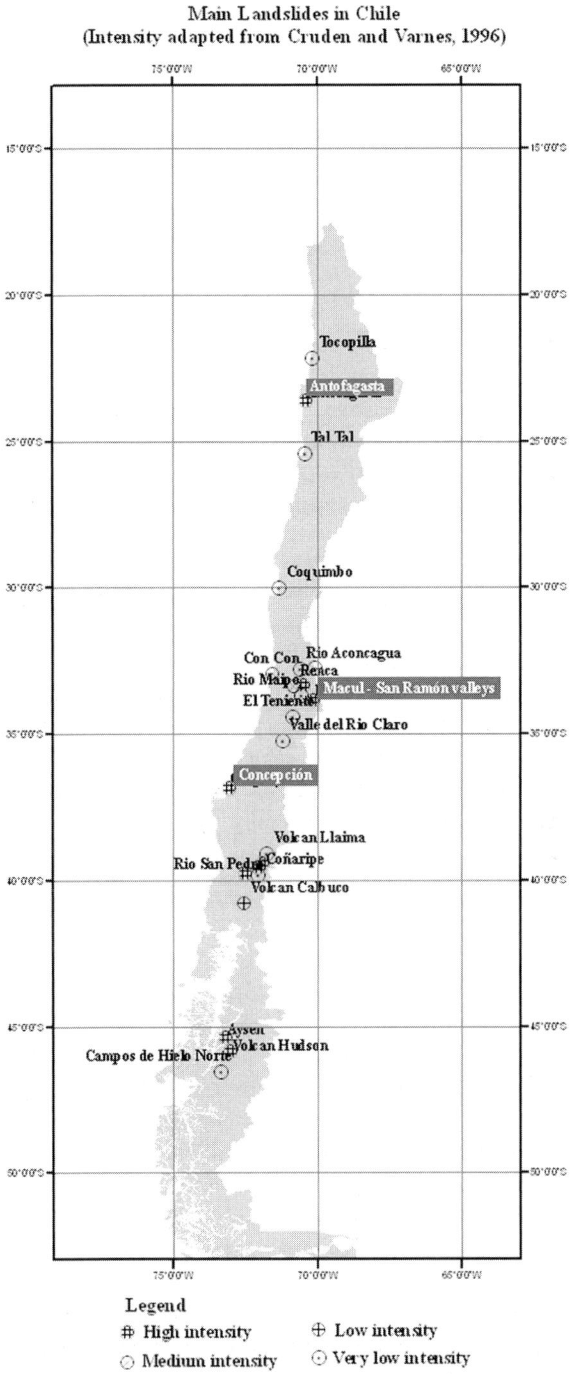

Figure 18.3 Intensity of landslides (adapted from Cruden and Varnes, 1996).

Table 18.1 Velocity of Mass Movements and Involved Volume

	Velocity (adapted from Cruden and Varnes, 1996)		
	0.03–300 m/min	0.03 m/min to 13 m/month	< 13 m/month
> 100,000	High	High	Medium
50,000–100,000	High	Medium	Low
5,000–50,000	Medium	Low	Very Low
< 5,000	Low	Very Low	Very Low

Table 18.2 Intensity of Landslides

Length (m) of mass displacement	**Volume m^3 = 1/6 π Dr Wr Lr (Dr = depth of rupture; Wr = width of displaced mass; Lr = length of rupture)**			
	> 100,000	10,000–100,000	1,000–10,000	< 1,000
> 1,000	High	High	Medium	Low
500–1,000	High	High	Medium	Low
100–500	High	Medium	Medium	Low
< 500	Medium	Low	Low	Very Low

In northern Chile, the Southeast Pacific anticyclone impedes the arrival of southern frontal systems. However, when the anticyclone becomes weak, especially during the El Niño (ENSO) events, which produce a warming of the southeastern Pacific Ocean, humid fronts can reach the coast, as happened on June 18, 1991. On this occasion, an anomalous propagation of a bad weather front with 5–14 mm/h rainfall took place. Rainfall intensity reached values as high as 24 mm/h to 42 mm/h in three or four hours (Garreaud and Rutllant, 1996). The rainfall triggered a sequence of debris flows along alluvial fans generated by small basins such as Salar del Carmen (drainage area (DA) = 33 km^2); La Negra (DA = 43 km^2); La Chimba (DA = 26 km^2); and La Cadena (DA = 21 km^2), which affected the city's activity (Sepúlveda et al., 2006). The most active areas were along the proximal zones of the alluvial fans (Vargas et al., 2000) where debris flow deposits reached 1 m thick. In Antofagasta, sandy deposits spread to the south and north of the city. Debris flows caused more than 100 fatalities and transported at least 7–8 $\times$ 10^{5}m^3 of sediment (Hauser, 1997). Due to the energy of the flows, many houses were dragged from their foundations, and schools, hospitals, and other public buildings were partially destroyed. Basic supplies such as energy, water, and gas were interrupted for several days. The total cost of this event caused losses of US$ 66 million, mainly from destruction of property and infrastructure (ONEMI, 1996). After this event, some mitigation measures were implemented, as large retaining dams were built in the valleys that descend toward the city.

Table 18.3 Hazards of Main Landslides ed in Chile

Type / Locality	Year	Length / Volume / Injury	Hazard
Lahars			
Trufultruful valley–Llaima volcano	1955, 1957	35 km length (1957)	High
Villarica volcano	1948	1000 Ha covered	High
Coñaripe–Villarica volcano	1964	$2^*10^6\,m^3$	High
Coñaripe–Villarica volcano	1971	$20^*10^6\,m^3$	High
Rio Colorado–Calbuco volcano	1961	$3{,}000\,m^3$	Medium
Río Huemules valley–Hudson volcano	1972	$752^*10^6\,m^3$	High
Hudson volcano	1991	$45^*10^6\,m^3$	High
Multirotational			
Río San Pedro valley	1960	$38^*10^6\,m^3$	High
Río San Pedro valley	1575	$100^*10^6\,m^3$	High
Cerro Colorado, Renca	1984	$2{,}5^*10^{12}\,m^3$	High
Avalanches			
Lo Valdés, Santiago	1953	$45^*10^6\,m^3$	High
Customs House Los Libertadores	1984	Inferred intensity	Very Low
Campo de Hielo Norte	1984, 1985	Inferred intensity	Very Low
Subsidence and slides			
La Africana mine, Santiago	1985	8,000 to 10,000 m^3	Medium
Rotational slides			
South slope of Villarrica volcano	1960	Intensity inferred	High
Road from Pelluco to Caihuin	1979	Intensity inferred	Low
Río Claro valley	1928	Inferred Intensity	Very Low
Chiguayante	2006	Inferred Intensity	Medium

(Continued)

Table 18.3 (*Continued*)

Type / Locality	Year	Length / Volume / Injury	Hazard
Laminar slides			
Flanks of Caburga lake	1960	Inferred Intensity	Very Low
Flanks of Panguipulli lake	1960	Inferred Intensity	Very Low
Debris flows			
Antofagasta	1991	$8^{*}10^{6}$ m^{3}/100 fatalities	High
Los Maitenes	1992	Inferred Intensity	Very Low
Macul valley, Santiago	1993	$2^{*}10^{6}$ m^{3} /26 fatalities	High
Pellaifa lake	1960	Inferred Intensity	Very Low
Concepción and surrounding areas	2005	US$ 2,000 million	High
Concepción and surrounding areas	2006	10 fatalities/95.000 injuries	High
Río Aconcagua valley	1980	Inferred Intensity	Very Low
Rockfalls			
Río Colorado	1987	2.83×10^{5} m^{3}	High

(Adapted from Varnes, 1978; Cruden and Varnes, 1996; Hutchinson, 1988; and Hungr et al., 2001).

3.1.2.2. Macul and San Ramon Valleys, May 3, 1993

Santiago is the capital and most populous city of the country, with more than 5.5 million inhabitants. It is located between the Maipo and Mapocho hydrological basins at the foot slopes of the San Ramon Range, an alluvial piedmont deposit (Fig. 18.4). The climate is Mediterranean, with mean annual rainfall varying approximately between 300 and 350 mm. The Macul valley is located in the southwest hillsides of the Andean Range, which reach elevations above 4000 m. From the Andean Range several valleys remain almost waterless throughout the year. During intense rainfall, the Macul and San Ramón valleys (Fig. 18.4) collect significant quantities of water, causing serious flooding in the urban area of Santiago. In this zone, snowfalls are normally above 2300 m. On May 3, 1993, the isotherm of 0°C rose to 3800 m

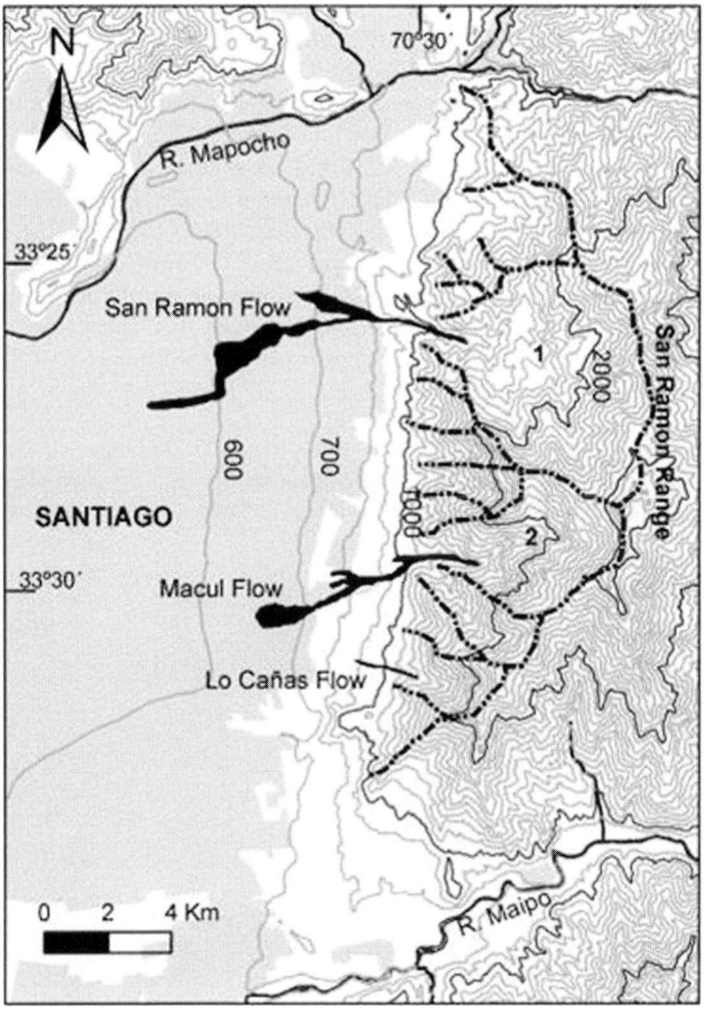

Figure 18.4 Macul and San Ramón debris flows (from Sepúlveda et al., 2006).

(Corvalán et al., 1997). A frontal system produced strong rainfall on the Andean Pre-Cordillera as a result of the warm conditions of a moderate El Niño event (Garreaud and Rutllandt, 1996). On May, 3, 1993, this phenomenon produced a rainfall of 67 mm/day (Lara, 1996). The increase in temperature and the conspicuous rains melted the snow and originated the forceful Macul and San Ramón valleys' debris flows that mobilized $2 \times 10^6 \, m^3$ of sediments. The viscous and dense flow transporting rock blocks of several meters flow to more than 30 km/h and the wave were more than 10 m thick (Naranjo and Varela, 1996), causing 26 fatalities and 1910 injuries, destroying 307 houses and damaging 5000 others, and affecting more than 28,000 people at an estimated damage cost of US$ 5 million (Corvalán et al., 1997; ONEMI, 1995). The cleanup took several weeks, and large dam-reservoirs were constructed to mitigate future alluvial-colluvial effects.

3.1.2.3. The Concepción Region, July 2006

During the winter of 2005 and 2006, in the Concepción region (Fig. 18.3), a long, intense rainy period took place that generated several debris and mudflows, completely isolating many areas and the city of Ralco located in the Andean Cordillera (Fig. 18.5). The Concepción region has a rainy temperate climate with an average rainfall of nearly 1800 mm/year. Rainfall, however, is locally affected by reliefs and by dense forest montane vegetation. Due to the destruction of roads, the isolation of these zones extended for many months and caused serious problems to forestry, agricultural, fishing, and cattle activities, with severe economic losses. One of the most critical floods occurred on July 12, 2006 in the city of Concepcion, claiming 10 victims and 95,000 injured persons (Concepción Regional Emergency Office, 2006).

Figure 18.5 Landslide damage to the road from Santa Bárbara to Ralco city (photo by Cecioni, 2006).

3.1.3. Climatic Forcing and Landsides

Debris flows in northern Chile are related to years of significant rainfall of short duration (one- or two-day storms). In central Chile, however, higher rainfall is not necessary (Rutlland and Fuenzalida, 1991; Garreaud and Rutllant, 1996) to reactivate the transport of debris, and the antecedent rainfall some days or weeks before seems to play a relevant role (Sepúlveda et al., 2006).

In northern and central Chile debris flow activity seems to be well correlated with El Niño events. Central-southern landslides in Chile are naturally triggered by rainy climate and the effect of "antecedent" rain (Sepulveda et al., 2006). Snow accumulation is also an important factor, especially in central Chile, as, for example, happened in the Macul event. Existing information on mechanisms and processes as well on the recurrence of events in relation to climatic dynamics through time, however, is rather preliminary in Chile.

3.2. Volcanic Hazards

In the Chilean Andean Range there are over 100 Pleistocene and Holocene stratovolcanoes as well as large volcanic fields and giant caldera complexes, of which 60 have documented Holocene eruptive activity. Volcanic areas are distributed in three sectors: 14°S to 28°S (CVZ: Central Volcanic Zone), 33°S to 46°S (SVZ: Southern Volcanic Zones), and 49°S to 55°S (AVZ: Austral Volcanic Zone) (Stern et al., 2007) (Fig. 18.6).

The volcanic activity results from the subduction of the Nazca and Antarctic plates below the South American Plate. Volcanoes in northern Chile (CVZ) and central-south of Chile (SVZ) occur where the angle of subduction of the Nazca Plate is relatively steep ($>25°$), at depth >90 km. These two segments are separated by the Pampean flat-slab segment where the angle of subduction decreases and becomes relatively flat ($<10°$) at depth >90 km. The volcanic arc segment of the SVZ is located between the Juan Fernández Ridge, volcanic oceanic islands, and the Chile Ridge, the tectonic divergent plate boundary between the Nazca and Antarctic plates. South of Chile Ridge, the Patagonian volcanism gap is observed between the SVZ and the AVZ (Stern et al., 2007). The southernmost volcanic segment of SVZ, between 38°S and 42°S, is the Liquiñe-Ofqui Fault System, which controls the location of the larger volcanic centers as well as minor eruptive centers. This area is the center of the greatest volcanic activity along the whole Andes (Fig. 18.7).

Geologic records show evidence of explosive eruptions that affected large areas, such as the volcanoes Antuco (10,000 year BP), Llaima (7600 year BP), and Chaitén (2008). Figure 18.8 shows the location of active volcanoes in these sectors, as well as the main population centers potentially exposed to volcanic hazardness. In northern Chile (CVZ), the Late Pleistocene and Holocene volcanism has been characterized mainly by andesitic stratovolcanoes and dacitic dome complexes, associated with pyroclastic flows, tephra fallout, debris avalanches, and block-and-ash flow deposits. Eruptions in this region could affect southwestern Bolivia or northwestern Argentina, as happened during the eruption of the Lascar volcano in 1993 (Gardeweg and

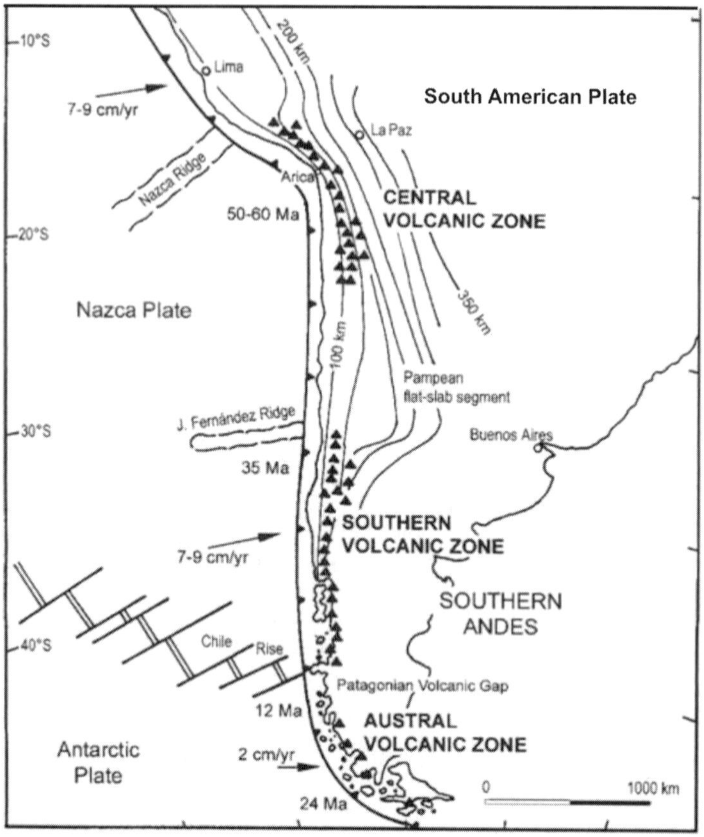

Figure 18.6 Active volcanic zones in the Andes Cordillera and subduction geometry indicated by depth (km) of Benioff zone (adapted from Stern et al., 2007).

Medina, 1994). Due to the high-altitude easterly winds, tephra from large eruptions could be carried west into Chile and affect local populations as well as altiplanic pasturelands, which are important feeding areas for camelids (Stern et al., 2007).

Volcanoes in central-south Chile (SVZ) form a continuous volcanic arc 1400 km long, extending from 33°S to 46°S (Stern et al., 2007), which includes more than 70 Pleistocene and Holocene composite stratovolcanoes and large volcanic fields, and at least nine caldera complexes, as well as hundreds of minor eruptive centers formed by scoria cones, lava flows, and maars. Postglacial (from 14,000 yr BP) activities in the central-south sector include the full range of Hawaiian, Strombolian, Vulcanian, Subplinian, and phreatomagmatic eruptions, with a Volcanic Explosivity Index (VEI) ranking from 0 to 6 (Stern et al., 2007). Lahars, lava flows, pyroclastic flows and surges, ashfalls, and voluminous debris avalanches have been the main types of volcanic activity in historic times (Fig. 18.8).

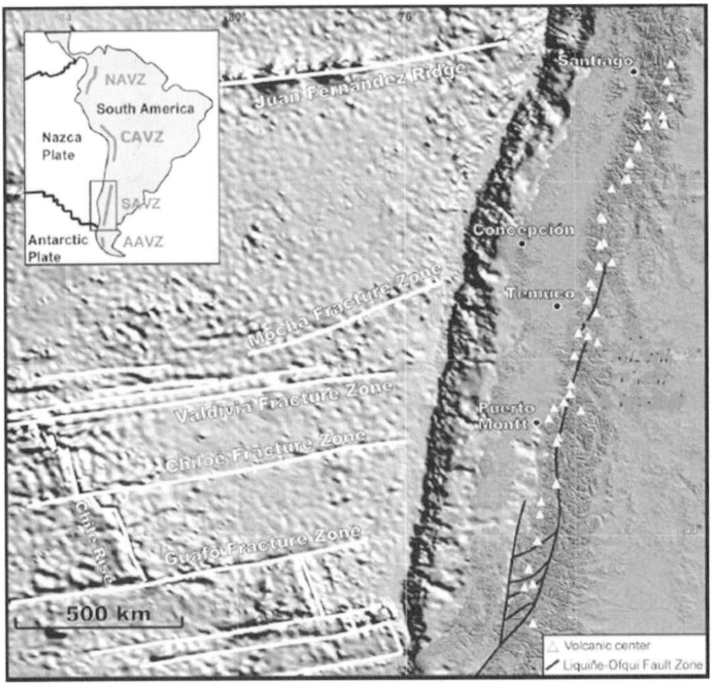

Figure 18.7 Volcanoes (white triangles) associated with the Liquiñe-Ofqui fault (black line), in the Southern Volcanic Zone (SVZ) (adapted from Lohmar et al., 2007).

The southernmost sector (AVZ) includes only six volcanic centers. The southernmost volcanic complex (Cook Island) consists of a group of postglacial domes, and all the other volcanoes are glaciated stratovolcanoes, with pre-Holocene and/or Holocene explosive activity. These volcanoes are far away from population centers; however, future eruptions similar to those in the Holocene could produce significant tephra fall in populated areas (Punta Arenas, Puerto Natales and Calafate) (Stern et al., 2007).

As is well known, volcanic hazards can be mitigated basically through land-use planning based on hazard and risk assessments and maps, and the establishment of monitoring, early warning, alert, and alarm systems linked to evacuation plans. In Chile eight volcanoes have monitoring systems. Indigenous populations, tourist areas, and some hydropower dams are exposed to volcanic hazards.

The riskiest areas are located in central-south Chile (SVZ) (Figs. 18.6, 18.7, 18.8). At least 40 volcanoes had Holocene activity, and 20 had historic eruptions. This sector, with an average of one eruption per year in historic times, is much more active than the others. Villarrica and Llaima are two of the most active volcanoes in the entire Andean arc, both with more than 80 reported episodes of activity since 1558. The 1932 Quizapu explosive eruption was the biggest in the last 100 years. Following is a description of some important eruptions of the Villarica volcano and the recent eruption of Chaitén volcano.

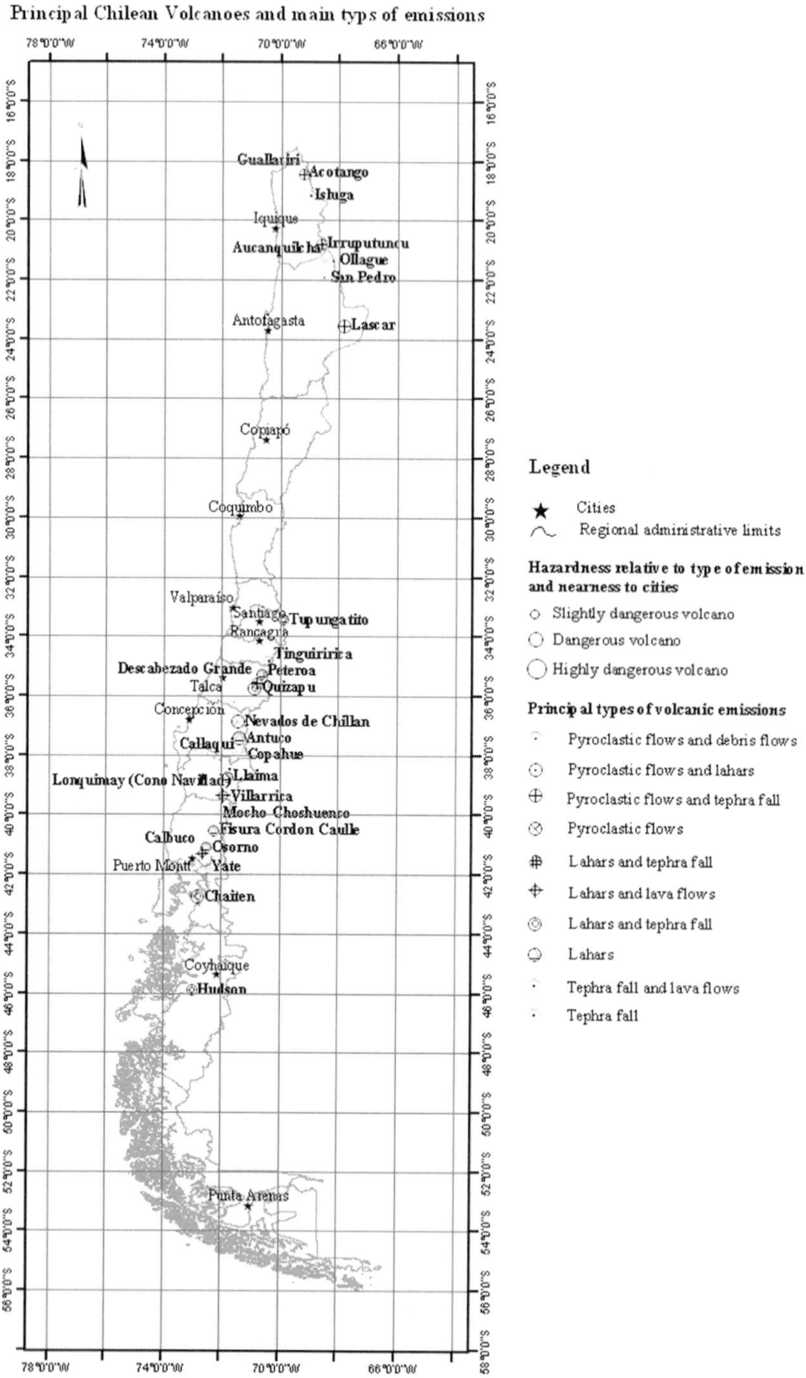

Figure 18.8 Principal Chilean volcanoes and hazards relative to different types of emissions and nearness to cities.

3.2.1. Villarrica Volcanic Eruptions

The Villarrica volcano (2847 m), one of the most active in the Andes, is considered the most dangerous volcano in Chile because of its proximity to the towns of Pucón and Villarrica, important tourist centers. This volcano has had more than 30 well-reported eruptions in historic time. Villarrica is a Late Pleistocene-Holocene compound stratovolcano, and its evolution has been divided into three stages: Villarrica 1(> 114,000 yr BP), Villarrica 2 (14,000–3700 yr BP), and Villarrica 3 (3700 yr to present) (Clavero and Moreno, 2004). In its early evolution (middle-Upper Pleistocene), an ancestral stratocone was built mainly with basaltic to basaltic andesite lavas and fallout deposits. At c.100.000 yr BP, this volcanic edifice partially collapsed, generating an elliptical caldera structure (6.5 km*4.2 km diameter), whose associated eruptions generated a series of pyroclastic flow deposits. During the Llanquihue Glaciation (95,000–14,000 yr BP), subglacial basaltic andesite lavas erupted, and dacitic domes and dikes were emplaced, although it is not clear if an edifice was formed. A second collapse (14,000 yr BP) with the generation of a series of piroclastic flows marks the beginning of an explosive phase, which has continued to the present. During postglacial eruptive activity (14,000 yr BP–3,700 yr BP), a new stratocone was formed through successive effusive and explosive eruptions. At 3700 yr BP, this stratocone partially collapsed, forming a smaller summit caldera, which collapsed, and a new stratocone was formed within the summit caldera. This cone has been built through successive effusive and explosive eruptions (3700 yr to present).

Historical activity at Villarrica has been essentially effusive and slightly explosive (strombolian). The volcano shows an almost permanent lava lake in the crater with occurrences of small explosions and emissions of SO_2 and HCl that exceed recommended limits and threaten the safety and health of visitors (Witter and Delmelle, 2004). Figure 18.9 shows the map of Villarrica volcano, indicating the distribution of lahars, lava flows, and the main limit of the principal risky and hazard areas (adapted from González-Ferrán, 1995). The most important eruptive events in recent times occurred in 1948–1949, 1963–1964, 1971–1972, and 1984. Descriptions of some major episodes follow.

April 1948: Signs of activity were first noted in April with an increase in explosive activity. On October 18, the first large eruption started with a strong explosion and the formation of a very large ash plume with incandescent pyroclastics and gases (3000 m high), continued emissions of lava flows, and lahar flows descending along the main river systems (Turbio, Pedregoso, Molco, Coñaripe, and Voipir rivers), with a mean velocity of 30 km/h. These flows transported large tree trunks and boulders (10–20 m^3), causing a 1-m rise in the water level of Lake Villarrica as well as generating a micro-tsunami. On December 29 the volcanic activity began again, and on January 1st, 1949, there was a very strong explosion with the formation of the new column of gases and pyroclastics that reached 8000 km high, emissions of lava, and lahar flows (Clavero and Moreno, 2004; González-Ferrán, 1995). The eruption affected a 200-km^2 area, destroying homes, the town of Coñaripe, a bridge, road, and highway, cultivated areas, and national parks; 100 human lives were lost.

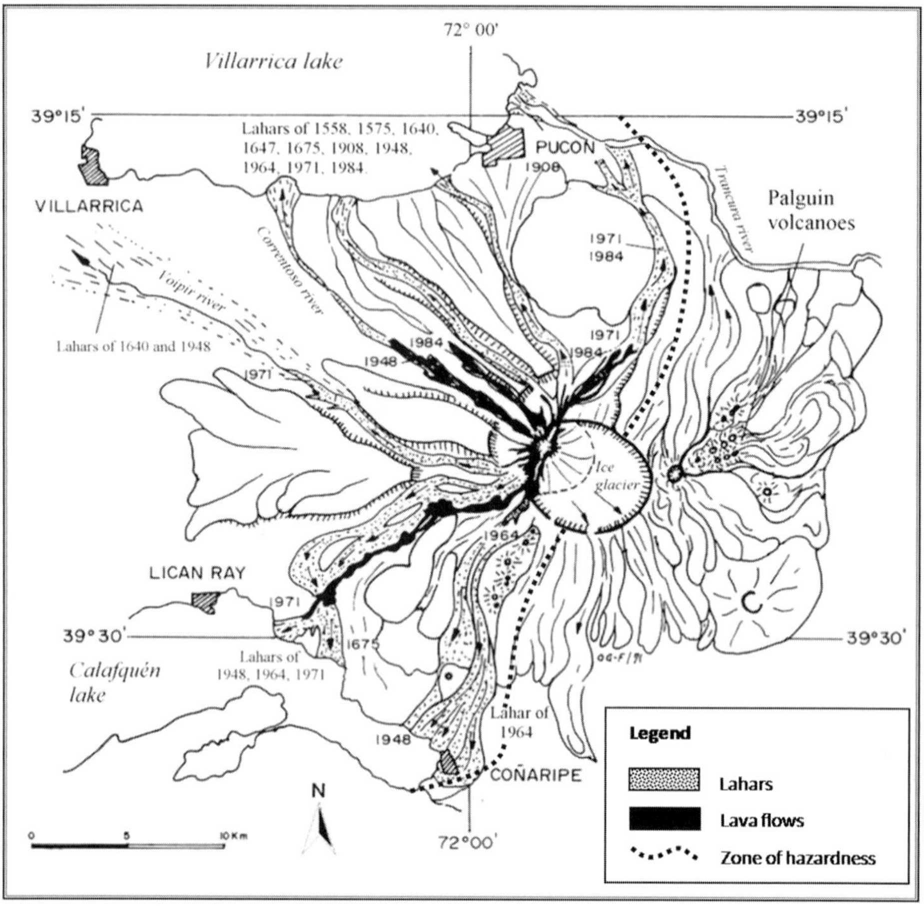

Figure 18.9 Map of Villarrica volcano, showing the distribution of lahars, lava flows and the main limit of the principal risky and hazardous areas (adapted from González-Ferrán, 1972).

October 29, 1971. The eruption began with violent emissions of lava without premonitory signs. On December 29, a 2-km-long fissure at the summit of the volcanic cone generated a 400-m-high glowing lava fountain. The lava reached 14 km into the air with a 200-m width and a 5-m mean thickness (30 million m^3 of total lava were ejected) (González-Ferrán, 1995). The eruption quickly melted the glacier on the volcano, generating lahars that flowed toward the north, west, and south, with estimated velocities of 30–40 km/h, dragging large quantities of tree trunks and rocky blocks up to 20 m^3. The biggest lahar was 10 m thick, more than 200 m wide, and 15-20 km long, reaching Lake Calafquén. The cities of Coñaripe, Pocura, Traitraico, Quilentué, Llauquén, Chaillupén, Licanray, and Lanalhue suffered heavy damage. In addition, the cities of Pucón and Villarrica were invaded by toxic gas clouds. About 200 persons died, and thousands of inhabitants had to be evacuated.

3.2.2. Chaitén Volcano

Chaitén volcano is a small, glacier-free caldera generated with a Holocene lava dome located 10 km NE of the town of Chaitén on the Gulf of Corcovado (Fig. 18.10). The caldera presents an ellipsoidal shape of 3.5 km width by 4 km length. The caldera is breached on the SW side by a river that drains to the bay of Chaitén. Inside the caldera a dome of rhyolithic composition of Holocene age has been raised (López-Escobar et al., 1993). The scarce available geological antecedents suggest that the caldera and its pyroclastic deposits were generated approximately 9300 yrs BP (Naranjo and Stern, 2004).

The eruptive event of May 2, 2008, consisted of a sequence of subplinian eruptions preceded by seismic activity that was perceived at Chaitén village located 10 km from the volcano. A pulsating white-to-gray ash plume rose to an estimated height of more than 21 km and drifted SSE (Fig. 18.11). On May 6, ONEMI and SERNAGEOMIN reported that the eruption became more forceful and generated a wider and darker gray ash plume, rising to an estimated altitude of 30 km. Anyone still in the towns of Chaitén and Futaleufú (30 cm of ash accumulated there), as well as anyone within 50 km of the volcano, was ordered to evacuate.

During May 2–6, the pyroclastic material generated a plume that was dispersed by the westerlies toward the east and southeast, through Palena Province (Chile) and the Argentinean territory. Pyroclastic waves of small volume and extension were generated at the base or the eruptive columns burned the surrounding vegetation around the crater. Water temperature in some creeks increased 10°C following the introduction of pyroclastic materials. During this period, intense to moderate rainfall became more frequent, triggering an increased number of mudflows and lahars. These affected the town of Chaitén and surrounding areas, because of the remobilization of big amounts of volcanic ash from the headwaters of the Blanco River (Fig. 18.12).

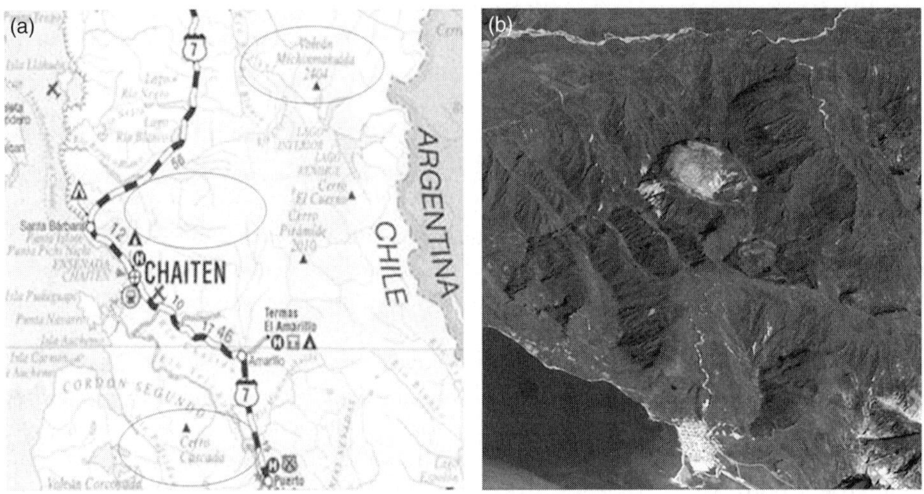

Figure 18.10 Location of Chaitén volcano. (A) Chaitén volcano is not represented on the map of Chaitén region; (B) Satellite view of the caldera of Chaitén volcano.

Figure 18.11 Sub-Plinian to Plinian eruption of Cahitén volcano (photo by *La Tercera* newspaper).

As a consequence of the eruption of May 21, a new dome was formed on the northern caldera. In the contact between the new and the older dome, occurred continuous pulses and explosions which generate two new craters (Servicio Nacional de Geología y Minería-SERNAGEOMIN).

Because of the violent eruption in May–June, the whole population of Chaitén (ca. 7000 inhabitants) had to be evacuated. The authorities demarked a "zero zone" of access 50 km from the volcano. Lahars and mudflows flooded the village, 5400 houses were destroyed, and a good part of the city became uninhabitable.

3.3. Seismic Hazards

The active Chilean margin of the South American Plate is constantly affected by subduction earthquakes owing to the interaction with Nazca, Antarctic, and Scotia plates. Chile's written history, which began with the arrival of the Spaniards some 500 years ago, is too short for anyone to evaluate the statistical recurrence of historical earthquakes. However, more than 10 events with magnitudes equal to or greater than magnitude 8 took place in Chile during

Fig. 18.12 Lahar, mudflows, and ash from the Chaitén volcano from mid-May to early June 2008 (Photo by A.B. Lockhart, USGS)-© Smithsonian Institution.

the twentieth century alone. Among these earthquakes is the 1960 event, which was the largest earthquake ever recorded since the beginning of instrumental seismology (Barrientos, 2007). Earthquakes and associated tsunamis have caused the biggest losses of lives and material in the country. Earthquakes have been studied on the basis of accelerograms and attenuation curves, and the difference between subduction and superficial intraplate earthquakes, which are important for the assessment of seismic risk (Ruiz and Saragoni, 2005a, 2005b).

In general, seismogenic zones in Chile are large, shallow (0–50 km) thrust earthquakes along the coast; large, deeper (70–100 km) tensional as well as compressional events within the subducting Nazca Plate; and very shallow seismicity (0–20 km) in a few places, such as the Cordillera region of central Chile and the southern extremity of the continent by the Magellan Strait. Deeper seismicity (150 to 650 km) occurs further to the east, beneath Bolivia and northwestern Argentina (Barrientos, S., 2007) (Fig. 18.13). The Chilean civil works procedures define three earthquake zones: interplate coastal earthquakes (depth < 50 km); intraplate earthquakess (depth > 70 km) inside the Nazca Plate; and superficial earthquakes (depth < 20 km) inside the continental crust on the western side of the Andean Cordillera (Alvarado, 1998). Intraplate earthquakes with normal mechanism produce higher superficial acceleration than subduction earthquakes with an inverse mechanism (Ruiz and Saragoni, 2005c).

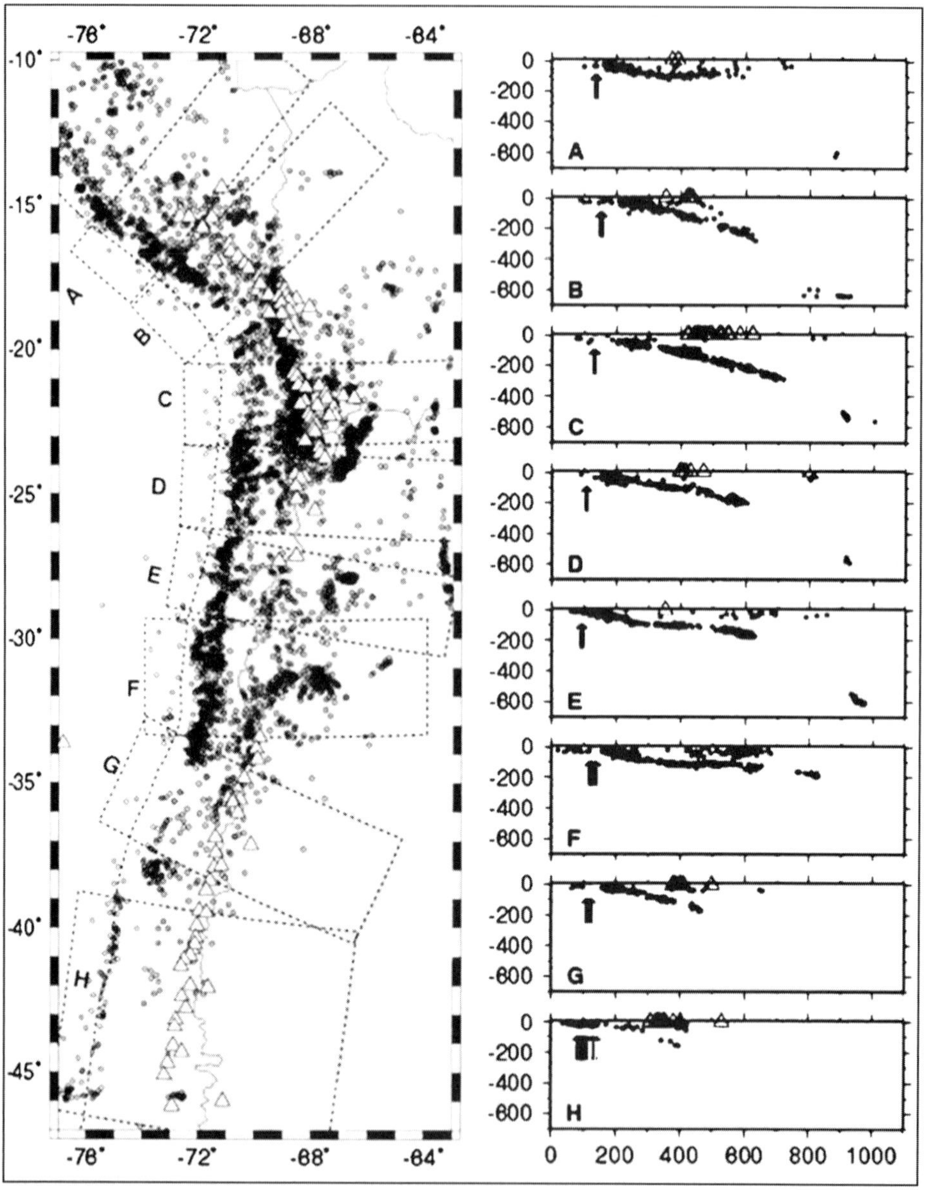

Figure 18.13 Seismicity (black circles) on the active margin of south South America (Engdahl and Villaseñor, 2003). Profiles show the Benioff zone and the distribution of seismicity with depth as a function of distance from the Peru-Chilean trench.

Subduction earthquakes are frequent in Chile, and the majority of them are normally. tsunamigenic. Violent earthquakes in the central zone of Chile, between 32° S and 33° S, have occurred with a periodicity of 83 ± 9 years (Comte et al., 1986).

Records during the last century registered high-magnitude intraplate earthquakes at Chillán (1939, Ms = 7.8); Calama (1950, Ms = 8.0) (Kausel, 1991), and Tarapacá (2005, Ms = 7.9). Other intraplate earthquakes have produced through the rupture of the continental crust, the fragile deformation due to tectonic dynamism controlled by subduction (Barrientos and Kausel, 1993). This type of earthquake is related to fault systems associated with the formation of the Andean Chain and with geologically active coastal faults (Armijo and Thiele, 1990; Hervé, 1987; Naranjo, 1987).

According to Cisternas et al. (2003), on the basis of C^{14} dating of ancient tsunamideposits, mega events such as the Valdivia tsunamigenic earthquake (Mw = 9.5) have a recurrence of around 500 years. In Chile, subduction earthquakes are more frequent, and they often are tsunamigenic. Therefore, earthquake and tsunami damages are not easily differentiated. The major earthquakes recorded are shown in Fig. 18.14.

At our present state of knowledge, we have only limited ability to predict earthquakes. Therefore, mitigation measures must be based mainly on geological, structural, and seismic micro-zoning studies and on implementing seismic building codes. Micro-zoning studies are not mandatory in Chile for land-use planning and construction of civil works.

3.3.1. Outline of Some Destructive Earthquakes

3.3.1.1. The Chillán 1939 Earthquake

On January 24, 1939 at 23:30 local time, a violent earthquake of M = 8 took place in Chillán. It was a product of down-dip tension along the subducting slab at a depth varying between 80 and 100 km (Beck et al., 1998) (Fig. 18.15). According to the Dirección General de Obras Públicas (the public works headquarters), 95% of the town of Chillán was destroyed, and about 30,000 people were killed (SUBDERE). The census of 1930 showed that the city had 39,511 inhabitants (INE, 1930). The earthquake also destroyed 95% of Concepción and produced a thousand unconfirmed fatalities.

3.3.1.2. The Concepción Earthquakes

Concepción was founded by the Spaniards in the southeastern part of the bay. On February 8, 1570, a violent earthquake and tsunami totally destroyed Concepción and its port. The city was reconstructed in the same place and became the headquarters of the colonial government. It was destroyed again by earthquake in 1657 and 1751. After the 1751 event, the population, shocked by the successive destructions, decided to relocate the city in its present emplacement away from the ocean. The new port of Talcahuano was constructed in the southwestern section of Concepción bay. On February 20, 1835, a new earthquake, described by Charles Darwin, destroyed Concepción, and the associated tsunami penetrated 4 km inland, totally destroying the port of Talcahuano and all the coastal villages.

3.3.1.3. The Valdivia Earthquake in 1960

The major historical earthquake registered with instruments was the Valdivia earthquake, which occurred on May 22, 1960. The quake and subsequent tsunami

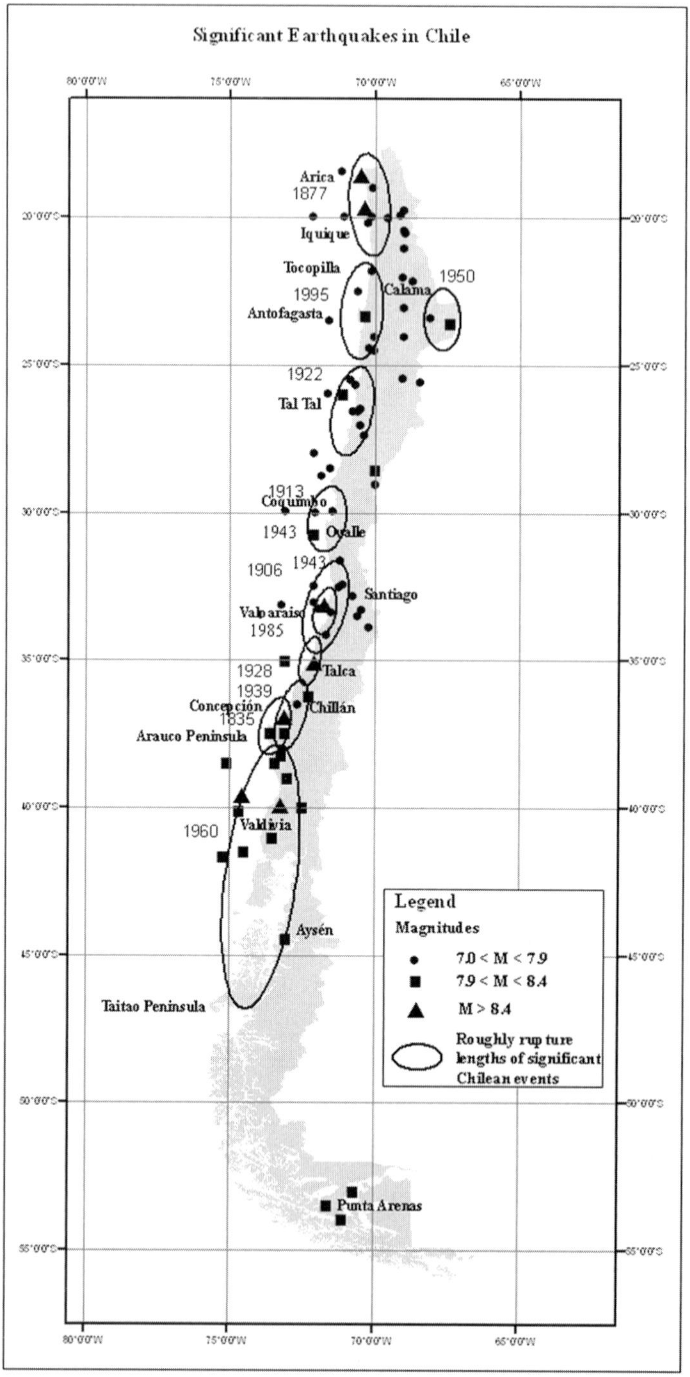

Figure 18.14 Principal earthquakes in Chile. The symbols indicate ranges of magnitude and the ellipses the approximate length of rupture.

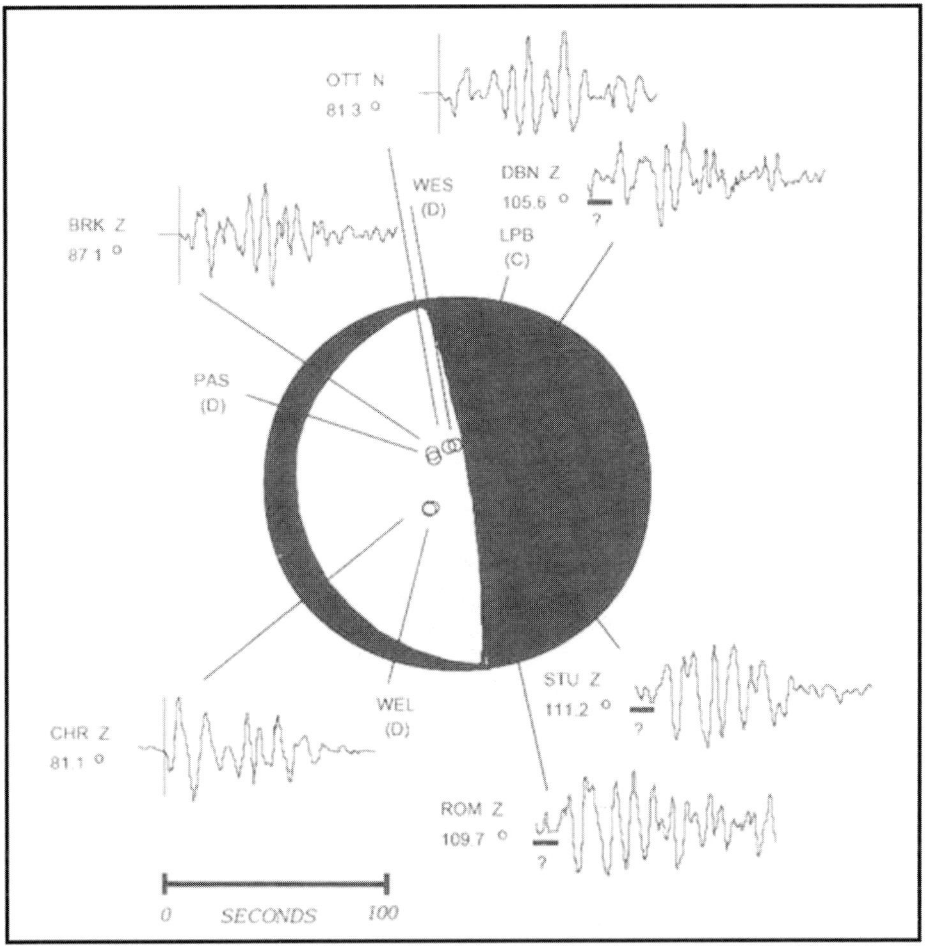

Figure 18.15 P-wave first motion focal mechanism and waveform for the Chillán 1939 earthquake (after Beck et al., 1998).

affected an extended region from Concepción to Puerto Montt, inhabited by around 2.5 million people. Over 2000 people were killed, and US$ 550 million in damages were recorded (Plafker and Savage, 1970). The tsunami spread all around the Pacific Ocean and caused another 61 deaths and US$ 24 million damages in Hawaii; 199 deaths and US$ 50 million damages in Japan; and about US$ 500,000 to US$ 1,000,000 in damages on the west coast of the United States (Lander and Lockridge, 1989).

The event had a magnitude moment of Mw = 9.6 with a rupture of nearly 1000 km length, 150 km width, and 20 m of fault displacement (Barrientos and Ward, 1990). The Chilean coast had raisings and permanent subsidence of several meters, and some Andean volcanoes erupted. Plafker and Savage (1970) reported coastal elevation changes produced by the 1960 event from Península de Arauco to

Península de Taitao, a 1000-km-long segment along the coast of southern Chile. These measurements were carried out in 1968 and were based on intertidal algal environment modification. Evidence from dead barnacles, balanus, and trees revealed extreme values of 6 m of uplift in Isla Guamblin and 2 m of subsidence in the city of Valdivia (Barrientos, 2007). Repeated measurements of triangulation as well as leveling were used in estimating the earthquake size: a nearly 1000-km-long dislocation with 20 to 40 m of fault displacement. Later, Plafker (1972) reanalyzed the static deformation and deduced a causative fault 120 km wide by 1000 km long, dipping 20°E with 20 m of slip. Assuming a rigidity modulus of 5×10^{10} Pa, the total seismic moment reached 1.2×10^{23} N m or Mw = 9.3. Kanamori and Cipar (1974) estimated a moment magnitude of 9.5.

The slip model of rupture is variable from nearly 40 m of displacement in the northern half of the rupture and only 15 m in the southern portion (Plafker and Savage, 1970; Barrientos and Ward, 1990). Postseismic displacements associated with this large event have been reported (Barrientos et al. 1992, in Barrientos, 2007), with observations made in 1991–1992 at several sites previously reported by Plafker and Savage (1970). These observations, together with Beck et al. (1998), concluded that this event was a product of down-dip tension along the subducting slab at a depth varying between 80 and 100 km.

This event confirmed the idea that earthquakes can cause free oscillations of the Earth (Benioff et al., 1961). For the first time it was possible to measure the lower frequency of oscillation of the Earth, with a period of 53 minutes (Cisternas et al., 2003).

3.4. Tsunami Hazards

Tsunamis are another and very dangerous consequence of the Chilean Nazca–South American active margin, especially between Arica and the Taitao Peninsula (Fig. 18.1, 18.2). Since 1966 the Servicio Hidrográfico y Oceanográfico de la Armada (SHOA; Navy Hydrographic and Oceanographic Service) has been responsible for seismic and sea-level monitoring to evaluate the possible approach of tsunamis, as well as sending out tsunamis alert and alarms. SHOA has set up three tsunami categories: (1) Distant field or trans-Pacific tsunamis generated more than 1,000 km away; (2) regional tsunamis generated less than 1000 km away; and (3) near field or local tsunamis, generated between the Chilean-Peru Trench and the Chilean coasts.

Information on distant field and regional tsunami information is provided by the Pacific Tsunami Warning System (PTWS) to SHOA, which sends out the alert. For near field tsunamis, no instrumentation is available or detecting tsunamigenic earthquakes in real time. The only instrument the SHOA has is the Buoy Dart (pressure sensor), which is useful only in the zone between Arica and Antofagasta (Fig. 18.16) To reduce the devastation and fatalities associated with near field tsunamis, it is necessary to follow three procedures: (1) risk evaluation; (2) mitigation measures; and (3) early alert systems (Bernard, 1998; Takehata, 1998; Bernard, 2002; Titov and González, 2002). Nowadays, the risk evaluation and the mitigation procedures are not satisfactorily developed, and the early alert system does not exists.

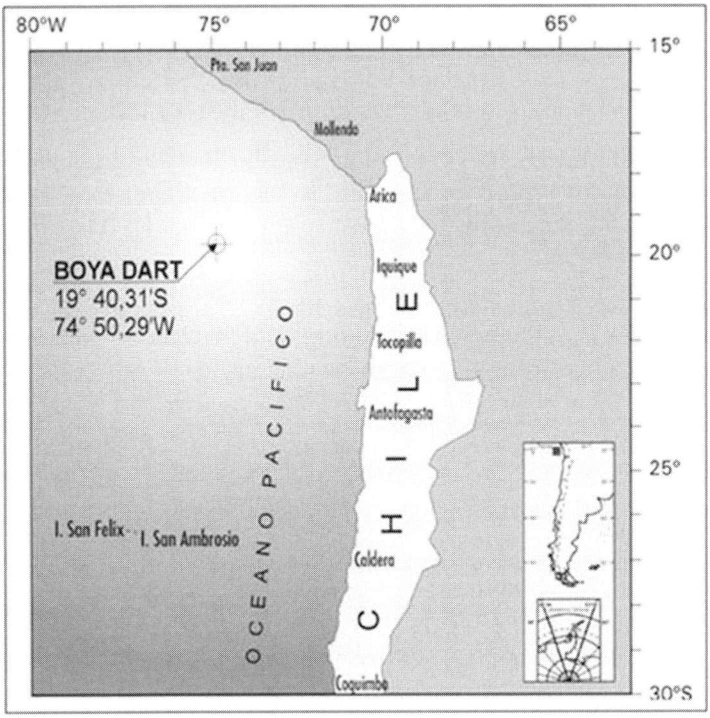

Figure 18.16 Location of buoy DARTof SHOA.

3.4.1. Main Tsunamis

October 28, 1562: On October 28, 1562, a strong tsunamigenic earthquake with an estimated magnitude of M = 8 to 8.5 hit the coastal area between 36°S and 38°S. Concepción was located in the southeast margin of Concepción Bay. The ocean retreated about 10 km and later completely flooded the city of Penco, leaving some crafts aground and destroying all of what had been saved from the earthquake. Some inhabitants survived by moving to higher ground. The tsunami affected 1200 km of coast and claimed numerous victims in several regions (Medina, 1956).

February 8, 1570: After eight years, another tsunamigenic earthquake occurred in the bay of the Concepción area, with an estimated magnitude of M = 8–8.5. The ocean went back about 10 km from the previous the tsunami wave, which killed about 2000 persons and destroyed the city (Medina, 1956).

December 16, 1575: Imperial, Valdivia, Villarrica, Osorno, and Castro (southern Chile) were destroyed by an earthquake in 1575. Shortly after the earthquake, a tsunami reached Valdivia, located on a river of the same name about 25 km upstream from its mouth. The water came rushing upstream, reversing the natural flow of the river. The rising water knocked over houses and uprooted trees. Two ships, riding at anchor in this port, were sunk. In Valdivia 22 persons died (Medina, 1956). The estimated magnitude was M = 8.5.

November 24, 1604: A tsunamigenic earthquake occurred in southern Peru and northern Chile. Many cities were destroyed, including Arequipa, San Marcos de Arica, and Ica. The tsunami flooded about 1200 km of the Chilean coast, causing many victims. Estimated magnitude was M = 8.7. In Arica, the tsunami wave was 16 m high.

May 13, 1647: An earthquake occurred in Santiago and surrounding coastal areas. There are anonymous reports of strong movements of the sea along the whole central coast, with waves sufficiently high to plunge coastal hills. The estimated magnitude was M = 8.7.

March 15, 1657: A strong earthquake occurred in Concepción again, destroying the city and surroundings. There were hundreds of victims of Spanish and natives, who were surrounding the city (Historia de la Compañía de Jesús 1595–1736). The estimated magnitude was M = 8.0.

July 8, 1730: The whole central area of the country was shocked by a strong earthquake of estimated magnitude M = 8.7. The tsunami affected 1000 km of coast, and the city of Valparaíso city was flooded and severely damaged. In Concepción the tsunami wave was 16 m high, which caused important damage.

May 25, 1751: For the third time an earthquake (M = 8.5 estimated) and a tsunami destroyed Concepción. The tsunami arrived at the city about half an hour after the earthquake, with an initial retreat of the ocean followed by three big waves.

April 11, 1819: A severe earthquake and later tsunami occurred in Copiapó, north Chile, and affected along 800 km of the coast. The estimated magnitude was M = 8.5.

November 19, 1822: The second earthquake of magnitude M = 8.3, preceded by several precursors and aftershock earthquakes, occurred in Valparaiso and destroyed most of the public buildings and about 700 houses. The tsunami arrived at the coast 15 minutes after the earthquake, and three waves of tsunami were registered with a height of 3.5 m.

February 20, 1835: For the fourth time in recorded history, the Concepción zone was destroyed by a strong earthquake (estimated magnitude M = 8.3) and tsunami, affecting hundreds of victims. After a half hour of the earthquake, the ocean began to retreat, leaving Talcahuano bay almost dry. The first tsunami wave that arrived was 9 m high, followed by other waves of 13 and 16 m height. The effects of the earthquake and tsunami were described by Charles Darwin.

October 5, 1859: Another earthquake of estimated magnitude M = 7.3–7.7 happened in the area of Copiapó, northern Chile, destroying about 300 houses and buildings. The level of the ocean fluctuated approximately 6 m.

August 13, 1868: Great earthquakes and tsunamis happened in the south of Peru and north of Chile, completely destroying Arica. After 20 minutes of the earthquake, the ocean retreated for about 2 km. The tsunami arrived after several minutes with two 15- to 18-m waves producing more than 200 victims. This tsunami affected practically the whole Pacific Ocean Basin and was registered in Peru, Australia, Alaska, Marquise islands, Chatman islands, New Zealand, Hawaii,

Japan, the Philippines, and the west coast of the United States. The estimated magnitude was M = 8.8.

May 9, 1877: Another catastrophic earthquake and tsunami happened in the north of Chile, destroying several cities. The tsunami began 5 minutes after the earthquake with waves between 10 and 23 m high, as registered in Mejillones (Antofagasta). The tsunami extended between Iquique and Antofagasta and also affected Hawaii, San Francisco, Acapulco (Mexico), Fiji, Samoa, New Zealand, Australia, and Japan. The estimated magnitude was M = 8.8.

November 10, 1922: The main destruction area in the north of Chile was between Copiapó and Coquimbo, causing more than 500 victims at Copiapó. The tsunami arrived rapidly at the coast between Huasco and Caldera. The sea-level elevation was gradual and reached 9 m over high tide in Chañaral, where it was very destructive. The greatest part of the damages was likely caused by the water retreat. The estimated magnitude was M = 8.4.

May 22, 1960: The major earthquake registered by instruments was called the Valdivia Earthquake (in the south of Chile). The earthquake and tsunami affected the whole south-center of Chile, causing more than 1000 victims. Several ships sank. Enormous deformation took place in the continental coastal crust with elevations of up to 6 m on Guamblin island and 5 m on Guafo island. Inland from the area surrounding Valdivia larger than 2-m subsidence was produced. The tsunami waves were propagated in the whole Pacific Ocean, causing serious damages and numerous victims. The height of the waves was between 15 m and 18 m. The calculated earthquake magnitude was M = 9.5.

July 30, 1995: An earthquake in Antofagasta also produced damages in Taltal, Mejillones, Calama, San Pedro de Atacama, and Tocopilla. The tsunami wave was 2.8 m high in Antofagasta, and it was also registered in the Marquise Islands and Hawaii with waves of 0.75 m. The calculated magnitude was M = 8.0.

3.5. The Concepción Tsunami Hazard: A Case Study

The seismotectonic region of Concepción extends from Constitución to south of Arauco Peninsula (Figs. 18.17 and 18.18). About one million people live in this high hazard zone. Five tsunamigenic earthquakes affected the area: in 1562, 1570, 1657, 1751, and 1835 (Lomnitz, 1970; CERESIS, 1997). Sesimic data analysis concludes that this area is considered a seismic gap (Campos et al., 2002; Barrientos, 2007) (Figs. 18.18 and 18.19).

Analysis of the vertical deformation of the Arauco Tectonic Block, including the associated islands, enabled the identification of three main fault systems: the Interface Fault (with parameters calculated for the co-seismic movement), the Lanalhue Fault Zone (NW-SE), and the forearc Bío Bío Fault Zone (Fig. 18.20) (Moreno, 2004, modified from Engdahl et al., 1998.).

Recent geodynamic studies (PISCO, SFB-267, and TIPTEQ projects) carried out in the active Peru-Chile margin have shown that the highest convergence

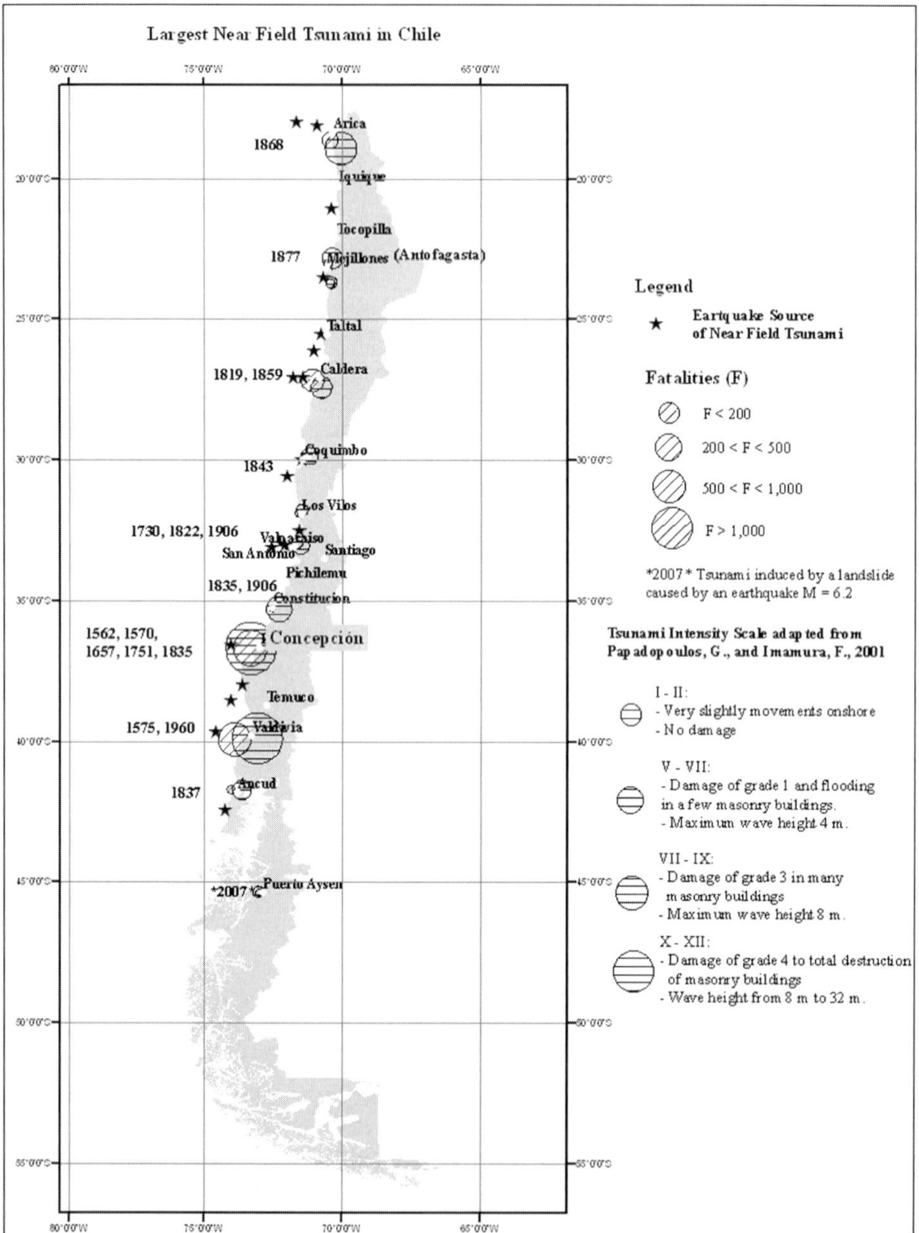

Figure 18.17 Greatest near field Tsunami in Chile generated by tsunamigenic earthquakes. The tsunami occurred in Aysén (2007) was caused by a landslide.

velocity is concentrated in the Concepción area and that coast and islands are increasing their altitude. This evidence suggests that this area has the highest accumulation of tectonic energy in the whole of Chile.

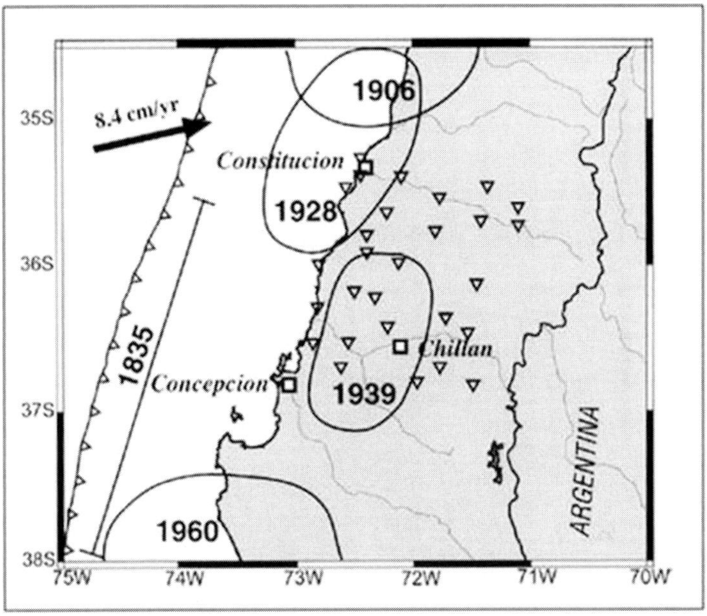

Figure 18.18 Source areas of the largest M $\geq$ 8 Chilean earthquakes of the last 95 years. The ellipsoidal patches define the estimated source areas for these events. The estimated fault length of the great 1835 earthquake is indicated by a solid line. The 1939 Chilean earthquake is not a thrust earthquake. Inverted triangles correspond to seismic stations deployed during the period March 1 to June 1, 1996 (from Campos et al., 2002).

Specifically, the superficial field of velocities (Fig. 18.21, left figure) indicates that the Arauco-Constitución seismic gap (segment "A" in the left figure) has the high velocity of the active margin of Chile (Klotz et al., 2001; Ruegg et al., 2002; Brooks et al., 2003; Moreno, 2004). Also, the N-S and W-E velocity models (Klotz et al., 2001; Ruegg et al., 2002; and Brooks et al., 2003) (Fig. 18.21, central and right figures) confirm the Arauco–Constitución anomaly and the high-convergence velocity associated with the Arauco Block, compared with other Chilean tectonic zones.

Isla Santa María at the active margin of south-central Chile is the result of earthquake-related uplift and deformation in the forearc since at least late Pleistocene time. Field mapping, dating of key depositional horizons, and analysis of seismic-reflection profiles reveal ongoing deformation in this sector of the Chilean forearc. The 30 km^2 island is located 12 km above the interplate seismogenic zone and 75 km landward of the trench. It is situated near the southern termination of the Concepción earthquake rupture segment, where Charles Darwin measured 3 m of co-seismic uplift during a M > 8 megathrust 1835 earthquake (Barrientos, et al., 1990; Melnick et al. 2006). Kaizuka et al. (1973) and Nelson and Manley (1992) calculated the tectonic uplift of this island at about 5 to 70 mm/yr. The 1835 co-seismic uplift was estimated to range from 1.5 to 3 m in this region (Melnick et al., 2006).

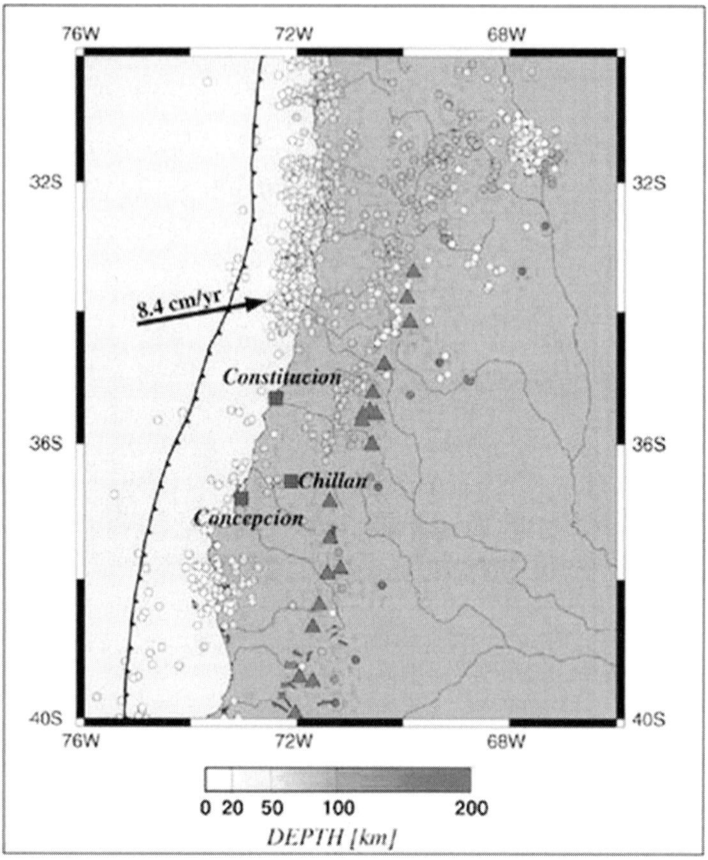

Figure 18.19 Seismicity of M > 5 in the Concepción–Constitución area since 1980 determined from the new seismic catalogue of Engdahl et al., 1998. This region looks less seismic than the rest of central Chile (from Campos et al., 2002).

On the basis of those studies and the historic periodicity of events, it can be concluded that the Concepción area could soon experience a significant and dangerous tsunamigenic earthquake. Moreover, the Bío Bío canyon could concentrate a west-generated, near field tsunami favoring its inland penetration along the Bío Bío River, thus increasing damages all along its banks (Pineda, 1999). For the Geodynamic Sector of Concepción, SHOA considered the characteristics of the last tsunamigenic earthquake that occurred in 1835. Figure 18.22 shows the flooding area (7.0 m)(Cecioni et al, 2004) that should be affected by a tsunami similar in intensity to that of 1835. The cities of Talcahuano and San Vicente would likely disappear, and the Tumbes Peninsula could be transformed into an island.

The historical occurrence of near field tsunamis, produced by tsunamigenic earthquakes, shows that actual return periods have varied significantly: 8 years (1562–1570), 87 years (1570–1657), 94 years (1657–1751), and 84 years (1751–1835). No subduction earthquake has occurred with focal mechanism located

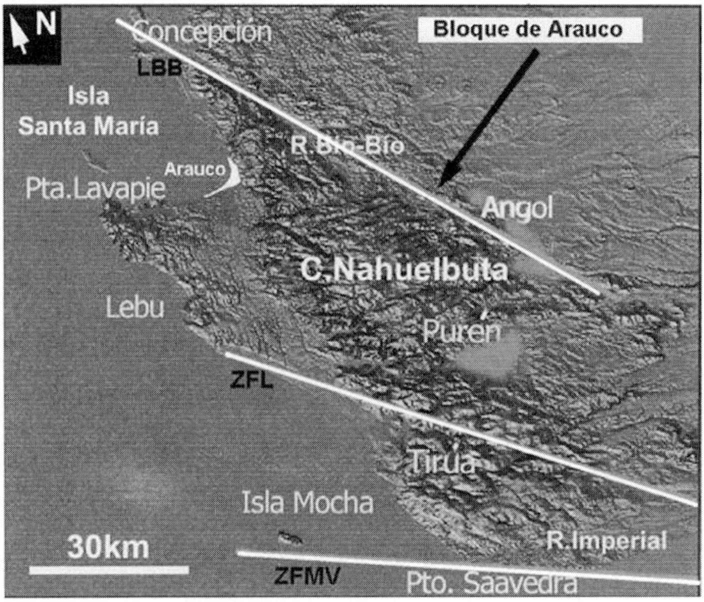

Figure 18.20 Arauco tectonic block, showing the three main fault systems (from Moreno, 2004) (modified from Engdahl et al., 1998).

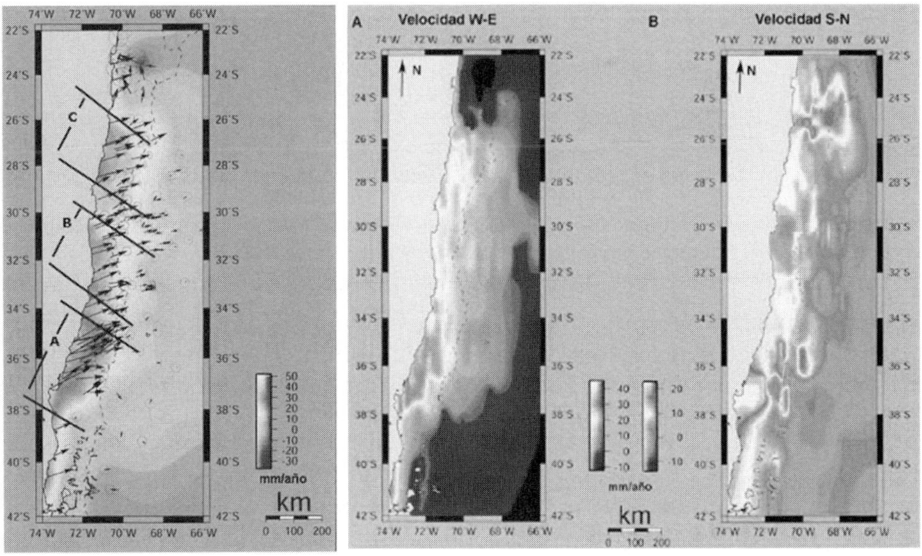

Figure 18.21 Superficial field of velocities indicates that the Arauco-Constitución seismic gap (segment "A" in the left figure) has the high velocity of the active margin of Chile. The N-S and W-E velocity models confirm the Arauco–Constitution anomaly and the high-convergence velocity associated with the Arauco Block, compared with other Chilean tectonic zones (adapted from Moreno, 2004).

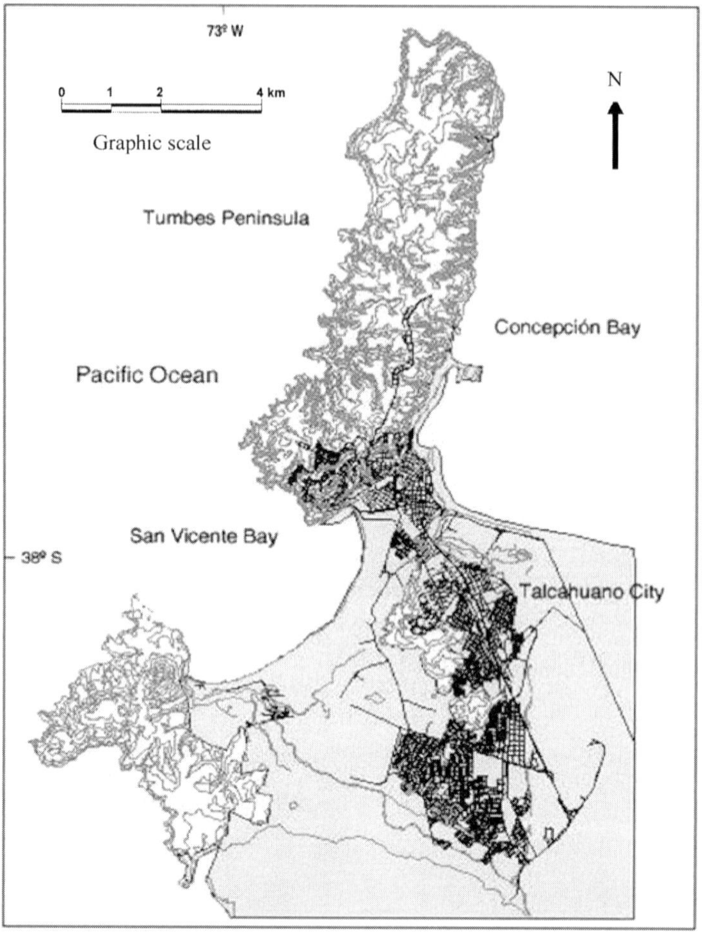

Figure 18.22 Integrated inundation Chart of Concepción and San Vicente bays (flooding area in clear grey color) (modified from charts of floods by tsunamis [SHOA]).

in the seismotectonic region of Concepción-Constitución since 1835 (174 years, nearly twice the longest historical return period). Consequently, it appears very likely that a tsunamigenic earthquake could occur in the near future. Considering that the coastal population in the area is about 1 million, this is not a reassuring scenario.

4. Final Remarks

Clearly, Chile has a high level of risk owing to a variety of geologic and geomorphic processes. It hazard is very high because practically all dangerous geologic processes affect the country. Exposure and risk are also considerable because many cities, some of them with large populations, are located in high-hazard zones. The country also has an added vulnerability due to the linear nature of essential infrastructure such as highways, located in hazard areas and often without alternative.

Not enough effort has been made to carry out detailed geologic and geomorphologic studies that could help mitigate the dangers of natural hazards in Chile. Unfortunately, such studies are not required in Chile for land-use planning. It is also necessary to design and implement programs to educate the population about hazard perception and preparedness.

Other structural and nonstructural mitigation measures for the hazards described are well-known. Seismic building codes have proven their effectiveness throughout the world. Stabilization works could solve the problem in the case of specific, well-identified landslides. The monitoring and prediction of volcanic eruptions and tsunamis is somewhat more advanced, though still far from satisfactory. Nevertheless, although not very appropriate to reduce material damage, monitoring and warning systems would reduce loss of lives. Instruments such as seismometers, accelerometers, geodetic GPS, and others should be installed. Data should be collected and transmitted in real time at institutions responsible for issuing alert and alarm warnings.

For the particular case of Concepción, where 174 years have elapsed since the last destructive tsunamigenic earthquake, very high deformation energy is being accumulated, and 1 million persons live in exposed coastal areas. In this situation, obviously it is extremely necessary to install monitoring, early warning, and alarm systems, as well as elaborate evacuation and self-protection plans. In all cases, detailed hazard and risk analyses, including zoning maps and their incorporation into rigorous land-use plans that take them into account, would be essential to reduce future exposure and help implement more sustainable land-use practices.

Acknowledgments

We give special thanks to Antonio Cendrero for his detailed review of the manuscript and invaluable comments. Edgardo Latrubesse and an anonymous reviewer also provided valuable suggestions that helped improve the original manuscript. We also thank the students of the Engineering Geology course (2007) for their dedication and professional passion during the investigation that furnished part of the material used for this chapter.

CHAPTER 19

Impact of Global Climate Change on Glaciers and Permafrost of South America, with Emphasis on Patagonia, Tierra del Fuego, and the Antarctic Peninsula

Jorge Rabassa[a]

[a]*CADIC-CONICET, and Universidad Nacional de la Patagonia-San Juan Bosco, C.C. 92, 9410 Ushuaia, Tierra del Fuego, Argentina*

Contents

1. Global Climate Change

Global climate change (GCC) can be recognized at the global level through rising mean annual or seasonal temperature, rising or diminishing regional precipitation, rising global sea level, and a general increase in the frequency and intensity of extreme meteorological events. The impact of GCC, being perhaps both beneficial and damaging according to different regions, has been observed along the entire South American Andes, but particularly in Patagonia, Tierra del Fuego,

Developments in Earth Surface Processes, Volume 13
ISSN 0928-2025, DOI 10.1016/S0928-2025(08)10019-0

and the Antarctic Peninsula. Its impact has been clearly shown since 1978, when the Andean glaciers started to retreat, its intensity increasing with time. The cited regions are characterized by their high vulnerability, deriving from their location in the mountain ranges where the regional snowline is very close to the summits, their extreme climates, and their high inherent variability. In southern South America and the Antarctic Peninsula, their geographical location with respect to the southern oceans and the Antarctic Circum-Polar Current is also very relevant to this problem. The problem of GCC has been amply discussed by the present author (Rabassa, 2007a, b) and several international media and website presentations, at least since 2004. Moreover, already in 1986 and much before GCC had become a daily issue in the world press, and our research group had predicted the total melting of one Argentine Patagonian glacier due to regional warming, the Castaño Overo ice-cone. Unfortunately, our prediction was all too precise: the total melt-out happened a few years later, during the 1990s (Bertani et al., 1986).

When discussing GCC, natural climatic variability is in fact being considered, together with anthropogenic perturbation of the atmosphere, both in the lower atmosphere or troposphere and in the stratosphere. The sun is the essential source for heat at the Earth's surface. The atmosphere, or more precisely, certain gas molecules in the atmosphere, retain a portion of the incoming solar energy before radiating it back into outer space. This mechanism is known as the greenhouse effect. Thus, the greenhouse effect provides higher global temperatures than expected according to the distance of our planet from the sun. If not for these conditions, our planet would be much colder than it is, with no oceans or free water, and permanent and extensive ice caps would descend from the poles and mountain chains, a similar situation as that which prevails today on Mars. The Earth's greenhouse effect, which has been active at least since Late Precambrian times (>600 million years ago), has had a great influence on the occurrence and development of life because it has allowed and preserved the existence of the oceans as such, as well as the evolution of terrestrial organisms, both plants and animals. The mild but efficient greenhouse effect in the Earth contrasts with its tremendous effect on Venus, where the thick atmosphere keeps a boiling surface where no free water is present, or with its almost total absence in Mars, where the atmosphere almost disappeared in a distant past and no significant greenhouse effect exists today, though it was certainly effective in ancient ages.

In contrast with the widely accepted anthropogenic origin of global warming due to increasing greenhouse gas content (data by the International Panel on Climate Change, a UN organization; IPCC, 2001, 2007), other scientists have suggested that higher global temperatures are related solely to variability in solar energy emission, denying human impact at such a level. Active investigations today will elucidate these matters in the near future.

Earth's climatic variability has been significant at all timescales in planetary history. Glaciations occurred several times in the geological record, in the Precambrian, the Ordovician, the Carboniferous and Permian, with the last three during the Paleozoic and much more recently during the Late Cenozoic, particularly during the last 10 million years. In contrast, hypertropical climates, extremely warm and wet climates, developed during Mesozoic times, especially during the

Jurassic and Cretaceous (IPCC, 2001, 2007; Berner, 1997). Then no polar ice caps existed, and dinosaurs were living in very high latitudes, probably even beyond the position of the polar circles, perfectly adapted to long polar nights or developing migration strategies.

This climatic variability has also been very significant in the studied regions during the Quaternary (roughly the last 2.5 million years), due to modifications in certain parameters of the Earth's orbit, such as its eccentricity, the obliquity of the terrestrial axis, and the equinoccial precession. These regular modifications are known as the Milankovitch cycles, after a Serbian astronomer and climatologist who predicted these changes in the early twentieth century. These variations have been particularly relevant to humankind during the Late Pleistocene (120 to 15 thousand radiocarbon years ago [^{14}C ka BP]), the Late Glacial (15–10 ^{14}C ka BP), and throughout the Holocene (the last 10 ^{14}C ka BP) until present times.

The atmospheric concentration of CO_2 has varied significantly during the history of the Earth, being related to changes in the extent of continental surfaces and terrestrial ecosystems, the orbit of the Earth, global climate, temperature of the oceans, and volcanic activity. But no major, large-scale changes have been generated by human activity until recent times, with perhaps the exception of agricultural development in the earliest Holocene (Ruddiman, 2005).

The Earth's natural climatic variability has had superimposed on it human-induced climatic modifications. This has occurred since the Industrial Revolution, particularly since 1850, because of the increase in CO_2 and other greenhouse gases emissions. Such gases are due to fossil fuel combustion, to the extensive development of agriculture and cattle-raising, and to the reduction of forest areas and other vegetated portions of the landscape, which act as natural sinks for excess CO_2.

Information provided by the IPCC shows that human influence on atmospheric changes is enormous, not only by the carbon dioxide content, but also methane and nitrous oxide. All the information presented in this chapter has been extensively discussed by IPCC (IPCC, 2001, 2007) and the reader is referred to their website (www.ipcc.ch). The content of greenhouse gases in the atmosphere increased dramatically after 1850, reaching concentrations that had never been attained in human history before, as shown by measurements in Arctic and Antarctic ice cores.

These gases, both of natural or human-made origin, have a large greenhouse effect due to their relative molecular size, physicochemical properties, and atmospheric concentration. Other gases are also important, such as tropospheric ozone and artificial halocarbonates (the so-called CFCs—chlorofluorocarbons).

The most important of the greenhouse gases is CO_2, although its total concentration in the atmosphere is only $<1\%$. The present CO_2 concentration, at around 380 ppm, is the largest during the last 20 million years since the middle Miocene and before the Alpine-Andean orogeny deeply modified global climate. This concentration was achieved with an increase of only 30% with respect to the atmospheric content of the last 1000 years. Growing CO_2 concentration was followed by a sharp increase in mean annual global temperature, reaching an historical maximum in 1998, the warmest year in the last millenium.

Global temperatures in the last several centuries have varied significantly due to changes in solar activity ("sun spots"), which experienced an extensive minimum between 1650 and 1720 (known as the Maunder Minimum) and again between 1790 and 1840 (the Dalton Minimum), diminishing the total solar radiation received in the Earth's surface. These minima were responsible for a period of lower global temperatures known as the Little Ice Age. After 1840, solar activity returned to normal conditions, under the present 11-year sun-spot cycle.

From 1850 through the twentieth century, global temperature increased constantly, with short cooling episodes occurring around 1910, 1945, and 1970. With regard to the global impact of these cool periods, the glaciers of the world advanced in the 1940s and the 1970s, with 1977 being the last year of general glacier advance in the Swiss Alps and 1978 the last advance in glaciers of Northern Patagonia, which have been receding since then (Brandani et al., 1986). In recent years, global mean annual temperature has increased considerably, with the only exception being 1992, following the explosion of Pinatubo Volcano in the Philippines. This event sent enormous amounts of volcanic ash into the stratosphere, partially blocking the incoming solar radiation and cooling the atmosphere.

In regional terms, the winter temperature increase for the last 50 years has been larger in the Arctic and Sub-Arctic regions and the Antarctic Peninsula, with the exception of the Labrador/Greenland area. The summer temperatures have gone up also elsewhere, particularly in the Antarctic Peninsula, northeastern Africa, and the Mediterranean Sea.

The impact of GCC may be considered as either harmful or beneficial in human terms, both socially and economically, regarding the different geographical areas involved or the kind of human activities considered. For example, among the beneficial impacts of GCC for South America, the displacement of the southern areas toward more benign climates and the southwestward widening of the agricultural frontier in the Argentine Pampas may be cited. The negative impacts on our continent are clearer and stronger, such as biodiversity and forest mass loss, degradation of ecotonal fringes, a higher frequency of extreme hydrological events such as flooding and drought, reducing or dissapearing permafrost conditions in the Andean ranges above the treeline, wetland and peatland dissecation, rising climatic snowline, and fast recession of mountain glaciers and snowfields, among many others (IPCC, 2001, 2007).

IPCC has studied these impacts and their regional distribution (IPCC, Third and Fourth Annual Reports, 2001, 2007). In addition, IPCC has prepared models that could predict the global and regional climatic and environmental conditions for the remaining portion of the twenty-first century. These mathematical models are of the probabilistic type, and they have been prepared following a large variety of global and regional, tentative scenarios, according to the possible, expected social, technological, and economic development of humankind. Each of these scenarios takes into consideration the cultural, political, and technological variables that would expand, maintain, or reduce greenhouse gas emissions in the future. These models predict possible consequences for the global and regional climate, and their influence on ecosystems, agroecosystems, and society. Once these consequences have been identified, processes to mitigate impact are proposed to the regional governments and international organizations.

The impact of climatic change is larger when regional vulnerability is higher. These conditions are particularly significant in Latin America and the polar zones (IPCC, 2001, 2007). In the case of the impacts on Latin America, IPCC opinion is as follows:

1. The adaptative capacity of human systems is low, particularly in respect to extreme climatic events, and vulnerability is very high.
2. A glacier volume loss and recession would negatively impact surficial runoff in those areas in which glacial melting is an important water source.
3. Floods and droughts will occur more frequently, with increasing sedimentary load and water-quality degradation.
4. An increase in tropical cyclone intensity would have an effect on life, property, and ecosystem hazards generated by strong rain storms, flooding, wind, and extreme events.
5. The yield of crops will decrease in most of Latin America, even if the favorable effects of CO_2 are taken into consideration. Subsistence agriculture in certain regions will be at risk.
6. The geographic distribution of infectious diseases will expand polewards and also to higher areas, and exposure to diseases such as malaria, dengue fever, and cholera will increase.
7. Coastal human population, economic activities, infrastructure, and manglar ecosystems will be negatively affected by rising sea level.
8. The rate of biodiversity loss will increase.

In the polar zones, and particularly in matters that apply to the Antarctic Peninsula, IPCC scientists indicated that (IPCC, 2001, 2007)

1. the natural systems in the polar zones are highly vulnerable to climatic change, and the present ecosystems have a low adaptative capacity.
2. the polar zones are expected to be among those with larger and stronger impact, particularly in the Antarctic Peninsula and the Southern Seas.
3. climate changes have an impact on lesser extension and thickness of sea ice, increasing coastal erosion, melting out of permafrost, changes in ice sheets and platforms, and the distribution of ice-related species.
4. some ecosystems may adapt to climate change by the replacement of species by migration and ecosystem composition, and perhaps by increase in total productivity. Ice marginal systems providing habitat for certain species are endangered.
5. the polar zones have climatic-change conduction mechanisms ,which, once triggered, would continue for centuries, even after the greenhouse gas concentration had been stabilized. Irreversible impact is expected on ice platforms, global oceanic circulation, and sea-level rise.

If a big portion of meltwater from Antarctic ice merges into the ocean, sea level will rise significantly. The West Antarctic Ice Sheet alone has enough water to raise the ocean by 5 m. A 1-m rise would be expected to cost the Netherlands alone US$ 10 billion to defend their coastal lowlands. Similarly, the cost of sea rise impact in Bangladesh could be measured in the lives of the millions who live on the fertile, coastal, deltaic lands (Shepherd et al., 2005).

2. Climate of the Twenty-first Century and Its Impact on Glaciers

According to the IPCC (2001, 2007), the climate of the twenty-first century will be characterized by increasing global mean annual temperatures and, as a consequence, a slow but constantly rising sea level. Mean annual temperature in the year 2100 would increase between a minimum of 1.4 °C and a maximum of 5.8 °C, above 1990 conditions, including the statistical errors of the method used. This temperature increase would push a global sea-level rise between 0.2 m and 0.7 m. A probabilistic analysis of the studied conditions indicates that the lower 95 % confidence bound is located at 0.9 °C and that the upper 95 % is at 4.8 °C (Jacoby et al., 2001), with a median of 2.5 °C and a highest probability density at 2.0 °C. Even these middle values are very important in terms of mean annual temperature variability, a climatic parameter that is known to have high stability. A similar increase in temperate latitudes during the Hypsithermal period (around 8000–6000 ^{14}C years BP, middle Holocene) is considered to be responsible for the global Flandrian transgression, the flooding of coastal areas all around the world, and the development of the Diluvium myth in many world cultures.

Today, there are many proofs of climate change, temperature elevation, and sea-level rising at all latitudes. In addition to all instrumental records, one of the most outstanding examples of global impact is its effects on glaciers and sea ice.

3. A Few Examples of Global Climate Change Impact on South American Glaciers

South American glaciers have been receding constantly since the mid-nineteenth century, after the end of the Little Ice Age, as has been the case for glaciers almost everywhere else in the world (Rabassa et al., 1984; Brandani et al., 1986; Oerlemans, 1994; Warrick et al., 1996).

3.1. Colombia

The small equatorial glaciers of Colombia, most of which are on high tectonic peaks or huge volcanoes along the northernmost Andes, have been rapidly receding since the 1980s. Multiple evidence of recession was identified during the glacier inventory of Sierra de Santa Marta, in the northernmost part of the country near the Caribbean coast, based on glacial landform identification on aerial photographs and topographic maps (Rabassa et al., 1993).

According to Ceballos et al. (2005), the Colombian glaciers have lost 50% of their area in the second half of the twentieth century. Glacier shrinkage continued constantly between 1990 and 2005, with an area loss of 10 to 50%, but no sensitive acceleration trend is yet evident. Glacier retreat in terms of length change was established about a mean of 15 m per year. Several peaks of the Colombian Andes have totally lost their glaciers during the last decades. In particular, the loss of ice mass has been more intense in Nevado Santa Isabel and Tolima, whereas the Nevado del Huila has had a smaller glacier loss.

3.2. Peru and Bolivia

Mark et al. (2005) have demonstrated that the Yanamarey Glacier in the Cordillera Blanca of Peru experienced increased meltwater runoff during the 2000–2004 period compared to 1998–1999, as demonstrated by the increased contribution to total annual mean discharge in the basin of 58% compared with 23% during the previous period due to regional warming. At the present rate of melting, the future of this glacier is uncertain. In general terms, the Cordillera Blanca underwent a general reduction in glacier volume throughout the twentieth century. Compared with the existing glacier inventory based on 1962–1970 aerial photographs (Ames et al., 1989), which listed a total glacier area of 723 km^2, a 1990 satellite-imagery survey has shown a reduction of the glacierized area down to less than 600 km^2, more than 20% of the surface area (Georges, 2004). Similarly, in Bolivia, the small Chacaltaya Glacier could disappear within the next decade, causing at least a 30% loss of total stream discharge within the basin where it is located and generating serious problems to the local agricultural communities (Ramírez et al., 2001).

3.3. The Central Andes of Argentina and Chile

Leiva et al. (2007) have amply discussed the glacier recession in the Central Andes. According to Escobar and Aceituno (1998) and Casassa et al. (2003), the meteorological stations in central Chile showed a 0.3 to 0.7 °C increase in mean annual temperature (MAT) during the last century. Likewise, the 0 °C annual isotherm has been pushed upward in the mountains by 150 m in winter and 250 m in summer during the last three decades due to regional warming. This has had significant impact on glacier nourishment. The Agua Negra Glacier (Province of San Juan, 30°10' S; 69°50' W) lost most of its thickness during that period (Leiva, 1999), though the frontal recession has been minimal. The glaciers of the Río Plomo valley (33°10' S; 70°15' W), included in the Río Mendoza Basin, have shown frontal recession of several kilometers between 1914 and 2005, with a minimum loss of ice mass of 1500 x 10^6 m^3 (Leiva et al., 1989; Espizúa, 1986; Leiva et al., 2007).

Casassa et al. (2003) have indicated that the glaciers of central Chile (32–34° S) have been receding for decades. Snowline data collected during the last 20 years show a clear tendency toward rising, based on direct observations and ski area data. Casassa et al. (2003) also stated that if the climate conditions continue with the same rates of change, glacier shrinkage will increase rapidly in forthcoming decades.

Finally, Brenning (2005) has demonstrated that rock glaciers at lower altitudes in the Andes of central Chile (33–35° S) are presently not in equilibrium with modern climate, due to regional warming. Since these ice bodies are significant water reservoirs, they will be seriously affected by GCC, losing water storage to the regional stream basins.

3.4. The Patagonian and Fuegian Glaciers

Regional snowline or permanent snowline is defined as the line that joins the lowest position on the landscape (particularly, the mountain landscape) at the end of the melting season (usually the beginning of fall) of the snow accumulated during the last winter. The equilibrium line is the position of a line that separates the area with net annual accumulation above (in terms of mass balance) from that of net annual ablation below. In some glaciers, this line may be located slightly below the regional snowline as a result of slushflow, recrystallization, and other translocations of snow and ice across the snowline, which extends the net accumulation area downslope. In the case of Patagonia and the Antarctic Peninsula, increasing mean annual temperature, particularly summer temperatures, has had a sizable effect on the position of the regional snowline, and thus, the equilibrium line, forcing a rise of more than 200 m for the last 20 years in Patagonia and Tierra del Fuego, and perhaps up to 100 m in some areas of the Antarctic Peninsula. This has forced a general recession of most of the Patagonian and Fuegian glaciers, mostly due to loss of accumulation area, rising temperatures at the glacier snout elevation, and increase of ice calving in lakes or the sea. This general recession of the Patagonian glaciers has been observed for more than 20 years (Aniya and Enomoto, 1986). These authors observed that between 1944 and 1984, there was a maximum recession of around 2.5 km at two calving glaciers. The amount of the glacier surface lowering ranged from 40 to 120 m during the last 40 years. In a more recent paper, Aniya (1999) estimated the Patagonian glaciers' contribution to sea-level change from increased melting. The total sea-level rise due to the Patagonian glaciers alone amounts to 1.93 ± 0.75 mm for the last 50 years, which is 3.6% of total changes in sea level. Analyses of climatic data provided by meteorological stations located around the icefield revealed a slight increase in air temperature and a decrease in precipitation over the last 40 to 50 years (Aniya, 1999).

The famous Perito Moreno Glacier (Parque Nacional Glaciares, Province of Santa Cruz, southernmost Patagonia) (Fig. 19.1) keeps advancing actively year after year, blocking Brazo Rico, a branch of Lago Argentino and generating an ice wall that later collapses when the accumulated water pressure in the southern side of the wall exceeds the ice resistance. When the wall breaks, it is a stunning event, which attracts tourists and naturalists from all over the world who come in large numbers to see it. This anomalous behavior is probably due not to climatic factors but to internal, glaciological forcing or to recurrent, small-magnitude seismic events, though large enough to induce glacier sliding. Skvarca (2006) has studied these anomalous glacier fluctuations. This glacier is presently at an almost steady-state condition, having reached its maximum extent in 1947, suggesting a recent increase in mass balance. It is highly probable that its Chilean counterpart, the Pío XI Glacier, acts in a similar way (Rivera and Cassassa, 1999). Rivera and Cassassa (1999) have estimated that the ice front in this glacier has advanced significantly in past decades probably due to variations in surging and regional equilibrium line

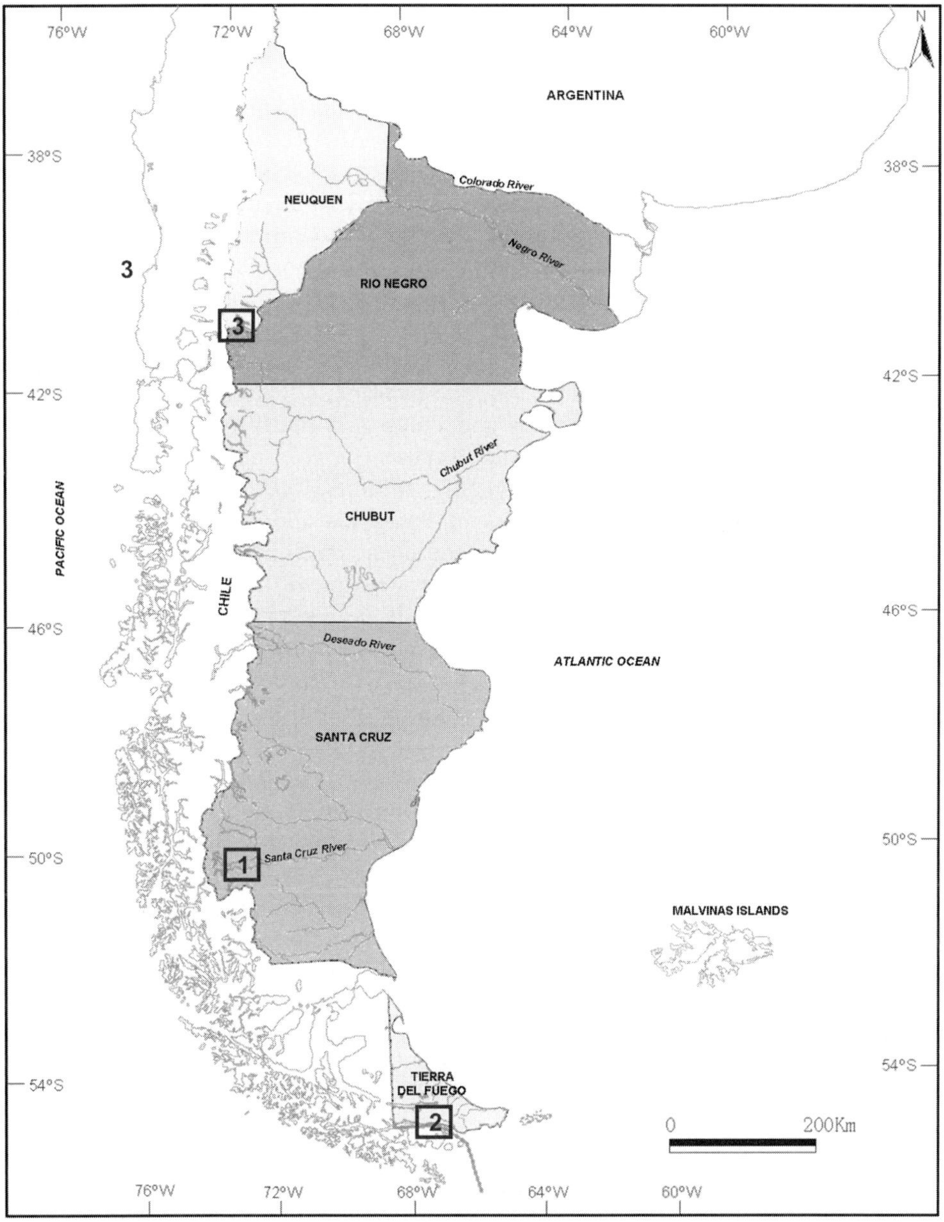

Figure 19.1 Location of the main areas with glaciers described in the chapter. 1: Upsala, Perito Moreno; 2: Monte Alvear, Martial; 3: Tronador (Rio Manso-Castaño Overo).

altitude (ELA) in relationship to glacier morphology. However, these authors observe that the constant rise in ELA will lead to a rapid decline of this glacier in the future. In the Torres del Paine National Park, Rivera and Cassassa (2004) have established that the glaciers in this park have lost a total of 62.2 km^2, or 8 % of the original ice area of 1945, with a maximum ice thinning of up to 7.6 m per year during the period under study.

The Upsala Glacier, the largest glacier in Argentina and one the biggest in South America and the Southern Hemisphere outside of Antarctica, is undergoing a clear, catastrophic recession both in its front and its thickness. Just in the last decades, its frontal recession has reached 8 km (Figs. 19.1, 19.2, 19.3, and 19.4). The floating portion of its tongue partially collapsed after the photograph shown in Figure 19.1 was taken, allowing a much deeper penetration of the boats sailing into this fjord-like branch of Lago Argentino. Between April 1999 and October 2001, the glacier front fluctuated seasonally within about 400 m, in contrast to the dramatic recession of previous years. During this period, the Upsala Glacier west terminus had a net advance of around 300 m. In addition, the available satellite images allowed the observation of recent calving speeds and confirmed the improved calving-rate/water-depth relationship (Skvarca et al., 2003).

A similar destiny may befall the smaller, mountain glaciers and discharge ice tongues from the surviving ice sheets in Patagonia and Tierra del Fuego: the Northern Patagonian Ice Sheet, the Southern Patagonian Ice Sheet, the Darwin Cordillera Ice Sheet ,and some other smaller ice caps in the Magellanic Archipelago. On the Argentine side of the Isla Grande de Tierra del Fuego, the Alpine glaciers of the Fuegian Andes are in rapid retreat. See, for example, the photographs corresponding to the Martial Glacier and the Alvear Este Glacier (Figs. 19.1 and Figs. 19.5 through 19.9). The front of the glacier continued receding very rapidly during 2007 and 2008, as shown in Figure 19.8. The cave in the left has almost disappeared, due both to ice melting and collapsing of the cave roof. The left margin of the glacier has strongly receded, losing contact with the rocky outcrops to the left of the picture and totally uncovering the ancient, blackish ice that was

Figure 19.2 Upsala Glacier, calving ice front in Brazo Norte, Lago Argentino, Glaciares National Park, Province of Santa Cruz, Argentina, 1981. (Photo: J. Rabassa).

Figure 19.3 A mountain glacier, tributary of Upsala Glacier from the western slope of the valley, seen from several kilometers from the south, 1981. Note the upper surface of Upsala Glacier front at the foreground of the photograph, illustrating the position of the ice front at that time, proximal to the ship from where this photo was taken. (Photo: J. Rabassa).

Figure 19.4 The same tributary, now seen from the north, 2004. The Upsala Glacier calving ice front had receded more than 8 km in that period, allowing boats to sail the Lago Argentino to locations that were not reachable in 1981.

exposed in 2005–2006 (Fig. 19.8). Similarly, the entire front has undergone a fast melting process, and new outcrops have become exposed (Fig. 19.8). These photographic documents prove that most of the lowest portions of the glacier will be gone by about 2015. Most likely, around the year 2020 AD, most of them will have vanished, generating a priceless loss from the point of view of

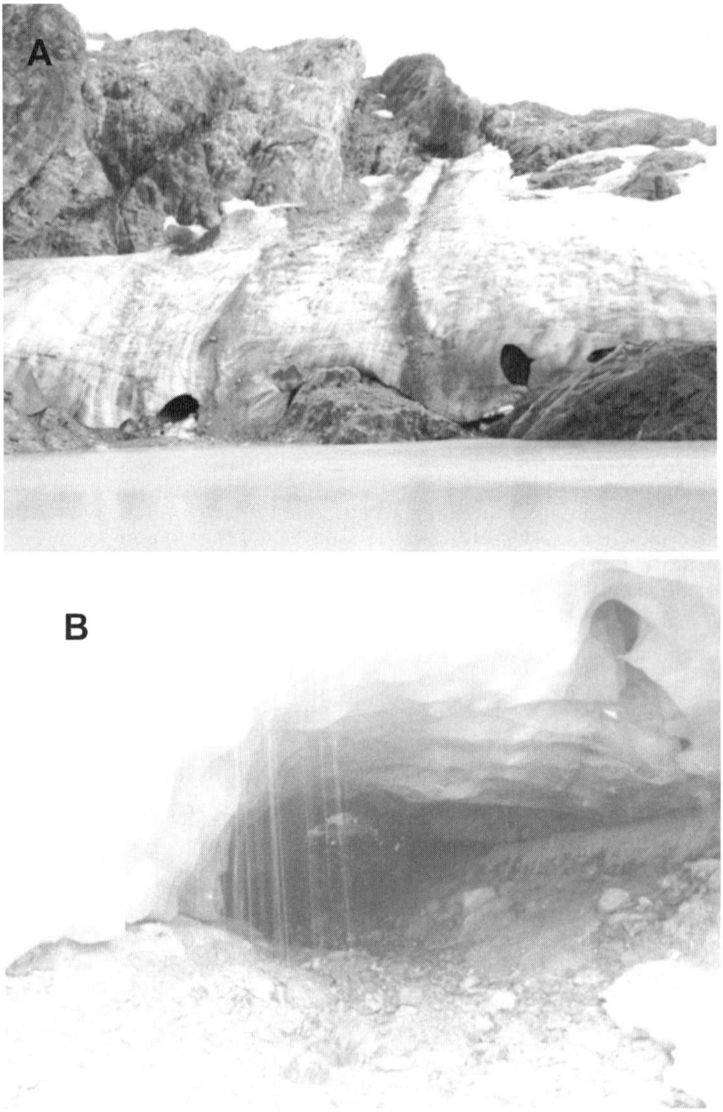

Figure 19.5 (A) Monte Alvear East Glacier, Fuegian Andes, lat. 54° S. Southern portion of the ice margin as seen in February 2004. Note the relatively small size of the ice cave at the left portion of the glacier front and the dimension of the rocky outcrop straight away to the right of it. (B) The ice cave shown in Figure A, February 2004. Note the little illuminated area at the far end of the cave. Photos: J. Rabassa.

environment, hydrology, water resources, Alpine wetlands, and scenic and tourism resources, as well as in terms of natural and cultural legacy.

In Northern Patagonia, the consequences have been similar. The Río Manso Glacier (Cerro Tronador, Nahuel Huapi National Park, Northern Patagonia; lat.

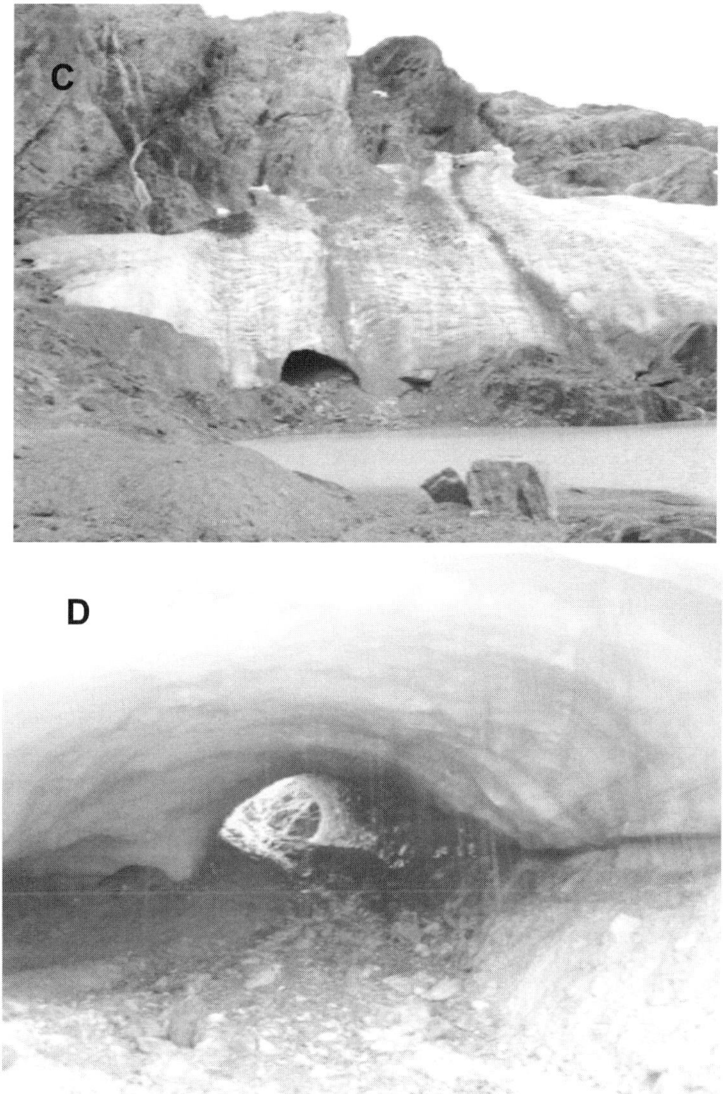

Figure 19.6 Monte Alvear East Glacier. Southern portion of the ice margin as seen in February 2005. Note the enlarged size of the ice cave and the emergence of a large rock boulder in the ice wall directly to the right of it, which was previously unobserved in the 2004 photograph, revealing a noteworthy ice front recession of several meters. Note also the exposure of dark ice remains beneath the whitish, more recent ice, to the left of the ice cave. This debris-rich, dark ice is a very old ice remnant, clearly pre-Little Ice Age, perhaps a thousand years old, or even extending perhaps from the Last Glacial Maximum (25,000 years ago). The actual age of these ice remains has not yet been investigated. Regrettably, it is likely that it will melt away before suitable sampling can be fully accomplished. The rocky outcrop to the right has increased in size as well, as the ice front has receded. (B). The ice cave shown in Figure 19.3, February 2005. Note the larger illuminated area at the far end of the cave. Photos: J. Rabassa.

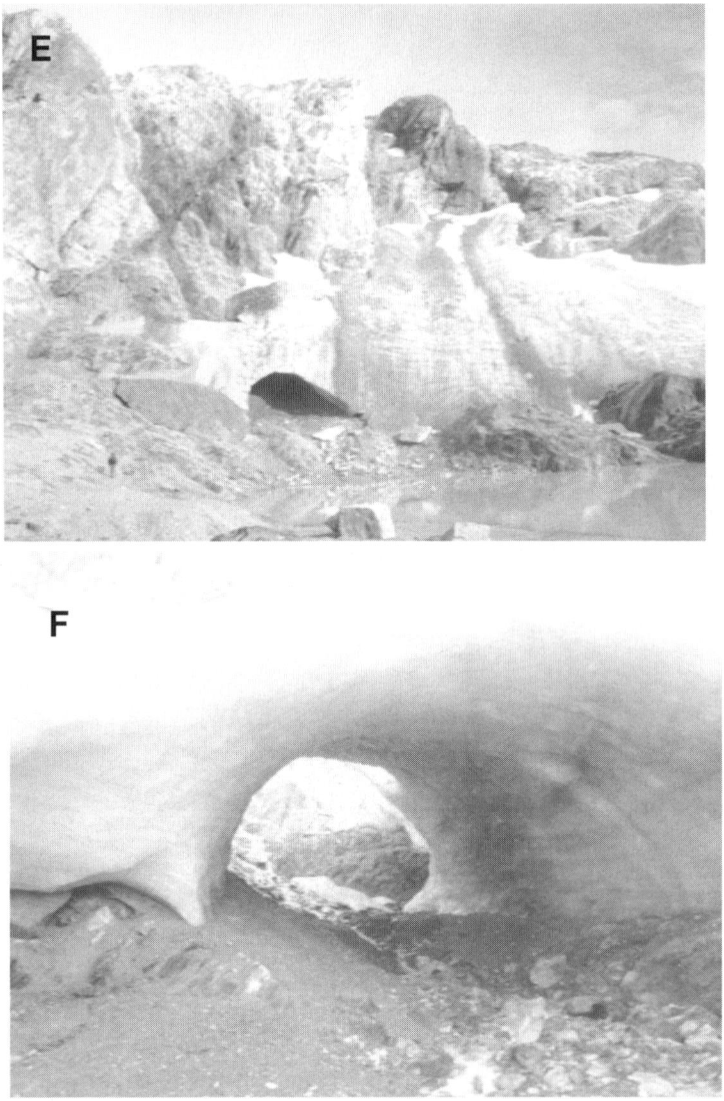

Figure 19.7 (A) Monte Alvear East Glacier. Southern portion of the ice margin as seen in February 2006. Note the much larger size of the ice cave at the left portion of the ice front, with sizable thinning of the ice at the top of the cave. The large rock boulder that was uncovered in 2005 has fallen down from the ice wall, and it is now on the ground (see Figure 19.6), indicating that the ice front retreat from the 2005 position is at least of several meters more. The dark ice remnant underneath the whitish ice, to the left of the ice cave, is more exposed than the year before. The rocky outcrop to the right has increased in size and exposition as well. (B) The ice cave shown in Figures 19.5 and 19.7, February 2006. Note the much larger extent of the ice cave. The ice tunnel is now much shorter and wider than in previous years. This site is of high interest for tourism, and trekking excursions are commercially offered to the "Alvear Ice Caves.". Photo: J. Rabassa.

Figure 19.8 Monte Alvear. (A) East Glacier. The remaining farthest end of the left ice cave, almost totally closed by infilling sediments and collapsing of the ice roof. (B) East Alvear Glacier. A general view of the receding ice front. See the large rock outcrops now fully exposed along the front, which were being uncovered in 2004–2005. Note that new outcrops have also been exposed in the last two years. Photograph taken by Natalia Martínez, Ushuaia, March 2008.

41°S, Fig. 19.1) has been the subject of detailed mapping, glaciological, and dendrochronological studies (Rabassa et al., 1978, 1984; Brandani et al., 1986). This glacier is a regenerated ice tongue, formed below a very tall icefall, where ice blocks drop from the upper, local ice cap. This lower tongue is covered in debris

Figure 19.9 The Martial Glacier, Fuegian Andes, Ushuaia, lat. 55° S, as seen from CADIC. 2004. Note the thinning in ice thickness, the exposure of large rock outcrops unseen before in the central part of the cirque glacier and its breaking apart in two or more, smaller ice bodies. Photo: J. Rabassa.

and has undergone a dramatic collapse during the last 30 years (see Figs. 19.10 to 19.12). In a nearby valley, the Lower Cone of the Castaño Overo Glacier was the topic of a graduation thesis in geography (Bertani et al., 1986), but it has already vanished as a permanent ice body, due to intense summer melting (Fig. 19.13). Thus, in only 20 years, a focus of scientific, geographical and glaciological studies is gone forever.

Finally, the case of Casa Pangue Glacier, Cerro Tronador, Chile, should be considered (Fig. 19.1 and Fig. 19.14). This is the largest glacier in Northern Patagonia, with a regenerated, lower ice tongue, forming also beneath huge icefalls in the western slope of Cerro Tronador, a dormant Pliocene volcano. This lower portion of the glacier is totally debris covered. When described in 1979 for the first time, debris cover was between 1.0 and 2.0 m thick, continuous and stable. It was so steady and firm that in the past it allowed the formation of in-transit moraines and soil development, on which a mature, full-grown, almost exact replica of the regional forest (the Valdivian Rain Forest) grew, perhaps since Little Ice Age times (Rabassa et al., 1981). This forest, which for decades and at a very slow rate moved downslope, disappeared sometime in the 1990s as the melting of the underlying ice made unstable the soil surface and the trees lost support, collapsed, fell down, and died. This puzzling ecosystem, unusual in the world, vanished forever as a result of a strong, regional warming trend. This was the first victim of GCC and global warming in the region, and it represents the extinction of a matchless, irreplaceable natural community. The appropriate question is who is going to pay for the loss of this breathtaking piece of natural patrimony? In neighboring areas, Larsen (2004) concluded that ice-thinning rates from 1995 to 2000 in the Patagonian glaciers

Figure 19.10 (A) Recent fluctuations of the Río Manso Glacier, Cerro Tronador, Nahuel Huapi National Park, lat. 41° S, Northern Patagonia, Argentina, in 1972. Debris covered, regenerated a lower ice tongue. (B) Río Manso Glacier, 1982. Note the significant retreat of the ice front from the valley marginal moraines, and the building of a small scale, abandoned lateral moraine on top of marginal kame deposits. Photos: J. Rabassa.

were more than double those of the previous two decades, thus extending the observation of these phenomena to a much larger region.

Rivera et al. (2006) have determined that 87% of the Patagonian glaciers studied showed signs of negative variation rate, ranging from a few meters per year to a maximum of 792 m per year at Martinelli Glacier in the Admiralty Sound, Chilean Tierra del Fuego, between 1992 and 2000. This situation

Figure 19.11 (C) Río Manso Glacier, 1998. Note the further recession of the ice from the 1982 position and the formation of a new, marginal lake, with many icebergs. (D) Río Manso Glacier, 2002. The marginal lake is now very extensive and the ice front has receded significantly toward the left of the image, and the dark ice remains at the right of the photograph have vanished. Photos: J. Rabassa.

continues with negative mass balances and significant ice thinning rates in all retreating glaciers. In this chapter, we estimated that 8.2% of the worldwide contribution of small glacier recession to global sea-level rise comes from Chilean glaciers alone. Only 6% of the analyzed Chilean glaciers showed evidence of net advance during some time in the twentieth century, particularly the Pío XI Glacier, which advanced until 1997 (Rivera and Casassa, 1999).

Figure 19.12 The Rio Manso Glacier. (E) Situation of the glacier in March 2008 as observed from the terminal moraine of the Little Ice Age. (F) View of the glacier valley showing the area abandoned by the glacier during the last decades. Photos: P. Milana.

The Patagonian and Fuegian glaciers have probably existed continuously for at least the last 100,000 years, since the beginning of the Last Glaciation. Human-induced climate change has gravely damaged these glaciers during the last 200 years, since the Little Ice Age. Their fading will generate enormous damage and will mean economic loss to Patagonian tourism, which today partially depend upon their survival and perpetuation.

The impact of global warming on snow accumulation in ski and winter sport resorts is clearly observable from the rising of the transient snowline in most sites in the Andean Ranges, projecting negative future perspectives and consequences. A dramatic example is the Club Andino Ski area in the city of Ushuaia, at

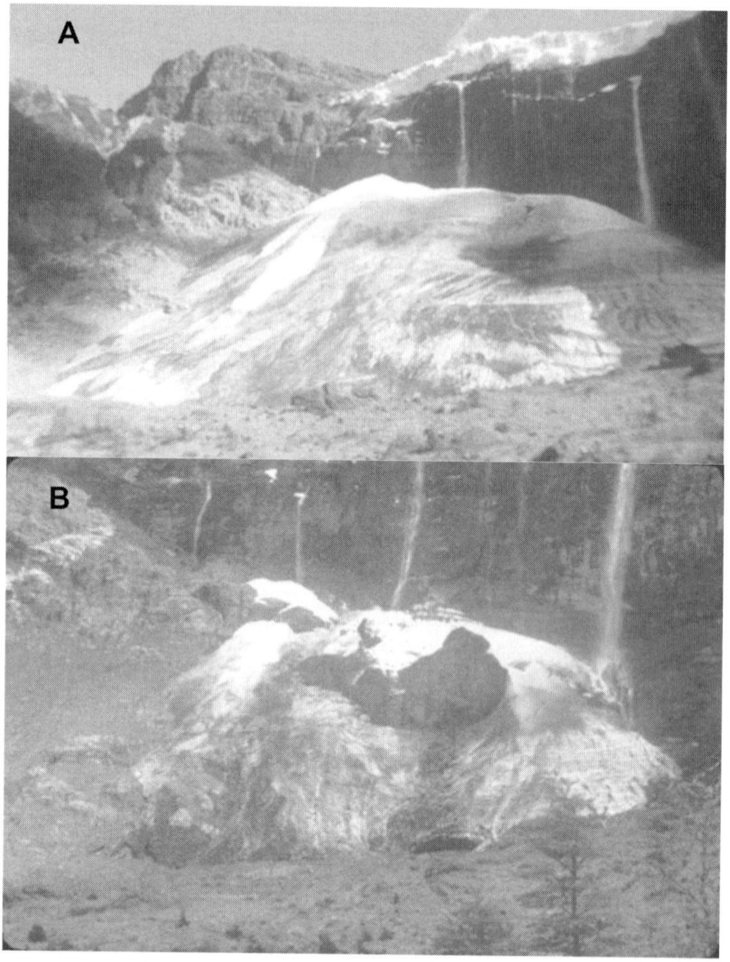

Figure 19.13 (A) The Castaño Overo Glacier, Cerro Tronador, Nahuel Huapi National Park, lat. 41° S, Northern Patagonia, Argentina, as seen in 1975. A regenerated lower ice cone, formed by ice avalanches from the upper glacier, seen at the top of the photograph. (B) The Castaño Overo Glacier, 1987. The same ice cone as in Figure 19.16. Note the large rocky outcrop that has been exposed in the central portion of the cone. This ice cone disappeared as a permanent ice body sometime during the 1990 decade. Photo: J. Rabassa.

about 200–250 m.a.s.l., which was used for decades until the end of the 1980s. At that time, it was inactivated and definitively dismantled due to the permanent lack of snow. Snowmaking technology is very expensive even in developed countries (Wang and Charneides, 2005). This technique is mostly beyond the economic possibilities of the operating agents of the South American ski areas, thus increasing the potential impact of GCC on these economic activities.

In other regions of Argentina, such as the central Argentine Andes, and particularly the irrigated vineyards of the Cuyo piedmont areas, as well as in other parts

Figure 19.14 The Casa Pangue Glacier, Cerro Tronador, Chile. This is the largest glacier in Northern Patagonia, with a regenerated, lower ice tongue, which is totally debris covered. This debris cover was between 1.0 and 2.0 m thick. In the past it allowed the formation of in-transit moraines, on which a mature, full-grown, almost exact replica of the regional forest (the Valdivian Rain Forest) developed perhaps during the Little Ice Age (Rabassa et al., 1981). These trees, which moved downslope with the ice for decades at a very slow rate, disappeared forever in the 1990s, as the melting ice made unstable the soil surface and the plants lost support. At the lower right, the late Jorge Suarez, distinguished mountaineer, friend, and colleague, climbing the ice ridge for scale. Photo: J. Rabassa.

of the world (central Chile; Sierra Nevada de Santa Marta, Colombia (Rabassa et al., 1993); Tibet; Eastern Africa, etc.), the seasonal melting of the glaciers and snowfields makes an important contribution to agricultural irrigation and provides fresh water resources in settled areas. This is the case for the city of Ushuaia, Tierra del Fuego, which depends entirely on glacier meltwater for its fresh water supply. Moreover, the glaciers of the Glaciares National Park of Southern Argentina have been declared to be part of World Mankind Heritage in the matching UNESCO program, which purportedly accords the glaciars perpetual recognition. Amazingly, humankind itself has doomed them in a very short time, probably before full scientific research can be completed.

3.5. The Antarctic Peninsula

The impact of GCC on the Antarctic Peninsula merits a special comment here. As a result of higher temperatures, the ice barriers or ice shelves of the Weddell Sea on the eastern side of the Antarctic Peninsula have partially collapsed in recent years, for the first time in 100,000 years, calving colossal icebergs referred to as as "ice islands," tens of kilometers long and thousands of square kilometers in surface area. Oceanographic studies show that the Larsen-B ice shelf had not experienced full recession and reformation at least since the

Last Glacial Maximum (LGM). The identification of ice rafted stones may mark the breakup event, which was first recorded in 1999 associated with countless icebergs and complete shelf breakdown (Domack et al., 2002). In the summer of 1995, 1600 km^2 disappeared from the Larsen A Ice Shelf. Several years later, in February 2002, 3200 km^2 vanished from theLarsen B Ice Shelf. Iceberg generation was largely increased after these events. Preliminary studies of sediment cores from the front of the Larsen B suggest that this breakup event was unprecedented in the time since an ice sheet last occupied the entire continental shelf, roughly 12,000 years BP (see www.nsidc.org/iceshelves/larsenb2002 and related links). The collapse of the ice shelf is attributed to recent warming trends (Domack et al., 2005). Similarly, the breakage of the Larsen ice shelf has led to very rapid glacier surge advances, locating more terrestrial ice in direct contact with sea water, increasing the total meltout runoff, and thus powerfully contributing to the rise in sea level (De Angelis and Skvarca, 2003). For more quantitative data, look for the Antarctic Glaciological Data Center (AGDC) at the National Snow and Ice Data Center (NSIDC) archives (www.nsidc.org). The ice barriers, such as is the case of the Larsen Ice Shelf, would probably be incapable of regeneration in any foreseeable period. Moreover, the unusual abundance of icebergs in the Drake Passage and the northernmost tip of the Antarctic Peninsula is seriously jeopardizing navigation in the area, affecting the intense tourist cruising in the area, mostly ships departing from the harbour of Ushuaia, Argentina. This situation has already resulted in two iceberg/cruiser incidents; in the one case the ship sank (fortunately with no casualties) during the first months of the 2007–2008 Antarctic tourism season (see the press notes at www.botellaalmar.-com.ar, December 2007). More of such dangerous events may be expected in the rest of the season and in future years. This will certainly augment the cost of ship insurance, thus increasing the cost of passenger tickets and largely affecting this booming economic activity.

In the Antarctic Peninsula, the climatic, regional snowline has also risen markedly in the last two decades, predominantly along the western coast of the Antarctic Peninsula. In just the last 15 to 29 years, it has increased as much as 100 m to 200 m, as is verified by the occurrence of recently exposed bedrock surfaces, which had been ice covered since at least the LGM, around 25,000 years ago. This has generated a noteworthy expansion in the exposed rocky areas at or near sea level, during the Austral summer, forcing a large expansion of the areas for colonization by penguins and other marine birds. This fact will positively stimulate an increase in their matching populations, probable migrations, and other ecological consequences of difficult forecast, since these birds compete with certain marine mammals, such as cetaceans and crab-eater seals, for the same food resources. But increasing the regional snowline elevation would also trigger a steady fading of many snowfields, small glaciers, and local ice caps in the lowlands as well as in adjacent islands, underfed by the decrease of their accumulation basins, increasing glacial sliding and marginal calving and collapse, augmented ice melting with marine salinity alteration, and partial permafrost melting.

Cook et al. (2005) recognized the distribution of receding glacier fronts in the Antarctic Peninsula for the last 60 years and identified a pattern associated with atmospheric warming and a southward displacement of the latitudinal separation between receding and advancing glaciers with time. According to these authors, the average change of the ice front position moved from +40 m/yr (advancing) in 1945 to -60 m/yr (receding) in 2004. The total proportion between advancing to receding glaciers also shifted from 61/39 (1945) to 32/95 (2004). The Sjogren Glacier, in the eastern Antarctic Peninsula, has retreated 13 km since 1993, when the Prince Gustav Ice Shelf collapsed. Moreover, all glaciers observed north of lat. 64° S receded during that period, proving that those ice bodies located at northernmost positions are more vulnerable to climate change. An accelerated sea-level rise of up to 0.2 mm per year has also been detected from the melting input of West Antarctica glaciers, which are pouring into the sea around 250 km^3 of ice annually. This figure is 60% more than the equivalent amount of accumulated snow in their nourishment areas (Thomas et al., 2004). At this rate, this region already accounts for more than 20% of the total global sea rise.

4. Permafrost

With regard to the permanent frozen grounds or permafrost, the steady decline in their thickness, the lowering of the frozen surface, and the consequent expansion of the active layer are projected to generate serious structural problems in high mountain buildings and roads, a general increase of surficial runoff, and probably an increase in the frequency and intensity of catastrophic mass-movement events. A project to measure the environmental and atmospheric variables in the permafrost zone of the Fuegian Andes above treeline is presently under development by CADIC and the Department of Geography, University of Santiago de Compostela, Spain.

5. Final Remarks

Global warming may force changes in the latitudinal position, power, and individuality of the Antarctic Circumpolar Current. These modifications, which are due to GCC and a vast input of fresh water into the southern oceans resulting from extensive ice melting may cause unforeseen oceanographic, climatic, and ecological alterations in the southern end of South America.

GCC impacts have long been thought to be greater in the high-latitude regions. Patagonia, Tierra del Fuego, and the Antarctic Peninsula may become outstanding examples of these circumstances, providing heartbreaking and voiceless proof of environmental damage induced by thoughtless human activities. The never-ending, escalating consumption of fossil fuels by the opulent industrial societies, mostly in the Northern Hemisphere, will cause loss of the invaluable Patagonian and Fuegian glaciers and snowfields and inflict serious damage on Antarctic environments, many thousands of kilometers away.

Acknowledgments

The author is particularly indebted to all the colleagues with whom he has shared fieldwork activities in the Patagonian and Fuegian Andes during the last three decades. He also thanks Professor Elena Chiozza (member of the Editorial Board of Ciencia Hoy, Buenos Aires) and Dr. Wagner Costa Ribeiro and colleagues of the Department of Geography, University of São Paulo, São Paulo, Brazil, for their invitation to publish preliminary results in previous papers, and to participate in the II Conferência Regional sobre Mudanças Globais, held in November 2005, as well for their generous contribution to finance travel and residence expenses. I am deeply grateful to Professor Edgardo Latrubesse as well for his invitation to contribute to this volume.

The preliminary version of this chapter was prepared while the author was an International Visiting Scholar of the University of Illinois at Urbana/Champaign (UIUC) at the Illinois State Geological Survey (ISGS), in April–June 2006. The author is deeply grateful to Dr. William W. Shilts, ISGS Director, Dr. E. Donald McKay III, Chief ISGS Scientist, and Dr. Richard Berg, ISGS Scientist, as well as to many other distinguished colleagues at ISGS, for the openhanded invitation and for providing a magnificent atmosphere for scientific research and for reviewing a first draft of this chapter.

Fieldwork in the Patagonian and Fuegian Andes on these subjects has been supported by several grants by the University of Comahue (Neuquén, Argentina), UNESCO World Glacier Inventory, National Geographic Society, University of Santiago de Compostela (Spain), Instituto Antártico Argentino, CONICET (the Argentine Research Council), and ANPCYT (the Argentine Agency for the Promotion of Science and Technology). The author is greatly indebted to his colleagues at CADIC, who have participated in most of his research projects since 1986.

The authour is greatly indebted to Dr. Juan Pablo Milana (Universidad Nacional de San Juan, San Juan, Argentina) and Natalia Martínez (Ushuaia) for kindly and generously providing some of their own unpublished glacier photographs to be included in this chapter.

CHAPTER 20

Possible Future Changes in Geomorphological Hazards in Latin America

Andrew S. Goudie

St. Cross College, Oxford, OX1 3LZ, United Kingdom

Contents

1. Introduction

Global warming, produced by the enhanced greenhouse effect, is a major issue in the twenty-first century and is likely to have a major effect on various geomorphological hazards (Goudie, 2006a). It will work in tandem with other anthropogenic activities, including changes in land cover and land use. This chapter provides a brief survey of how future climate change could modify the operation of some geomorphological hazards in Latin America.

The impact of change in any area depends on a variety of factors. The first of these is threshold reliance. Some landforms and land-forming processes are prone to change across crucial climatic thresholds. For example, the melting of components of the cryosphere is strongly temperature dependent, and permafrost can only exist where mean annual temperatures are negative. Thus as temperatures rise, permafrost in the Andes will move upward in altitude, and the depth of summer thaw will change (Couture and Pollard, 2007). Likewise, the mass balance of glaciers is largely controlled by the relative significance of ablation and snow nourishment, and these in turn depend on temperatures and precipitation amounts.

Developments in Earth Surface Processes, Volume 13
ISSN 0928-2025, DOI 10.1016/S0928-2025(08)10020-7

A second factor is the extent to which global climate change is compounded by other more local human actions. In Amazonia, global climate changes may cause droughts to intensify, but could occur concurrently with regional climate changes caused by land-use change (e.g.. albedo modification, alteration of surface roughness, and reduction in transpiration of moisture).

A third factor relates to the susceptibility of landforms to change. Some landforms are robust, while others are not. For example, muddy and sandy coastlines will plainly be more prone to erosion than hard-rock coastlines, and heavily lithified reefs will be more resistant than less lithified ones (Woodroffe, 2008).

Fourth, the severity of climate change, and thus its likely impact, will also vary spatially. For instance, reduction in soil moisture and stream flows may be especially great in areas that are currently relatively dry, so that stream networks will shrink (Goudie, 2006a,b). Warming will have an especially strong impact on river behavior in areas where winter precipitation currently falls as snow. In the Andes future temperature increases may be enhanced at higher altitudes (Vuille et al. 2008).

1.1. General Trends in Future Climates

Christensen et al. (2007) have summarized the likely climatic trends that may develop in South America. They suggest that all of South America is very likely to becomer warm during this century, with the greatest degree of warming occurring over the central part of the continent (Christensen et al., 2007, Fig. 11.15). Many areas may see mean annual temperatures rise by 3 to 4°C.

With regards to precipitation, the annual mean precipitation is projected to decrease over northern South America near the Caribbean coast, as well as over large parts of northern Brazil, Chile, and Patagonia, while it is projected to increase in Colombia, Ecuador, and Peru, around the equator, and in southeastern South America. It is also possible, as in other parts of the world, that precipitation events may become more intense. Detailed scenarios for southern South America are provided by Labraga and Villalba (in press) and Nunez et al. (2008).

Two issues are of particular concern with regard to precipitation trends. The first of these relates to the El Niño-Southern Oscillation (ENSO), which is important with respect, inter alia, to the occurrence of floods and debris flows in the Atacama and Peruvian deserts (Bendix et al., 2000; Magilligan and Goldstein, 2001; Houston, 2006). The situation here is far from clear, for as An et al. (2008, p. 3) have put it "dynamical understanding of ENSO responses to global warming is still in a toddling stage!" Some models suggest a weakening of ENSO activity under a global warming scenario, while others do not.

The other big issue relates to what may happen to Amazonia and whether this area will become subject to substantial regional drought and to forest dieback. Were this to happen, the regime of the Amazon, the nature of its channels, and its sediment budgets could be transformed. This is still a matter of considerable debate, but there are those who argue that global climate change, combined with regional climate change promoted by deforestation, could produce substantial drying, especially in the east and southeast of Amazonia, but only to a lesser extent in northwestern Amazonia (Malhi et al., 2008; Betts et al., 2008). In general, changes in precipitation amounts and intensities may have implications for changes in lake levels, erosion rates, desertification, slope instability, and the like. Given that two-thirds of Latin America is arid or semiarid, including large portions of Argentina, Bolivia, Chile, Peru, northeastern Brazil, Ecuador and Colombia, any increase in temperature or slight decrease in precipitation could have a disproportionate effect on river flows and on drainage densities, as is the case in Africa (de Wit and Stanaewicz, 2006). There is also the possibility that tropical cyclones, which at present are relatively unimportant in South America, could become more significant, as presaged by Hurricane Catarina, which hit southern Brazil in March 2004 (Pezza and Simmonds, 2005). For Brazil, Favis-Mortlock and Guerra (1999) used the Hadley Centre HADCM2GCM and an erosion model (WEPP—Water Erosion Prediction Project) to estimate future erosion rates. They found that by 2050 the increase in mean annual sediment yield in their area in the Mato Grosso would be 27 percent.

2. Glaciers and Permafrost

Glaciers occur in the Andes and southward into Patagonia and Tierra del Fuego. Observations over recent decades have shown that most South American glaciers have been suffering from intensified retreat (see, for example, Thompson, 2000; Kaser and Osmaston, 2002), and many may disappear in the next two decades (Ramirez et al., 2001; Magrin et al., 2007). Although an increase in glacier melting initially increases runoff, the eventual disappearance of glaciers will cause abrupt changes in stream flow because of the lack of a glacial buffer or reservoir during the dry season (Bradley et al., 2006; Vuille et al., 2008). The disappearance of glaciers can cause slope instability (Vilímek et al., 2005) by reducing their buttressing effect and by exposing rock slopes to severe weathering that produces rock falls. It is also possible that melting glaciers will produce large proglacial lakes that may subsequently fail and produce down valley floods (Harrison et al., 2006; Vilímek et al., 2005).

The southern Andes contain the largest Southern Hemisphere permafrost environment outside Antarctica, and this is associated with many rock glaciers of considerable size, especially in the semiarid Andes of Chile and Argentina. This permafrost will be subject to retreat and thinning under warmer conditions, which could have implications for ground stability, as well as diminution of water supplies downstream as ground ice melts out.

3. Sea Level and Coastal Environments

In a warmer world, sea levels will rise through a combination of the steric effect and the melting of the cryosphere. There has been a considerable diversity of views about how much sea-level rise is likely to occur by 2100. In general, estimates have tended to be revised downward through time (Pirazzoli, 1996) and have now settled at best estimates of just under 50 cm by 2100. This implies rates of sea-level rise of around 5 mm per year, which compares with a rate of about 1.5 to 2.0 mm during the twentieth century (Miller and Douglas, 2004). However, should Greenland melt at a faster rate than is currently predicted, then the amount of rise will be greater. Low-lying coastlines, including beaches, lagoons, deltas, barriers, spits and swamps and marshes may be especially sensitive, especially if they are in areas that are also subsiding. Bird (1993, p. 4) identifies three areas in South America where subsidence is greater than 2 mm per year: the gulf of La Plata, the Amazon Delta and the Orinoco Delta. Some parts of the Brazilian coast showed relative sea-level rising at 4 mm per year between 1960 and 2000 (Magrin et al., 2007, Table 13.2).

Mangrove swamps are widespread in South America and stretch from 33°S on the Atlantic coast to 3° 40′S on the Pacific coast, where their extent is restricted by the cold Humboldt Current. They are ecologically important biomes and also offer some degree of protection against coastal erosion. Many mangrove communities are affected by rises in sea level. Species zonation may change, with trees to seaward being inhibited by extended submergence times, while those on the landward margins might be able to extend, provided that suitable habitat was available and assuming that their migration was not impeded by sea walls and other anthropogenic structures. A key issue is the extent to which the rate of sediment deposition in the swamps can keep pace with the rate at which sea level is rising. In the great river deltas, on high oceanic islands with considerable runoff from rainfall or in those situations where large quantities of marine sediments can be trapped, mangroves may be able to grow up at a rate greater than that of sea-level rise. Some mangroves are relatively slow growing, and this may make them susceptible to fast rates of sea-level changes. On the other hand there may be some expansion in the range of mangroves as sea-surface temperatures rise (Hogarth, 1999).

Coral reefs, because of their low-lying nature, may be subject to overwashing, inundation, and salt water incursion as sea level rises (Woodroffe, 2008), particularly if, because of pollution, they are unable to grow upwards at an adequate rate to keep up with sea-level rise. They may also suffer from increasing ocean acidification. Sea water absorbs a proportion of the extra carbon dioxide being released into the atmosphere by the burning of fossil and biomass. As this combines with water it produces carbonic acid, which will cause it to become more acidic (i.e., it will have a lower pH than now). Several centuries from now, if we continue to add carbon dioxide to the atmosphere, ocean pH will be lower than at any time in the past 300 million years (Doney, 2006). This will be harmful to organisms such as corals, which depend on the presence of carbonate ions to build their hard parts out of calcium bicarbonate (Orr et al., 2005). Reefs also play an important role in coastal protection against storms so that "we can anticipate that decreasing rates of reef of coral accretion, increasing rates of bio-erosion, rising sea-levels, and intensifying storms may combine to jeopardize a wide range of coastal barriers" (Hoegh-Guldberg et al., 2007, p. 1742).

4. Conclusions

The incidence of natural hazards in Latin America is likely to change in coming decades as a result of both human activities (land-use change, etc.) and climate change. These will sometimes work in synergy. A challenge for geomorphologists and hydrologists is to identify areas that are likely to be especially sensitive to these influences, so that appropriate mitigation strategies can be adopted.

CHAPTER 21

A Latin American Perspective on Geomorphologic Hazards and Related Disasters

Edgardo M. Latrubesse

University of Texas at Austin, Department of Geography and the Environment, GRG 334; 1 University Station A 3100, Austin TX 78712 USA

The Latin American region has often been treated as a single entity because of its relative uniformity in language and its predominant historical connection to the Spanish and Portuguese empires. It is commonly referred to as Latin America and is considered a relatively cohesive territorial unit in the social, political, and economic sciences—a concept that has gradually penetrated into disaster analysis as well (Stillwell, 1992). This perspective has resulted in too broad a synthesis and an oversimplification of diverse issues concerning Latin American countries. A similar situation exists for developing countries or tropical countries (Alcantara Ayala, 2002). However, historical heritage, cultural aspects, socioeconomic models, relief, and natural resources among the different countries in Latin America are too strong and diverse to support "regional" theories and synthesis.

Latin America is a very diverse region that can be perceived in significantly diverse ways by individuals. For example, it can be considered to encompass a coincidence of Andean cultures (Quechua and Aymara), Portuguese and Spanish conquerors, and colonial cities; the longest mountain chain of the world (Andes); the biggest fluvial system (the Amazon); the driest desert (Atacama); the largest and the most complex forest (Amazon); the mythic Patagonia; the Caribbean coast of Colombia and Venezuela; and a megalopolis such as Mexico, Buenos Aires, Caracas, Rio de Janeiro, or São Paulo, among many other "perceptions." Geomorphologic hazards and related disasters can also be part of these perceptions. Catastrophes such as Hurricane Mitch (Chapters 1 and 4), Nevado del Ruiz, which destroyed Armero in Colombia (Chapter 7), Vargas-La Guaira disaster in Venezuela (Chapter 6), the Pisco earthquake (Chapter 9), and the floods in Santa Fe, Argentina (Chapter 16) demonstrate the extreme natural processes and the irregular social and economic development seen across South America.

We encounter similar generalizations when we compare the different developing countries or "tropical" countries (Latrubesse et al., 2005). Large populations in developing economies, the chaotic growth of urban areas, and the consequent sharp

Developments in Earth Surface Processes, Volume 13
ISSN 0928-2025, DOI 10.1016/S0928-2025(08)10021-9

increase in water and power demands are some of the common problems ofl tropical countries. The resources available and the management strategies adopted to tackle geomorphic disasters, however, may be entirely different from country to country. These differences eventually affect the overall economic growth of the country. For example, Brazil, with a total of 8.5 million km^2 of area and around 190 million people, is considered one of the largest agricultural producers in the world because of its widespread agricultural area and its intensive water management practices. On the other hand, millions of people in India live on rather rudimentary agriculture and face scarce availability of ground and surface water. This has obviously resulted in a much lower rate of growth in the agricultural sector in India and has also affected the water and power demands in many parts of the country.

The tremendous impact of agriculture in Brazil in recent decades has no equal in any other world region (Chapter 11). The country's large-scale urbanization and agricultural expansion have resulted in significant changes in land use during the last few decades. First to undergo these changes were the southeastern states (São Paulo, Paraná, and others) during the first coffee period and. more recently, the Cerrado (Brazilian savannahs). The Brazilian Cerrado, for example, is an environmental catastrophe that both the national and international communities underestimated. From a geomorphological point of view, it produced widespread erosion and soil loss. However, the Cerrado area, cultivated to answer the worldwide demand for biofuel and agrofuel, is presently considered the world's largest area of cultivation for sugar cane. Brazil's agricultural advances in the early twenty-first century have resulted in repeating old models and policies based on the agriculture of monocultures. After the coffee and soy rush, today is the time for the sugar cane rush, and Brazil is expected to feed probably 10% of the world's fuel with bio- or agrofuel.

The impact of land-use changes in Argentina has also been dramatic. Because of the scarce historical socioeconomic role and underpopulation of the southern region, the shameful desertification of Patagonia by overgrazing (Chapter 17) and the environmental degradation by oil companies have been just cited as an "anecdotic" episode in the environmental policy of Argentina instead of being a priority national issue. This environmental disaster is being repeated with the destruction of the Chaco and the Sub-Andean forest along the socially and economically underdeveloped northwest and northern parts of the country.

Flood disasters in Latin America are another concern when we compare different countries and regions. In Asia, floods are recurrent and catastrophic, and even though some of the largest rivers of the world drain through South America, floods are not as significant in this part of the world as in Asia. With the exception of the Paraná in Argentina, floods occur mainly in small- to medium-size basins. Central America is an extreme example of how catastrophic flash floods in small fluvial basins may be triggered by hurricanes and tropical storms and affect thousands of people in poorly developed countries, as in Guatemala where, for example, 40,000 people were killed in 1949. In Asia, the alluvial plains of the Yangtze, Ganges, and Brahmaputra rivers are severely affected by floods. More than 500 million people in Nepal, India, Bhutan, and Bangladesh (more than the entire population of South America) are affected by floods (Latrubesse et al., 2005).

In China, in 1998 floods affected more than 200 million people (Hoyois et al., 2006), and the Yangtze River Basin floods have caused damages of more than US$ 20 billon (Yi and Li, 2001). The rivers in these countries are crucial for irrigation and for domestic and industrial consumption uses, but at the same time when flooded they are responsible for huge losses of life and property. Bangladesh, for example, is considered to be the most frequently flooded country in the world followed by India. Flood damage increased 40-fold in India from the 1950s to the 1980s (Centre for Science and Environment—CSE, 1992), although part of this increase may be attributed to the improved techniques used to assess damages (Mirza et al., 2001).

In contrast to Asia, floods in South America, affect only a minor proportion of the population because the risky areas are sparsely populated. Nonetheless, these floods generate socioeconomic catastrophes relating to production losses in agriculture and cattle farming, as in the Argentinean Pampas. In Argentina, floods constitute a major natural hazard. Floods in the Paraná Basin and along the flat plain rivers of the productive Pampean region spell economic calamity to this country, which has been economically dependent on agriculture and cattle farming since historic times.

In Brazil, floods and landslides affect the cities, the megalopolis, and the industrial complexes located on or surrounded by a hilly relief (*morros*), particularly in southeastern Brazil (see Chapters 12 and 13). Large urban centers are affected mainly by floods in small to medium-size basins such as São Paulo, Blumenau, the suburban areas of Buenos Aires, or cities such as Rio Branco in southwestern Amazonia and Santa Fe in the Salado Basin of Argentina. In all these cases the catastrophes are related to urban centers located in hazardous places and in areas where the programs for basin management are very deficient. In many cases, poor river engineering works that fail to use a multidisciplinary approach have contributed to increasing flooding that has occurred due to changes in the rivers' natural conditions.

Because of their location in the Andes chain and the Circum-Pacific ring of fire, countries such as Guatemala, Nicaragua, Mexico, Costa Rica, El Salvador, Venezuela, Colombia, Ecuador, Peru, Bolivia, Chile, and Argentina are susceptible to earthquakes and some of them to volcanic eruptions as well (the latter include mainly Mexico, Guatemala, Colombia, Ecuador, Peru, and Chile) (see Chapters 2, 4, 5, 8, 9, and 18). With the exception of the Vargas floods in Venezuela (1999), the most devastating disasters in terms of human loss in South America have occurred in Peru, Colombia, Chile, and Argentina and were related to endogenic processes (earthquakes and volcanoes). In Central America and Mexico, the main disasters involving human loss are related to both exogenic processes (mainly floods and landslides) and endogenic (volcanoes and earthquakes) (Chapters 2 and 4).

A natural disaster can be defined as the product of an extreme relationship between physical phenomena and a society's structure and organization (ECLAC-UNEP, 2000). Concepts such as environmental risks and disasters have the advantage of including natural and human dimensions (Smith, 1996). Nevertheless, the human dimension of disasters can also be a direct variable, although it is not necessarily the main one needed to define a disaster. The concept of a regional environmental vulnerability implies evaluating the area's susceptibility or resistance to disasters that may be caused by natural phenomena. It is a fact that environmental vulnerability must be taken into account in all future regional, national, and local activities. In environmental disasters, the human dimension can also act as a geomorphologic agent, triggering or strengthening geomorphologic processes that may have serious socioeconomic consequences and affect a region's environmental value. As we have stated, many of the disasters in South America have been minimal in terms of human losses (e.g., the disasters that affect the regional economies that are highly dependent on agricultural products). Drought and floods in Bolivia and floods in Argentina or environmental disasters in Patagonia and Brazil are also included in this list.

It has been estimated that nearly 88,448,000 people were affected or killed by disasters in South America between 1974 and 2003 as a consequence of 538 disasters (32% of the total recorded for the American continent); the economic losses from these disasters have been set at US$ ~54 billion (Guha Sappir et al., 2004). Floods constituted the main disaster in South America, followed by droughts and earthquakes. However, these economic values either underestimate or do not include several environmental disasters such as the Patagonia desertification, the expansion of the agricultural frontier (and loss of biodiversity) in Brazil, or the glacier recession in the Andes. Between 1902 and 2007, 465 disasters associated with natural hazards occurred in Mexico and Central America, causing more than 160,000 fatalities, affecting about 32 million people, and generating damages estimated at approximately US$ 43 billion (Chapter 4).

The vulnerability of Latin American societies to natural disasters is high because of inadequate public policies. In general terms, all the Latin American countries have poorly developed management strategies for handling natural disasters. Public policies and laws seem to be insufficient to guarantee disaster prevention and mitigation because the political instability in several of the Latin American countries since the late 20th century. This factor has also contributed to a poor implementation of long-term and sustained national or regional programs on disasters, and low respect by legal fulfillment.

A geomorphic perspective on natural disasters that focuses on their prevention and mitigation is still in the infancy stage in Latin American countries. There is a strong need to recognize the role of geomorphology and geomorphologists in Latin America. In the middle of this negative scenario, geomorphologists continue to await the call to contribute their expertise to society and play a more relevant role in their countries.

REFERENCES

Ab'Saber, A.N., 2003. Os Domínios de Natureza no Brasil: potencialidades paisagísticas. Ateliê Editorial, São Paulo, p. 159.

Ab'Saber, A.N., 1973. A organização natural das paisagens inter e subtropicais brasileiras. S. Paulo, USP, IGEOG, Geomorfologia 41, p. 39.

Abraham, E., Maccagno, P., Tomasini, D., 2003. Experiencia argentina vinculada a la obtención y evaluación de indicadores de desertificación. In Abraham, D., Tomasini, E., Maccagno, P. (eds.), Desertificación: Indicadores y puntos de referencia en América Latina y el Caribe. IADIZA—LaDyOT, pp. 81–105

Abreu de Castilhos, J., Gre, J.C.R., 1997. Praias da Ilha de Santa Catarina: Caracerizacao morfologica e problemas de erosao costeira. VI Congresso da ABEQUA, Curitiba, pp. 388–392.

Aceituno, P., 1988. On the functioning of the Southern Oscillation in the South American sector. Part I: Surface Climate. Monthly Weather Review 116, pp. 505–524.

Addad, J., Martins-Neto, M.A., 2000. Deforestation and coastal erosion. A case from east Brazil. Journal of Coastal Research 16 (2), pp. 423–431.

AIS Asociación Colombiana de Ingeniería Sísmica, INGEOMINAS and Universidad de Los Andes, 1998. Estudio general de amenaza sísmica de Colombia. Publicación especial de Ingeominas. Bogotá, Colombia.

Albalá-Bertrand, J.M., 1993. Political Economy of Large Natural Disasters: With Special Reference to Developing Countries. Oxford Univ. Press, London, p. 259.

Alcántara-Ayala, I., 2002. Geomorphology, natural hazards, vulnerability and prevention of natural disasters in developing countries. Geomorphology 47, pp. 107–124.

Alcántara-Ayala, I., 2004a. Hazard assessment of rainfall induced landsliding in Mexico. Geomorphology, 61, pp. 19–40.

Alcántara-Ayala, I.. 2004b. Flowing mountains in Mexico—Incorporating local knowledge and initiatives to confront disaster and promote prevention. Mountain Research and Development 24 (1), pp. 10–13.

Alcántara-Ayala, I.; Lopez-Mendoza, M.; Melgarejo-Palafox, G.; Borja-Baeza, R.C.; Acevo-Zarate, R. 2004. Natural hazards and risk communication strategies among indigenous communities—Shedding light on accessbility in Mexico's mountains, Mountain Research and Development 24 (4), pp. 298–302.

Alcántara-Ayala, I., 2008. On the historical account of disastrous landslides in Mexico: The challenge of risk management and disaster prevention. Adv. Geosci. 14, pp. 159–164.

Alcántara-Ayala, I., 2009. Geomorphosite management in areas sensitive to natural hazards. In Reynard, E., Coratza, P., Regolini, G. (eds.), Geomorphosites: Assessment, Mapping and Management, München, Pfeil Verlag (in press).

Alexander, D., 1997. The study of natural disasters, 1977–97: Some reflections on a changing field of knowledge. Disasters 21 (4), pp. 284–304.

Allan, N.J.R., Knapp, G.W., Stadel, C. (eds.). 1988. Human Impact on Mountains. Rowan & Littlefield, Totowa, NJ.

Alley, R.B., Clark, P.U., Huybrechts, P., Joughin, I., 2005. Ice-sheet and sea-level changes. Science, 310, pp. 456–460.

Allmendinger, R., Strecker, M., Eremchuck, J., Francis, P. 1989. Neotectonic deformation of the southern Puna Plateau, Northwestern Argentina. Journal of South American Earth Sciences Volume 2, pp. 11–130.

Allmendinger, R., Jordan, T., Kay, S., Isacks, B., 1997. The evolution of the Altiplano-Puna plateau of the Central Andes. Ann. Rev. Earth and Planetary Sciences 25, pp. 139–174.

Almeida, Y., 1995. Variabilidad del nivel medio del mar. Vértice cartográfico. Etapa 2. Servicio autónomo de Geografía y Cartografía Nacional, Cracas, Venezuela.

Alonso R.N., Wayne W.J., 1992. Riesgos geológicos en el norte Argentino. Congreso Geológico Boliviano, Boletín 27, pp. 213–216.

Altez, R., 2005. Historia sin memoria: la cotidiana recurrencia de eventos desastrosos en el Estado Vargas–Venezuela. Revista Geográfica Venezolana, Mérida, Venezuela, pp. 313–342

Alva-Hurtado, J., 1985. Mapa de licuación de suelos en el Perú. Earthquake Mitigation Program in the Andean Region, 14, pp. 303–313.

Alvarado, P., 1998. Sismicidad superficial de Los Andes Centrales (33°–35°S, 69,5°–70,5°W). M. Sc. Thesis of Geophysic, Departamento de Geofísica, Facultad de Ciencias Físicas y Matemáticas, Universidad de Chile, Santiago, Chile.

Alvarado, G.E., Vega, E., Chávez, J., Vásquez, M., 2004. Los grandes deslizamientos (volcánicos y no volcánicos) de tipo debris avalanche en Costa Rica. Rev Geol Am Central, 30, pp. 83–99.

Ames, A., Dolores, S., Valverde, A., Evangelista, P., Javier, D., Gavnini, W., Zúñiga, J., Gómez, V., 1989. Glacier Inventory of Peru. HIDRANDINA, Huaraz, Peru, internal report.

Amigo, A., 1965. El sobrepastoreo en la región patagónica. Causas que lo originan y soluciones que se proponen. Proyectos especiales 14. Consejo Nacional de Desarrollo, pp. 27–54

Amsler, M.L., Prendes, H.H., 2000. Transporte de sedimentos y procesos fluviales asociados. In Paoli, C., Schreider, M.I. (eds.), El río Paraná en su tramo medio. Contribución al conocimiento y prácticas ingenieriles en un gran río de llanura. Centro de Publicaciones, Universidad Nacional del Litoral, Santa Fe, Argentina, pp. 233–306.

Amsler, M., Ramonell, C., Toniolo, H.A., 2005. Morphologic changes in the Parana River channel (Argentina) in the light of the climatic variability during the 20th century. Geomorphology 70 (3–4), pp. 257–278.

An, S-I., Kug, J-S., Ham, Y-G., Kang, I-S. 2008. Successive modulation of ENSO to the future greenhouse warming. Journal of Climate 21, pp. 3–21.

Andersen, R.J., Santos, N., Díaz, H.F., 1993. An analysis of flooding in the Paraná/Paraguay river basin. LATEN Dissemination, Note N° 5, The World Bank, Latin American Technical Dep. Environ. Div. September 1993, p. 19.

Andrada, L., 2002. Territorio y ganadería en la Patagonia Argentina: Desertificación y rentabilidad en la Meseta Central de Santa Cruz. Economía, Sociedad y Territorio III (12), pp. 675–706.

Andrioli, C.P., Barros, M.T.L., 2005. Análise hidrometeorológica preliminar da precipitação ocorrida em São Paulo entre os dias 25 e 26 de maio de 2005. In XVI Simpósio Brasileiro de Recursos Hídricos, João Pessoa, 2005.

Angulo, R.J., Lessa, G.C., de Souza, M.C., 2006a. A critical review of mid– to late–Holocene sea-level fluctuations on the eastern Brazilian coastline. Quaternary Science Reviews 25, pp. 486–506.

Angulo, R.J., Soares, C.R., Marone, E., Souza, M.C., Odreski, L.L.R., Noernberg, M.A., 2006b. Paraná. In D. Muhe (ed.), Erosao e Progradacao do Litoral Brasileiro, Programa de Geologia e Geofísica Marina, Ministerio do Meio Ambiente, pp. 347–400.

Aniya, M., Enomoto, H., 1986. Glacier variations and their causes in the Northern Patagonia Icefield, Chile, since 1944. Arctic and Alpine Research 18 (3), pp. 307–316.

Apoluceno, D.De M., Dominguez, J.L.M., Bittencourt, A.C.S.P., Cruz, F.C., 1997. Erosao da linha da costa na regiao de Ilheus (BA) e sua relacao com a instalacao do Porto do Malhado. VI Congresso da ABEQUA, Curitiba, pp. 383–387.

Aragón-Durand, F., 2007. Urbanisation and flood vulnerability in the peri–urban interface of Mexico City. Disasters 31 (4), pp. 477–494.

Arnold, M., Chen, R.S., Deichmann, U., Dilley, M., Lerner-Lam, A.L., Pullen R.E., Trohanis, Z., (eds.), 2006. Natural Disaster Hotspots: Case Studies. World Bank, Disaster Risk Manag. Ser. 6, p. 184.

Arntz, W.E., Farbach, E., 1996. El Niño. Experimento Climático de la Naturaleza. Fondo de Cultura Económica, México, p. 309.

Araujo, M., Tello, G., Perez, A., Perez, I., Puigdomenech, C., 1999. Shallow seismicity in the north–western part of Argentina and its relations with tectonics. Fourth ISAG, Germany, pp. 42–46.

Armijo, R., Thiele, R., 1990. Active faulting in northern Chile: Ramp stacking and lateral decoupling along a subduction plate boundary. Earth and Planetary Science Letters 98, pp. 40–61.

ASAGAI, 2004. Argentinean Association of Applied Geology and Engineering. Special Issue N° 4, Buenos Aires.

Astroza, M., Sandoval, M., Kausel, E. 2005. Estudio comparativo de los efectos de los sismos chilenos de subducción del tipo intraplaca de profundidad intermedia. Congreso Chileno de Sismología e Ingenería Antisísmica. IX jornadas. Concepción, Chile.

Atwater, B.F., Nuñez, H.J., Vita-Finzi, C., 1992. Net Late Holocene emergence despite earthquake–induced submergence, south–central Chile. Quaternary International 15/16, pp. 77–85.

Aubreville, A., 1949. Climats, forêts et désertification de l'Afrique tropicale. Société d'Edition de Géographie Maritime et Coloniale Paris, vol. 1, p. 381.

Audemar, F., De Santis, F., Montes, L., Lugo, M., Singer, A., 1989. El Alud Torrencial del 06–09–87 del río Limón al Norte de Maracay, Estado Aragua. Memorias 50 Aniversario de la Escuela de Geología, Minas y Geofísica, Facultad de Ingeniería, Universidad Central de Venezuela. GEOS No. 29, pp. 250–260.

Audemard, F.A., Singer, A., 1996. Active fault recognition in northwestern Venezuela and its seismogenic characterization: Neotectonic and paleoseismic approach. In Enghdal, E.R., Castano, J., Berrocal, J. (eds.), Proceedings of the 1994 Regional Seismological Assembly in South America. Geofísica Internacional, México 35(3), pp. 245–255.

Audemard, F.A., 1999. Morpho–structural expression of active thrust fault systems in the humid tropical foothills of Colombia and Venezuela. Zeitschrift für Geomorphologie 118, pp. 1–18.

Audemard F.A., Pantosti, D., Machette, M., Costa, C., Okumura, K., Cowan, H., Diederix, H., Ferrer, C., SAWOP Participants, 1999. Trench investigation along the Merida section of the Boconó fault (central Venezuelan Andes), Venezuela. Tectonophysics 308, pp. 1–21.

Audemard, F.A., Machette, M., Cox, J., Dart, R., Haller, K., 2000. Map and database of Quaternary faults in Venezuela and its offshore regions. U.S. Geological Survey Open–File Report 00–018, Golden.

Audemard, F.E., Audemard, F.A., 2002. Structure of the Mérida Andes, Venezuela: relations with the South America–Caribbean geodynamic interaction. Tectonophysics 345, pp. 299–327.

Audemard, F.A., 2005. Paleoseismology in Venezuela: Objectives, methods, applications, limitations and perspectives. Tectonophysics 408 (1–4), pp. 29–61.

Audemard, F.A., Romero, G. Rendón, H., Cano, V., 2005. Quaternary fault kinematics and stress tensors along the southern Caribbean from microtectonic data and focal mechanism solutions. Earth–Science Reviews 69 (3–4), pp. 181–233.

Audemard, F.A., Singer, A., Soulas, J–P., FUNVISIS' Neotectonics Section, 2006. Quaternary faults and stress regime of Venezuela. Revista de la Asociación Geológica Argentina 61, pp. 480–461.

Audemard M, F.A., Gómez, J.C., Tavera, H.J., Orihuela G, N., 2005. Soil liquefaction during the Arequipa M_w 8.4, June 23, 2001 earthquake, southern coastal Peru. Engin. Geol. 78, pp. 237–255.

Audemard, F., 2009. Falla de Boconó en ambientes peri-glaciales de los Andes de Mérida Venezuela. In: Proyecto Multinacional Andino (Ed.). Atlas de Deformaciones Cuaternarias de los Andes. SERNAGEOMIN, Publicación Multinacional No 7: 260-272, Santiago.

Auer, V., 1956. The Pleistoceneof Fuego-Patagonia. Part I: The ice and interglacial age. Annales Academiae Scientiarum Fennicae, Series A III (45), pp. 1–226.

Auer, V., Cappannini, D., 1957. La erosión en la región de los lagos San Martín y Tar. IDIA 111. INTA, Buenos Aires.

Avelar, A.S., 1996. Investigação histórica e geotécnica do movimento de massa do Soberbo (RJ). Tese de M.Sc., COPPE/UFRJ, Rio de Janeiro, RJ, Brasil, p. 122.

Avelar, A.S., Coelho Netto, A.L., 1992. Fraturas e desenvolvimento de unidades geomorfológicas côncavas no médio vale do rio Paraíba do Sul. Rev. Bras. Geociências, 22 (2), pp. 222–227.

Avelar, A.S., Lacerda, W.A., 1997. Causas iniciais do movimento de massa do Soberbo, RJ—Brasil. 2° Simpósio Panamericano sobre Escorregamentos e 2ª Conferência Brasileira sobre Estabilidade de Encostas—COBRAE, Rio de Janeiro, RJ, Brasil, vol. 1, pp. 3–12.

Ayesa, J.D., Bran, C., Lopez, A., Cingolani, S., Clayton., Sbriller, D., 1995. Evaluación del estado actual de la desertificación en la transecta Río Negro. In Lucha contra la desertificación en Patagonia. Evaluación del estado actual de la desertificación en áreas representativas de la Patagonia. Informe Final Fase I. INTA—GTZ.

Aysan, Y.F., 1993. Vulnerability assessment. In Merriman, P.A., Browitt, C.W.A. (eds.), Natural Disasters: Protecting Vulnerable Communities. Thomas Telford, London, pp. 1–14.

Azambuja, E., Zeni, C.A., Bressani, L.A., Nanni, A.S., 2001. Avaliação dos fluxos de detritos em Alto Feliz no Rio Grande do Sul. 3ª Conferência Brasileira sobre Estabilidade de Encostas—COBRAE, Rio de Janeiro, RJ, Brasil, vol. 1, pp. 17–24.

Azarevich, M., Novara, M., Armella, M., Deus, R., Sanchez, M.C., 1999. Geomorfología de la región de Tilcara, Jujuy. XIV Congreso Geológico Argentino, Asociación Geológica Argentina, Salta, Actas 1, pp. 31–34.

Balay, M.A., 1961. El Rio de la Plata entre la atmósfera y el mar. Servicio de Hidrografía Naval, Buenos Aires, Publ. H–621, p. 153.

Bandieri, S., 1990. Contribución al estudio de la apropiación privada de tierras en Patagonia: La singularidad del caso neuquino, primera aproximación a los actores sociales. Revista de Historia, Fac. de Humanidades. U.N.Comahue.

Bangs, N., Cande, S.C., 1997. Episodic development of a convergent margin inferred from structures and processes along the southern Chile margin. Tectonics 16 (3), pp. 489–503.

Barata, F.E., 1969. Landslides in the Tropical Region of Rio de Janeiro, Proc. of the 7th ICSMFE, México, vol. 2, pp. 507–516.

Barazangui, M., Isacks, B., 1976. Spatial distribution of earthquakes and subduction of the Nazca Plate beneath South America. Geology 4, pp. 686–692.

Barbería, E., 1987. El arrendamiento de la tierra pública en Santa Cruz. 1884–1911. Primeras Jornadas de Historia Argentino Americana. Universidad del Centro de la Provincia de Buenos Aires, Tandil.

Barbería, E., 1995. Los dueños de la tierra en la Patagonia Austral. Universidad Federal de la Patagonia Austral, Buenos Aires.

Barkun, N., 1974. Disaster and the Millennium. Yale Univ. Press, New Haven, CT, p. 246.

Barreira, C.C.M.A., 2002. Vão do Paranã. A Estruturação de Uma Região. 1. ed. Brasilia: Ministério da Integração Nacional–Coleção Centro–Oeste de Estudos e Pesquisas, vol. 1, pp. 320.

Barrientos, S., Kausel, E., 1993. Características de la sismicidad superficial en la Zona Central de Chile. VI Jornada Chilena de Sismología e Ingeniería Antisísmica, August 9–13, 1993, Universidad de Chile, Santiago, vol. 1, pp. 3–9.

Barrientos, S., Ward, S., 1990. The 1960 Chile earthquake: inversion for slip distribution from surface deformation. Geophysical Journal International 103 (3), pp. 589–598.

Barros, V., Doyle, M., 1996. Precipitation trends in Southern South America to the East of the Andes. Report N° 26 April 1996. Center for Ocean Land atmosphere Studies. In Kinter III, S.S., Schneider, E.K. (eds.), Proceedings of the Workshop on Dynamics and Statistics of Secular Climate Variations. ICTP Trieste Italy. December 4–8, 1995, Calvelton, USA, pp. 76–86.

Barros, V., Frumento, O., Camilloni, I., Rivero, M., 1997. Energia eólica. Subs. de Energía, Praepra, p. 400.

Barros V., Castañeda, M.E., Doyle, M., 2000. Recent precipitation trends in southern South America to the east of the Andes: An indication of a mode of climatic variability. Capitulo del libro Southern Hemisphere Paleo and Neoclimates.Concepts, Methods, Problems. Springer.

Barros, W.T., Bogssian, F., Lopes, P.C.C., 1992. O escorregamento do Soberbo, Publicação especial da 1a Conf. Bras. sobre Estabilidade de Encostas–COBRAE, Rio de Janeiro, p. 14.

Bastias, H., 1986. Fallamiento Cuaternario en la Región Sismotectónica de Precordillera. Ph.D. Thesis (Unpublished), San Juan University, Argentina, p. 160.

Bastias, H., Weidmann, N., Perez, A., 1984. Dos zonas de fallamiento Plio-Cuaternario en la Precordillera de San Juan. IX Congreso Geológico Argentino, Actas, vol. 2, pp. 329–341.

Bastias, H., Tello, G., Perucca, L., Paredes, J., 1993. Peligro Sísmico y Neotectónica. 12° Congreso Geológico Argentino y 2° Congreso de Exploración de Hidrocarburos. Geología y Recursos Naturales de Mendoza. Ramos, V. (ed.), Relatorio, VI (l), pp. 645–658.

Bastias, H., Paredes, J., Perucca, L., Tello, G., 1995. Mapa Neotectónico preliminar de la República Argentina. Universidad Nacional de San Juan, Mapa inédito. Escala 1:50.000. Argentina.

Bastias, H., Uliarte, E., Paredes, J., Sanchez, A., Bastias, J., Ruzycki, L., Perucca L., 1990. Neotectónica de la provincia de San Juan. 11° Congreso Geológico Argentino. Relatorio de Geología y Recursos Naturales de la provincia de San Juan, San Juan, pp. 228–245.

Bates, D.C., 2002. Environmental refugees: Classifying human migrations caused by environmental change. Population and Environment 23 (5), pp. 465–477.

Bates, R.L., Jackson, J.A., 1987. Glossary of Geology. American Geological Institute, Alexandria, VA, USA.

Beck, S.L., Barrientos, S., Kausel, E., Reyes, M., 1998. Source characteristics of historic earthquakes along central Chile subduction zone. Journal of South American Earth Science 11 (2), pp. 115–129.

Bejerman, N., 1993. Mapa de peligro de deslizamiento del tramo La Calera–Dique San roque de la ruta provincial E-55. Provincia de Córdoba, Argentina. Asociación Argentina de Geología Aplicada a la Ingeniería, Publicación Especial (3), San Juan, Argentina, pp. 53–70.

Bejerman, N., Giraud, C., 1993. Mapa de posibilidad de deslizamiento en un tramo de la ruta provincial A–73 (Departamento de Punilla-Provincia de Córdoba, Argentina. Asociación Argentina de Geología Aplicada a la Ingeniería, Publicación Especial (3), San Juan, Argentina, pp. 71–79.

Belinazzi Jr., R., Bertolini, F., Lombardi Neto, F., 1981. A ocorrência de erosão rural no Estado de São Paulo. 2° Simp. Erosão, São Paulo, pp. 117–137.

Beltramone C.A., 2005. Dinámica de las vertientes en la ladera occidental de la Sierra Chica de Córdoba. Revista de la Asociación Geológica Argentina, Buenos Aires, 60 (1), pp. 9–15.

Bendini, M., P. Tsakoumagko, Destéfano, D., 1990. El trabajo trashumante en la provincia del Neuquén. Universidad Nacional del Comahue, Neuquén.

Bendix, J., 2000. Precipitation dynamics in Ecuador and northern Peru during the 1991/92 El Niño: A remote sensing perspective. Int. J. Remote Sensig 21 (3), pp. 533–548.

Bendix, J., Bendix, A., Richter, M., 2000. El Niño 1997/1998 in Nordperu: Anzeichen eines Őkosystem– Wandels? Petermann's Geographische Mitteilungen 144, pp. 20–311.

Benson, C., 2003. The Economy–wide Impact of Natural Disasters in Developing Countries. University of London.

Bernard, E., 1998. Program aims to reduce impact of tsunamis on Pacific status. Eos Trans. AGU 79 (22), 258, pp. 262–263.

Bernard, E., 2002. The U.S. National Tsunami Hazard Mitigation Program. In Ewing, L., and Wallendorf, L. (eds.), Proceedings of Solutions to Coastal Disasters. ASCE, San Diego, CA, February 24–27, 2002, pp. 964–971.

Berner, R.A., 1997. The rise of plants and their effect on weathering and atmospheric CO_2. Science 276, pp. 544–546.

Bernex, N., 2002. Riesgos naturales y desastres en el Perú, una responsabilidad social. In Alegre, M., Bernex, N., Carrasco, A., Castro–Pozo, H., Chirinos, C., Felipe–Morales, C., Iturregui, P., Corp, M., Masson, L., Portilla, A., Ríos, M. (eds.), El Medio Ambiente en el Perú. Año 2001. Lima, Peru, pp. 155–170.

Bers, G., 2004. List of major natural disasters, 1960–1987. Natural Hazards 31 (1), pp. 97–99.

Berri, G.J., Ghietto, M.A., García N.O., 2002. The influence of ENSO en the flows of the Upper Parana River of South America over the last 100 years. Journal of Hydrometeorology 3 (1), pp. 57–65.

Berrocal, J., Espinoza, A.F., Galdos, J., 1978. Seismological and geological aspects of the Mantaro landslide in Peru. Nature 275, pp. 533–536.

Bertani, L., Brandani, A., Rabassa, J., 1986. Fluctuations of Castaño Overo Glacier in Northern Patagonia since the beginning of the XVIth Century. In Proceedings of the International Symposium on Glacier mass–balance, fluctuations and runoff, Alma–Ata, USSR, 1985. Data of Glaciological Studies, No. 57, pp. 192–196; Russian version: pp. 67–72. Academy of Sciences of the USSR, Moscow.

Bertiller, M., 1993. Conclusiones. In Paruello, J.M., Bertiller, F. (eds.), Secuencias de deterioro en distintos ambientes patagónicos. Su caracterización mediante el modelo de estados y transiciones. INTA—GTZ, Buenos Aires, p. 110.

Bertiller, M.B., Elissalde, O.E., Rostagno, C.M., Defossé, G.E., 1995. Environmental patterns and plant distribution along a precipitation gradient in western Patagonia. Journal of Arid Environment 29, pp. 85–97.

Bertoni, J., 1959. O espaçamento dos terraços em culturas anuais, determinadfo em função das perdas por erosão. Campinas, Bragantia 18, pp. 113–140

Bertoni, J., Lombardi Neto, F., 1985. Conservação do solo. Ed. Livroceres, Piracicaba, p. 392.

Bertoni, J.C., and Tucci, C.E.M., 2003. Urbanización. In Tucci, C.E.M., and Bertoni, J.C. (eds.), Inundação Urbana na América do Sul. World Meteorological Organization, Global Water Partnership, ABRH, Porto Alegre, RS, Brazil, pp. 1–9.

Bessa, O., Angulo, R.J., 2001. Variacoes da morfologia costeira na Praia Brava de Caiobá, Estado de Paraná. VIII Congresso da ABEQUA, Mariluz, Imbé, pp. 509–511

Betts, R.A., Malhi, Y., Roberts, J.T., 2008. The future of the Amazon: New perspectives from climate, ecosystem and social sciences. Philosophical Transactions of the Royal Society, 363 B, pp. 1729–1735.

Bezada, M., 2000. El Evento Torrencial Catastrófico de Vargas, Diciembre de 1999. Conferencia Caracas, Marzo 2000. Universidad Pedagógica Experimental Libertador, Caracas.

Bezada, M., Pacheco, H., Gil, R., González, O., 2001. La destrucción de Carmen de Uria en diciembre de 1999: Una catástrofe natural o una sumatoria de errores humanos. IV Congreso Venezolano de Geografía, 2001, Caracas.

Bischoff, S.A., García, N.O., Vargas, W.M., Jones, P.D., Conway D., 1999. Climatic variability and Uruguay river flows. Water International 25 (3), p. 24.

Bisigato, A.J., Bertiller, M.B., 1997. Grazing effects on patchy dryland vegetation in northern Patagonia. Journal of Arid Environment 36, pp. 639–653.

Bird, R.M., 1987. A postulated tsunami and its effect on cultural development in the Peruvian Early Horizon. Amer. Antiq. 52 (2), pp. 285–303.

Bird, E.C.F., 1993. Submerging Coasts. John Wiley and Sons, Chichester.

Bitrán, D., 2000. Evaluación del impacto socioeconómico de los principales desastres naturales ocurridos en la República Mexicana, durante 1999. Cuadernos de Investigación 50, CENAPRED, SEGOB, p. 194.

Bitschene, P.R., 1995. Environmental impact and hazard assessment of the August 1991 eruption of Mt. Hudson (Patagonian Andes). In Bitschene, P.R., Mendia, J. (eds.), The August 1991 Eruption of the Hudson Volcano (Patagonian Andes): A Thousand Days After. Univ. Nacional de la Patagonia San Juan Bosco, Serv Nac Geol, pp. 2–15.

Blaikie, P., Cannon, T., Davis, I., Wisner, B., 1994. At Risk: Natural Hazards, People's Vulnerability and Disasters. Routledge, London, p. 275.

Blanchard–Boehm, D.R., 2004. Natural Hazards in Latin America: Tectonic forces and storm fury. The Social Studies 95 (3), pp. 93–105.

Blanco, F.A., Burgos, E.A., Mejía, M., 2002. Estudio de amenaza por lahar en El Salvador: revision de casos historicos y calibracion de herramientas para la evaluacion de amenaza. Universidad Centroamericana José Simeón Cañas, El Salvador, p. 158.

Bodenbender, G., 1894. El terremoto argentino del 27 de octubre de 1894. Boletín de la Academia Nacional de Ciencias en Córdoba, Córdoba, tomo XIV, pp. 293–329.

Bommer J.J., Rodriguez C.E., 2002. Earthquake–induced landslides in Central America. Engineering Geology 63 (3), pp. 189–220.

Bonilla, M. 1988. Minimum earthquake magnitude associated with coseismic surface faulting. Bulletin Association of Engineering Geologists 25, pp. 17–29.

Bonnot, D., 1988. Évolution géodynamique plio–quaternaire du bassin intracordillérain du Callejon de Huaylas et de la Cordillère Blanche, Perou. Géodynamique 3, pp. 57–83.

Bordas, N.P., Lanna, A.E., Leprun, J.C., Semmelmann, P.R., 1987. Diagnóstico preliminar dos riscos de assoreamento no Brasil. VII SBRH, ANAIS.

Bourgeois, J., Petroff, C., Yeh, H., Titiv, V., Synolakis, C.E., Benson, B., Kuroiwa, J., Lander J., Norabuena, E., 1999. Geologic setting, field survey and modeling of the Chimbote, northern Peru, Tsunami of 21 February 1996. Pure Appl. Geophys. 154, pp. 513–540.

Bourgois, J., Bigot–Cormier, F., Bourles, D., Bracuher, R., Dauteuil, O., Witt, C., Michaud, F., 2007. Tectonic record of strain buildup and abrupt coseismic stress release across the northwestern Peru coastal plain, shelf, and continental slope during the past 200 kyr. J. Geophys. Research 112: doi 10.1029/2006JB004491.

Bovis, M.J., Dagg, B.R., 1992. Debris flow triggering by impulsive loading: mechanical modelling and case studies. Canadian Geotech. Journ. 29, pp. 345–352.

Bowler, J., 1973. Clay dunes: The occurrence, formation and environmental significance. Earth Sci. Rev. 9, pp. 315–338.

Bradley, R.S., Vuille, M., Diaz, H.F., Vergara, W. 2006. Threats to water supplies in the tropical Andes. Science 312, pp. 1755–1756.

Bragagnolo, N., 1994. Uso dos solos altamente suscetíveis a erosão. In Pereira, V.P., Ferreira, M.E., Pessôa da Cruz, M.C. (eds.), Solos altamente suscetíveis à erosão, Soc. Bras. Ci. Solo/Fac. Ci. Agrárias. e Veterinárias, UNESP, Jaboticabal, pp. 5–16.

Brandani, A., Rabassa, J., Boninsegna, J., Cobos, D., 1986. Glacier fluctuations during and since the Little Ice Age and forest colonization: Monte Tronador and Volcán Lanín, Northern Patagonian Andes. In Proceedings of the International Symposium on Glacier Mass–balance, Fluctuations and Runoff, Alma–Ata, USSR, 1985; Data of Glaciological Studies, no. 57, pp. 196–205; Russian version: pp. 72–83. Academy of Sciences of the USSR, Moscow.

Brandt, C.H., 1828. Journey of a vogaye to Peru: A passage across the Cordillera of the Andes, in the winter of 1827, performed on foot in the snow; and a Journey across the Pampas. London.

Brenning, A., 2005. Geomorphological, hydrological and climatic significance of rock glaciers in the Andes of Central Chile (33–35° S). Permafrost and Periglacial Processes, 16, pp. 231–240.

Bridges, L., 2000. El último confín de la tierra. Editorial sudamericana, Capítulo 16, p. 511.

Bridges, T., 1879. Southern Misión: Tierra del Fuego. South American Missionary Magazine, July 1, pp. 151–156.

Brooks, N., Nicholls, R., Hall, J., 2006. Sea level rise: Coastal impacts and responses. Final Draft, WBGU, p. 46.

Brooks, B., Bevis, M., Smalley Jr., R., Kendrick, E., Manceda, R., Lauría, E., Maturana, R., Araujo, M., 2003. Crustal motion in the Southern Andes (26°–36°S): Do the Andes behave like a microplate? Geochemistry, Geophysics, Geosystems (G^3) 4(10). doi:10.1029/2003GC000505.

Brooks, W.E., Willett, J.C., Kent, J.D., Vasquez, V., Rosales, T., 2005. The Muralla Pircada—An ancient Andean debris flow retention dam, Santa Rita B archaeological site, Chao valley, northern Peru. Landslides 2, pp. 117–123.

Bruce, J.P., 1994. Natural disaster reduction and global change. Bulletin of the American Meteorological Society 75(10), pp. 1831–1835.

Brush, S.B., 1976. Man's use of Andean ecosystems. Human Ecology 4, pp. 147–166.

Burton, I., Kates R.W., 1964. The perception of natural hazards in resource management. Nat. Resour. J. 3, pp. 412–441.

Burton, I., Kates, R.W., White, G.F., 1968. The human ecology of extreme geophysical events. Department of Geography. Natural Hazards Research Working Paper No. 1, University of Toronto.

Byers, W.G., 2004. Behavior of railway bridges and other railway infrastructure in three strong to great earthquakes in 2001. Transport. Research Record, 1863, pp. 37–44.

Cadavid, M.F., Hermelin, M., 2005. El evento del 29 y 30 de mayo de 2000 en La Estrella y Sabaneta (Antioquia). In Hermelin, M. (ed.), Desastres de origen natural en Colombia 1979–2004. Universidad EAFIT, OSSO–Univalle, Medellín, Colombia, pp. 181–198.

Cadavid, M.F., Hermelin, M., De Greiff, P., 2004. Lessons learned from the May 2000 catastrophic event in the town of La Estrella, Antioquia, Colombia. Journal of Human Security and Development 1 (1), pp. 77–92.

Cahil, T., Isacks, B.L., 1992. Seismicity and shape of the subducted Nazca Plate. Journal Geophysical Research 97, pp. 17503–17529.

Carey, M., 2005. Living and dying with glaciers: People's historical vulnerability to avalanches and outburst floods in Peru. Global Planet. Change 47, pp. 122–134.

Carvalho, L.M.V., Jones, C., Liebmann, B., 2004. The South Atlantic convergence zone: Intensity, form, persistence, and relationships with intraseasonal to interannual activity and extreme rainfall. J. Climate 17, pp. 88–108.

Calle, B., González, H., 1980. Geología y geoquímica de la plancha 166 Jericó. Ingeominas Technical Report 1822. Medellín, Colombia.

Calvache, M.L., 2006. Nevado del Ruiz: Reactivación y erupción en 1985. Boletín de vías 32 (98), pp. 29–38.

Campagnoli, F., 2005. The Brazilian lands: Rates of potential production of sediments. Sediment Budget Symposium/VII IAHS Scient. Assembly, Foz do Iguaçu, PR, Brazil, p. 7.

Campbell, C.J., 1974. Ecuadorian Andes in Mesozoic–Cenozoic orogenic belt, data for orogenic studies. Geological Society of London Special Publication, 4, pp. 725–732.

Campos, J., Ruiz, S., Pérez, A., Ruiz, J., Kausel, E., Thiele, R., Saragoni, R., Sepúlveda, S., 2005. Terremotos corticales de las melosas 1958, Chusmiza 2001 y Curicó 2004: un análisis comparativo de los terremotos de northridge 1994 y kobe 1995. Nuevos antecedentes para el peligro sísmico en Chile. Congreso Chileno de Sismología e Ingeniería Antisísmica. IX jornadas, Concepción, Chile.

Cande, S.C., Leslie, R.B., Parra, J.C., Hobart, M., 1987. Interaction between the Chile Ridge and Chile Trench: Geophysical and geothermal evidence. Journal Geophysical Research 92, pp. 495–520.

Cannon, T., 1993. A hazard need not a disaster make: Vulnerability and the causes of "natural" disasters. In Merriman, P.A., C.W.A., Browitt, C.W.A. (eds.), Natural Disasters: Protecting Vulnerable Communities, Thomas Telford, London, pp. 92–105.

Cannon, T., 1994, Vulnerability analysis and the explanation of "natural" Disasters. In Varley, A. (ed.), Disasters, Development and Environment, New York, John Wiley and Sons, pp. 13–30.

Cannon, T., 2003. Vulnerability Analysis, livelihoods and disasters components and variables of vulnerability: Modelling and analysis for disaster risk management. Indicators for Disaster Risk Management, OPERATION ATN/JF–7907–RG, Inter–American.

Canziani, P.O., 2003. Tropospheric–stratospheric coupling: Extratropical synoptic systems in the lower stratosphere. Quart. J. Roy Meteorol. Soc., p. 129.

Cárdenas, C., 2005. Erupción de 1985 del volcán Nevado de Ruiz: El despertar del león dormido. In Hermelin, M. (ed.), Desastres de origen natural en Colombia 1979–2004. Universidad EAFIT, OSSO–Univalle, Medellín, Colombia, pp. 39–53.

Cardona, O.D., 1999. The earthquake of Armenia, Colombia, January 25, 1999. Lessons in seismic engineering and disaster prevention. International seminar on natural disaster reduction for roads in Latin American countries. Cartagena, Colombia, pp. 87–109.

Cardona, O.D., 2001. Estimación Holística del Riesgo Sísmico Utilizando Sistemas Dinámicos Complejos. Polytechnical University of Catalonia, Barcelona. (http://www.desenredando.org/public/varios/2001/ehrisusd/index.html).

Cardona, O.D., 2003b. Indicadores para la Medición del Riesgo: Fundamentos para un Enfoque Metodológico. IDB/IDEA Program of Indicators for Disaster Risk Management. National University of Colombia, Manizales (http://idea.unalmzl.edu.co).

Cardona, O.D., 2004a. The need for rethinking the concepts of vulnerability and risk from a holistic perspective: A necessary review and criticism for effective risk management. In Bankoff, G., Frerks, G., and Hilhorst, D. (eds.), Mapping Vulnerability: Disasters, Development and People. Earthscan Publishers, London.

Cardona, O.D., 2004b. Dimensionamiento Relativo del Riesgo y de la Gestión: Metodología Utilizando Indicadores a Nivel Nacional. IDB/IDEA Program of Indicators for Disaster Risk Management. National University of Colombia, Manizales (http://idea.unalmzl.edu.co).

Cardona, O.D., 2004c. Resultados de la Aplicación del Sistema de Indicadores en Doce Países de las Américas. IDB/IDEA Program of Indicators for Disaster Risk Management. National University of Colombia, Manizales (http://idea.unalmzl.edu.co).

Cardona, O.D., 2005. Sistema de indicadores para la gestión del riesgo de desastre: Informe técnico principal. IDB/IDEA Program of Indicators for Disaster Risk Management. National University of Colombia, Manizales (http://idea.unalmzl.edu.co).

Cardona, O.D. 2006. A system of indicators for disaster risk management in the Americas in measuring vulnerability to hazards of natural origin: Towards disaster resilient societies. In Birkmann, J. (ed.). United Nations University Press, Tokyo.

Cardona, O.D., Hurtado, J.E., Duque, G., Moreno, A., Chardon, A.C., Velásquez L.S., Prieto, S.D., 2003. La Noción de Riesgo desde la Perspectiva de los Desastres: Marco Conceptual para su Gestión Integral. IDB/IDEA Program of Indicators for Disaster Risk Management. National University of Colombia, Manizales (http://idea.unalmzl.edu.co).

Carlotto, V., Tintaya, D., Cárdenas, J., Carlier, G., Rodríguez, R., 2006. Fallas transformacionales Permo–Triásicas: La Falla Patacancha–Tamburco (sur del Perú). XIII Congreso Peruano de Geología Resúmenes Extendidos.

Carr, M.J., Stoiber, R.E., 1990. Volcanism, in The Geology of North America, Volume H, The Caribbean Region. In Dengo, G., Case, J.E. (eds.). Geol. Soc. America, Boulder, CO, pp. 375–391.

Carr, M.J., Feigenson, M.D., Patino, L.C., Walker, J.A., 2003. Volcanism and Geochemistry in Central America: Progress and Problems, Inside the Subduction Factory. Geophysical Monograph. 138, pp. 153–179.

Carreño R., 1999. Identificación y evaluación preliminar de peligros asociados a terrenos inestables en algunos puntos regionales del centro y norte de Nicaragua. COSUDE–INETER, Managua, p. 22.

Carreño C.R., Kalafatovich C.S., 2006. The Alcamayo and Cedrobamba debris flow (January, March, and April 2004) in Machupicchu area–Peru. Landslides, 3, pp. 79–83.

Carrillo, E., Audemard, F.A., Beck, C., Cousin, M., Jouanne, F., Cano, V., Castilla, R., Melo, L., Villemin, T. 2006. A Late Pleistocene natural seismograph along the Boconó fault (Mérida Andes, Venezuela): The moraine–dammed Los Zerpa paleolake. Bulletin de la Societé Géologique de France 177, pp. 3–17.

Carvalho, A.P., 1994. Solos do arenito Caiuá. In Pereira, V.P., Ferreira, M.E., Pessoa da Cruz, M.C. (eds.), Solos altamente suscetíveis à erosão. Soc. Brasil. Ci. Solo/Fac. Ci. Agrárias e Veterinárias, UNESP, Jaboticabal, pp. 3–16.

Casertano, L., 1963. General characteristics of the active Andean volcanoes and summary of their activities during recent centuries. Bulletin of the Seismological Society of America 53, pp. 1415–1433.

Castañeda, M.E. Barros, V.R., 1994. Las tendencies de la precipitación en el Cono Sur de América al este de los Andes. Meteorológica 19, pp. 23–32.

Casassa, G., Rivera, A., Escobar, F., Acuña, C., Carrasco, J., Quintana, J., 2003. Snow line rise in Central Chile in recent decades and its correlation with climate. EGS–AGU–EUG Joint Assembly, April 6–11, 2003, Nice, France. Abstracts volume, Abstract No. 14395.

Castagnoli, N., 1966. Perdas anuais provocadas pela erosão no Estado de São Paulo. I Congr. Panamericano Conserv. Solo, São Paulo, Anais, pp. 147–153.

Castellanos, A. 1945. Cuatro Lecciones sobre terremotos. Parte B: El terremoto de San Juan. Asociación Cultural de Conferencias de Rosario. Publicación 6:77-243, Rosario.

Castañeda, M.E., Barros, V.R., 1994. Las tendencies de la precipitación en el Cono Sur de América al este de los Andes. Meteorológica 19, pp. 23–32.

Cazes–Boezio, G., Robertson, A.W., Mechoso, C.R., 2003. Seasonal dependence of ENSO teleconnections over South America and relationships with precipitation in Uruguay. J. Climate 16, pp. 1159–1176.

CBHAT, 1999. Plano da Bacia do Alto Tietê. Comitê da Bacia Hidrográfica do Alto Tietê, Relatório Síntese, São Paulo.

Ceballos J.P., 2006. Peligro de Aluviones en la localidad de Barreal, Departamento Calingasta, San Juan. Ph.D. Dissertation. Universidad Nacional de San Juan, p. 120. Unpublished.

Ceballos, J., Escátegui, C., Ramírez, J., 2005. Colombia. In Haeberli, W., et al., Fluctuations of Glaciers, 1995–2000, IUGG–UNEP–UNESCO. Zurich, Switzerland, vol. 8, pp. 40–41.

Cecioni, A., Pineda, V., Cendrero, A., Hurtado, M., Panizza, M., 2004. Environmental quality assessment in the coast of Concepción, Chile. In Martín–Duque, J.F., Brebbia, C.A., Godfrey A.E., Díaz de Terán, J.R. (eds.), Geo–Environment: Monitoring, Simulation and Remediation of the Geological Environment. WIT Press. Ashurst Lodge, Ashurst, Southampton, UK

CENICAFÉ Centro Nacional de Investigaciones de Café., 1989. Estudio climático de Risaralda y Quindío. Chinchiná, Colombia. Boletín Técnico, 15.

Central Pacific Hurricane Center–NOAA, 2006. Tropical cyclone climatology. Available at: http://www.pch.noaa.gov/cphc.

CENTRO, 2004. Informe final IAI– 2004 ENSO–argentina, CENTRO estudios sociales y ambientales.

CEPAL, 1984. The Natural Disasters of 1982–1983 in Bolivia, Ecuador and Peru. ECLA, Economic Commision for Latin America, E/CEPAL/G 1274, p. 80.

CEPAL Comisión económica para América Latina., 1999. El terremoto de enero de 1999 en Colombia: Impacto socioeconómico del desastre en la zona del Eje Cafetero. Technical Report. Méjico DF, Méjico. Available at: http://www.cepal.org/. Accessed in July 2006.

CEPAL, 2003. Las inundaciones dze 2003 en Santa Fe, Argentina. Evaluación del Impacto Socioeconómico y Ambiental. http://www.eclac.org/analisis/default.asp?idioma=FR&unbisCategory =13&unbisNumber

CERESIS, 1997. Catálogo sísmico regional de parámetros focales para América del Sur y mapa con el resultado del cálculo del peligro sísmico. Centro Regional de Sismología para América del Sur.

Charrier, R., Pinto, L., Rodríguez, M.P., 2007. Tectonostratigraphic evolution of the Andean Orogen in Chile. In Moreno, T., and Gibbons, W. (eds.), The Geology of Chile. The Geological Society, London, pp. 21–114.

Chatelain, J.L., Tucker, B., Guillier, B., Kaneko, F., Yepez, H., Fernandez, J., Valverde, J., Hoefer, G., Souris, M., Deuperier, E., Yamada, T., Bustamante, G., Villacis, C., 1999. Earthquake risk management pilot project in Quito, Ecuador. GeoJournal 49 (2), pp. 185–196.

Chayle W., Agüero C., 1987. Características de remoción en masa en la cuenca del Rio Grande (Qda de Humahuaca) Jujuy. Instituto de Geol. y Minería. Universidad Nacional de San Juan, Revista No. 8, pp. 97–122. Salta, Argentina.

Chayle W., Wayne W.J., 1995. Impact of erosion, mass wasting, and sedimentation on human activities in the Río Grande Basin, Jujuy Province, Argentina. Environmental & Engineering Geosciences, 1, (4), pp. 403–416.

Chen, J.L., Wilson, C.R., Tapley, B.D., 2006. Satellite gravity measurements confirm accelerated melting of Greenland Ice Sheet. Science 313, pp. 1958–1960.

Chibnik, M., 1994. Risky Rivers: The Economcs and Politics of Floodplain Farming in Amazonia. The University of Arizona Press.

Christensen, J.H., 16 others 2008. Regional climate projections. In Solomon, S., Qin D., Manning, M., Chen, Z., Marquis M., Avery, K.B, Tignor M., Miller H.L. (eds.), Climate Change 2007: The Physical Science Basis. Cambridge University Press, Cambridge, pp. 848–940.

CIAT Centro Internacional de Agricultura Tropical., PNUMA Programa de las Naciones Unidas para el Medio Ambiente, 1998. Atlas de Indicadores Ambientales y de Sustentabilidad para América Latina y el Caribe. Cali, Colombia.

Cibils, A., 2001. El clima de Río Gallegos: más de 60 años de registros de temperatura ambiente y precipitaciones. INTA–EEA Santa Cruz, Río Gallegos. Unpublished.

Cisternas, A., Kausel, E., Comte, D., 2003. El mayor sismo registrado: Chile, 1960. 3er Congreso Nacional Ingeniería Sísmica. Coloquios de la AEIS. Madrid, España.

Clapperton, C., 1993. Quaternary geology and geomorphology of South America. Dept. of Geography, University of Aberdeen, Scotland.

Clarín newspaper, 1991. Buenos Aires, Argentina.

Clarke, C., Munasinghe, M., 1995. Economic aspects of disasters and sustainable development: An introduction. In Munasinghe, M., Clarke, C. (eds.), Disaster Prevention for Sustainable Development, Economy and Policy Issues. IDNDR and the World Bank, Washington, DC, pp. 1–10.

Claure, B., 2008. Bolivia: El Niño has bigger bite with climate change. Inter Press Service News Agency, Monday, March 24, 2008.

Clavero, J.E., Sparks, R., Huppert, H., 2002. Geological constraints on the emplacement mechanism of the Parinacota debris avalanche, northern Chile. Bulletin of Volcanology 64, pp. 40–54.

Clifton, H.E., 1988. Sedimentologic relevance of convulsive geologic events. In Clifton, H.E. (ed.), Sedimentologic consequences of convulsive geologic events. Geological Society of America special paper 229. GSA, Boulder, CO, USA, pp. 1–5.

Cobbing, E.J., 1978. The Andean geosyncline in Peru, and its distinction from Alpine geosynclines. J. Geol. Soc. Lond. 135, pp. 207–218.

Codignotto J., 2004. Erosion costera. In Gonzalez, M., Bejerman, NJ (eds.), Peligrosidad geológica en Argentina. ASAGAI, Bs. Ar, Special Issue N° 4, pp. 90–111.

Coe, J.A., Godt, J.W., Baum, R.L., Bucknam, R.C., Michael, J.A., 2004. Landslide susceptibility from topography in Guatemala. In Lacerda, W.A., Ehrlich, M., Fontura, S.A.B., and Sayao, A.S.F. (eds.), Landslides: Evaluation and Stabilization. Taylor & Francis, vol. 1, pp. 69–78.

Coelho Netto, A.L., Alvera, A.S., Fernandez, M.C., Lacerda, W.A., 2007. Landslide susceptibility in a mountainous geoecosystem, Tijuca Massif, Rio de Janeiro: The role of morphometric subsivision of the terrain. Geomorphology, 87, pp. 120–131.

Coelho Netto, A.L., 1985. Surface hydrology and soil erosion in a tropical mountainous rainforest drainage basin, Rio de Janeiro, Ph.D. Thesis, Katholieke University Leuven, Belgium.

Coelho Netto, A.L., 1996. Produção de sedimentos em bacias fluviais florestadas do Maciço da Tijuca: Respostas aos eventos extremos de Fevereiro de 1996. Anais do II Encontro Nacional de Engenharia de Sedimentos, ABRH, Rio de Janeiro, vol. 1, pp. 209–217.

Cohen, J.C.P., Silva Dias, M.A.F., Nobre, C.A., 1995. Environmental conditions associated with Amazonian squall lines: A case study. Mon. Wea. Rev. 123, pp. 3163–3174.

Coira, B., Kay, S.M., 1993. Implications of Quaternary volcanism at Cerro Tuzgle for crustal and mantle evolution of the Puna Plateau, Central Andes, Argentina. Contr Mineral Petr, 113, pp. 40–58.

Coira, B., Kay, S., Viramonte, J., 1993. Upper Cenozoic magmatic evolution of the Argentine Puna. A model for changing subduction geometry. International Geology Review 35 (8), pp. 677–720.

Coleman, A., 1949. Mi vida de ferroviario inglés en la Argentina: 1887–1948: Al vaciarse un lago cordillerano, provoca una gran inundación, Bahía Blanca, Argentina, Capítulo III, pp. 241–250.

Comissão de Solos, 1960. Levantamento de reconhecimento dos solos do estado de São Paulo. Min. Agric., SNPA, Rio de Janeiro, p. 634.

Comte, D., Eisenberg, A., Lorca, E., Pardo, M., Ponce, L., Saragoni, G.R., Singh, S.K., Súarez, G., 1986. The central Chile earthquake of 3 March, 1985: A repeat of previous great earthquakes in the region? Science 233, pp. 449–453.

Cook, A.J., Fox, A.J., Vaughan, D.G., Ferrigno, J.G., 2005. Retreating glacier fronts in the Antarctic Peninsula over the past half-century. Science 308, pp. 541–544.

Coronato, A., 1994. Geomorfología glacial de valles de los Andes Fueguinos y condicionantes físicos para la instalación humana. Ph.D. Dissertation, Universidad de Buenos Aires, p. 318. Unpublished.

Coronato, A., 1995. The last Pleistocene glaciation in tributary valleys of the Beagle Channel, Southernmost South America. Quaternary of South America and Antarctic Peninsula, Rotterdam, 9, pp. 173–182.

Coronato, A., Roig C., 1999. Peligro geomorfológico en ambientes de génesis reciente: Valles de Tierra Mayor y Río Olivia. I Congreso Argentino de Geomorfología y Cuaternario, Santa Rosa, Actas, pp. 123–132.

Coronato, A., Escobar J., Mallea C., Roig C., Lizarralde M., 2003. Características geomorfológicas de ríos de montaña colonizados por Castor canadensis en Tierra del Fuego, Argentina. Ecología Austral 13, Buenos Aires, pp. 15–26.

Correa, I.D., Acosta, S., Bedoya, G., 2007. Análisis de las causas y monitoreo de la erosión litoral en el Departamento de Córdoba. Corporación Autónoma regional de los valles del Sinú y del San Jorge, Universidad EAFIT, Area Ciencias del Mar, p. 128.

Correa, I.D., González, J.J., 2000. Coastal erosion and village relocation: A Colombian case study. Ocean and Coastal Management 43, pp. 51–64.

Correa, I.D., Vernette, G., 2004. Introducción al problema de la erosión litoral en Urabá (Sector Arboletes–Turbo), costa Caribe colombiana. Boletín de Investigaciones marinas y costeras, INVEMAR 33, Santa Marta, 1.

Cortés, J., 2000. Fallas cuaternarias oblicuas al frente montañoso en la Cordillera Frontal de Mendoza (34°– 34° 30'). Revista de Cuaternario y Ciencias ambientales. Publicación Especial N° 4. Asociación Geológica Argentina, pp. 57–62.

Cortés, C., Franchi, M., Nullo, F., 1987. Evidencias de neotectónica en Las Sierras de Aguilar y del Tanque, Cordillera Oriental y Puna Jujeñas, Argentina. X Congreso Geológico Argentino, Actas, vol. 1, pp. 239–242.

Corvalán, N., Kovacic, I., Muñoz, O., 1997. Quebrada de Macul: El aluvión del 3 de mayo 1993 causas y consecuencias. Revista Geográfica de Chile Terra Australis 42. Instituto Geográfico Militar, pp. 139–167.

Costa, C., 1996. Análisis geotectónico en las sierras de San Luis y Comechingones: Problemas y métodos. XIII Congreso Geológico Argentino, Vol. 2, pp. 285–300.

Costa, C., 1999. Tectónica Cuaternaria en las Sierras Pampeanas. In Caminos, R. (ed.), Geología Argentina, Segemar. Anales, 29 (24), Sección 2B, pp. 779–784.

Costa C., Rockwell, T., Paredes, J., Gardini, C., 1999. Quaternary deformations and seismic hazard at the Andean Orogenic Front (31°–33°, Argentina): A paleoseismological perspective. 4° International Symposium on Andean Geodynamics, IRD Paris, Extended Abstracts, pp. 187–191.

Costa, C., Machette, M., Dart, R., Bastias, H., Paredes, J., Perucca, L., Tello, G., Haller, K., 2000. Map and Database of Quaternary Faults and Folds in Argentina. U.S.Geological Survey Open–File Report 00–0108, p. 75.

Costa, C., Gardini, C., Diederix, H., 2000b. The Montecito anticline: A Quaternary growing structure in the Precordilleran foothills of northern Mendoza, Argentina. 9° Congreso Geológico Chileno Actas 1, pp. 758–762.

Costa, C., Murillo, V., Sagripanti, G., Gardini, C., 2001. Quaternary intraplate deformation in the southeastern Sierras Pampeanas, Argentina. Journal of Seismology, 5, pp. 399–409.

Costa, C.H., Audemard, F.A.; Becerra, F.H., Lavenu, A., Machette, M.N., París, G., 2006a. An overview of the main Quaternary deformation of South America. Revista de la Asociación Geológica Argentina 61, pp. 461–479.

Costa, C., Smalley, R., Schwartz, D., Stenner, H., Ellis, M. Ahumada, E., Velasco, M., 2006b. Paleoseismic observations of an onshore transform boundary: The Magallanes–Fagnano fault, Tierra del Fuego, Argentina. Revista Asociación Geológica Argentina 61, pp. 647–657.

Costa, C., 2009a. Falla La Rinconada, Argentina. In: Proyecto Multinacional Andino (Ed.). Atlas de Deformaciones Cuaternarias de los Andes. SERNAGEOMIN, Publicación Multinacional No 7: 104-110 Santiago.

Costa, C., 2009-b. Anticlinal Montecito, Argentina. In: Proyecto Multinacional Andino (Ed.). Atlas de Deformaciones Cuaternarias de los Andes. SERNAGEOMIN, Publicación Multinacional No 7: 273-277, Santiago.

Couto, E.G., 1990. O uso da terra e o garimpo na bacia do rio S. Lourenço, Mato Grosso. Reflexos no ambiente. FEMA–MT/UFMT/CCA, p. 206.

Couture N.J., Pollard, W.H., 2007. Modelling geomorphic response to climatic change. Climatic Change, 85, pp. 407–431.

Cowan, H., Machette, M., Amador, X., Morgan, S., Dart, R., Bradley L–A., 2000. Map and database of Quaternary faults in the vicinity of Managua, Nicaragua. U.S. Geological Survey Open–File Report 00–0437. Golden.

Cruz, E.S.A., Vilela, C.L., Coelho Netto, A.L., 1999. Spatial distribution of the 1996 landslides scars at the Tijuca Massif and subsequent erosion: The influence of topography and vegetation changes. 1999/IGU–Gertec Meeting on Geomorphic Responses to Vegetation Changes: Problems & Remedial Work, Bananal (SP), Brazil, vol. 1, pp. 89–102.

Cruz, O., 1974. A Serra do Mar e o litoral na área de Caraguatatuba. São Paulo, USP, Inst. Geogr., ser. Teses e Monogr. 11, p. 181.

Cruz, O., 2000. Studies on the geomorphic processes of overland and mass movements in the Brazilian geomorphology. São Paulo, Rev. Brasil. Geoc. 30 (9), pp. 504–507.

CSE. 1992. Floods, Flood Plains and Environmental Myths. Centre for Science and Environment–CSE, New Delhi.

DAEE/IPT, 1986. Orientações para o combate à erosão no Estado de São Paulo: bacia do Peixe–Paranapanema. Inst. Pesq. Tecnol. S. Paulo, Rel. 24.739, vol. 6.

DAEE/IPT, 1988. Orientações para o combate à erosão no Estado de São Paulo: bacia do Baixo Tietê. Inst. Pesq. Tecnol. S.Paulo, Rel. 26.339, vol. 3.

DAEE/lPT, 1989. Controle de erosão: bases conceituais e técnicas. S. Paulo, Secret. Energia e Saneamento/Inst. Pesq. Tecnol., pp. 2.

DAEE/IPT, 1990. Orientações para o combate à erosão no Estado de São Paulo: bacia do Pardo–Grande. Inst. Pesq. Tecnol. S. Paulo, Rel. 28.194, vol. 3.

DAEE/IPT, 1992. Orientaõees para o combate a erosão no Estado de São Paulo: bacia do Médio Tietê. Inst. Pesq. Tecnol. S. Paulo, Rel. 29.004, vol. 3.

DAEE/IPT, 1997. Mapa de erosão do Estado de São Paulo (coord. Gouveia, M.I., Femandes K., Almeida F.G.S., Souza, L.A.P.). Dep. Águas Energ. Eletr. e Inst.Pesq.

DANE Departamento Administrativo Nacional de Estadística., 2005. Censos de población y vivienda. Available at: http://www.dane.gov.co/. Accessed in July 2006.

Darwin, C., 1845. Voyage of the Beagle. 2nd edition, John Murray, Albermarle Street, London,434p.

Davis, L., 2002. Natural disasters: Revised edition. Facts On File, Inc., New York, USA, p. 420.

De Angelis, H., Skvarca, P., 2003. Glacier surge after ice shelf collapse. Science, 299, pp. 1560–1562.

De Cserna Zoltan, 1990. Téctónica, mapa IV.2.1, T. II, Atlas Nacional de México, Instituto de Geografía, UNAM, México.

Degg, M.R., Chester, D.K., 2005. Seismic and volcanic hazards in Peru: Changing attitudes to disaster mitigation. Geographic J. 171 (2), pp. 125–145.

Degg, M.R., Doornkamp, J.C., Sufflebotham, E.J., 1999. Earthquake hazard atlas: 11. Peru. International Underwriting Association of London, London.

Defant, M., Durmmond, M., 1990. Derivation of some modern arc magmas by melting of young subducted lithosphere. Nature 347, pp. 662–665.

De la Cruz–Reyna, S., Tilling, R.I., 2008. Scientific and public responses to the ongoing volcanic crisis at Popocatepetl Volcano, Mexico: Importance of an effective hazards–warning system. Journal of Volcanology and Geothermal Research 170 (1–2), pp. 121–134.

Delpino, D., 1993. Peligrosidad del volcanismo de retroarco del sur de Mendoza. Primeras Jornadas Nacionales de Vulcanología, Medio Ambiente y Defensa Civil. Mendoza.

Delpino, D., Bermúdez, A., 2002. La erupción del volcán Copahue del año 2000. Impacto social y al medio natural. Provincia de Neuquén. Argentina. XV Congreso Geológico Argentino, Calafate, Actas 3, pp. 365–370.

Del Cogliano, D., Perdomo, R., Hormaechea, J.L., 2000. Desplazamiento entre placas tectónicas en Tierra del Fuego, Actas de la XX Reunión Científica de la AAGG, Mendoza.

Del Valle, H., 1998. Land degradation status in the Patagonian region: Desertification categories. Proceedings Third Latin American Space Days, Mexico, pp. 21–29.

Del Valle, H., Beeskow, A., Coronato, F., Rostagno, C., 1995. Evaluación del estado actual de la desertificación en tres segmientos de la provincia del Chubut. In: Lucha contra la desertificación en Patagonia. Evaluación del estado actual de la desertificación en áreas representativas de la Patagonia. Informe Final Fase I. INTA–GTZ.

Del Valle, H., Elissalde, N., Gagliardini, D., Milovich, J., 1998. Status of desertification in the Patagonian Region: Assessment and mapping from satellite imagery. Arid Soil Research and Rehabilitation 12, pp. 95–122.

Del Valle, H., Rostagno C.C., Bouza, P.P., 2000. Los médanos del S de Península Valdés, su dinámica y los cambios asociados en los suelos y la vegetación. XVII Congreso Argentino de la Ciencia del Suelo, Mar del Plata.

Demant, A., 1978. Características del eje neovolcánico transmexicano y sus problemas de interpretación. Universidad Nacional Autónoma de México: Instituto de Geología, Revista 2, pp. 172–187.

Demets, C., Gordon, R., Argus, D., Stein, S., 1990. Current plate motions. Geophys. J. Int. 101, pp. 425–478.

De Silva, S., Francis, P., 1991. Volcanoes of the Central Andes. Berlin: Springer-Verlag, p. 216

De Silva, S.L., Zielinski, G.A., 1998. Global influence of the AD 1600 eruption of Huaynaputina, Peru. Nature 393, pp. 455–458.

DesInventar., 2003. Disasters database for Colombia, 1914–2002. Available at: http://www.desinventar.org/desinventar.html. Accessed in July 2006.

Devoli, G., Morales, A., Høeg, K., 2007. Historical landslides in Nicaragua—Collection and analysis of data. Landslides 4, pp. 5–18.

De Wit, M., Stanciewicz J., 2006. Changes in surface water supply across Africa with predicted climate change. Science 311, pp. 1917–1921.

Dewey, J.W., Algermissen, S.T., 1974. Seismicity of the Middle America Arc–Trench System near Managua, Nicaragua. Bulletin of the Seismological Society of America, 64, pp. 1033–1048.

Digregorio, J., and Uliana, M., 1980. Cuenca Neuquina. In Turner, J. (ed.), Geología Regional Argentina.

Academia Nacional de las Ciencias II, Córdoba, pp. 985–1032.

DNP Departamento Nacional de Planeación., 1997. Sistema de indicadores sociodemográficos para Colombia v 2.0. Available at: http://www.dnp.gov.co/. Accessed in July 2006.

Dolan, R., Davis, R.E., 1994. Coastal storm hazards. Journal of Coastal Research, Sp. Issue 12, pp. 103–114.

Dolan, R., Hayden, B., 1981. Templates of change: Storms and shoreline hazards. Oceanus, 23 (4), Woods Hole Oceanographic Institution, pp. 32–37.

Dolan, R., Hayden, B., Bosserman, K., Lisle, L., 1987. Frequency and magnitude data on coastal storms. Journal of Coastal Research 3, 2, pp. 245–247.

Dolique, F., Anthony, E.J., 2005. Short–term profile changes of sandy pocket beaches affected by Amazon derived mud, Cayenne, French Guiana. Journal of Coastal Research 21, 6, pp. 1195–1202.

Domack, E., Gilbert, R., Berger, G., Tewksbury, D., McCallum, S., Backman, E., McMullen, K., Duran, D., McCloskey, A., Rubin, A., 2002. Marine geological survey off the Larsen B Ice Shelf: Cruise NBP–01–07 (December 2001 to January 2002). In Antarctic Peninsula Climate Variability: An Historical and Paleoenvironmental Perspective, Hamilton College, New York, April 3–6, 2002.

Domack. E.R., Duran, D., Leventer, A., Ishman, S., Doane, S., McCallum, S., Amblas, D., Ring, J., Gilbert, R., and Prentice, M., 2005. Stability of the Larsen Ice B Shelf on the Antarctic Peninsula during the Holocene epoch. Nature 436 (4), pp. 681–685.

Domínguez, E., Rabassa, J., 1981. Estabilidad de la ladera del valle del río Ñireco, San Carlos de Bariloche, Río Negro. VIII Congreso Geológico Argentino, San Luis, Argentina, Actas II, pp. 277–286.

Domínguez E., Rabassa J., Cabral D., 1981. Estudio del aluvión de Melipal San Carlos de Bariloche, Prov. de Río Negro. VIII Congreso Geológico Argentino, San Luis, Argentina, Actas II, pp. 221–233.

Doney, S.C., 2006. The dangers of ocean acidification. Scientific American 294 (3), pp. 38–45.

Dorbath, L., Cisternas, A., Dorbath, C., 1990. Assessment of the size of large and great historical earthquakes in Peru. Bull. Seism. Soc. Amer. 80(3), pp. 551–576.

Dorbath, L., Dorbath, C., Jiménez, E., Rivera, L., 1991. Seismicity and tectonic deformation in the eastern Cordillera and Sub Andean zone of Central Perú. Journal of South American Earth Sciences 4, pp. 13–24.

Dragani W.C., D'Onofrio, E.E., Grismeyer W., Fiore M.E., 2006. Tide gauge observations of the Indian ocean tsunami, December 26, 2004, in Buenos Aires coastal waters, Argentina. Continental Shelf Research, 26, pp. 1543–1550.

Dregne, H.E., 1976. Desertification: Symptoms of a crisis. In Desertification, Process, Problems, Perspectives. University of Arizona Press.

Dregne, H.E., 1986. Desertification of arid lands. El–Baz, F., Hassan, M.H.A. (eds), In Physics of desertification., Martinus, Nijhoff, Dordrecht, The Netherlands.

Duarte, G.M., 1995. Depósitos cenozóicos costeiros e a morfologia do extremo sul de Santa Catarina, T. Doutorado, Inst. de Geociências, Univ. São Paulo, p. 234.

Dumont, J.F., 1993. Lake patterns as related to neotectonics in subsiding basins: The example of the Ucamara Depression, Peru. Tectonophysics 222, pp. 69–78

Dumont, J.F., 1996. Neotectonics of the Subandes–Brazilian craton boundary using geomorphological data—The Marañon and Beni basins. Tectonophysics 257, pp. 137–151.

Dumrauf, C., 1992. Historia de Chubu: Colección historia de nuestras provincias. Plus Ultra, Buenos Aires.

Dykes, A.P., Welford, M.R., 2007. Landslides in the Tandayapa valley, northern Andes, Ecuador: implications for landform development in humid and tectonically active mountain ranges. Landslides 4, pp. 177–187.

ECLAC, 1999. Central America: Assessment of the damage caused by hurricane Mitch, 1998: Implications for economic and social development and for the environment.

ECLAC–UNEP.2000. Panorama of the environmental impact of the recent natural disasters in Latin America and the Carribbean, Economic Commission for Latin America and the Caribbean–E-CLAC and the United Nations Environment Program, p. 51.

Elissalde, N., Pappalardo, J., Mühleman, M., 1995. Evaluación del estado actual de la desertificación en la transecta Chubut. In Lucha contra la desertificación en Patagonia: Evaluación del estado actual de la desertificación en áreas representativas de la Patagonia. Informe Final Fase I. INTA—GTZ.

El Masri, S., Tipple, G., 2002. Natural disaster, mitigation and sustainability: The case of developing countries. International Planning Studies 7 (2), pp. 157–175.

Ellenberg, H., 1979. Man's influence on tropical mountain ecosystems in South America. Journal of Ecology 67, pp. 401–416.

EM-DAT: The OFDA/CRED International Disaster Database. Université Catholique de Louvain. Brussels, Belgium. At http://www.cred.be/emdat.

Emery, K.O., Aubrey, D.G., 1991. Sea Levels, Land Levels and Tidal Gauges. Springer-Verlag, p. 226.

EMPLASA, 2001. Memória Urbana: A grande São Paulo até 1940. EMPLASA, Arquivo do Estado, Imprensa Oficial, São Paulo.

Endlichter, W., Santana, A., 1988. El invierno de 1995: Un fenómeno climático muy severo en la Patagonia Austral. Anales del Instituto de la Patagonia, serie Ciencias Naturales, Chile, 25, pp. 77–88.

Enever, A., 2002. Fifty dead in shock Bolivian flood. BBC News. http://news.bbc.co.uk/2/hi/americas/1833002.stm.

EOSDIS Science Operations Office. Impacts of El Niño and La Niña 1997–2000, Goddard Space Flight Center Greenbelt, Maryland http://outreach.eos.nasa.gov/EOSDIS–CD–03/start.htm

Escosteguy L., Geuna S., Fauqué L., 1999. Las avalanchas de rocas del Moncol, Cordillera Principal, Neuquén. XIV Congreso Geológico Argentino, Salta, Actas II, pp. 67–70.

Escosteguy L., Cobos J., Anselmi G., 2005. Geomorfologia de la región de tres lagos, sector extrandino de la provincia de Santa Cruz. Actas XV Congreso Geológico Argentino (CD), La Plata.

Espinosa, A., 2001. Erupciones históricas de los volcanes Colombianos (1500–1995). Academia Colombiana de ciencias exactas, físicas y naturales. Colección Jorge Alvarez Lleras 16. Bogotá, Colombia.

Espizúa, L., 1986. Fluctuations of the Río Plomo Glaciers. Geografiska Annaler 68A, 4, pp. 317–327.

Espizua L.E., 2005. Megadeslizamientos Pleistocénicos en el valle del Rio Mendoza. Actas XV Congreso Geológico Argentino (CD), La Plata.

Espizua L.E., Bengochea J.D., 2002. Landslide hazard and risk zonation mapping in the Río Grande Basin, Central Andes of Mendoza. Argentina Mountain Research and Development 22(2), pp. 177–185.

Espizua L.E., Bengochea J.D., Aguado C., 1993. Mapa de riesgo de remoción en masa en el valle del río Mendoza. VII Congreso Geológico Argentino; II Congreso de Exploración de Hidrocarburos, Mendoza, Actas VI, pp. 323–332.

Esteves, L.S., Toldo, E.E., Almeida, L.E.S.B., Nicolodi, J.L., 2001. Erosao na costa do Río Grande do Sul entre 1975–2000. VIII Congresso da ABEQUA, Mariluz, Imbé, pp. 511–513.

Evans, S.G., Bent A.L., 2004. The Las Colinas landslide, Santa Tecla: A highly destructive flowslide triggered by the January 13, 2001, El Salvador earthquake. In Rose, W.I., Bommer, J.J., Lopez, D.L., Carr, M.J., Major, J.J. (eds.), Natural Hazards in El Salvador. Geological Society of America, Boulder, CO, Special paper 375, pp. 25–37.

Evans, S.G., Alcántara-Ayala, I., 2007. Disasters resulting from landslides, snow avalanches, and geotechnical failures in North America (Canada, United States, and Mexico) 1841–2006, first assessment. In Turner, A.K., and Schuster, R.L. (eds), Landslides in North America. Association of Engineering Geologists, Special Publication 22, pp. 3–23.

FAO and UNEP, 1984. Metodología provisional para la evaluación y representación cartográfica de la desertización. FAO/UNEP. Rome, Italy.

Farber, D., Hancock G., 2005. Uplift and topography formation of the Cordillera Blanca, central Peruvian Andes. Geophysical Research Abstracts vol. 7, 10,374.

Faria, A.P., 1996. Os processos erosivos e suas variações nas escalas temporal e espacial: revisão e análise. Rio de Janeiro, Rev. Brasil. Geogr. 58 (1/4), pp. 37–54

Fauque, L., 1987. Avalanchas de rocas en la Qda de Segovia y zonas vecinas de la vertiente occidental de las Sierras del Famatina, prov. de La Rioja, Argentina. X Congreso Geológico Argentino, San Miguel de Tucuman, Actas III, pp. 33–336.

Fauque, L., Strecker, M.R., 1998. Large rock avalanche deposits (Struzstrome, Struzstroms) at Sierra Aconquija, Northern Sierra Pampeanas, Argentina. Eclogae Geologie Helvetica 81, pp. 579–592.

Fauque, L., Tchilinguirian, P., 2002. Villavil rockslides, Catamarca Province, Argentina. In Evans, S.G., and DeGraff, J.V. (eds.). Catastrophic rockslides: Effects, occurrence, and mechanisms. Geological Society of America Reviews in Engineering Geology 15, pp. 303–324.

Fauque, L., Gonzalez M.A., 2004. Qué características presentan los movimientos actuales con antecedentes históricos en las Sierras Pampeanas? In Gonzalez M.,and Bejerman, N.J. (eds.). Peligrosidad geológica en Argentina. ASAGAI, Buenos Aires, Special Issue No. 4, pp. 408–410.

Fauqué L., Cortés J.M., Folguera A., Etcheverría M., 2000. Avalanchas de rocas asociadas a neotectónica en el valle del río Mendoza, al sur de Uspallata. Revista de la Asociación Geológica Argentina, 4, Buenos Aires, No. 55, pp. 419–423.

Fauqué L.E., Rosas M., Copplecchia M., Hermanns R., Etcheverria M., Tejedo A., Wilson C.G.J., 2005a. Laderas afectadas por deformaciones gravitacionales profundas en el valle del Río Cuevas, provincia de Mendoza. Actas XVI Congreso Geológico Argentino (CD), La Plata.

Fauqué L.E., Baumann V., Di Tommaso I., Rosas M., Hermanns R., Gonzalez M.A., Copplecchia M., Wilson C.G.J, 2005b. Evidencias de paleoendicamientos en la cuenca del Río Mendoza, provincia de Mendoza, Argentina. Actas XVI Congreso Geológico Argentino (CD), La Plata.

Favis-Mortlock, D.T., Guerra, A.J.T., 1999. The implications of general circulation model estimates of rainfall for future erosion: A case study from Brazil. Catena 37, pp. 329–394.

Federici, P.R, aRodolfi, G., 2004. Geomorphological features and evolution of the Ensenada de Atacames (Provincia de Esmeraldas, Ecuador). Journal of Coastal Research 20, 3, pp. 700–708.

Fernandez, D.S., 2005. The giant paleolandslide deposits of Tafí del Valle, Tucuman Province, Argentina. Geomorphology 70, pp. 97–111

Fernandez, D.S., Lutz M.A., 2003. Procesos de remoción en masa y erosión fluvial en la quebrada del río Los Sosa, provincia de Tucumán. Revista de la Asociación Geológica Argentina 2, Buenos Aires, No. 58, pp. 255–266.

Fernandes, N.F., Amaral, C.P. 1996. Movimentos de massa: Uma abordagem geológico–geomorfológica, In Guerra, A.J.T., Cunha, S.B. (eds.), Geomorfologia e Meio Ambiente. Ed. Bertrand, Rio de Janeiro, Brasil, Cap 3, pp. 393.

Ferreira Centeno J.P., Giambasti M., 1992. Estudio geológico de San Carlos Minas. Aporte para su relocalización. Secretaría de Minería, Dirección de ecología, promoción e industrias mineras, Córdoba. Unpublished report.

Feruglio, E., 1949. Descripción geológica de la Patagonia, Tomo I. Dirección General de YPF, Buenos Aires.

FICH-UNL-INA, 1998. Estudios hidrológicos, hidráulicos y sedimentológicos de un sector del tramo inferior del Río Salado destinados a la concreción de una nueva conexión entre las ciudades de Santa Fe y Santo Tomé, en la Provincia de Santa Fe, Dirección Prov. de Vialidad de Santa Fe. Informe Final, Santa Fe, Argentina, p. 186.

Fidalgo, F., Riggi. J.C., 1965. Los rodados patagónicos en la meseta de Guenguel y alrededores. Rev. Asociación Geológica Argentina 3 (20), Buenos Aires, pp. 273–325.

Fidalgo, F., 1972. Consideraciones sobre los bajos situados al norte de la provincia de Santa Cruz. V° Congreso Geológico Argentino, Córdoba, Actas V, pp. 123–137.

FNUAP. Fondo de las Naciones Unidas de Asistencia de Población, www.fnaup.org.

Foghin, S., 2001. Algunas consideraciones meteorológicas acerca de la catástrofe del Estado Vargas en Diciembre de 1999. Aula y Ambiente año 1 N° 1. UPEL, Caracas, pp. 93–98.

Fonoseca, A.M.M.C.C., 1969. Relato sobre causas e problemas das encostas da Guanabara, Anais da 1ª Semana Paulista de Geologia Aplicada, São Paulo, vol. 1, pp. 1–10.

Forbes, 1861. Informe sobre el terremoto de Mendoza. Revista de la Junta de Estudios Históricos de *Mendoza* (10): 111-120, Argentina.

Foster, M.S., Terborgh, J., 1998. Impact of rare storm event on an Amazon forest. Biotropica 30(3), pp. 470–474.

Franca, C.A.S., de Mesquita, A.R., in press. The December 26th 2004 tsunami recorded along the southeastern coast of Brazil. Natural Hazards.

Francis, P., 1982. The Cerro Galan caldera, Argentina. Earthquake Inf. Bulletin 14, pp. 124–133.

Francou, B., et al. 2005. Glaciers of the tropical Andes: Indicators of the global climate variability. In: Huber, U., et al. (eds.), Global Change and Mountain Regions: A State of Knowledge Overview. Springer, New York, pp. 197–204.

Fraser, B., 2007. Earthquake highlights mental-health issues in Peru. The Lancet, 370(9590), pp. 815–816.

Freitas, P.L., Castro, A.F., 1983. Estimativas das perdas de solo e nutrientes pela erosão no estado do Paraná. Campinas, Soc. Bras. Ci. Solo, Bol. Inform. 8(2), pp. 43–52.

Frenguelli, J., 1946. Las grandes unidades físicas del Territorio Argentino. Geografía de la República Argentina. Sociedad Argentina de Estudios Geográficos, Buenos Aires, 3, p.114.

Fritz, S.C., Baker, P.A., Lowenstein, T.K., Seltzer, G.O., Rigsby, C.A., Dwyer, G.S., Tapia, P.M., Arnold, K.K., Ku, T.L., Lou, S., 2004. Hydrologic variation during the last 170,000 years in the southern hemisphere tropics of South America. Quaternary Research 61, pp. 95–104.

Fuenzalida, R., Harambour, S., 1984. Evidencias de subsidencia y solevantamiento en la Península Brusnwick, Magallanes. Comunicaciones Universidad de Chile, 34, pp. 117–120.

Fuerza Aérea Venezolana, 2000. Análisis de las Situaciones Meteorológicas que afectaron el Área Norte Costera del País durante el mes de diciembre de 2000. Servicio Meteorológico, Ministerio de la Defensa, Venezuela.

Funning, G.J., Barkea, R.M.D., Lamb, S.H., Minaya E., Parsons, B., and Wright, T.J., 2005. The 1998 Aiquile, Bolivia earthquake: A seismically active fault revealed with InSAR. Earth and Planetary Science Letters, 232, pp. 39–49.

Furian, S., 1995. Morphogénèse/Pedogénèse en milieu tropical humide de la Serra do Mar, Brésil: contribution de l'altération et de la pedogénèse a une dynamique actuelle. França, Universidade de Caen, U.V. Sciences de la Terre et de l'Amenagement (tese Dr.).

Gade, D.W., 1999. Nature and Culture in the Andes. University of Wisconsin Press, Madison.

Gallardo E.F., 1988. Geología del Cuaternario en la confluencia de los Rios Calchaquie y Santa Maria (Salta). Revista de la Asociación Geológica Argentina 4, (43), pp. 435–444.

Gambino, P., Vargas, D., 2002. Evaluación del impacto del trazado del gasoducto TGS—Puerto Madryn sobre la calidad de los suelos. Congreso Argentino de la Ciencia del Suelo. Puerto Madryn. 16 al 19 de abril de 2002. Actas XVIII

Garcia-Acosta, V., 2007. Risks and disasters in the history of the Mexico Basin: Are they climatic or social? Medieval History Journal 10 (1–2), pp. 127–142.

García, M., 1988. Eventos catastróficos del 13 de Noviembre de 1985. Boletín de vías 15 (65), pp. 7–106.

García, N.O., 1994. South America Climatology, Quaternary International. INQUA/Elsevier Science LTD., Amsterdam, vol. 21, pp 7–27.

García, N.O., Vargas W.M., 1996. The spatial variability of runoff and precipitation in the Río de la Plata Basin. J.Hydrol. Sc. 41, pp. 279–299.

García, N.O., Vargas, W.M., 1998. The temporal climatic variability in the Rio de la Plata Basin displayed by the river discharges. Climatic Changes 38, pp. 359–379.

García, N.O., 2000. About the climate variability and the runoff in the Río de la Plata basin. Preprints of 6th International Conference on Southern Hemisphere Meteorology and Oceanography, Santiago, Chile, pp. 279–281.

García, N.O., Mechoso, C.R., 2005. Variability in the discharge of South American rivers and in climate. Hydrological Science Journal 50 (3), pp. 459–478.

García, N.O., Krepper, C.M., 2006. Spatial–Temporal Str4ucture of the Interanual Variability of Precipitation Over "Río de la Plata" Basin. Proceedings 8th International Conference on Southern Hemisphere of Meteorology and Oceanography (Electronic Support). American Meteorological Society, Foz de Iguazu, Brasil..

García, N.O., Pedraza, R.A., Krepper, C.M., 2007. Temporal variation of rainfall frequency and intensity in the Northeastern Argentina. Variability or Climatic change? Proceeding 7th EMS Annual Meeting / 8th European Conference on Applications of Meteorology. October 1–5, 2007, El Escorial, Espain.

Gardini C.E., Sales D.A., Ricci W.R., 2003. Deslizamiento de la ladera sobre la ruta provincial No. 9. Localidad de Nogoli–Prov. de San Luis. ASAGAI, Buenos Aires, 19, pp. 23–30.

Garga, V.K., de la Torre, M., 2002. Emergency remediation of instability at Caudalosa tailings dam, Peru: A case history. Can. Geotech. J. 39, pp. 1193–1200.

Garreaud, R.D., 1999. Cold air incursions over subtropical and tropical South America: A numerical case study. Monthly Weather Review 127, pp. 2823–2852.

Garreaud, R.D., 2000. Cold air incursions over subtropical South America: mean structure and dynamics, Mon. Wea. Rev., 128, pp. 2544–2559.

Garreaud, R., and Aceituno, P., 2001. Interannual rainfall variability over the South American Altiplano. Journal of Climate 14, pp. 2779–2789.

Garreaud, R., Rutllant, J., 1996. Análisis meteorológico de los aluviones de Antofagasta y Santiago de Chile en el período 1991–1993. Atmósfera, No. 9, pp. 251–271.

Gavidia, J., 1991. The economic and social effects of recent natural disasters in Ecuador and Nicaragua. In Organizational Behaviour and Industrialization: Lessons from Transnational Corporations in the Manufacturing and Agrobased Industry Sectors. UNCRD meeting report series Num. 45. Nagoya. Japan, pp. 66–79.

Georges, C., 2004. The 20th century glacier fluctuations in the tropical Cordillera Blanca (Peru). Arctic, Antarctic and Alpine Research 36, pp. 100–107.

Ghosh, P., Garzione, C.N., Eiler, J.M., 2006. Rapid uplift of the Altiplano revealed through ^{13}C–^{18}O bonds in paleosol carbonates. Science 311, pp. 511–515.

Giardini, D., Grunthal, G., Shedlock, K., Zhang, P., 1999. The GSHAP global seismic hazard map. Ann. Geofis. 42, pp. 1225–1230.

Gibs, R.J., 1967. The geochemistry of the Amazon River system: Part I. The factors that control the salinity, composition and concentration of suspended solids. Geological Society of American Bulletin 78, pp. 1203–1232.

Giesecke, A., 1996. Strengthening existing adobe housing in earthquake-prone regions in the Andean countries. J. Conting, Crisis Manag. 4 (4), pp. 241–244.

Giovanni, M.K., Beck, S.L., Wagner, L., 2002. The June 23, 2001 Peru earthquake and the southern Peru subduction zone. Geophys. Resear. Letters, 29 (21): DOI:10.1029/2002GL015774,2002.

Glantz, M.H., Orlovsky, N.S., 1983. Desertification: A review of the concept. Desertification Control Bulletin 9, pp. 15–22.

Goldbrunner, A., 1984. Atlas climatológico de Venezuela, 1951–1970. Servicio de Meteorología de la Fuerza Aérea Venezolana.

Goldbrunner, A., 1976. El clima de Venezuela y su clasificación. Instituto Pedagógico Universitario de Caracas. Mimeografiado.

Goldbrunner, A., 1961. Las causas meteorológicas de las lluvias de extraordinaria magnitud en Venezuela. Ministerio de Obras Publicas, Caracas.

Golluscio, R., Mercau, J., 1994. Cambios en la biodiversidad ante distintos grados de desertificación provocada por el pastoreo. Actas del Taller Internacional sobre recursos fitogenéticos desertificación y uso sustentable. Centro Regional Patagonia Sur INTA, Buenos Aires, pp. 60–71.

González, H., Núñez, A., 1991. Mapa geológico generalizado del Departamento del Quindío escala 1:100 000. Memoria explicativa. Ingeominas Technical Report. Bogotá, Colombia.

González, J., Correa, I.D., 2001. Late Holocene evidence of coseismic subsidence in theSan Juan Delta, Pacific coast of Colombia. Journal of Coastal Research 17, (2), pp. 459–467.

González Bonorino, F., Rabassa, J., 1973. La Laguna Carri Lafquen Grande y el origen de los bajos patagónicos. Revista Asociación Geológica Argentina 3, (27), Buenos Aires, pp. 314–318.

González Bonorino, G., 2002. Erosión y acreción litoral durante el Holoceno, con especial referencia a las puntas Bustamante y Dungeness. Geología y Recursos Naturales de Santa Cruz. Relatorio XV Congreso Geológico Argentino. El Calafate, I–20, pp. 317–324. Buenos Aires.

Gonzalez, M.A., 1979. Condiciones de inestabilidad del cierre de la Laguna del Atuel, provincia de Mendoza Revista de la Asociación Geológica Argentina, Buenos Aires, No. 34, pp. 151–155

González, M., González Díaz, E., Sepúlveda, E., Regairaz, M., Costa, C., Cisneros, H., Bea, S., Gardini, C., Pérez, I., Pérez, M., 2002. Carta de Peligrosidad Geológica 3369—II Mendoza. Provincias de Mendoza y San Juan. Boletín No. 324. Segemar, p. 178.

González Díaz, E.F., 1972. La participación del proceso de remoción en masa (mass-wasting) en el llamado "aluvión" de Chumbica (SE de la Provincia de Catamarca). V Congreso Geológico Argentino, Actas V, pp. 101–122.

González Díaz, E.F., 2003. El englazamiento en la región de la caldera de Caviahue–Copahue (Provincia del Neuquén): Su reinterpretación. Revista de la Asociación Geológica Argentina 3 (58), Buenos Aires, pp. 356–366.

González Díaz, E., 2004. La extensa zona de deslizamientos de la escarpa oriental de la pampa de Salamanca, Chubut, entre los 45° 00' S y 45° 45' S. Revista de la Asociación Geológica Argentina, 4 (59), Buenos Aires, 4, pp. 743–762.

González Diaz, E.F., 2005. Geomorfología de la región del volcán Copahue y sus adyacencias (centro–oeste del Neuquén). Revista de la Asociación Geológica Argentina 1 (60), Buenos Aires, pp. 72–87.

González Díaz, E.F., Mon, R., 1996. El origen de las algunas de Yala, prov de Jujuy, noroeste de la Argentina (24°05′de LS y 65° 28′de LO). XIII Congreso Geológico Argentino, 4, Buenos Aires, 4 pp. 209–217.

González Díaz, E.F., Folguera, A., 2005. El reconocimiento de avalanchas de rocas y deslizamientos de bloques rocosos prehistoricos en el área andina de Neuquen (37° 15′–37° 30′). **RAGA, 3** (60), pp. 446–460.

González Díaz, E.F., Folguera, A., Costa C.H., Wright, E, 2006. La avalancha de rocas de Picun Challa: ratificación de una única glaciación en la región de Cavihue–Copahue y sus adyacencias (Neuquén). Revista de la Asociación Geológica Argentina 1 (61), Buenos Aires, pp. 19–30.

González Díaz, E.F., Fauque L., Costa C., Giaccardi A.D., Palomera D.A., Pereyra F. 1997. La avalancha de rocas del Potrero de Leyes, Sierras Pampaeanas australes. Sierrra Grande de San Luis, Argentina (32° 30′ lat S). Revista de la Asociación Geológica Argentina 1 (52), Buenos Aires, pp. 93–107.

González Díaz, E.F., Fauque L., Costa C., Giaccardi A.D. 1998 La avalancha de rocas de Las Cañas: otra manifestación de los fenómenos en el flanco oeste de la sierra Grande de San Luis. Revista de la Asociación Geológica Argentina 3 (53), Buenos Aires, pp. 325–336.

González Díaz, E.F., Costa C., Giaccardi, A.D., Bierman P., Caffee M., 1999. Edades de cosmonuclídeos ^{10}Be y ^{26}Al en avalanchas de rocas de la Sierra de San Luis: Implicancias paleosísmicas. XIV Congreso Geológico Argentino, Salta, Actas I, p. 81.

González Díaz, E.F., Fauqué, L.E., Giaccardi, A., Costa, C., 2000. Las lagunas de Varvar Co Campos y Varavar Co Tapia (N del Nauquén, Argentina): Su relación con avalanchas de rocas. Revista de la Asociación Geológica Argentina, 3 (55), Buenos Aires, pp. 147–164.

González Díaz, E., Giaccardi, A., Costa, C., 2001. La avalancha de rocas del río Barrancas (cerro Pelán), norte del Neuquén: su relación con la catástrofe del río Colorado (29/12/1914). Revista de la Asociación Geológica Argentina 4, No. 56, pp. 466–480.

González Díaz, E.F., Costa C.H., Giaccardi A.D., 2003. El complejo deslizamiento de Ailinco–Cerro Papas–Las Olletas (Departamento Minas, norte del Neuquén, Argentina). Revista de la Asociación Geológica Argentina 2 (58), Buenos Aires, pp. 194–200.

González Díaz, E.F., Folguera A., Hermanns R., 2005. La avalancha de rocas del cerro Los Cardos (37°10′S, 70°53′O) en la región norte de la provincia del Neuquén. Revista de la Asociación Geológica Argentina, Buenos Aires 1 (60), pp. 207–220.

González-Ferrán, O., 1972. Villarrica Volcanic Eruption, Nov–Dec. 1971. Smithsonian Institution Cent. for Short–lived Phenomena. 1971 Ann. Rep. Event 95–71, pp. 134–138.

Gonzalez-Ferran, O., 1995. Volcanes de Chile. Santiago: Instituto Geografico Militar, p. 635.

Gonzalez-Ferran O., Lahsen, A., Brown, F., 2003. 1932 Quizapu Eruption: The Impact on the atmosphere and environment in South America. Volcanism and the Earth's Atmosphere (IAVCEI, IAMAS). Abstract.

Gordillo, S., G. Bujalesky, G., Pirazzoli, P., Rabassa, J., Saliège, J., 1992. Holocene raised beaches along the northern coast of the Beagle Channel, Tierra del Fuego, Argentina. Palaeogeography, Palaeoclimatology, Palaeoecology, 99, pp. 41–54.

Goring, D.G., 2002. Response of New Zealand waters to Peru tsunami of 23 June 2001. N.Z. J. Mar. Freshw. Res. 36, pp. 225–232.

Goudie, A.S., 2006a. The Human Impact on the Natural Environment (6th edition). Oxford, Blackwell.

Goudie, A.S., 2006b. Global warming and fluvial geomorphology. Geomorphology 79, pp. 384–394.

Groeber, P., 1916. Informe sobre las causas han producio las cescientes del Río Colorado (territorios del Neuquen y La Pampa) en 1914, Ministerio de Agricultura de la Nación (Argentina), Direccíon General de Minas, Buenos Aires, Geologia e Hidrologia, Serie B (Geología), Bolletín No. 11, p. 28.

Groeber P., 1925. La región de Copahue y su glaciación diluvial. Sociedad Argentina de Estudios Geográficos (GAEA), Buenos Aires, 3, pp. 92–110.

Groeber P., 1933. Confluencia de los Río Grande y Barrancas (Mendoza y Neuquen), Ministerio de Agricultura de la Nacíon (Argentina), Dirección de Minas y Geología, Buenos Aires, Boletín. No. 38, p. 72.

Groeber P., 1947. Observaciones a lo lago del meridiano 70. Hojas Domuyo, Mari Mahuida, Huarhuar Co y parte de Epu Lauken. Revista de la Asociación Geológica Argentina 2, Buenos Aires, pp. 347–408.

Gretener, P.E., 1984. Reflections on the "rare event" and related concepts in geology. In Berggren, W.A., and Van Couvering, J.A. (eds.), Catastrophes and Earth History. Princeton University Press, Princeton, NJ, USA, pp. 77–89.

Groeber, P., 1953. Glacial, tardío y post–glacial en Patagonia. Rev. Museo de La Plata, 1, pp. 79–103

Grubb, P.J., 1970. The impact of man on the Cerro Antisana, Ecuador. Journal of Applied Ecology 7 (2), pp. 7–8.

Guerrard, J., 1992. Soil Geomorphology: An Interaction of Pedology and Geomorphology. London, Univ. Press pub, Chapman's Hall, 1992

Guglielmo, R., 1996. Les grandes métropoles du mond et leurs crise. Ed. Armand Colin, Paris.

Guha Sapir, D., Hargitt, D., Hoyois, P., 2004. Thirty Years of Natural Disasters, 1974–2003: The Numbers. Centre for Research on the Epidemiology of Disasters, UCL presses universitaires, Lovain, pp. 190.

Guidicini, G., Iwasa, O.Y., 1977. Tentative correlation between rainfall and landslides in a humid tropical environment. Symp. on Landslides & Other Mass Movement, Praga, IAEG Bulletin, 16, pp. 13–20.

Guilderson, T.P., Burckle, L., Hemming, S., Peltier, W.R., 2000. Late Pleistocene sea level variations derived from the Argentine shelf. Geochemistry, Geophysics and Geosystem 1, AGU, ISSN 1525–2027, p. 15.

Gusiakov, V.K., 2005. Tsunami generation potential of different tsunamigenic regions in the Pacific. Marine Geology 215, pp. 3–9.

Gutscher, M.-A., Spakman, W., Bijwaard, H., Engdahl, E.R., 2000. Geodynamics of flat subduction: seismicity and tomographic constraints from the Andean margin. Tectonics 19, pp. 814–833.

Hagen, R., Vergara, H., Naar, D., 1996. Morphology of San Antonio submarine canyon on the Central Chile forearc. Marine Geology 129, pp. 197–205.

Halcrow and Partners. 1999. Plan Maestro Integral de la Cuenca de Río Salado. Ministerio de Obras y Servicios Públicos de la Provincia de Buenos Aires. La Plata, Argentina.

Haller, M., Mendía, J., Ostera, H., 1991. Mapa preliminar de riesgo en la vertiente argentina del Volcán Peteroa. VI Congreso Geológico Argentino, Actas II, pp. 355–358.

Hanuš, V., Vaněk, J., Špičák, A., 2000. Seismically active fracture zones and distribution of large accumulations of metals in the central part of Andean South America. Mineral. Deposita 35, pp. 2–30.

Harden, C.P., 2004. The 1993 Landslide dam at La Josefina in Southern Ecuador: A review of "Sin plazo para La Esperanza." Engineering Geology 74 (1–2), pp. 157–161.

Harden, C.P., 2006. Human impacts on headwater fluvial systems in the northern and central Andes. Geomorphology 79, pp. 249–263.

Hare, K., 1977. Connections between climate and desertification, Environmental Conservation, 4 (2), p. 82.

Harmon, R., Rapela, C. (eds.), 1991. Andean magmatism and its tectonic setting. Geological Society of America, Special Paper, 265, Boulder, CO.

Harrington, H., 1944. El sismo de San Juan del 15 de enero de 1944. Corporación para la Promoción del Intercambio S.A., p. 79. Unpblished report.

Harrington, H.J., 1945. Las corrientes de barro (mudflows) de "El Volcán," Quebrada de Humahuaca, Jujuy. Revista de la Asociación Geológica Argentina 2 (1), Buenos Aires, pp. 148–165.

Harrison S., Glasser, N., Winchester, V., Haresign, E., Warren, C., Jansson, K., 2006. A glacial outburst flood associated with recent mountain glacier retreat Patagonian Andes. The Holocene 16, pp. 611–620.

Hastenrath, S., Polzin, D., Francou, B., 2004. Circulation variability reflected in ice core and lake records of the southern tropical Andes. Climatic Change 64, pp. 361–375.

Heilbron, M., 1995. O segmento central da Faixa Ribeira: síntese geológica e ensaio de evolução geotectônica. Tese de Livre Docência, Departamento de Geologia, UERJ, p. 110.

Hellin, J., Haigh, M., Marks, F., 1999. Rainfall characteristics of hurricane Mitch. Nature 399, p. 316.

Herd, D., Youd, T.L., Meyer, H., Arago, J.L., Person, W.J., Mendoza, C., 1981. The great Tumaco Colombia earthquake of December 1979. Science 211 (4481), pp. 441–445.

Herd, D.G., 1982. Glacial and volcanic geology of the Ruiz-Tolima volcanic complex, Cordillera Central, Colombia. Publicaciones geológicas especiales del Ingeominas 8, pp. 1–48.

Hermanns, R.L.M., Strecker, M.R., 1999a. Tipos de frentes de montaña susceptibles a colapsos catastróficos y distribucion spacial de grandes deslizamientos de la ladera en el árido NW de la Argentina. XIV Congreso Geológico Argentino, Salta, Actas I, pp. 54–56.

Hermanns R., Strecker M., 1999b. Structural and lithological controls on large Quaternary bedrock landslides in NW–Argentina. Geological Society of America Bulletin 111 (6), pp. 934–948.

Hermanns, R., Trauth, M.H., McWilliams M., Strecker, M.R., 2000. Tephro-chronologic constraints on temporal distribution of large landslides in NW-Argentina. Journal of Geology, 108, pp. 35–52.

Hermanns, R.L., Niedermann, S., Villanueva, A., Sosa Gomez, J., Strecker, M.R., 2001. Neotectonics and catastrophic failure of mountain fronts in the southern intra-Andean Puna Plateau. Geology 29, pp. 619–623.

Hermanns, R.L., González Diaz F.E., Folguera, A., Mardones, M., 2003. Large massive rock slope failures, landslide dams, related valley evolution, and their association with the tectonic setting in the Argentine and Chilean Andes between 36 and 38° S. Actas X Congreso Geológico Chileno: 5, Concepción, Chile.

Hermann, R.L., Folguera, A., Penna, I., Naumann, R., Niedermann, S., 2006. Morphologic characterization of giant flood deposits downriver landslide dams in the Northern Patagonian Andes. Geophysical Research Abstracts, 8, EGU06–A–09181.

Hermelin, M., 2005. Introducción. In Hermelin, M. (ed.), Desastres de origen natural en Colombia 1979–2004. Universidad EAFIT, OSSO-Univalle, Medellín, Colombia, pp. 11–16.

Hermelin M. 1993. Informe—Reconocimiento de las partes altas de la cuenca del río Pedral. Technical report, Gobernación de Antioquia. Medellín, Colombia.

Hermelin, M., Velásquez, E., Mejía, O., 1992. Erosional and depositional features produced by a convulsive event, San Carlos, Colombia, September 21, 1990. Bulletin of the International Association of Engineering Geology 45, pp. 89–95.

Hernádez, E., and Valbuena, J., 2001. Las Amenazas Naturales, los Riesgos y la vulnerabilidad Ambiental del poblamiento del Eje del Chama–Mocotíes. Universidad de los Andes Mérida, Venezuela.

Hervé, M., 1987. Movimiento normal de la Falla Paposo, zona de Falla de Atacama, en el Mioceno, Chile. Revista Geológica de Chile No. 31, pp. 31–36.

Hewitt, K., 1997. Risks and Disasters in Mountain Lands. In Messerli, B., Ives J.D. (eds.), Mountains of the World: A Global Priority. Parthenon Publishing, New York, pp 371–408.

Hildreth, W., Drake, R.E., 1992. Volcán Quizapu, Chilean Andes. Boletín of Volcanology 54, pp. 93–125.

Hirtz, N., Pérez, H., Grizinik, M., Tejedo, A., Blachakis, A., Stronati, M., Cavallaro, S., 2000. Uso de mapas temáticos en la confección de la carta geoambiental de Comodoro Rivadavia, Chubut, Argentina. Revista ASAGAI 15, Buenos Aires, pp. 41–55.

Hoegh-Guldberg, O., and 16 others, 2007. Coral reefs under rapid climate change and ocean acidification. Science 318, pp. 1737–1742.

Hogarth, P.J., 1999. The Biology of Mangroves. Oxford University Press, Oxford.

Hongn, F., Tubía, J., Aranguren, A., Mon, R., 2002. La monzodiorita Las Burras: un plutón Mioceno en el Batolito de Tastil, Cordillera Oriental Argentina. Actas del XV Congreso Geológico Argentino 2, pp. 128–133.

Hormann, P.K, Pichler, H., Zeil, W., 1973. New data on the young volcanism in the Puna of NW. Argentina. Geol Rundschau 62, pp. 397–405

Horne, F., 1998. Dinámica del agua en los mallines de Patagonia. III° Congreso Chileno de Ingeniería Agrícola. Chillán, Chile.

Hoyois, P., Scheuren, J.M., Below, R., Guha-Sapir, D., 2006. Annual disaster Statistical review: number and trends, 2006. Centre for Research on the Epidemiology of Disasters, UCL presses univesitaires, Lovain, pp. 54.

Houston, J., 2006. The great Atacama flood of 2001 and its implications for Andean hydrology. Hydrol. Processes 20, pp. 591–610.

Houston J., 2006. The great Atacama flood of 2001 and its implications for Andean hydrology. Hydrological Processes 20, pp. 591–610.

Hubbard, B., Heald, A., Reynolds, J.M., Quincey, D., Richardson, S.D., Zapata Luyo, M., Santillan Portill, N., Hambrey, M.J., 2005. Impact of a rock avalanche on a moraine-dammed proglacial lake: Laguna Safuna Alta, Cordillera Blanca, Peru. Earth Surf. Process. Landforms 30, pp. 1251–1264.

Huerta, G., and Sarmiento, R., 1989. La rentabilidad de la producción lanera. Rev. Presencia A 4 (19), pp. 47–48.

Huerta, G., 1991. Una aproximación a la Unidad Ganadera Ovina en condiciones patagónicas. Serie Técnica. INTA, AEE, Bariloche.

Humboldt, A., 1816. Viaje a las Regiones Equinocciales del Nuevo Continente. Traducción de la edición francesa: Alvarado, L., Rol, E., and Nucete-Sardi, J. Ediciones Monte Avila, Caracas, 1985.

IBGE. Anuários estatísticos e Censos Agropecuários de 1960 a 2003.

IBGE, 2006. Anuário estatístico do Instituto Brasileiro de Geografia e Estatística: www.ibge.gov.br

IGAC Instituto Geográfico Agustín Codazzi., 1982. Atlas regional andino. Bogotá, Colombia.

IGAC Instituto Geográfico Agustín Codazzi, 1980. Mapa físico del Departamento del Tolima, escala 1:400 000. Bogotá, Colombia.

Igarzabal A.P., 1971. Remoción en masa en la Quebrada del Toro (Salta). Opera Lilloana 21 p. 60. Universidad Nacional de Tucumán, Tucumán.

Igarzabal A.P., 1978. Los flujos densos de la Quebrada de Escoipe. VII Congreso Geológico Argentino, Neuquen, Actas II, pp. 109–117.

Igárzabal, A., 1999. Cuaternario: 1. Cuaternario de La Puna. Geología Argentina. Anales 29 (23), pp. 683–714.

Ihmlé, P.F., Gómez, J.M., Heinrich, P., Guibourg, S., 1998. The 1996 Peru tsunamigenic earthquake: broadband source process. Geophys. Resear. Letters 25 (14), pp. 2691–2694.

Inbar, M., Llerena, C.A., 2000. Erosion processes in high mountain agricultural terraces in Peru. Mount. Resear. Develop. 20 (1), pp. 72–79.

INCYTH (Instituto Nacional de Ciencia y Tecnología Hídrica)—Centro Regional Litoral, 1992. Estudio de delimitación de áreas de riesgo hídrico en Santa Fe. Informe Final, Santa Fe, Argentina, p. 64.

INDEC, 2001. Censo nacional de población, hogares y vivienda. http://www.indec.mecon.gov.ar / webcenso/provincias_2/provincias.asp

INDECI (Instituto Nacional de Defensa Civil), 2004. Plan Nacional de Prevención y Atención de Desastres. Sistema Nacional de Defensa Civil, Lima, Peru, p.196.

INGEOMINAS Instituto Colombiano de Geología y Minería, 1994. El sismo de Páez, Cauca—Evaluación de emergencia. Technical report. Bogotá, Colombia.

INGEOMINAS, 1986. El sismo de Popayán del 31 de marzo de 1983. Technical report. Bogotá, Colombia.

INGEOMINAS, 1985. Mapa de riesgos volcánicos potenciales del Nevado del Ruiz. Technical report. Medellín, Colombia.

INMET—Instituto Nacional de meteorologia: www.inmet.gov.br

INPRES 1993. La verdadera dimensión del problema sísmico en la provincia de San Juan. Publicación Técnica N° 18, San Juan, p. 46.

INPRES, 2006. Listado de terremotos históricos, on line catalog (www.inpres.gov.ar).

IPCC, 2001. Third Assessment Report (TAR). Geneva, IPCC. Vol. 1: Climate Change 2001: Working Group I: The Scientific Basis; Vol. 2: Climate Change 2001: Working Group II: Impacts, Adaptation and Vulnerability; Vol. 3: Climate Change 2001: Working Group III: Mitigation; Vol. 4: Climate Change 2001: Synthesis Report. Published by IPCC and Cambridge University Press, UK. Available from www.ipcc.ch

IPCC, 2007. Fourth Assessment Report. Geneva, IPCC. Available from www.ipcc.ch

IPCC, 2007. Climate Change 2007: The Physical Science Basis. Cambridge University Press, Cambridge, UK, p. 996.

IPCC, 2008a. Fourth Assessment Report, Working Group I, Summary for Policymakers: http://195.70.10.65/pdf/assessment–report/ar4/wg1/ar4–wg1–spm.pdf.

IPCC, 2008b. Fourth Assessment Report, Working Group II Report: http://195.70.10.65/ipccreports/ar4–wg2.htm.

Isla, F.I., 1989. The Southern Hemisphere sea level fluctuation. Quaternary Science Revs. 8, pp. 359–368.

Isla, FI., Bértola, G.R., 2005. Litoral bonaerense. In: de Barrio, R., Etcheverry, R.O., Caballé, M. Llambías, E. (Eds.) Geología y Recursos Minerales de la Provincia de Buenos Aires. Relatorio XVI Congreso Geológico Argentino, La Plata, pp. 265–276.

Isla, F.I., Bujalesky, G., 2004. El maremoto de los Yaganes. Revista Nexos, UNMDP, Mar del Plata, pp. 29–33.

Isla, F.I., Cortizo, L.C., 2005. Patagonian cliff erosion as sediment input to the continental shelf. Actas XVI Congreso Geológico Argentino, La Plata, tomo IV, pp. 773–778.

IPCC, 2001. Climate Change 2001: Impacts, Adaptation, and Vulnerability. Contribution of Working Group II to the Third Assessment Report of the Intergovernmental Panel on Climate Change, p. 967.

Iverson, R.M., Delinger, R.P., Lahusen, R.G., Logan, M., 2000. Two phase debris flows across 3–D terrain: model predictions and experimental tests. pp. 521–529. In Wickzorek, G.F., Naeser, N.D. (eds.), Debris–flow Hazards Mitigation: Mechanics, Prediction and Assessment. Ed. Balkema, Teipei, Taiwan, pp. 608.

Jakson, L., 2007. Cosmogenic isotopic chronology of rock avalanches on the Montalban watershed on Venezuelan Andes. Unpublished data. Canada Geological Survey.

Jansen, R.C., 2001. Distribuição dos sistemas radiculares em encostas florestadas e sua influência sobre a infiltração. Tese de M.Sc. Prog. de Pós-Grad. em Geografia/UFRJ, Rio de Janeiro, p. 147.

Jaschek, E., Sabbione, N., Sierra, P., 1982. Reubicación de sismos localizados en territorio argentino. Serie Geofísica, tomo XI, N°1, Publicaciones Observatorio de la Universidad de La Plata, pp. 1920–1963.

Jibson, R.W., Crone, A.J., 2001. Observations and recommendations regarding landslide hazards related to the January 13, 2001 M–7.6 El Salvador Earthquake, USGS Open-File Report 01–141.

Jobbagy, E.G., Paruelo, J.M., Leon, R.J.C., 1996. Vegetation heterogeneity and diversity in flat and mountain landscapes of Patagonia. Journal of Vegetation Science 7, pp. 599–608.

Jordan, T., Isacks, B., Allmendinger, R., Brewer, J., Ramos, V., Ando, C., 1983. Andean tectonics related to geometry of subducted Nazca plate. Geological Society of America Bulletin 94, pp. 341–361.

Jordan, E., 1999. Glaciers of Bolivia. In Williams, R.S., Ferrigno, J.G. (eds.), Satellite Image Atlas of Glaciers of the World. Geological Survey Professional Paper 1386, USA

Jordan, T., and Allmendinguer, R., 1986. The Sierras Pampeanas of Argentina. A modern analogue of Rocky Mountains foreland deformation: American Journal of Science 286, pp. 737–764.

Jordan, T., Gardeweg, M., 1987. Tectonic evolution of the late Cenozoic Central Andes. In Ben Avraham Z. (ed.), Mesozoic and Cenozoic Evolution of the Pacific Margins, Oxford University Press, New York, pp. 193–207.

Jordan, T., Isacks, B., Allmendinguer, R., Brewer, J., Ramos, V., Ando, C., 1983. Andean tectonics related to geometry of subducted Nazca Plate. Geological Society of America Bulletin 94, pp. 341–361.

Kahn, M.E., 2005. The death toll from natural disasters: The role of income, geography and institutions. The Review of Economics and Statistics 87 (2), pp. 271–284

Kantor M., 1916. El problema de las inundaciones en Andalgalá (prov. de Catamarca). Revista del Museo de la Plata, Buenos Aires.

Kaser, G., Osmaston, H., 2002. Tropical Glaciers. Cambridge University Press, Cambridge.

Kates, R.W., 1962. Hazard and Choice, Perception in Flood Plain Management. Research Paper No. 78. Department of Geography, University of Chicago.

Kausel, E., 1991. The influence of large thrust and normal earthquakes in the assessment of the seismic hazard, Workshop. New horizons in Strong Ground Motion: Seismic Studies and Engineering Practice, Santiago, Chile.

Kay, S., Mpodozis, C., Ramos, V., Munizaga, F., 1991. Magam source variations for mid-late Tertiary magmatic rocks associated with a shallowing subduction zone and a thickening crust in the central Andes (28° to 33° S). In Harmon, R.S., Rapela, C.W. (eds.), Andean Magmatism and its Tectonic Setting, Geol. Soc. Am. Spec. Pap. 265, pp. 113–137.

Keane, D., 2007. Bolivia declares El Nino flood disaster. USA Today, March 1, 2007. http://www.usatoday.com/weather/news/2007–03–01–bolivia–floods_N.htm

Keating, P.L. 2007. Fire ecology and conservation in the high tropical Andes: Observations from northern Ecuador. Journal of Latin American Geography 6 (1), pp. 43–62.

Keefer, D.K., 1994. The importance of earthquake-induced landslides to long-term slope erosion and slope-failure hazards in seismically active regions. Geomorphology 10, pp. 265–284.

Keefer, D.K., 2002. Investigating landslides caused by earthquakes–a historical review. Surveys Geophys. 23, pp. 473–510.

Keefer, D.K., Moseley, Me., 2004. Southern Peru desert shattered by the great 2001 earthquake: implications for paleoseismic and paleo-El Niño-Southern Oscillation. Proceed. Nat. Acad. Sc. 10, 1(30), pp. 10878–10883.

Kendrick, E., Bevis, M., Smalley Jr, R., Cifuentes, O., Galbán, F., 1999. Current rates of convergence across the Central Andes: Estimates from continuous GPS observations. Geophysical Research Letters 26, pp. 541–544.

Kendrick, E., Bevis, M., Smalley Jr. R., Brooks, B.,Vargas, R.B., Lauría, E., Fortes, L. 2003. The Nazca-South America Euler vector and its rate of change. Journal of South American Earth Sciences, 16, pp. 125–131.

Kertzman, F.F., Queirozneto, J.P., 1993. Análise estrutural acplicada ao estudo da degradação dos solos. V Simp. Nac. Geogr. Fís. Aplic., S. Paulo, Anais, pp. 501–504.

Kilt E., 1939. Derrumbamientos, deslizamientos y torrentes en caminos de montaña de la República Argentina. Universidad de Buenos Aires, Serie B, Cientifico-Tecnológica No. 16, Buenos Aires.

Killeen, T.J., Douglas, M., Consiglio, T., Jørgensen, P.M., Mejia, J., 2007. Dry spots and wet spots in the Andean hotspot. Journal of Biogeography 34, pp. 1357–1373.

King, G., III, 2007. Narcissism and effective crisis management: A review of potential problems and pitfalls. J. Contingencies & Crisis Manag. 15 (4), pp. 183–193.

Klein, A.H. Da F., Calliari, L., 1997. Concheiros do Albardao: Variaçoes espaço–temporais da morfologia praial. VI Congresso da ABEQUA, Curitiba, pp. 401–406.

Klepeis, K. 1994. The Magallanes and Deseado fault zones: Major segments of the South American-Scotia transform plate boundary in southernmost South America, Tierra del Fuego. Journal of Geophysical Research 99, pp. 22,001–22,014.

Klohn, E., 1963. The February 1961 eruption of Calbuco volcano. Bulletin of the Seismological Society of America; 53: 1435-1436.

Klotz, J., Khazaradge, G., Enferman, D., Reigber, C., Perdono, R., Cifuentes, O., 2001. Earthquake cycle dominates contemporary crustal deformation in central and southern Andes. Journal of Earth and Planetary Science Letters 193, pp. 437–446.

Knapp, G., 1991. Andean Ecology: Adaptive Dynamics in Ecuador. Westview Press, Boulder, CO..

Konagai, K., Johannson, J., Mayorcap, P., Yamamoto, T., Miyajima, M., Uzuoka, R., Pulido, N.E., Duran, F.C., Sassa, K., Fukuoka, H., 2002, Las Colinas landslide caused by the January 13, 2001 off the coast of El Salvador earthquake. Journal of Japan Association for Earthquake Engineering 2 (1), pp. 1–15

Konagai, K., Johansson, J., Mayorca, P., Uzuoka, R., Yamamoto, T., Miyajima, M., Pulido, N., Sassa, K., Fukuoka, H., Duran, F., 2004, Las Colinas landslide: Rapid and long-traveling soil flow caused by the January 13, 2001, El Salvador earthquake. GSA Paper 375—Natural Hazards in El Salvador. In William, I., Rose (eds.). Geological Society of America, pp. 39–53.

Kousky, V.E., 1979. Frontal influences on northeast Brazil. Mon. Wea. Rev. 107, pp. 1140–1153.

Kovda, V., 1980. Land Aridization and Drought Control. Westview Press, Boulder, CO.

Krepper, C.M., Scian, B., Pierini, G., 1982. Time and space variability of raifall in central-east Argentina. J.Climat,e 2, pp. 39–47.

Krepper, C.M., García, N.O., Jones, P.D., 2003. Interannual variability in the Uruguay River Basin. International Journal of Climatology 23, pp. 103–115.

Krepper, C.M., García, N.O., 2004. Spatial and temporal structures of trends and interannual variability of precipitation over the La Plata Basin. Quaternary International 114. pp. 11–21.

Krepper, C.M., García, N.O., Jones, P.D., 2006. Paraguay River Basin response to seasonal rainfall. International Journal of Climatology 26, pp. 1267–1278.

Krepper, C.M., García, N.O., Jones, P.D., 2007. Low-Frequency Response of Upper Paraná River Basin and Vulnerability. International Journal of Climatology published online by Wiley Inter-Science on July 2007, p. 10.

Kruse, E., Laurencena, P., 2005. Aguas Superficiales relación con el régimen subterráneo y fenómenos de anegamiento. In Barrio, R., Etcheverry, R., Caballé, M., Llambías, E. (eds.), Geologia y Recursos Minerales de la Provincia de Buenos Aires. Relatorio del XVI Congreso Geológico Argentina, La Plata, pp. 313–326.

Krühn T., 1915. Informe sobre un viaje de estudios en la prov. de Catamarca y parte de Tucumán. Boletín de Instrucciones Públicas, Buenos Aires, tomo XV, pp. 50–51.

Kulikov, E.A., Rabinovich, A.B., Thompson, R.E., 2005. Estimation of tsunami risk for the coasts of Peru and northern Chile. Natural Hazards 35, pp. 185–209.

Kuroiwa, J., 2002. Sustainable cities, a regional seismic scenario and the 6–23–2001 Arequipa Peru earthquake. Natural Hazards Review 3, pp. 158–162.

Laity, J. 1994. Landforms of aeolian erosion. In: Abrahams, A., Parsons, A. (ed.), Geomorphology of Desert Environments. Chapman & Hall, London, pp. 506–535.

Labraga, J.C., Villalba, R., 2009. Climate in the Monte Desert: Past trends, present conditions, and future projections. Journal of Arid Environments 73, pp. 154–163.

Lacerda, W.A., 1997. Stability of natural slopes along the tropical coast of Brazil. Almeida, M.S.S. (ed.), Symp. on Recent Developments in Soil and Pavement Mechanics, COPPE–UFRJ. Balkema, Rio de Janeiro, pp. 17–40.

Lacerda, W.A., Avelar, A.S., 2003. Flume tests on sand subjected to seepage with the influence of hidden barriers. Proc. of the International Workshop on Occurrences and Mechanisms of Flows in Natural Slopes and Earthfill, Sorrento, Italy, vol. 1, pp. 136–144.

Lacerda, W.A., Sandroni, S.S., 1985, Movimentos de massas colluviais. Mesa Redonda sobre Aspectos Geotecnicas de Taludes, ABMS, Rio de Janeiro, pp. iii–1—iii–19.

Lacerda, W.A., Schilling, G.H., 1992. Rain enduced creep-rupture of Soberbo Road Landslide. Proceedings of the 6th International Symp. on Landslides, New Zealand, Christchurch, vol. 1, pp.145–152.

Lafone, Quevedo, 1880. Inundación en Andalgalá. La Nación newspaper. Buenos Aires

Lander, J.F., Whiteside, L.S., 1997. Caribbean tsunamis: an initial history. Tsunami workshop, June11–13, Mayaguez, Puerto Rico. http://www.cima.–uprm.edu/–tsunami/Lander/J_Lander.html.

Landsea, C.W., Gray, W.M., 1992. The strong association between Western Sahelian monsoon rainfall and intense Atlantic hurricanes. J. Climate 5, pp. 435–453.

Lanfredi, N.W., Pousa, J.L., D'Onofrio, E.E., 1998. Sea-level rise and related potential hazards on the Argentine coast. Journal of Coastal Research 14 (1), pp. 47–60.

Lapido, O., Pereyra, F., 2002. Riesgos geológicos. In Haller, M. (ed.), Geología y Recursos Naturales de Santa Cruz, Relatorio del XV Congreso Geológico Argentino, VI–1, pp. 831–845.

Larsen, J., 2004. Glaciers and Sea Ice Endangered by Rising Temperatures. Earth Policy Institute, 1350 Connecticut Ave. NW, Suite 403, Washington, DC 20036. Available at: www.earth–policy.org.

Lavell, A., 1994. Prevention and mitigation of disasters in Central America: Vulnerability to disasters at the local level. In Varley, A. (ed.), Disasters, Development and Environment. John Wiley and Sons, Chichester.

Lavenu, A., Winter, T., Dávila, F., 1995. A Pliocene-Quaternary compressional basin in the Inter-andean Depression. Central Ecuador. Geophysical Journal International 121, pp. 279–300.

Lavenu, A., 2006. Andean neotectonics between 1°N and 47°S (Ecuador, Bolivia and Chile): A review. Revista de la Asociación Geológica Argentina, No. 61, pp. 504–524.

Laya, H., Kruger, H., Sanchez, L., Silenzi, J.C., 1984. Resultado de observaciones directas sobre acciones eólicas en la Patagonia. Interpretación e intento de evaluación. Seminario Metodología regional del proceso de desertificación. Universidad Nacional del Comahue. 24 de octubre al 4 de noviembre de 1984. Neuquén.

Latrubesse, E.M., Stevaux, J.C., Santos, M.L., Assine, M.L., 2005. Grandes sistemas fluviais: geologia, geomorfologia e paleoidrologia. In Oliveira, P.E., Gouviea Souza, R,C., Suguio, K., and Oliveira, A.M.S. (eds.), Quaternário do Brasil. Holos Editora, Ribeirão Preto, pp. 276–297.

Latrubesse, E.M., Silva, S.A., Moraes, M.J., 2001. Floods in the city of Rio Branco, Brazil: A case study on the impacts of human activities on flood dynamics and effects. In Marchettti, M., Rias, V.

(eds.), Geomorphology and Environmental Impact Assessment. A.A. Balkema Publishers, Lisse., pp. 163–175.

Latrubesse, E.M., Stevaux, J.C., and Sinha, R., 2005. Tropical rivers. Geomorphology, 70, pp. 187–206.

Laurance, W.F., Williamson, G.B., Delamônica, P., Oliveira, A., Lovejoy, T.E., Gascon, C., Pohl, L., 2001. Effects of a strong drought on Amazonian forest fragments and edges. J. Trop. Ecol. 17, pp. 771–785.

Lavallée, Y., de Silva, S.L., Salas, G., Byrnes, J.M., 2006. Explosive volcanism (VEI 6) without caldera formation: insight from Huaynaputina volcano, southern Peru. Bull. Volcanol. 68, pp. 333–348.

Leanza, A., 1958. Geología Regional. La Argentina, Suma de Geografía I, Editorial Peuser, Buenos Aires, pp. 217–349.

Leiva, J.C., 1999. Recent fluctuations of the Argentinian glaciers. Global and Planetary Change 22, pp. 169–177.

Leiva, J.C., Lenzano, L., Cabrera, G., Suarez, J., 1989. Variations of the Río Plomo Glaciers, Andes Centrales Argentinos. In Oerlemans, J. (ed.), Glacier Fluctuations and Climatic Change. Kluwer Academic Publishers, pp 143–151.

Leiva, J.C., Espizúa, L., Iturraspe, R., Masiokas, M., Norte, F., Villalba, R., 2007. La respuesta de los glaciares argentinos al clima de los siglos 20 y 21. Manuscript.

Lemos, A., 1883. Apuntes de viaje a través de Los Andes. London.

León, R., Brand, D., Collantes, M., Paruelo, J., Soriano, A., 1998. Grandes unidades de vegetación de la Patagonia extra andina. Asoc. Arg. De Ecología. Ecología austral 8, pp. 125.144.

León, R., Facelli, J.M., 1981. Descripción de una coenoclina en el SO del Chubut. Revista de la Facultad de Agronomía 2, pp. 163–171.

León, R.J.C., Aguiar, M.R., 1985. El deterioro por uso pasturil en estepas herbáceas patagonicas. Phytocoenologia 13, pp. 181–196.

Leopold, L.D., Wolman, M.G., Miller, J.P., 1964. Fluvial Processes in Geomorphology. W.H. Freeman, San Francisco.

Leprun, J.C., 1981. A erosão, a conservação e o manejo do solo no Nordeste Brasileiro: Balanço, diagnóstico e novas linhas de pesquisa. Recife, SUDENE, Div. Rec. Renováveis, série Recursos do solo 15, pp. 105.

Lepsch, I.E., 2002. Formação e conservação dos solos. Oficina de Textos, São Paulo, pp. 178.

Liu, J., 2008. Interannual to decadal variability of rainfall in tropical South America. M.S. Thesis, Duke University.

Lliboutry, L., 1999. Glaciers of the Wet Andes. In Williams, R.J., Jr., Ferringo, J.G. (eds), Satellite Image Atlas of Glaciers of the World, South America, United States Geological Survey, Professional Papers 1386–I, online version 1.01. http://pubs.usgs.gov/prof/p1386i/index.html.

Lomnitz, C., 1970. Mayor earthquakes and tsunamis in Chile during the period 1535 to 1955. Geological Rundshau 59, pp. 938–960.

Lomnitz, C., Flores, J., Novaro, O., Seligman, T.H., Esquivel, R., 1999. Seismic coupling of interface modes in sedimentary basins: A recipe for disaster. Bulletin of the Seismological Society of America 89 (1), pp. 14–21.

Loos, P.A., 1926. Los terremotos del 17 de diciembre de 1920 en Costa de Araujo, La valle, La central, Tres Porteñas, etc. Contribuciones Geofísicas, Tomo I, No. 2. Observatorio Astronómico de la Universidad Nacional de La Plata.

Loos P.A., 1928. El terremoto Argentino-Chileno del 14 de abril de 1927. Contribuciones Geofísicas, Tomo II, N° 2. Observatorio Astronómico de la Universidad Nacional de La Plata.

Loos, P.A., 1929. Relatos del Dr Lukenheimer: Observaciones del sismo 1929 en Villa Atuel. Observatorio de La Plata.

Lopardo, R.A., Soeane, R., Paoli, C.U., 2003. A scientific-technological view on the disasters caused by floods. Academia Nacional de Ciencias Exactas Fisicas y Naturales, Buenos Aires, Argentina, p. 11.

López Bermúdez, F., 1996. La degradación de tierras en ambientes áridos y semiáridos, Causas y consecuencias. In Lasanta, T., García Ruiz, J.M. (eds.). Erosión y recuperación de tierras en áreas marginales. Instituto de Estudios Riojanos. Sociedad Española de Geomorfología. Geoforma Ediciones. Logroño, pp. 51–72.

López Bermúdez, F., 1997. Desertificación y degradación del suelo: Un desafío de fin de milenio. En Sustentabilidad, Desarrollo Económico, Medio Ambiente y Biodiversidad. David M. Rivas, Coord. Editorial Parteluz, S. L. Madrid, pp. 129–155.

Los Andes newspaper, 1890, Mendoza, Argentina.

Lukenheimer, F., 1930. El terremoto sudmendocino del 30 de mayo de 1929. Contribuciones geofisicas 3 (2). Observatorio de La Plata.

Lünkenheimer F., 1929. El terremoto surmendocino del 30 de mayo de 1929. ContribucionesgGeofísicas 3 (2). Observatorio Astronómico de la Universidad Nacional de La Plata

Maack, R., 1968. Geografia física do estado de Paraná. Curitiba, Banco Desenv. Paraná, Univ. Fed. Paraná, IBPT, pp. 350.

Macharé, J., Benavente, C., Audin, L., 2009. Sistema de Fallas de la Cordillera Blanca. In: Proyecto Multinacional Andino (Ed.). Atlas de Deformaciones Cuaternarias de los Andes. SERNAGEOMIN, Publicación Multinacional No 7: 208-214, Santiago.

Madueño Luján, R.A., 2003. Optimal use of the hydropower potential of the upper catchment area of Tablachaca River in Peru. Abstract, Master's Thesis, École Polytechnique Fédérale de Lausanne, France.

Malfait, B.T., Dinkelman, M.G., 1972. Circum-Caribbean tectonic and igneous activity and the evolution of the Caribbean plate. Bulletin of the Geological Society of America 83, pp. 251–272.

Magaña, V., Amador, J.A., Medina, S., 1999. The midsummer drought over Mexico and Central America. J. Climate 12, pp. 1577–1588.

Magilligan, F.J., Goldstein, P.S., 2001. El Niño floods and culture change: A late Holocene flood history for the Rio Moquegua, southern Peru. Geology 29 (5), pp. 431–434.

Magnoli, D., Araujo, R., 1992. A Nova Geografia, 2nd ed., São Paulo: Moderna.

Magrin, G., García., C.G., Cruz Choque D., Giménez J.C., Moreno A.R., Nagy, G.J., Nobre C., Villamizar, A., 2007. Latin America. In Parry,M.L., Canziani, O.F., Palutikof, J.P., van der Linden P.J., Hanson, C.E. (eds.), Climate Change 2007: Impacts, Adaptation and Vulnerability. Cambridge University Press, Cambridge, pp. 581–615.

Maia, L.P., Freire, G.S.S., Lacerda, L.D., 2005. Accelerated dune migration and aeolian transport during El Niño events along the NE Brazilian coast. Journal of Coastal Research 21 (6), pp. 1121–1126.

Malhi, Y., Roberts, T.M., Betts, R.A., Killeen, T.J., Li, W., Nobre, C.A., 2008. Climate change, deforestation and the fate of the Amazon. Science, 319, pp. 169–172.

Manners, R.B., Magilligan, F.J., Goldstein, P.S., 2007. Floodplain development, El Niño, and cultural consequences in a hyperarid Andean environment. Annals Assoc. Amer. Geogr. 97 (2), pp. 229–249.

Manuel–Navarrete, D., Gomez, J.J., Gallopin, G., 2007. Syndromes of sustainability of development for assessing the vulnerability of coupled human–environmental systems. The case of hydrometeorological disasters in Central America and the Caribbean. Global Environmental Change—Human and Policy Dimensions 2, pp. 207–217.

Marchant, M., Cecioni, A., Figueroa, S., González, H., Giglio, S., Hebbeln, J.K., Lamy,F., Mohtadi, M., Pineda, V., and Romero, O., 2007. Marine geology, oceanography and climate. In Moreno, T., and Gibbons, W. (eds.), The Geology of Chile. The Geological Society, London, pp. 289–308.

Marcuzzi J.J., Wayne W.J., Alonso R.N., 1994. Geologic hazards of Salta Province, Argentine. 7° International Congress. International Association of Engeneering Geology, Proceedings 3, pp. 2039–2048. Lisbon.

Marengo, J., Hastenrath, S., 1993. Case Studies of Extreme Climatic Events in the Amazon Basin. J.Climate 6, pp. 617–627.

Marengo, J.A., Liebmann, B., Kousky, V.E., Filizola, N.P., Wainer, I.C., 2001. Onset and end of the rainy season in the Brazilian Amazon Basin. J.Climate 14, pp. 833–852.

Marengo, J., 1992. Interannual variability of surface climate in the Amazon basin. Int. J. Climatol. 12, pp. 853–863.

MARNR, 2000. Análisis de la precipitación extrema del mes de diciembre de 1999 en la región norte costera de Venezuela. Dirección de Hidrología y Meteorología, Caracas.

Martin, L., Dominguez, J.M.L., Bittencourt, A.S.S.P., 2003. Fluctuating Holocene sea levels in Eastern and Southeastern Brazil: Evidence from multiple fossil and geometric indicators. Journal of Coastal Research 19 (1), pp. 101–123.

Martínez, C., 2001. El efecto de ensenada en los procesos litorales de las ensenadas de Valparaíso, Algarrobo y Cartagena, Chile Central. Unpublished Mg. Thesis, Escuela de Post Grado, Universidad de Chile. Santiago, p. 152.

Martínez, J.M., Parra, E., París, G., Forero, C.A., Bustamante, M., Cardona, O.D., Jaramillo, J.D., 1994. Los sismos del Atrato medio, 17 y 18 de Octubre de 1992, noroccidente de Colombia. Revista Ingeominas, 4, pp. 35–76.

Martos, L., 1995. Fallamiento cuaternario en el borde oriental de Precordillera entre el río San Juan y el río de Los Pozos. Unpublished Doctoral Thesis, Universidad Nacional de San Juan, San Juan, Argentina, pp. 250.

Martos, L., 1999a. Dinámica morfoestructural del paisaje cuaternario en el piedemonte oriental de la Precordillera Oriental. San Juan: Décimo Cuarto Congreso Geológico Argentino Actas, vol. I, pp. 275–278.

Martos, L., 1999b. Cronoestratigrafía de los niveles pedemontanos cuaternarios en el este de Precordillera Oriental. San Juan: Décimo Cuarto Congreso Geológico Argentino Actas, vol. I, pp. 279–282.

Marques, J.Q.A., Bertoni, J., Barreto, G.B., 1961. Perdas por erosão no Est. de São Paulo. Campinas, Bragantia 20, pp. 1143–1182.

Maskrey, A., 1993a. Los Desastres no son Naturales. Tercer Mundo Editores, Santa Fé de Bogotá, Colombia.

Maskrey, A., 1993b. Vulnerability accumulation in peripheral regions in Latin America: The challenge for disaster prevention and management. In Merriman, P.A., Browitt, C.W.A. (eds.), Natural Disasters: Protecting Vulnerable Communities. London: Thomas Telford, pp. 461–472.

Maza J., Ruiz, A., 2006. Inundaciones Urbanas. Comunicación Nacional de Cambio Climático: vulnerabilidad de la Patagonia y sur de las provincias de Buenos Aires y La Pampa. Fundación e Instituto Torcuato Di Tella, Buenos Aires, pp. 157–182.

Mazzoni, E., 2001. Distribución espacial y caracterización geomorfológica de pequeñas cuencas endorreicas de la Patagonia austral. Anales del Instituto de la Patagonia Serie Ciencias Naturales, 29, pp. 5–24.

Mazzoni, E., Vazquez, M., Ruiz Posse, E., 2002. Dinámica geomorfológica de depósitos eólicos asociados a "bajos sin salida" de la Patagonia austral. IV Jornadas Nacionales de Geografía Física, Mar del Plata.

Mazzoni, E., Vazquez, M., 2004. Ecosistemas de mallines y paisajes de la Patagonia Austral, Provincia de Santa Cruz. INTA, Buenos Aires, p. 63.

Mazzoni, E., 2006. Geomorfología y evolución geomorfológica de escoriales volcánicos y sus mallines asociados en diferentes de Patagonia extrandina. Tesis doctoral. Universidad Nacional del Sur, Bahía Blanca.

McCalpin, J. 1996. Paleoseismology. Academic Press, London, p. 588.

McEntire, D.A., Fuller, C., 2002. The need for a holistic theoretical approach: an examination from the El Niño disasters in Peru. Disast. Prev. Manag, Intern. J. 11 (2), pp. 128–140.

McNulty, B., and Farber D., 2002. Active detachment faulting above the peruvian slab. Geology 30, pp. 567–570.

McPhaden, M.J., Zebiak, S.E., Glantz, M.H., 2006. ENSO as an integrating concept in Earth Science. Science, 314, pp. 1740–1745.

McQuarrie, N., Horton, B.K., Zandt, G., Beck, S., DeCelles, P.G., 2005. Lithospheric evolution of the Andean fold-thrust belt, Bolivia, and the origin of the Central Andean Plateau. Tectonophysics 399, pp. 15–37.

Meckelein, W., 1980. The problem of desertification within deserts. In Desertification in Extremely Arid Environments, IGU Working Group on Desertification in and around Arid Lands, 24th International Geographical Congress, Tokyo.

Medina, J.T., 1956. Colección de Documentos Inéditos para la Historia de Chile, Segunda Serie, vol. 2, pp. 1573–1580. Fondo Histórico y Bibliográfico.

Meglioli, A., 1982. Glacial Geology and Geochronology of Southernmost Patagonia and Tierra del Fuego, Argentina and Chile. Tesis doctoral. Lehigh University, Pennsylvania, USA.

Meigs, A., Krugh, W., Schiffman, C., Vergés, J., Ramos, V., 2006. Refolding of thin-skinned thrust sheets by active basement-involved thrust faults in the eastern Precordillera of western Argentina. Revista Asociación Geológica Argentina 4 (61), pp. 589–603.

Meis, M.R.M., Silva, J.X., 1968. Considerações geomorfológicas a propósito dos movimentos de massa ocorridos no Rio de Janeiro. Rev. Bras. de Geografia, ano 30 (1), pp. 55–73.

Mendoza, B., Jauregui, E., Diaz-Sandoval, R., Garcia-Acosta, V., Velasco, V., Cordero, G., 2005. Historical droughts in central Mexico and their relation with El Niño, Journal of Applied Meteorology, 44 (5), pp. 709–716.

Mendoza, B., Velasco, V., and Jauregui, E., 2006. A study of historical droughts in southeastern Mexico, Journal of Climate, 19 (12), pp. 2916–2934.

Mensching, H.G., 1996. Procesos geomorfológicos en la Patagonia como indicadores de la desertificación. Guia de campo. Proyecto Prodesar. INTA-GTZ, Bariloche.

Mercer, J., 1976. Glacial history of southernmost South America. Quaternary Research 6, pp. 125–166.

Mérega, J.L., 2000. La desertificación. Publicaciones online, Fundación Fundasur. http://www.fundasur.org.ar/ladesertificación.htm.

Mertes, L., 1985. Floodplain development and sedimentat transport in the Solimões–Amazon River, Brazil. Máster's Thesis, University of Washington.

Methol, E., 1967. Rasgos geomorfológicos de la meseta de Somuncurá, Río Negro. Consideraciones acerca de los orígenes de los Pequeños Bajos sin salida. Asoc. Geol. Arg. Revista 4 (22), pp. 295–271.

Miers J., 1826. Travels in Chile and La Plata. Baldwin, Cradock and Joy. London, p. 640.

Mikkan R.A., 1992. Geomorfologia de la zona de influencia de la presa Potrerillos. Boletín de Estudios Geográficos, Mendoza 35 (88), pp. 91–113.

Mileti, D.S., Bolton, P.A., Fernandez, G., Updike, R.G., 1991. The Eruption of Nevado del Ruiz Volcano, Colombia, South America. National Academy Press, Washington, DC, USA.

Miller, L., Douglas, B.C., 2004. Mass and volume contribution to twentieth-century global sea level rise. Nature 428, pp. 406–409.

Minaya, E.R, Aliaga, P., Avila, G., Chirinos, C., Díaz, J., Gonzáles, M., Párraga, L., Valero, D., 2005. Probabilidad de ocurrencia de un sismo distante con effectos secundarios en Bolivia. http://www.observatoriosancalixto.org/OBSERVATORIO%20SAN%20CALIXTO–2.pdf

Mingorance, F., 1991. Análisis y evaluación tectónica asociada a la falla Sampacho, Provincia de Córdoba, Argentina Central. Revista Geofísica, 35, pp. 13–39.

Mingorance, F., 1998. Evidencias de paleoterremotos en la falla activa La Cantera–cinturón de empuje de la Precordillera, San Juan, Argentina. X Congreso Latinoamericano de Geología Actas, vol. II, pp. 161–166.

Mingorance, F., 2000a. Peligro de desplazamiento superficial de falla en el núcleo urbano del Gran Mendoza, Argentina. 9° Congreso Geológico Chileno, Actas I, pp. 81–85.

Mingorance, F., 2000b. Caracterización de la geometría de la zona de fallamiento activo La Cal, Mendoza, Argentina. 9° Congreso Geológico Chileno, Actas I, pp. 800–804.

Mirza, M.Q., Warrick, R.A., Ericksen, N.J., Kenny, G.J., 2001. Are the floods getting worse in the Ganges, Brahmaputra and Meghna basins? Environmental Hazards, pp. 37–48.

Mojica, J., Colmenares, F., Villaroel, C., Macía, C., Moreno, M., 1985. Características del flujo de lodo ocurrido el 13 de Noviembre de 1985 en el Valle de Armero (Tolima, Colombia). Historia y comentarios de los flujos de 1595 y 1845. Geología Colombiana 14, pp. 107–140.

Molion, L.C.B., Moraes, J.C., 1987. Oscilacao sul e descarga de rios na America do Sul Tropical. Rev. Bras. de Engennharia, Caderno do Recurso Hídricos, 5, pp. 53–63.

Mon, R., Mena, R., Marret, R., Hongn, F., 1988. Rasgos de Geotectónica en la Puna Austral (República Argentina). V Congreso Geológico Chileno, Actas I, pp. A69–A83.

Moore, P.D. 1990. Soils and ecology: temperate wetlands. In: Williams, M. (ed), Wetlands: a threatened landscape, pp. 95–114. Basil Blackwell, Oxford, UK, 419p.

Mooser, F., 1972. The Mexican volcanic belt: Structure and tectonics. Geofisica Internacional, 12, pp. 55–70.

Monbeig, P., 1952. Pionniers et planteurs de São Paulo. Libr. Armand Colin, Paris, p. 376.

Montes de Oca, I., 1997. Geografía y Recursos Naturales de Bolivia (3rd edition). EDOBOL, La Paz, Bolivia.

Moreiras, S.M., 2002. Ocurrencia de Procesos Gravitaciones en la Ruta Internacional Chile–Argentina–Paso de La Cumbre–entre los 69° y los 69° 30' Longitud Oeste. Actas Simposio Internacional de Geología Ambiental para planificación del uso del territorio (CD). Puerto Varas, Chile.

Moreiras, S.M., 2003. Remoción en masa en un sector del valle del Río Mendoza. Mendoza, Argentina (Landslide in a sector of the Rio Mendoza Valley). Revista de la Universidad Nacional de La Rioja, 4–1 (2), pp. 2–10.
Moreiras, S.M., 2004. Zonificación de Peligrosidad y Riesgo de procesos de remoción en masa en el valle del Río Mendoza, Argentina. Unpublished Doctoral Thesis, Universidad Nacional de San Juan, San Juan, Argentina, pp. 190.
Moreiras, S.M., 2004a. Landslide hazard and risk zonation along the Mendoza River. Universidad Nacional de San Juan. Ph.D. Dissertation, pp. 193. Unpublished.
Moreiras, S.M., 2004b. ¿Qué características tienen los procesos de Remoción en masa en Cordillera Frontal y Precordillera Mendocina? In Gonzalez, M., Bejerman, N.J. (eds.), Peligrosidad geológica en Argentina. Asociación Argentina de Geología Aplicada a la Ingeniería, Buenos Aires, publicación Especial No. 4, pp. 378–386.
Moreiras, S.M., 2004c. Landslide Incidence Zonification along the Rio Mendoza Valley. Mendoza, Argentina. Earth Surface Processes and Landforms, 29, pp. 255–266.
Moreiras, S.M., 2004d. Zonificación de peligrosidad y riesgo de procesos de remoción en masa en el valle del Río Mendoza Peligrosidad geológica en Argentina. Asociación Argentina de Geología Aplicada a la Ingeniería, Buenos Aires, publicación Especial No. 4, pp. 423–427.
Moreiras, S.M., 2005a. Landslide Susceptibility Zonation in the Rio Mendoza Valley, Argentina. Geomorphology, 66 (1–4), pp. 345–357.
Moreiras, S.M., 2005b. Avalancha de rocas pleistocena Tigre Dormido, valle del Río Mendoza, Mendoza Province. XVI Congreso Geológico Argentino, La Plata, Actas III, pp. 491–498.
Moreiras, S.M., 2005c. Climatic effect of ENSO associated with landslide occurrence in the Central Andes, Mendoza Province, Argentina. Landslides, 2/1, pp. 53–59.
Moreiras, S.M., 2005d. Zonificación de la peligrosidad de procesos de remoción en masa en el valle del Río Mendoza, provincia de Mendoza, mediante a un modelo probabilístico. Actas XVI Congreso Geológico Argentino, IV (CD). La Plata.
Moreiras, S.M., 2006a. Frequency of debris flows and rockfall along the Mendoza River Valley (Central Andes), Argentina. Quaternary International, Special Issue: Holocene Environmental Catastrophes in South America. (In press).
Moreiras, S.M., 2006b. Chronology of a Pleistocene rock avalanche probable linked to neotectonic, Cordon del Plata (Central Andes), Mendoza—Argentina. Quaternary International, 148, pp. 138–148.
Moreiras, S.M., 2006c. Estudio geomorfológico de la avalancha de rocas Piedras Blancas, provincia de Mendoza, Argentina. IV Congreso Argentino de Geomorfología y Cuaternario, Acta I, 23-31, Cordoba, Argentina.
Moreiras, S.M., 2006d. Climatic role in landslide occurrence in the Central Andes, Argentina. Proceedings of Symposium on Climate Change: Organizing the Science in the American Cordillera: 18. Mendoza.
Moreno, H., 1976. The upper Cenozoic volcanism in the Andes of southern Chile (from 40deg 00min to 41deg 30min lat S). In Gonzalez–Ferran O (ed.), Proc Symp Andean & Antarctic Volcanology Problems (Santiago, Chile, September 1974). Rome: IAVCEI, pp. 143–171.
Moreno H., Naranjo J.A., 1991. The southern Andes volcanoes (33 deg–41 deg 30 min S), Chile. 6th Geol Cong Chile, Excur PC–3, p. 26.
Moreno, M., 2004. Dinámica del Antearco Externo en la Zona del Bloque de Arauco, 37°–38° S, Octava Región. Chile. Thesis of Geology, Departamento Ciencias de la Tierra, Facultad de Ciencias Químicas, Universidad de Concepción.
Morton, R.A., González, J.L., López, G.I., Correa, I.A., 2000. Frequent non–storm washover of barrier islands, Pacific Coast of Colombia. Journal of Coastal Research 16 (1), pp. 82–87.
Morton, R.A., Richmond, B.M., Jaffe, B.E., Gelfenbaum, G., 2006. Reconnaissance investigation of Caribbean extreme wave deposits—Preliminary observations, interpretations, and research directions. USGS Open–File Report 2006–1293, p. 41.
Motoki A., Orihashi, Y., Naranjo, J., Hirata, D., Skvarca, P., Anma, R., 2006. Geologic reconnaissance of Lautaro Volcano, Chilean Patagonia. Revista Geológica de Chile 33 (1), pp. 177–187.
Mountain Research Initiative (MRI), 2006. Global Change and Mountain Regions Research Strategy. GLOCHAMORE. A joint project of the Mountain Research Initiative, UNESCO–MAB and IHP, and the EU framework programme 6. Bern, Switzerland.

Moura, A.D., Shukla, J., 1981. On the dynamics of droughts in Northeast Brazil: Observations, theory and numerical experiments with a general circulation model. J. Atmos. Sci. 38, pp. 2653–2675.

Movia, C., 1972. Formas de erosión eólica en la Patagonia. Photointerprétation 6/3: 14–19. Ed.Technip, Paris, pp. 14–19.

Movia, C. 1984a. Tipología de mallines. In Rabassa, J., Brandani, A., Capua, O.,and Ottonello, E (eds.), Curso de campo en sistemas ecogeomorfológicos: Guía de excursiones. Dpto. de Postgrado. Universidad Nacional del Comahue, Neuquén.

Movia, C. 1984b. Metodologías aplicables a la evaluación de la desertificación en Patagonia. In: Metodología regional del proceso de desertificación. Universidad Nacional del Comahue, Neuquén, pp. 154–191.

Muratori, A..M., 1984. Erosão no Noroeste do Paraná: uma proposta metodológica de estudo sistemático através do uso da fotografias aéreas. Curitiba, Univ. Fed. Paraná, Dep. Geociências, p. 143. (dissertaçao de Mestrado).

Muñoz, J.F., Fernández, B., Varas, E., Pastén, P., Gómez, D., Rengifo, P., Muñoz, J., Atenas, Mesenia, Jofre, J.C., 2007. Chilean water resources. In Moreno, T., Gibbons, W. (eds.), The Geology of Chile. The Geological Society, London, pp. 215–230.

Nágera, J.J., 1939. Geografía Física de la República Argentina. In Manito, O., and Nágera, J.J. (eds.), Geografía Física de las Américas y de la República Argentina. Ed. Kapeluz, Buenos Aires, p. 232.

Naranjo J., Polanco, E., 2004. The 2000 AD eruption of Copahue Volcano, Southern Andes. Revista Geológica de Chile 31 (2), pp. 279–292.

Naranjo, J.A., 1987. Interpretación de la actividad cenozoica superior a lo largo de la Zona de Falla de Atacama, norte de Chile. Revista Geológica de Chile, 31, pp. 43–55.

Naranjo, J.A., and Clavero, J.E., 2005. A rare case of grass flow induced by the M8.4 Arequipa earthquake, June 2001, in the Altiplano of Northern Chile. Quatern. Research, 64, pp. 242–248.

National Oceanic and Administration, 2006. http://www.csc.noaa.gov/hurricane_tracks.

Nester, P.L., Gayó, E., Latorre, C., Jordan, T.E., Blanco, N., 2007. Perennial stream discharge in the hyperarid Atacama Desert of northern Chile during the latest Pleistocene. Proc. Nat. Acad. Sci. 104 (50), pp. 19724–19729.

Nicholls, R.J., Lowe, J.A., 2004. Benefits of mitigation of climate change for coastal areas. Global Environmental Change, 14, pp. 229–244.

Nieto–Ferreira, R., Rickenbach, T.M., Herdies, D.L., Carvalho, L.M.V., 2003. Variability of South American convective cloud systems and tropospheric circulation during January–March 1998 and 1999. Mon. Wea. Rev., 131, pp. 961–973.

NOAA–NESDIS, 2000. Venezuelan rainfall December 15–17 1999. online consult: http://pr.water.usgs.

Noy Meir, I., 1973. Desert ecosystems: Environment and producers. Anual Review of Ecology and Systematics, 4, pp. 25–41.

Nunes, A.J.C., 1969. Landslides in soils of decomposed rock due to intense rainstorms, Proc. of the 7th ICSMFE, Mexico, 3, pp. 1123–1134.

Nuñez, M.N., Solomon, S.A., Cabré, M.F., 2008. Regional climate change experiments over southern South America. II: Climate change scenarios in the late twenty-first century. Climate Dynamics, DO1: 10. 1007/S 00 382–008–0449–8.

Oerlemans, J., 1994. Quantifying global warming from the retreat of glaciers. Science 264, pp. 243–245.

O'Hare, G., Rivas, S., 2005. The landslide hazard and human vulnerability in La Paz city, Bolivia. The Geographical Journal 171, pp. 239–258.

Okada, K., 1995. Los recursos fitogenéticos y el rol del IPGRI. In: Montes, L., Oliva, G. (eds.), Taller internacional sobre los recursos filogenéticos, desertificación y uso sustentable. Río Gallegos, pp. 26–33.

Okal, E.A., Borrero, J.C., Synolakis, C.E., 2006. Evaluation of tsunami risk from regional earthquakes at Pisco, Peru. Bull. Seism. Soc. Amer. 96 (5), pp. 1634–1648.

Oliva,G., Rial, P., González , L., Mazzoni, E., 1995. Evaluación del estado actual de la desertificación en la Transecta Santa Cruz. In Del Valle, H., Eiden, G., Mensching, H., Goergen, J. (eds.), Lucha contra la desertificación en Patagonia. INTA–GTZ, Bariloche, pp. 65–85.

Oliva, O., Cibils, A., Borelli, P., Humano, G., 1998. Stable States in relation to grazing in Patagonia: A 10–year experimental trial. Journal of Arid Environments 40, pp. 113–131.
Oliveira, A.M.S., 1990. Depósitos tecnogênicos associados a erosão atual. Congr. Brasil. Geol. Eng., Salvador, Anais 1, pp. 411–418.
Oliveira, A.M.S., 1994. Depósitos tecnogênicos e assoreamento de reservatórios. Exemplo do reservatório de Capivara, rio Paranapanema, SP/PR. São Paulo, USP, FFLCH, Dep. Geogr., vol. 2, pp. 211, 25 mapas, 6 perfis, 1 graf. (tese Dr.)
Oliveira, A.M.S., Queiroz Neto, J.P., 1992. Produção de sedimentos e depósitos tecnogênicos no Planalto Ocidental Paulista. VI Simp. Nac. Controle Erosão, Presidente Prudente.
Oliveira, A.M.S., Queiroz Neto, J.P., 1993. Erosão acelerada no Planalto Ocidental Paulista: desequilíbrio, descontinuidade e epiciclo. V Simp. Geogr. Fis. Aplic., São Paulo, Anais, pp. 461–463.
Oliveira, A.M.S., Queiroz Neto, J.P., 1994. Depósitos tecnogênicos induzidos pela erosão acelerada no Planalto Ocidental Paulista. S. Paulo, Bol. Paul. Geogr. 73, pp. 91–123.
Oliveira, A.M.S., Queiroz Neto, J.P., Brannstrom, C., 2001. Impactos ambientais da ocupação do Planalto Ocidental do estado de São Paulo, Brasil. V Reun. Quat. Ibérico (REQUI) e I Congr. Quat. Paises Ling. Port. (CQPLI), Lisboa, Portugal.
Oliveira, A.M.S., Queiroz Neto, J.P., Carlstrom, F.C., Salomão, F.X.T., Kertzmen, F.F., 1991. Depósitos tecnogênicos no Planalto Ocidental Paulista: O exemplo de Andradina e Bauru. III Congr. ABEQUA, Belo Horizonte, Anais: pp. 88–95.
Oliveira, A,M.S., Queiroz Neto, J.P., Ponçano, W., 1995. Deforestation and sediment production in the Western Plateau of the São Paulo State, Brazil. Reg. Conf. on Global Change, Braz. Acad. Sci., São Paulo, Abstracts: pp. 114–115.
Oliveira, A.M.S., Queiroz Neto, J.P., Ponçano, W., 1997. Produção de sedimentos e depósitos tecnogênicos no Planalto Ocidental Paulista. VI Simp. Nac. Controle de Erosão, Presid. Prud., Anais, pp. 3.
Oliveira, R.R., Avelar, A.S., Leão, O.M.R., Freitas, M.M., Coelho Netto, A.L., 1996, Degradação da floresta e desabamentos ocorridos em Fevereiro de 1996 no Maciço da Tijuca, RJ. Anais do XLVII Congresso Nacional de Botânica, Nova Friburgo (RJ), vol. 1, pp. 353–353.
Olson, R.S., Sarmiento, J.P., Olson, R.A., Gawronski, V.T., and Estrada, A., 2000. The marginalization of disaster response institutions: The 1997–1998 El Niño experience in Peru, Bolivia, and Ecuador. Boulder, Colorado, USA: Natural Hazards Research and Applications Information Center, Special Publication 36.
Orche, E., Vargas, J., and Lara, F., 2003. Informe Técnico de CYTED sobre las operaciones mineras y la seguridad en el distrito aurífero de Tipuani (Bolivia). CYTED.
Orche, E., 2003. El deslizamiento de Chima (Bolivia) del 31 de marzo de 2003. In Villas Boas, R., and Aranibar, A.M., (Eds.), Pequeña minería y minería artesanal en Iberoamérica. CYTED–CETEM–CONACYT, Río de Janeiro, Brasil, pp. 201–272.
Orche, E., 2003. Colapso del cerro minero Pucaloma (Chima, Bolivia) ¿Una tragedia evitable? Jornadas "Seguridad, calidad y medio ambiente en la Ingeniería de minas: nuestro futuro." Barcelona, Spain.
Ordaz, M. Reyes, C., 1999. Earthquake hazard in Mexico City: Observations versus computations. Bull. Seismol. Soc. Am. 89 (5), pp. 1379–1383.
Orfeo, O., Stevaux, J.C., 2002. Hydraulic and morphological characteristics of middle and upper reaches of Paraná River (Argentina and Brazil). Geomorphology, pp. 309–322.
Orr, J.C., Pantoja, S., Pörtner, H.O., 2005. Introduction to special section: The ocean in a high-CO_2 world. Journal of Geophysical Research 110 (c), doi: 10 1029/2005 JC 003086.
Ortlieb, L., 2000. The documented historical record of El Niño events in Peru: An update of the Quinn record (sixteenth through nineteenth centuries). In Diaz, H., Markgraf, V. (eds.), El Niño and the Southern Oscillation, Multiscale Variability and Global and Regional Impacts. Cambridge University Press, pp. 207–295.
Ortlieb, L., Macharé, J., 1993. Former El Niño events: records from western South America. Global and Planetary Change 7, pp. 181–202.
Ota, Y., Paskoff, R., 1993. Holocene deposits on the coast of north-central Chile: Radiocarbon ages and implications for coastal changes. Revista Geológica de Chile 20, pp. 25–32.

Pacheco, H., Suárez, C., 2004. Mediciones fotogramétricas para determinar variaciones de la posición de la línea de costa en el cordón litoral de la laguna de Unare, Estado Anzoátegui, Venezuela. Acta Científica Venezonala, Ciencias de la Tierra, 55, pp. 97–106.

Paerregaard, K., 1997. Linking Separate Worlds. Urban Migrants and Rural Lives in Peru. Explorations in Anthropology, Oxford, United Kingdom, p. 292.

PAHO, 2000. Crónicas de desastres. Fenómeno El Niño 1997–1998. Programa de Preparativos para Situaciones de Emergencia y Coordinación del Socorro en casos de Desastre. Panamerican Health Organization/World Health Organization, Washington, DC, p. 288.

PAGES, 2006. Reconstructing past regional climate variations in South America over the late Holocene: an international symposium. Abstracts volume, October 4–7, 2006, Malargüe, Mendoza, Argentina, p. 91.

Paniagua-Pérez, S., 2002. Síntesis de algunos volcanes activos y peligrosos de América Central: Prevención, preparación y mitigación, Balboa, Ancón, Panamá: CEPREDENAC, p. 296.

Paolini L., Villalba R., Grau R., 2005. Precipitation variability and landslide occurrence in a subtropical mountain ecosystem of NW Argentina. Dendrochronologia, 22 (3), pp. 175–180.

París, G., Marín, P.A., 1979. Generalidades acerca de la geología del Departamento del Cauca. Ingeominas technical report, Bogotá, Colombia.

Paruelo, J.M., Bertiller, M., Schlichter, T., Coronato, F. (eds.), 1993. Secuencias de deterioro en distintos ambientes patagónicos. Su caracterización mediante el modelo de estados y transiciones. INTA–GTZ. Buenos Aires, p. 110.

Paruelo, J.M., Lauenroth, W.K., 1996. Relative abundance of functional type in grassland and shrubland of North America. Ecol. Applic. 6, pp. 1212–1224.

Paz, C., Buffoni, H., 1986. Manejo de las vegas en el sur de Santa Cruz. Presencia INTA 2 (8), pp. 40–44.

Pedrosa, M.G.A., Soares, M.M., Lacerda, W.A., (1988). Mechanism of movements in colluvial slopes in Rio de Janeiro. Proc. 5th Intern. Symp. on Landslides, Lausane, vol. 2 , pp. 1211–1216.

Pelayo A., Wiens, D., 1989. Seismotectonics and relative motions in the Scotia Sea Region, JGR 94, No B6, pp. 7293–7320.

Pelayo, A., Wiens, D. 1989. Seismotectonics and relative plate motions in the Scotia Sea region. Journal of Geophysical Research 94, pp. 7293–7320.

Pellerin, J., Gama, A.M.R.C., Nascimento, E.E.D.J., Valdati, J., Paisani, J.C., Pontelli, M.E., 1996. Zoneamento e efeitos morfodinâmicos da enxurrada do dia 23.12.95 nas bacias dos rios Figueira e Pinheirinho, SC. Revista Sociedade e Natureza, Uberlândia, MG, No. 15, pp. 195–200.

Pennington, W.D., 1981. Subduction of the eastern Panama Basin and seismotectonics of northwestern South America. Journal of Geophysical Research 86, pp. 10753–10770.

Peralta, C. 1995. Algunas características sociales de la Patagonia. In Consorcio DHV–Swedforest. Desertificación en Patagonia. Cap. 4.

Pereira de Queiroz, M.I., 2004. Seigneurs ruraux et pouvoir local dans la vie politique brésiliènne. Paris, Ed. Maison dês Sci. De l'Homme, Coll. Brasília, pp. 166.

Pereyra F.X., 1995. Esquema geomorfológico del sector norte del valle del río cuevas, entre Puente de Inca y Las Cuevas, Provincia de Mendoza. Revista de la Asociación Geológica Argentina 1–4 (50), Buenos Aires, pp. 103–110..

Pereyra F., Lapido O., González Díaz E., 2002. Geomorfología. In Haller, M. (ed.), Geología y Recursos Naturales de Santa Cruz, Relatorio del XV Congreso Geológico Argentino, Buenos Aires, I–21, pp. 325–352.

Pérez, O., Aggarwal, Y., 1981. Present–day tectonics of southeastern Caribbean and northeastern Venezuela. Journal of Geophysical Research 86, pp. 10791–10805.

Pérez, A., 1983. Algunos aspectos del clima. In Van der Hammen, T., Pérez, A., Pinto, P. (eds.), La Cordillera Central Colombiana, transecto Parque Los Nevados 1. J. Cramer, Berlin, Germany, pp. 38–47.

Perucca, L., 1990. Sistema de Fallamiento La Dehesa–Maradona–Acequión, San Juan, Argentina. XI Congreso Geológico Argentino, San Juan, Tomo II, pp. 431–434.

Perucca L.P., 1995. Actividad tectónica cuaternaria en al sierra de la Punilla. San Juan-La Rioja. Argentina. Universidad Nacional de San Juan, Ph.D. Dissertation, San Juan, pp. 250.

Perucca, L., Bastias, H., 2005. El terremoto argentino de 1894. Fenómenos de licuefacción asociados a sismos. In Aceñolaza, F. (ed.), INSUGEO, Serie Correlación Geológica, 19, pp. 55–70.

Perucca, L., Bastias, H., 2006. Regiones sismotectónicas en el centro-oeste argentino. Provincias de La Rioja, San Juan y Mendoza. In Aceñolaza, F. (ed.), INSUGEO, Serie Correlación Geológica 16, pp. 7–16.

Perucca, L.P., Moreiras, S.M., 2003. Avalanchas de rocas holocenas y fenómenos de licuefacción asociados a paleoterremotos en el Rio Acequión, provincia de San Juan, Argentina. Actas II° Congreso Nacional de Cuaternario y Geomorfología, San Miguel de Tucumán, pp. 137–146.

Perucca L.P., Moreiras, S.M., 2006. Liquefaction phenomena associated to historic earthquakes in San Juan and Mendoza provinces, Argentina. Quaternary International, Special Issue: Holocene Environmental Catastrophes in South America 158, 96-109.

Perucca, L., Navarro, C., 2005. El terremoto de Jagüé, departamento Vinchina, provincia de La Rioja. In Cabaleri N., Cingolani, C.A., Linares, E., López de Luchi, M.G., Ostera, H.A., Panarello, H.O. (eds.), XVI Congreso Geológico Argentino, Tomo 4, pp. 501–504.

Perucca, L., Paredes, J., 2003. Fallamiento cuaternario en la zona de La Laja y su relación con el terremoto de 1944, Departamento Albardón, San Juan, Argentina. Revista Mexicana de Ciencias Geológicas 20 (1), pp. 20–26.

Perucca, L., Paredes, J., 2004. Descripción del Fallamiento activo en la provincia de San Juan. Tópicos de Geociencias. Un volumen de Estudios Sismológicos, Geodésicos y Geológicos en Homenaje al Ing. Fernando Séptimo Volponi, pp. 269–309.

Perucca L., Paredes J., Tello. G., and Bastias, H., 1999. Fallamiento Activo en el área norte del Sistema de Fallamiento El Tigre, San Juan-La Rioja, Argentina. Revista de la Asociación Geológica Argentina 3 (54), pp. 206–214.

Petrinovic, I.A., 1999. La caldera de colapso del cerro Aguas Calientes, Salta; República Argentina; Evolución y Esquema estructural. In Colombo, F., Queralt, I., and Petrinovic, I. (eds.), Geología de los Andes Centrales Meridionales: El Noroeste Argentino, Acta Geológica Hispánica 34, pp. 243–255.

Pettiford, L. 1995. Towards a redefinition of security in Central America—The case of natural disasters. Disasters 19 (2), pp. 148–155.

Pezza, A., Simmonds, I., 2005. The first South Atlantic hurricane: Unprecedented blocking, low shear and climate change. Geophysical Research Letters 32 (15), doi: 10.1029/2005GL023390.

Pezza, A.B., Simmonds, L., 2005. The first South Atlantic hurricane: Unprecedented blocking, low shear and climate change. Geophysical Research Letters 32, L15712, doi: 10. 1029/2005 GL023390

Pichler, E.,1957, Aspectos geológicos dos escorregamentos de Santos, Boletim da Soc. Bras. de Geologia 6 (2), pp. 69–77.

Piedrahíta, I., Hermelin, M., 2005. La avenida torrencial del río Tapartó, abril de 1993. In Hermelin, M. (ed.), Desastres de origen natural en Colombia 1979–2004. Universidad EAFIT, OSSO–Univalle, Medellín, Colombia, pp. 109–120.

Pielke, R.A., Jr., Rubiera, J., Landsea, C., Fernández, M.L., Klein, R., 2003. Hurricane vulnerability in Latin America and the Caribbean: Normalized damage and loss potentials. Natural Hazards Review 4 (3), pp. 101–114.

Pineda, V., 1999. El Cañón Submarino del Bío-Bío: Aspectos Dinámicos y Ambientales. Ph.D. Thesis, Universidad de Concepción, Concepción, Chile.

Pinto, S.A.F., 1983. Utilização de técnicas de sensoriamento remoto para a caracterização de erosão do solo no SW do Estado de S. Paulo. São José dos Campos, INPE (dissertação de Mestrado), p. 128.

Pinto S.A.F., 1991. Sensoriamento remoto e integração de dados aplicados ao estudo da erosão de solos: contribuição metodológica. S. Paulo, USP, FFLCH, Dep. Geogr., p. 130, ilustr. (Tese Dr.)

Pirazzoli, P.A., 1986. Secular trends of relative sea-level (RSL) changes indicated by tide-gauge records. Journal of Coastal Research SI 1, pp. 1–26.

Pirazzoli, P.A., 1996. Sea Level Changes: The Last 20,000 Years. John Wiley and Sons, Chicester.

Pivel, M.A.G., Speranski, M., Calliari, L.J., 2001. A erosao praial na costa atlántica uruguaia. VIII Congresso da ABEQUA, Mariluz, Imbé, pp. 517–519.

Plafker, G., Savage, J.C., 1970. Mechanism of the Chilean earthquakes of May 21 and 22, 1960. Geological Society of America Bulletin 81, pp. 1001–1030.

Plafker, G. 1976. Tectonic aspects of the Guatemala Earthquake of 4 February 1976. Science 193, pp. 1201–1208.

Plessis-Fraissard, M., 2007. Planning roads for rural communities. Transp. Res. Record 1989, pp. 1–8.

Plunket, P., Urunuela, G., 2006. Social and cultural consequences of a late Holocene eruption of Popocatepetl in central Mexico. Quaternary International 151, pp. 19–28.

Plunket, P., and Urunuela, G., 2008. Mountain of sustenance, mountain of destruction: The prehispanic experience with Popocatepetl volcano. Journal of Volcanology and Geothermal Research, 170 (1–2), pp. 111–120.

PNUD ARG 02/006, 2004. Excedentes Hídricos en la Cuenca del Río Quinto, PNUD/Ministerio de Infraestructura, Vivienda y Servicios Públicos, Gobierno de la Provincia de Buenos Aires, La Plata, Argentina, p. 42.

Pochat, V., 2002. Las mayores crecidas en la cuenca del Plata, Anexo 4.1/Appendix 4.1, in http://www.oas.org/usde/publications/Unit/oea26b/ch10.htm

Polanski J., 1966. Flujos rápidos de escombros rocosos en zonas áridas y volcánicas. Editorial Universitaria de Buenos Aires, Buenos Aires, p. 67.

Polk, M.H., Young, K.R., Crews–Meyer, K.A., 2005. Biodiversity conservation implications of landscape change in an urbanizing desert of southwestern Peru. Urban Ecosystems 8, pp. 313–334.

Ponçano, W., Christofoletti, A., 1987. Procedimento estimativo para taxas de erosão pluvial em regiões tropicais. 4^0 Simp. Nac. de Controle de Erosão, Martilia, Anais.

Pousa, J.L., Tosi, L., Kruse. E., Guaraglia, D., Bonardi, M., Mazzoldi, A., Rizzetto, F., Proctor R., 1920. Narraciones del viaje para la Cordillera de los Andes y Residencia en Lima y otros partes del Perú en los años 1823 y 1824. La cultura argentina.Vaccaro. Buenos Aires, Argentina, p. 540.

Pousa, J., Tosil, L, Kruse, E., Guaraglia, D., Bonardi, M., Mazzoldi, A., Rizzetto, F. Schnack, E., 2006. Coastal processes and environmental hazards: The Buenos Aires (Argentina) and Venetian (Italy) littorals. Environmental Geology, 51 (8), 1307-1316.

Prefeitura Municipal de Rio Branco, 1986. Primeiro Plano Diretor de Desenvolvimento Urbano de Rio Branco.

Prentice, C., Mann, P., Taylor, F., Burr, G., Valastro, S., 1993. Paleoseismicity of the North American–Caribbean plate boundary (Septentrional fault, Dominican Republic). Geology 21, pp. 49–52.

Quarantelli, E.L., 1996. Basic themes derived from survey findings on human behavior in the Mexico City earthquake. International Sociology 11 (4), pp. 481–499.

Queiroz Neto, J.P., 1978. Les problèmes de l'érosion accelerée dans l'état de São Paulo, Brasil. Geo–Eco–Trop. 2, pp. 205–220.

Queiroz Neto, J.P., 2003. Agricultura brasileira, a pesquisa de solos e a sustentabilidade. In Ribeiro, W.C. (eds.), O Patrimônio Ambiental do Brasil: avaliação e perspectiva, São Paulo, EDUSP/Impr. Of.do Estado São Paulo, pp. 49–74.

Rabassa, J., 2007a. El impacto del cambio climático en los glaciares patagónicos y fueguinos. Ciencia Hoy, Buenos Aires, 17 (97), pp. 50–57.

Rabassa, J., 2007b. Global climate change and its impact on the glaciers and permafrost of Patagonia, Tierra del Fuego and the Antarctic Peninsula. In Leite, P., et al., A contribution to understanding the regional changes of global climate in South America, II Regional Conference on Global Change in South America, November 2005, Institute of Advanced Studies, University of São Paulo, São Paulo, Brazil, pp. 21–40.

Rabassa, J., Rubulis, S., Suarez, J., 1978. Los glaciares del Monte Tronador, Parque Nacional Nahuel Huapi, Río Negro, Argentina. Anales de Parques Nacionales 14, pp. 261–316

Rabassa, J., Rubulis, S., Suarez, J., 1981. Moraine in-transit as parent material for soil development and the growth of Valdivian Rain Forest on moving ice: Casa Pangue Glacier, Mount Tronador (lat. 41° 10' S), Chile. Annals of Glaciology 2, pp. 97–102.

Rabassa, J., Brandani, A., Boninsegna, J., Cobos, D., 1984. Cronología de la "Pequeña Edad del Hielo" en los glaciares Río Manso y Castaño Overo, Cerro Tronador, Provincia de Río Negro. IX Congreso Geológico Argentino, San Carlos de Bariloche, Actas III, pp. 624–639.

Rabassa, J., De Lasa, L., Siciliano, F., 1993. Glacier and Permanent Snow Field Inventory in Sierra Nevada de Santa Marta, Colombia. Quaternary of South America & Antarctic Peninsula, A.A. Balkema Publishers, Rotterdam, 8 (1990): pp. 119–134.

Ramirez, E., and 8 others, 2001. Small glaciers disappearing in the tropical Andes: A case study of Bolivia: Glacier Chacaltya (16°S). Journal of Glaciology, 47, pp. 187–194.

Ramos, V. 1999. Las provincias geológicas del territorio argentino. Geología Argentina, Instituto de Geología y Recursos Minerales. Buenos Aires, Anales 29 (3), pp. 41–96.

Ramos, V.,1999. Plate tectonic setting of the Andean Cordillera. Episodes 22, pp. 183–190.

Ramos, V., Alonso, R., Strecker, M., 2003. Neotectónica cuaternaria en Lomas de Olmedo, Salta, Argentina. X Congreso Geológico Chileno, Actas CD Rom.

Ramos, V., 2008. The basement of the Central Andes: The Arequipa and Related Terranes. Annual Review of Earth and Planetary Sciences 36, pp. 289–324.

Ramírez, E., Francou, B., Ribstein, P., Descloitres, M., Guerin, R., Mendoza, J., Gallaire, R., Pouyaud, B., Jordan, E., 2001. Small glaciers disappearing in the tropical Andes: A case study in Bolivia: Glaciar Chacaltaya (16°S). Journal of Glaciology 47, pp. 187–194.

Rao, G.V., Cavalcanti, I., Hada, K., 1996. Annual variations of rainfall over Brazil and water vapor characteristics of South America. J. Geophys. Res. 101, pp. 36350–26551.

Rein, B., 2007. How do the 1982/83 and 1997/98 El Niños rank in a geological record from Peru? Quatern. Intern. 161, pp. 56–66.

Rein, B., Lückge, A., Reinhardt, L., Sirocko, F., Wolf, A., Dullo, W.C., 2005. El Niño variability off Peru during the last 20,000 years. Paleoceanogr. 20: doi:10.1029/2004/PA001099.

Rein, B., Lückge, A., Sirocko, F., 2004. A major Holocene ENSO anomaly during the medieval period. Geophys. Res. Letters 31, doi: 10.1029/2004GL020161.

Restrepo, I.C., Ojeda, G.Y.,Correa, I.D., 2007. Geomorfología de la plataforma somera del Departamento de Córdoba, Costa Caribe Colombiana. Boletín de Ciencias de la Tierra 20, pp. 39–52.

Rhyner, K., 2006. Cries in the dark: Reconstruction after Hurricane Mitch in Honduras. Open House International 31 (1), pp. 31–38.

Rial, P., Vazquez, M., González, L., 1999. Evaluación de la capacidad productiva de la meseta central de Santa Cruz, Argentina, mediante sistemas de información geográfica. VII Conferencia iberoamericana sobre S.I.G. Mérida, Venezuela.

Riller, U., Petrinovic, I., Ramelow, J., Greskowiack, J., Strecker, M., Oncken, O., 2001. Late Cenozoic tectonism, caldera and plateau formation in the central Andes. Earth Planetary Science Letter 188, pp. 299–311.

Ribeiro, A.G., Arantes, A.O., Macedo, D., Franco, J.B.S., 1997. O papel dos recursos hídricos na sustentabilidade do sistema agroalimentar no Domínio dos Cerrados do Brasil Central. Estudo de caso: a bacia do médio/baixo Paranaíba no estado de Minas Gerais. In Shiki, S., Graziano da Silva, J., and Ortega, A.C. (eds), Agricultura, meio ambiente e sustentabilidade do cerrado brasileiro, Uberlândia, Dep. Econ./Geogr., UFU/Inst.Econ. UNICAMP/ EMBRAPA, pp. 267–310.

Richey, J.E., Nobre, C., Deser, C., 1989. Amazon river discharge and climate variability: 1903 to 1985. Science 246, pp. 101–103.

Rivera, A., Cassassa, G., 1999. Volume changes on Pío XI Glacier, Patagonia: 1975–1995. Global and Planetary Change 22, pp. 233–244.

Rivera, A., Cassassa, G., 2004. Ice elevation, areal and frontal changes of glaciers from National Park Torres del Paine, Southern Patagonia Icefield. Arctic, Antarctic and Alpine Research 36 (4), pp. 379–389.

Rivera, A., Bown, F., Acuña, C., Casassa, G., 2006. Historical glacier variations in Chile. In PAGES-Past Global Changes, 2006. Reconstructing past regional climate variations in South America over the late Holocene: an international symposium. Abstracts volume, October 4–7, 2006, Malargüe, Mendoza, Argentina, pp. 31–32.

RJEHM, 1987. Revista de Junta de Estudios históricos de Mendoza. Segunda etapa, No. 11, Tomo I. Mendoza.

Robertson, A.W., Mechoso, C.R., García, N.O., 2001. Interannual prediction of the Paraná River. Geophysical Research Letters, 22, pp. 4235–4238.

Robinson, D.P., Das, S., Waters, A.B., 2006. Earthquake rupture stalled by a subducting fracture zone. Science 312, pp. 1203–1205.

Rodríguez, E.E., 1999. Sismo del eje cafetero: Comportamientos del suelo en el sismo. Noticreto 51, pp. 22–24.

Rohl, E., 1949. Los diluvios en las montañas de la cordillera de la Costa. Boletín de la Academia de Ciencias Físicas, Matematicas y Naturales 22 (38), Caracas, pp. 34–59.

Romero, C.C., Baigorria, G.A., Stoosnijder, L., 2007. Changes of erosive rainfall for El Niño and La Niña years in the northern Andean highlands of Peru. Climate Change 85, pp. 343–356.

Ronchail, J., 1989. Advecciones polares en Bolivia: caracterización de los efectos climáticos. Bulletin de l'Institut Francais d'Etudes Andines 18, pp. 65–73.

Rosales, C., Meyer, H., 2005. La catástrofe de 1999 en el Eje Cafetero. In Hermelin, M. (ed.), Desastres de origen natural en Colombia 1979–2004. Universidad EAFIT, OSSO–Univalle, Medellín, Colombia, pp. 169–185.

Rosengaus-Moshinsky, M., 2006. Objective evaluation of cyclonic activity, Ingeniería Hidráulica en México 21 (2), pp. 27–41.

Rostagno, C., Del Valle, H., 1988. Mounds asociated with shurbs in aridic soils of Northeastern Patagonia: Characteristics and probable genesis. Caten 15, pp. 347–359.

Rostagno, C., Coronato, F., Del Valle, H., Puebla, D., 1999. Runoff and erosion in five land units of closed basin of northeastern Patagonia. Arid Soil Research and Rehabilitataion 13, pp. 281–292.

Ruddiman, W.F., 2005. How did humans first alter global climate? Scientific American, March 2005, p. 8.

Ruegg, J.C., Campos, J., Madariaga, R., Kausel, E., De Chabalier, J.B., Armijo, R., Dimitrov, D., Barrientos, S., 2002. Interseismic strain accumulation in south central Chile from GPS measurements, 1996–1999. Journal of Geophysical Research Letters 29 (11).

Ruiz, S., Saragoni, G.R., 2005a. fórmulas de atenuación para la subducción de Chile considerando los dos mecanismos principales de sismogénesis y los efectos del suelo. Congreso Chileno de Sismología e Ingeniería Antisísmica. IX jornadas. Concepción, Chile.

Ruiz, S., Saragoni, G.R., 2005b. Estudio comparativo de los terremotos de subducción chilenos con los terremotos de subduccion del norte, centro y sur de América. Congreso Chileno de Sismología e Ingeniería Antisísmica. IX jornadas. Concepción, Chile.

Ruiz, S., Saragoni, G.R., 2005c. fórmulas de atenuación para la subducción de Chile de terremotos interplaca tipo thrust considerando los efectos del suelo y las asperezas. Congreso Chileno de Sismología e Ingeniería Antisísmica. IX jornadas. Concepción, Chile.

SA and DS (Secretaría de Ambiente y Desarrollo Sustentable, Dirección de Conservación del Suelo). 2000.

Programa de Acción Nacional de Lucha Contra la Desertificación. Sitio Web de la Secretaría de Ambiente y Desarrollo Sustentable de Argentina: http://www.medioambiente.gov.ar/areas/dcs.

Sagripanti, G., Costa, C., Giaccardi, A., Aguilera, D., Schiavo, H., Campanella, O., Dolso, M., 1999. Nuevos datos sobre la actividad cuaternaria de la falla Las Lagunas. XIV Congreso Geológico Argentino Actas, vol. 1, pp. 42.

Saint-Amand, P., 1961. Los terremotos de Mayo, Chile 1960. Technical article 14, NOTS TP 2701, Michelson Laboratories, China Lake, California, p. 39.

Sala, J., Gonzalez, N., Kruse, E., 1983. Generalización Hidrológica de la Provincia de Buenbos Aires. Hidrologia de Grandes Llanuras. Actas del Coloquio de Olavaria. UNESCO–CONAPHI 2, pp. 973–1009.

Salas, J., 2008. Risk assessment of socio-natural disasters in Central America and the Caribbean. Open House International, 33 (1), pp. 8–21.

Salati, E., Marques, J., Molion, I.C., 1978. Origem e distribuição das chuvas na Amazônia. Interciências 3, pp. 200–205.

Sales D., Origlia D., Yenes Ortega M., Gardini C., 2002. Deslizamiento rotacional en el cerro Inti Huasi. Departamento Pringles, Provincia de San Luis. ASAGAI 18, pp. 106–112.

Salfity, J., 1985. Lineamientos transversales al rumbo andino en el noroeste argentino. IV Congreso Geológico Chileno, vol. 2, pp. 119–137.

Salinas L., Varela S., Navas M., Sánchez E., 1995. Protección geológico-ambiental-cartas de erosión y remoción de áreas cordilleranas próximas a rutas. Departamento Iglesia, Provincia de San Juan, República Argentina. ASAGAI Special Issue (3), pp. 341–357.

Salomao, F.X.T., 1999. Controle e Prevenção dos Processos Erosivos. In Erosão e Conservação dos Solos: Conceitos, temas e aplicações. Bertrand Brasil, Rio de Janeiro, pp. 340.

Salomao, F.X.T., 1994. Processos erosivos lineares em Bauru, SP: regionalização cartográfica aplicada ao controle preventivo urbano e rural. USP, FFLCIH, Dep. Geogr., São Paulo, pp. 200, 2 mapas (tese Dr.).

Salomao, F.X.T., Queiroz Neto, J.P., Alarsa, C., 1993. Análise estrutural da cobertura pedológica aplicada à elaboração de cartas geotécnicas e de capacidade de uso das terras. V Simp. Geogr. Fis. Aplica., S. Paulo, Anais pp. 451–459.

Salomón, J.N., 1969. El alto valle del río Mendoza. Estudio de geomorfología. Boletín de Estudios Geográficos, Mendoza 16 (62), pp. 1–50.

Salomone, J., Schenkel, R., 2002. Impacto de las lenguas de erosión patagónicas en los componentes del suelo, la vegetación y la productividad. XVIII Congreso Argentino de la Ciencia del Suelo, Puerto Madryn, Chubut, Argentina.

Sánchez M.C., Lucena L.L., Rivelli F.R., Salfity J.A, 2005. Geología y geomorfología de la cuenca del río Pascha–Incamayo, Qda del Toro, Salta. Actas XVI Congreso Geológico Argentino, Cafayate.

Sandweiss, D.H., Maasch, K.A., Burger, R.L., Richardson, J.B., III, Rollins, H.B., Clement, A., 2001. Variation in Holocene El Niño frequencies: climate records and cultural consequences in ancient Peru. Geology 29 (7), pp. 603–606.

Santana, E., Dumont, J.F., King, A., 2001. Los efectos del fenómeno El Niño en la ocurrencia de una lata tasa de erosión costera en el sector de Punta Gorda, Esmeraldas. Acta Oceanográfica del Pacífico, INOCAR, Ecuador, 11, No. 1, pp. 1–8.

Santana, M.A., Queiroz Neto, J.P., 1995. Fatores responsáveis pela sensibilidade a erosão linear em Marilia. Rev. Dep. Geografia 9, pp. 21–33.

Santana, M.A., 1991. Avaliação dos fatores responsáveis pela fragilidade dos solos a erosão na região de Marilia (SP). USP, FFLCIH, Dep. Geografia, São Paulo, p. 104. (dissert. Mestrado).

Santos, A.R., 2004. A grande barreira da Serra do Mar: da trilha dos Tupiniquins à rodovia dos Imigrantes. O Nome da Rosa—CTE, São Paulo, p. 122.

Sarmiento, F.O., 1987. Antología Ecológica del Ecuador: desde la selva hasta el mar. Casa de la Cultura Ecuatoriana. Quito, pp. 382.

Sarmiento, F.O., 1994. Human impacts on the cloud forests of the Upper Guayllabamba River, Ecuador. Ecological Studies 110, pp. 284–295.

Sarmiento, F.O., 2002. Anthropogenic landscape change in highland Ecuador. The Geographical Review 92(2), pp. 213–234.

Sarmiento, F.O. (ed.). 2003. Las Montañas del Mundo: Una prioridad global con perspectivas latinoamericanas. Editorial Abya-Yala, Quito.

Sarmiento, F.O., 2005. Ethnoecology of the tropical Andes: Avian indicators of landscape change in highland Ecuador. Pirineos 160, pp. 87–119.

Sassa, K., 1985. The mechanism of debris flow, Proc. of the 11th ICSMFE, San Francisco, vol. 3, pp. 1173–1176.

Schilling, G.H. 1993. Instrumentação e análise dos movimentos da encosta do Soberbo—Alto da Boa Vista, R.J. Tese de Mestrado, COPPE/UFRJ, Rio de Janeiro, p. 237.

Schilling, G.H., Lacerda, W.A., Barros, W.T., 1992. Instrumentação da encosta do Soberbo, 1ª Conferência Brasileira sobre Estabilidade de Encostas–COBRAE, Rio de Janeiro, vol. 1, pp. 307–325.

Schimmelmann A., Lange C.B., Meggers B.J., 2003. Palaeoclimatic and archaeological evidence for a similar to 200-yr recurrence of floods and droughts linking California. Mesoamerica and South America over the past 2000 years, Holocene 13, (5), pp. 763–778.

Schneider. H., Schwab, M., Schlunegger, F., 2007. Channelized and hillslope sediment transport and the geomorphology of mountain belts. Int. J. Earth Sci. doi: 10.1007/s00531–006–0164–9.

Schaffers, A., 2002. Paleotsunamis in the Caribbean. Field evidences and dating from Aruba, Curacao and Bonaire. Essener Geographische Arbeite,no. 33, pp. 185.

Schnack, E.J., 1985. Argentina. In Bird, B., Schwartz, M. (eds.), The Worlds Coastlines. van Nostrand–Reinhold Co., pp. 69–78.

Schnack, E.J., 1993. The vulnerability of the east coast of South America to sea-level rise and possible adjustment strategies. In Warrick, R.A., Barrow, E.M., Wigley, M.L., (Eds.), Climate and Sea Level Changes: Observations, Projections and Implications, Cambridge Univ. Press, pp. 336–348.

Schnack E.J, Pousa J.L, Isla F.I., 1998. Erosive processes on the sandy coastline of Argentina. Vechtaer Studien zur Angewandten Geographie und Regionalwiseenschaft, Band 20, S. 133–136.

Schnack E.J., Pousa J.L., 2004. Episodic processes (Storm surges and tsunamis). In Isla, F. (ed.), Coastal zones and Estuaries, from Encyclopedia on Life Support Systems EOLSS–UNESCO, Eolss Publishers Co.Ltd., Oxford, UK, www.eolss.net.

Schubert, C., 1979. El Pilar fault zone, northeastern Venezuela: Brief review. Tectonophysics 52, pp. 447–455.

Schubert, C., 1980a. Late Cenozoic pull-apart basins, Boconó fault zone, Venezuelan Andes. Journal of Structural Geology 2, pp. 463–468.

Schubert, C., 1980b. Morfología neotectónica de una falla rumbo-deslizante e informe preliminar sobre la falla de Boconó, Andes merideños. Acta Científica Venezolana 31, pp. 98–111.

Schubert, C., 1982. Neotectonics of the Boconó fault, western Venezuela. Tectonophysics 85, pp. 205–220.

Schubert, C., 1984. Basin formation along Boconó–Morón–El Pilar fault system, Venezuela. Journal of Geophysical Research, 89, pp. 5711–5718.

Schubert, C., 1994. Tsunamis in Venezuela: Some observations on their occurrence. In Finkl, C., Jr. (ed.), Coastal Hazards: Perception, Susceptibility and Mitigation. Journal of Coastal Research, Special Issue 12, pp. 189–195.

Schuster R.L., Highland L.M., 2001.U.S. Geological Survey Open-File Report 01–0276.

Schuster, R.L., Nieto, A., O'Rourke, T.D., Crespo, E., Plaza, G., 1996. Mass wasting triggered by the March 1987 Ecuador earthquakes. Engineering Geology 42 (1), pp. 1–23.

Schuster, R.L., Salcedo, D.A., Valenzuela, L., 2002. Overview of catastrophic landslides of South America in the twentieth century. In Evans, S.G., DeGraff, J.V. (eds.), Catastrophic Landslides: Effects, Occurrence, and Mechanisms. Geological Society of America reviews in engineering geology XV, Boulder, CO, USA. pp. 1–34.

Schwartz, D., 1988. Paleoseismicity and neotectonics of the Cordillera Blanca fault zone, Northern Peruvian Andes. Journal of Geophysical Research 9, pp. 4712–4730.

Schwartz, D., Stenner, H., Costa, C., Smalley, R., Ellis, M., Velasco, S., 2002. Rupturas asociadas a los sismos Ms 7.8 de 1949 en Tierra del Fuego: Investigaciones Paleosismológicas iniciales. XV Congreso Geológico Argentino, Calafate, Actas 1, pp. 136–138.

Schurr, B., Asch, G., Rietbrock, A., Kind, R., Pardo, M., Heit, B., 1999. Seismicity and average velocities beneath the Argentine Puna plateau. Geophysical Research Letter, 26, pp. 3025–3028.

SDS and PA Secretaría de Recursos Naturales y Desarrollo Sustentable, 1998. Taller de Identificación y Evaluación de Indicadores de Desertificación. Resultados alcanzados. http://www.medioambiente.gov.ar/ areas/direcs/default.htm

Seggiaro, R., 1994. Petrología, geoquímica y mecanismos de erupción del Complejo volcánico Coranzulí. Tesis Doctoral, Universidad Nacional de Salta, pp. 150.

SEMETFAV, 1999. Datos de precipitación de la Estación Meteorológica del Aeropuerto de Maiquetía. Servicio Meteorológico de la Fuerza aérea de Venezuela.

Serbín, A. 2007. Paz, Conflicto y Sociedad Civil en América Latina y el Caribe. Coordinadora Regional de Investigaciones Económicas y Sociales. Centro Internacional de Investigaciones para el Desarrollo (IRDC): Ottawa. 403pp.

Serra-Filho, R., Cavalli, A.C., Guillaumon, J.R.; 1975. Levantamento da cobertura vegetal natural e do reflorestamento no Estado de São Paulo. Boletin Técnico do Instituto Florestal 11, São Paulo, pp. 1–53 (2nd ed.)

Serrano, J.S., 2007. Vulnerability, poverty and socio–natural disasters in Central America and the Caribbean. Informes de la Construcción 59 (508), pp. 29–41.

Sharma, S., Candia-Gallegos, M., 1992. Seismic hazard analysis of Peru. Engineer. Geol. 32, pp. 73–79.

Shedlock, K.M., 1999. Seismic hazard map of North and Central America and the Caribbean. Ann. Geofis. 42, pp. 977–997.

Shedlock, K.M., Tanner, J.G., 1999. Seismic hazard map of the western hemisphere. Ann. Geofis. 42, pp. 1199–1214.

Shedlock, K.M., Giardini, D., Grunthal, G., Zhang, P., 2001. The GSHAP global seismic hazard map. Seismol. Res. Lett. 71, pp. 679– 686.

Shepherd, J.B., 1993. Seismic hazard in the eastern Caribbean. In McGuire, R.K. (ed.), The Practice of Earthquake Hazard Assessment. IASPEI, Denver, pp. 51–55.

Shepherd, A. Wingham, D., Payne, T., Skvarca, P. 2003. Larsen Ice Shelf has progressively thinned. Science, 302, pp. 856–859.

Sheridan, M.F., Bonnard, C., Carreno, R., Siebe, C., et al, 1999, 30 October 1998 Rockfall/ Avalanche and Breakout Flow of Casita Volcano, Nicaragua. Landslide News, June, pp. 2–4.

Siame, L., Bourles, D., Sebrier, M., Bellver, O., Castano, J., Araujo, M., Perez, M., Raisbeck, G., Yiou, F., 1997. Cosmogenic dating from 20 to 700 ka of a series of alluvial fan surfaces affected by El Tigre fault, Argentina. Geology 25, pp. 975–978.

Siame, L., Sébrier, M., Bellier, O., Bourlés, D., 2006. Can cosmic ray exposure dating reveal the normal faulting activity of the Cordillera Blanca fault, Peru? Revista de la Asociación Geológica Argentina 61, pp. 536–544.

Sieh, K., 1996. The repetition of large-earthquake ruptures. Proceedings National Academy of Sciences USA, 93, Washington, DC, pp. 3764–3771.

Silgado E., 1978. Historia de los sismos más notables en el Perú (1513–1974). Instituto Geológico Minero y Metalúrgico. Lima. Boletín 3.

Sillitoe, R.H., 2003. Iron oxide–copper–gold deposits: An Andean view. Mineralium Deposita 38, pp. 787–812.

Silva Dias, M.A.F., Nieto-Ferreira, R., 1992. Application of a linear spectral model to the study of Amazonian squall lines during GTE-ABLE 2B. J. Geophys. Res. 97, 20,405–20,219.

Silver, P.G., Beck, S.L., Wallace, T.C., Meade, C., Myers, S.C., James, D.E., Kuehnel, R., 1995. Rupture characteristics of the deep Bolivian earthquake of 9 June 1994 and the mechanism of deep-focus earthquakes. Science 268, pp. 69–73.

Silveira, G.C., Lacerda, W.A., 1992. Caracterização mineralógica dos solos residuais e coluvionares do escorregamento da encosta do Soberbo, 1ª Conferência Brasileira sobre Estabilidae de Encostas–COBRAE, Rio de Janeiro, vol. 1, pp. 479–490.

Silveira, C.S., Barbosa, M.C., Coelho Netto, A.L., 2004. A tracer experiment as an alternative methodology to understand infiltration pathway in a tropical rainforest soil. Solos e Rochas 28 (3), São Paulo, pp. 261–270.

Singer, A., 1977. Síntesis regional de la evolución Geomorfológica de la cordillera de la Costa en el área de Caracas. V Congreso Geológico Venezolano. Guía Excursión 1.

Singer, A., Rojas, C., Lugo, M., 1983. Inventario Nacional de Riesgo Geológico. Estado Preliminar. FUNVISIS, Caracas, p. 128.

Skewes, M.A., Stern, C.R., 1979. Petrology and geochemistry of alkali basalts and ultramafic inclusions from the Palei-Aike volcanic field in southern Chile and the origin of the Patagonian plateau lavas. J Volc Geotherm Research 6, pp. 3–25.

Skvarca, P., 2006. Anomalous century fluctuations of Glaciar Perito Moreno, southern Patagonia. In PAGES, 2006. Reconstructing past regional climate variations in South America over the late Holocene: an international symposium. Abstracts volume, October 4–7, 2006, Malargüe, Mendoza, Argentina, p. 32.

Skvarca, P., Raup, B., De Angelis, H., 2003. Recent behaviour of Glaciar Upsala, a fast-flowing calving glacier in Lago Argentino, southern Patagonia. Annals of Glaciology 36 (1), pp. 184–188.

Sladen, A., Konca, A.O., Perfettini, H., Audin, L., Tavera, H., Avouac, J., Simons, M., Helmberger, D.V., 2007. Source characteristics of the 2007 Pisco earthquake (Mw = 8.0), central Peru, derived from seismic waveforms inversion and tsunami modeling. Eos Trans., Amer. Geophys Union 88 (52), p. 1055.

Slemmons, D., 1977. Faults and earthquake magnitude. In State of the art of assessing earthquake hazards in the United Status. Report 6. United States Army Engineers. Waterway Station. Miscellaneous Paper S–73–1, p. 129.

Slemmons, D., 1981. A procedure for analyzing fault controlled lineaments and the activity of faults. In O'Leary, D., Earle, J. (eds.), Proc. III Conference on Basement Tectonics. Bas. Tect. Comm. Publ., vol 3, pp. 33–49.

Smith, K, 1996. Environmental Hazards: Assessing Risk and Reducing Disaster. Printed and bound in Great Britain by Biddles Ltd, Guildford and King's Lynn. A volume in the Routledge Physical Environment Series. Edited by Keith Richards. University of Cambridge. Routledge London and New York, p. 389.

Sneyers, R., 1975. Sur l'analyse statistique des series d'observations. Technical Note No 143, WOM, Geneva, Switzerland.

Solberg, S., Hale, D., Benavides, J., 2003. Natural disaster management and the road network in Ecuador: Policy issues and recommendations. Inter–American Development Bank, Sustainable Development Department, Washington, DC.

Solis, N., Rivero, A., 1994. Remoción en masa en la localidad de Tilcara, prov. de Jujuy, Rep. Arg. Proccedings of 3 Simposio Latinoamericano de riesgos geológicos urbanos, Salta, pp. 262–277.

Solis, N., Orozco, O., 1996. Riesgo geológico en la localidad de Tilcara. Prov. de Jujuy, Republica Argentina. Memorias del Congreso Geológico de Bolivia, La Paz, pp. 927–934.

Solís, N., Chayle, W., Ramírez, A., 2004. ¿Que características tiene la remoción en masa en el noroeste argentino? In Gonzalez, M., and Bejerman, N.J. (eds.), Peligrosidad geológica en Argentina. Asociación Argentina de Geología Aplicada a la Ingeniería, Buenos Aires, publicación Especial No 4, pp. 366–377.

Soriano, A., 1956. Aspectos ecológicos y pasturiles de la vegetación patagónica relacionados con su estado y capacidad de recuperación. Revista de Investigaciones Agrícolas 10, pp. 349–386.

Soriano, A., Movia, C., 1986. Erosión y desertización en la Patagonia. Interciencia, 11, pp. 77–83.

Soulas, J.–P., 1985. Neotectónica del flanco occidental de los Andes de Venezuela entre 70°30′ y 71°00′W (Fallas de Boconó, Valera, Piñango y del Piedemonte). \tVI Congreso Geológico Venezolano, Caracas, vol. 4, pp. 2690–2711.

Soulas, J.–P., 1986. Neotectónica y tectónica activa en Venezuela y regiones vecinas. Proceedings VI Congreso Geológico Venezolano, Caracas, vol. 10, pp. 6639–6656.

Speck, N., Sourrouille, E., Wijnhoud, S., Munist, E., Monteith, N., Vofkheimer, W., Menéndez, J., 1982. Sistemas Fisiográficos de la Zona Ingeniero Jacobacci–Maquinchao (Provincia de Río Negro). INTA, Buenos Aires.

Sruoga P., 2004. Volcán Planchón-Peteroa (35°15′S/70°35¨O): Estratigrafía eruptiva y evaluación de su peligrosidad. In Gonzalez, M., Bejerman, N.J. (eds.), Peligrosidad geológica en Argentina. Asociación Argentina de Geología Aplicada a la Ingeniería, Publicación Especial No. 4, pp. 487–500.

Sruoga, P., Schonwandt, D., 2004. Volcanismo. Peligrosidad Geológica en Argentina. Metodologías de Análisis y mapeo. Estudio de casos. Revista de la Asociación Argentina de Geología Aplicada a la Ingeniería. Publicación Especial No. 4. Capítulo 8, pp. 469–475.

Sruoga, P., Guerstein, P., Bermúdez, A., 1993. Riesgo Volcánico. In Ramos, V. (ed.), Geología y Recursos Naturales de Mendoza. Parte III: Geología Ambiental y Riesgo Geológico, pp. 659–667.

Sruoga, P., Llambías, E., Fauqué, L., 2000. Geocronología t evolución geoquímica del Volcán Maipo (34° 10'S). IX Congreso Geológico Chileno, Actas II, pp. 72–76.

Stern, C.R., 2004. Active Andean volcanism: Its geologic and tectonic setting. Revista Geológica de Chile 31 (2), pp. 161–206.

Stern, C.R., Moreno, H., López–Escobar, L., Clavero, J., Lara, L.E., Naranjo, J.A., Parada, M.A., Skewes, A., 2007. Chilean volcanoes. In Moreno, T., and Gibbons, W. (eds.), The Geology of Chile. The Geological Society, London, pp. 146–178.

Sternberg, H.R., 1949. Enchentes e movimentos de massa do solo no vale do Paraíba em dezembro de 1948. Influencia da explotação destrutiva das terras. Rev. Bras. Geogr., Rio de Janeiro 11, p. 223.

Stipanicic, P., Methol, E., 1972. Macizo de Somún Curá. In Leanza, A. (ed.), Geología Reginal Argentina. Academia Nacional de Ciencias, Córdoba, pp. 581–599.

Stillweel, H.D., 1992. Natural hazards and disasters in Latin America. Natural Hazards 6, pp. 131–159.

Súnico, A., Bouza, P., and del Valle, H., 1996. Erosion of subsuface horizonts in northeastern Patagonia, Argentina. Arid Soil Research and Rehabilitation, 10, pp. 359–378.

Suárez, G., Molnar, P., Burchfield, C., 1983. Seismicity, fault plane solutions, depth of faulting and active tectonics of the Andes of Perú, Ecuador and Southern Colombia. Journal of Geophysical Research 89, pp. 10403–10428.

Suayter, L.E., 1997. Zonificación general del riesgo geológico y geotectónico en la prov. de Tucumán. Escala aproximada 1:100.000. ASAGAI, 10 10, pp. 109–117.

Taboada, A., Rivera, L.A., Fuenzalida, A., Cisternas, A., Philip, H., Bijwaard, H., Olaya, J., Rivera, C., 2000. Geodinamics in northern Andes: Subductions and intracontinental deformations (Colombia). Tectonics 19, pp. 787–813.

Takahashi, T., 2000. Initiation and flow of various types of debris-flow, pp. 15–28. In Wickzorek, G.F., Naeser, N.D. (eds.), Debris-flow hazards mitigation: Mechanics, prediction and assessment. Ed. Balkema, Teipei, Taiwan, pp. 608.

Takehata, H., 1998. The new tsunami warning system of the Japan Meteorological Agency. Science of Tsunami Hazard, 16(1), pp. 39–50.

Tang, B.H., Neelin, J.D., 2004. ENSO Influence on Atlantic hurricanes via tropospheric warming. Geophys. Res. Lett. 31, L24204.

Tanner, J.G., Shepherd, J.B., 1997. Seismic hazard in Latin America and the Caribbean. Final Report to the International Development Research Centre, Ottawa, Canada, Seismic Hazard in Latin America and the Caribbean. Instituto Panamericano de Geografia y Historia, Mexico, D.F, vol. 1, p. 142.

Tanner, J.G., Shedlock, K.M., 2004. Seismic hazard maps of Mexico, the Caribbean and Central and South America. Tectonophysics 390, pp. 159 –175.

Tobin, G.A., Montz, B.E., 1997. Natural Hazards: Explanation and Integration. Guilford Press, New York, p. 388.

Tarras-Wahlberg, N.H., Lane, S.N., 2003. Suspended sediment yield and metal contamination in a river catchment affected by El Niño events and gold mining activities: The Puyango River Basin, southern Ecuador. Hydrol. Processes 17, pp. 3101–3123.

Tavera, H., Buforn, E., 2001. Source mechanism of earthquakes in Peru. J. Seismology, 5, pp. 519–539.

Tavera, H., Fernández, E., Bernal, I., Antayhua, Y., Agüero, C., Salas, H., Rodríguez, S., Vilcapoma,L., Zamudio, Y., Portugal, D., Inza, A., Carpio, J., Ccallo, F., Valdivia, I., 2006. The southern region of Peru earthquake of June 23, 2001. J. Seismology 10 (2), pp. 171–195.

Tavera, H., Perfettini, H., Audin, L., Bernal, I., Farber, D., Bondoux, F., Sladen, A., Avouac, J., 2007. Overview of the August 15th, 2007 Pisco Earthquake (Mw = 8.0), Central Peru. Eos Trans., Amer. Geophys. Union 88 (52), pp. 1054.

Tejedo, A., Pereyra, F., Anielli, C., Jones, M., 1999. Carta de peligrosidad geológica 4566–III Comodoro Rivadavia. Boletín No. 305, SEGEMAR, Buenos Aires.

Tello, G., Perucca, L., 1993. El sistema de fallamiento Precordillera Oriental y su relación con los sismos históricos de 1944 y 1952. XII Congreso Geológico Argentino, San Juan, Argentina, Actas, vol. 3, pp. 258–265.

Thomas, R., and 17 authors, 2004. Accelerated sea-level rise from West Antarctica. Science, 306, pp. 255–258.,

Thompson, L.G., 2000. Ice core evidence for climate change in the Tropics: Implications for our future. Quaternary Science Reviews 19, pp. 19–35.

Thornburg, T.M., Kulm, L.D., 1987. Sedimentation in the Chile Trench: Depositasional morphologies, lithofacies and stratigraphy. Geological Society of American Bulletin, 98, pp. 33–52.

Thouret, J.C., Dávila, J., Eissen, J.P., 1999. Largest explosive eruption in historical times in the Andes at Huaynaputina volcano, AD 1600, southern Peru. Geology 27 (5), pp. 435–438.

Thouret, J.C., Finizola, A., Fornari, M., Legeley-Padovani, A., Suni, J., Frechen, M., 2001. Geology of El Misti volcano near the city of Arequipa, Peru. Geol. Soc. Amer. Bull. 113 (12), pp. 1593–1610.

Thouret, J.C., Juvigné, E., Gourgaud, A., Boivin, P., Dávila, J., 2002. Reconstruction of the AD 1600 Huyanaputina eruption based on the correlation of geologic evidence with early Spanish chronicles. J. Volcan. & Geotherm. Res. 115, pp. 529–570.

Thouret, J.C., Romero, M., Wörner, G., Gerbe, M.C., Finizola, A., Fornari, M., Gonzales, K., 2005. Ubinas: The evolution of the historically most active volcano in southern Peru. Bull. Volcan. 67, pp. 577–589.

Thouret, J.C., 1999. Urban hazards and risks: Consequences of earthquakes and volcanic eruptions—An introduction. GeoJournal 49 (2), pp. 131–135.

Thouret, J.C., Gourgaud, A., Calvache, M.L., 1989. The eruption of the Nevado del Ruiz on the 13th of November 1985. In Van der Hammen, T., Diaz–Piedrahita S., Alvarez, V.J. (eds.), La

Cordillera Central Colombiana, transecto Parque Los Nevados. J. Cramer, Berlin, Germany, vol. 3, pp. 217–255.

Thrauth, M.H., Alonso, R.A., Haselton, K.R., Hermanns, R.L., Strecker, M.R., 2000. Climate change and mass movements in the NW Argentine Andes. Earth Planet. Sc. Letters 179, pp. 243–256.

Titov, V., González, F., 2002. Modeling solutions for short-term inundation forecasting for tsunamis. In Underwater Ground Failures on Tsunami Generation, Modeling, Risk and Mitigation, NATO Advanced Workshop, May 23–26, 2001, Istanbul, Turkey, pp. 227–230.

Titov, V., Rabinovich, A.B., Mofjeld, H.O., Thomson, R.E., González, F.I., 2005. The global reach of the 26 December 2004 Sumatra tsunami. Science 309, pp. 2045–2048.

Tomazelli, L.J., Vilwock, J.A., Dillenburg, S.A., Bachi, F.A., Dehnhardt, B.A., 1997. A erosao costeira e a transgressao marinha atual na costa do Rio Grande do Sul. VI Congresso da ABEQUA, Curitiba, pp. 415–419.

Torres, M.A., 1985. Estratigrafía de la ladera occidental del Cerro Amarillo y la Qda La Yesera, Dpto de Calafate, Salta. Revista de la Asociación Geológica Argentina 40, Buenos Aires, pp. 141–157.

Trauth M.H., Strecker M.R., 1999, Formation of landslide-dammed lakes during a wet period between 40,000–25,000 yr B.P. in northwestern Argentina. Palaeogeography Palaeoclimatology Palaeoecology 153 (1–4), pp. 277–287.

Trauth M.H., Alonso R.A., Haselton K.R., Hermanns R.L., Strecker M.R., 2000, Climate change and mass movements in the northwest Argentine Andes. Earth and Planetary Sciences Letters 179 (2), pp. 243–256.

Tricart, J., 1965. Príncipes et méthodes de la géomorhologie. Ed. Masson, Paris, p. 496.

Tricart, J., 1966. As descontinuidades nos fenômenos de erosão. Not. Geomorf. 6 (12), pp. 3–4.

Troll, C. (ed.) 1968. Geo–Ecology of the Mountainous Regions of the Tropical Americas. Ferd. Dummlers Verlag: Bonn.

Tutiven Ubilla, I., 1998. Variaciones morfológicas y batimétricas de la línea de costa en el estuario del río Chone, producidas por los eventos ENSO. Bull. Inst. Fr. Études Andines 27 (3), pp. 557–563.

UNCOD. 1977. Conferencia de las Naciones Unidad sobre Desertificación. Nairobi, Kenia.

UNCED. 1992. Earth Summit '92. The UN Conference on Environment and Development. UNCED, Rio de Janeiro, Brasil.

UNDP. 2005. Crisis Prevention and Recovery. UNDP Publications on Disasters. p. 161.

UNDP. 2006. Reducing Disaster Risk: A Challenge for Development. UNDP Publications on Disasters. p. 213.

UN/ISDR, 2004. Living with risk, A global review of disaster reduction initiatives, Inter–Agency Secretariat of the International Strategy for Disaster Reduction, http://www.unisdr.org/eng/about_isdr/bd–lwr-2004-eng.htm.

UN/ISDR, 2008. Climate Change and Disaster Risk Reduction, Briefing Note 01, http://www.eird.org/publicaciones/Climate-Change-DRR.pdf

UN-ISA Universidad Nacional, Interconexión Eléctrica SA., 1984. Estudio geomorfológico de las cuencas de los ríos Nare y San Carlos. Technical Report. Medellín, Colombia.

United Nations Development Program (UNDP), 2007. 2007/2008 Human Development Report.

United Nations, 2008. UN Office for the Coordination of Humanitarian Affairs. Bolivia: Flash Appeal 2008, pp. 38.

Urbani, F., 2002. Geología del Macizo del Ávila y adyaciencias, estado Vargas, Miranda, Aragua y Distrito Capital. III Coloquio sobre Microzonificación Sismica y III Jornadas Venezolanas de Sismología Histórica. Funvisis, Caracas

USGS (United States Geological Survey), 2004. Shuttle Radar Topography Mission Data. Available at the Global Land Cover Facility, http://www.landcover.org. Accessed in July 2006.

USGS/NEIC, 2002. National Earthquake Information Center, World Data Center A for Seismology. Global Earthquake Search. United States Geological Survey, National Earthquake Information Center, http://wwwneic.cr.usgs.gov/neis/epic/epic_global.html.

Uyeda, S., Kanamori, H., 1979. Back-arc opening and mode of subduction. Journal Geophysical Research 84, pp. 1049–1061.

Vanoni, V.A., 1977. Sedimentation engineering. New York, ASAE Manual and Reports of Engineering Practices, 54, p. 743.

Van Wyk de Vries, B., Kerle, N., Petley, D., 2000. Sector collapse forming at Casita volcano, Nicaragua. Geology 28 (2), pp. 167.170.

Vargas, G., Ortlieb, L., Rutlland, J., 2000. Aluviones históricos en Antofagasta y su relación con eventos El Niño/Oscilación del Sur. Revista Geológica de Chile 27 (2), pp. 157–176.

Varley, A., 1991. Disasters: Vulnerability and response. Disasters 15 (3), pp. 285–287.

Veblen, T.T., Young, K.R., Orme, A.R. (eds.), 2007. The Physical Geography of South America. Oxford University Press, New York, p. 361.

Velásquez A., and Rosales, C., 1999. Escudriñando en los Desastres a todas las escala. Concepción, metodología y análisis de desastres en América Latina, Desinventar. Cali, Colombia.

Velásquez, A., 2005. Los terremotos del Atrato medio-Murindó en Octubre de 1992. In Hermelin, M. (ed.), Desastres de origen natural en Colombia 1979–2004. Universidad EAFIT, OSSO-Univalle, Medellín, Colombia, pp. 91–108.

Velásquez, E., Hermelin, M., 2005. El evento de San Carlos (Antioquia) de 1990. In Hermelin, M. (ed.), Desastres de origen natural en Colombia 1979–2004. Universidad EAFIT, OSSO-Univalle, Medellín, Colombia, pp. 77–90.

Velíz, R., 2001. Amenazas y Riesgos a las Inundaciones en el Valle del Chama. Municipio Santos Marguina, Estado Mérida. Universidad de los Andes. Escuela de Ingeniería Forestal. Mérida, Venezuela.

Venencio, M.V., García, N.O., 2005. Impacto de las sequías en la recarga natural al acuífero libre. Revista Brasileira de Recursos Hídricos 10, (1), pp. 49–60.

Verdaguer, J.A., 1929. Historia eclesiástica de Cuyo. Tomo I. Editorial Planeta

Vergara, W., Deeb, A.M., Valencia, A.M., Bradley, R.S., Francou, B., Zarzar, A., Grunwaldt, A., Haeussling, S.M., 2007. Economic impacts of rapad glacier retreat in the Andes. EOS, Transactions, American Geophysical Union, 88, pp. 261–268.

Vergés, J., Ramos, V., Meigs, A., Cristallini, E., Bettini, B.H., Cortés, J.M., 2007. Crustal wedging triggering recent deformation in the Andean thrust front between 31° S and 33° S: Sierras Pampeanas-Precordillera interaction. Journal of Geophysical Research 112, B03S15, doi: 10.1029/2006JB 004287.

Victor, M.A., 1974. A cobertura vegetal primitiva. S. Paulo, Soc. Brasil. Silvicult., p. 48.

Vieira, B.C, Vieira, A.C.F, Fernandes, N.F., Amaral, C.P., 1997. Estudo comparativo dos movimentos de massa ocorridos em Fevereiro/1996 nas bacias do Quitite e Papagaio (RJ): um enfoque geomorfológico, 2nd Panam Symp. on Landslides, Rio de Janeiro, 1, pp. 165–174.

Vilímek, V., Klimeš, J., Zapata, M.L., 2005a. Glacial lake outburst floods in the area of Huaras Cordillera Blanca, Peru. Studia Geomorph. Carpatho-Balcanica 39, pp. 115–124.

Vilímek, V., Luyo Zapata, M., Klimeš, J., Patzelt, Z., Santillán, N., 2005b. Influence of glacial retreat on natural hazards of the Palcacocha Lake area, Peru. Landslides 2, pp. 107–115.

Vilímek, V., Klimeš, J., Vicko, J., Carreño, R., 2006. Catastrophic debris flows near Machu Picchu village (Aguas Calientes), Peru. Environ. Geol., 50, pp. 1041–1052.

Vilímek, V., Zvelebil, J., Klimeš, J., Patzelt, Z., Astet, F., Kachlik, V., Hartvich, F., 2007. Geomorphological research of large-scale slope instability at Machu Picchu, Peru. Geomorphology 89, pp. 241–257.

Villamizar, J., 2006a. Desolación y Muerte por Inundaciones del Río Montalban. Relato, Cronista Oficial del Municipio Campo Elias. Estado Mérida, Venezuela, p. 4.

Villamizar, J., 2006b. Inundación de Ejido por el Río La Portuguesa. Relato del Cronista Oficial del Municipio Campo Elias. Estado Mérida, Venezuela, pp. 3.

Viramonte, J., Gallisky, M., Araña, V., Aparicio, A., García–Cacho, L., Escorza, C., 1984. El finivulcanismo básico de la depresión de Arizaro, provincia de Salta. IX Congreso Geológico Argentino, Actas III, pp. 234–253.

Villamor, P., Berryman, K., 1999. La tasa de desplazamiento de una falla como aproximación de primer orden en las estimaciones de peligrosidad sísmica. 1° Congreso Nacional de Ingeniería Sísmica, Murcia, España, 1a, pp. 153–163.

Volponi, F., 1976. El riesgo sísmico en territorio Argentino. Anales de la Sociedad Científica Argentina. Serie II, Ciencias Aplicadas No. 42, Buenos Aires, pp. 36–44.

Volponi, F., Quiroga, M., Robles, A., 1978. El terremoto de Caucete del 23 de noviembre de 1977. Instituto Sismológico Zonda. Universidad Nacional de San Juan, p. 81.

Vuille, M., Bradley, R., Werner, M., Keimig, F., 2003. 20th Century climate change in the tropical Andes: Observations and model results. Climatic Change 59, pp. 75–99.

Vuille M., Francou, B., Wagnon, P., Juen, I., Kaser, G., Mark, B.G., Bradley, R.S., 2008. Climate change and tropical Andean glaciers: Past, present and future. Earth-Science Reviews 89, pp. 79,–96.

Wallace, R., 1981. Active faults, paleoseismology and earthquake hazards in the Western United States. In Simpson, D., Richards, P. (eds.), Earthquake Prediction, American Geophysical Union, M. Ewing Series 4, pp. 209–216.

Wallace, R. (ed.). 1986. Active Tectonics. National Academy Press, Washington, DC, p. 266.

Walker, I., 1991. Algumas consideração sobre um programa dezoneamento da Amazônia. In Val, A. Figliuolo, R., Feldberg, E. (eds.), Bases científicas para estratégias de preservação e desenvolvimento da Amazônia: fatos e perspectivas. INPA, Manaus, pp. 37–46.

Warner, J., Oré, M.T., 2006. El Nino platforms: participatory disaster response in Peru. Disasters 30 (1), pp. 102–117.

Wang, J., Charneides, B., 2005. Global Warming's Increasingly Visible Impacts. Environmental Defense, New York, p. 43.

Warren, C., Sugden, D., 1993. The Patagonian Icefields: A glaciological review. Arctic Alpine Res. 25, pp. 316–331.

Warrick, R.A., Le Provost, C., Meier, M.F., Oerlemans, J., Woodworth, P.L., 1996. Changes in sea level. In Houghton, J.T., Meira Filho, L.B, Callander, N., Kattenberg, A. (eds.), The Science of Climatic Change. Cambridge University Press, Cambridge, pp. 359–405 .

Wayne W.J., 1987. Aluviones al acecho. El Tribuno newspaper, April 12, Salta, Argentina, pp. 6–7.

Wayne W.J., Alonso R.N., 1991. Catastrophic debris flows, eastern slope of the Andes, northern Argentina. Geological Society of America, Programs and Abstracts 24 (7), pp. 165–166.

Wayne, W., 1994. Mass wasting as a geological hazard in the province of Salta, Argentina, p. 28. Unpublished report.

Wayne W.J., 1999. The Alemania rockfall dam: A record of a mid-Holocene earthquake and catastrophic flood in northwestern Argentina. Geomorphology 27, pp. 295–306.

Wells, C., Coppersmith, K., 1994. New empirical relationships among Magnitude, Rupture Length, Rupture Width, Rupture Area and Surface Displacement. Bulletin of the Seismological Society of America 84, pp. 974–1002.

Westgate, K.N., O'Keefe, P., 1976. Some Definitions of Disaster. Disaster Research Unit Occasional Paper No. 4. Department of Geography, University of Bradford.

Westoby, M., Walker, B., Noy-Meir, I., 1989. Opportunistic management for rangelands not al equilibrium. Journal of Range Management 42 (4), pp. 266–274.

White, G.F. (ed.), 1961. Papers on Flood Problems. Research Papers No. 70, Department of Geography, University of Chicago.

White, G.F., 1964. Choice of Adjustments to Floods, Research Paper No. 93. Department of Geography, University of Chicago.

White, G.F., 1973, Natural hazards research. In R.J. Chorley (ed.), Directions in Geography. Methuen, London, pp. 193–216.

Wilches-Chaux, G., 2005. El terremoto, la avalancha y los deslizamientos de la cuenca del río Páez, 1994. In Hermelin, M. (ed.), Desastres de origen natural en Colombia 1979–2004. Universidad EAFIT, OSSO–Univalle, Medellín, Colombia, pp. 121–133.

Willys, B., 1914. El norte de la Patagonia. Naturaleza y riquezas. Reedición 1988. Editorial Universitaria de Buenos Aires (EUDEBA), p. 500.

Winchester, P., 1992. Power, Choice and Vulnerability: A Case Study in Disaster Management in South India. James & James, London, p. 225.

Winslow, M.A., 1982. The structural evolution of the Magallanes Basin and neotectonics in the southernmost Andes, In Craddock, C. (ed.), Antarctic Geoscience. University of Wisconsin Press, Madison, pp. 143–154.

Winslow , M., Prieto, X., 1991. Evidence of Active Tectonics along the Strait of Magellan. VI Congreso Geológico Chileno. Resúmenes Ampliados, pp. 654–655.

Wischemeier, W.S., Smith, D.D., 1978. Predicting Rainfall Erosion Losses: A Guide to Conservation Planning. Washington, DC, USDA, Agric. Handbook 537, pp. 58.

Withney, R., Bastias, H., 1984. The Tigre fault of the San Juan Province, Argentina—The late boundary of the Andes uplift. Geological Society of America Program with Abstract, 16 (6), p. 693.

WMO, 1980. Expert Meeting on Meteorological Aspects of Desertificatiion, December 1980. Report of the Meeting, point 2.4. Geneva, WMO.

Wolman, M.G., and Miller, J.P., 1960. Magnitude and frequency of forces in geomorphic processes, Reprint J. Geol. 68, pp. 54–57.

Woodroffe, C.D., 2008. Reef–island topography and the vulnerability of atolls to sea-level rise. Global and Planetary Change 62, pp. 77–96.

Woodworth, P.L., Blackman, D.L., Foden, P., Holgate, S., Horsburgh, K., Knight, P.J., Smith, D.E., Macleod, E.A., Bradshaw, E., 2005. Evidence for the Indonesian Tsunami in British tidal records, Weather 60 (9), pp. 263–267.

Yeats, R., Schwartz, D., 1990. Paleoseismicity: Extending the record of earthquakes into prehistoric time. Episodes. 13, pp. 9–12.

Yeats, R., Prentice, C., 1996. Paleoseismology. Special Section, Journal of Geophysical Research, 101 (B3), pp. 5847–6294.

Yepez, H. 2002. Riesgo Volcanico del Distrito Metropolitano de Quito. In: Memorias del Seminario Internacional Gestión de Riesgos y Prevención de Desastres. Facultad Latinoamericana de Ciencias Sociales (FLACSO), Cooperazione Internazionale (COOPI), European Commission Humanitarian Aid Office (ECHO), Quito, Ecuador, pp 13-15.

Yin, H., Li, C., 2001. Human impact on floods and flood disasters on the Yangtze River. Geomorphology 41, pp. 105–109.

Yong, C., Ling, C., Guendel, F., Kulhanek, O., Juan, L., 2002. Seismic hazard and loss estimation for Central America. Natural Hazards 25 (2), pp. 161–175.

Young, K.R., 1994. Roads and the environmental degradation of tropical montane forests. Conserv. Biol. 8 (4), pp. 972–976.

Young, K.R., Lipton, J.K., 2006. Adaptive governance and climate change in the tropical highlands of western South America. Climatic Change 78, pp. 63–102.

Zhou, J, Lau, K.M., 1998. Does a monsoon climate exist over South America? Journal of Climate 11, pp. 1020–1040.

Zimmerer, K.S., Young, K.R., 1998. Nature's Geography: New Lessons for Conservation in Developing Countries. University of Wisconsin Press, Madison.

Zinck, A., 1986a. Característica y Fragilidad de los Suelos en Ambiente de Selva Nublada: El Ejemplo de Rancho Grande, Cordillera de la Costa Venezuela. In Hubert, O. (ed.), La Selva Nublada de Rancho Grande. Parque Henry Pittier. Fondo Editorial Acta Científica Venezolana, Caracas, pp. 31–61.

Zinck, A., 1986b. Propiedades y Estabilidad Mecánica de los Suelos en Ambiente de Selva Nublada El Ejemplo de Rancho Grande, Cordillera de la Costa Venezuela. In Hubert, O. (ed.), La Selva Nublada de Rancho Grande. Parque Henry Pittier. Fondo Editorial Acta Científica Venezolana, Caracas, pp. 91–105.

Zúniga, R., Suárez, G., Ordaz, M., García Acosta, V., 1997. Seismic Hazard in Latin America and the Caribbean. Volume II: Mexico, vol. 2. IDRC, Ottawa. p. 82.

Index

A

D

E

O

P

R

S

T

U

V

W

Y